APPLIANCE
SERVICE
HANDBOOK

APPLIANCE SERVICE HANDBOOK

Second Edition

GEORGE MEYERINK

Professor Emeritus
Lake Michigan College

PRENTICE HALL, Englewood Cliffs, New Jersey 07632

Library of Congress Cataloging-in-Publication Data

Meyerink, George
    Appliance service handbook / George Meyerink. — 2nd ed.
      p.   cm.
    Includes index.
    ISBN 0–13–038902–1
    1. Household appliances, Electric—Maintenance and repair. 2. Gas
appliances—Maintenance and repair.   I. Title.
TK7018.M42   1988   87–18455
683′.83′0288   dc19   CIP

Editorial/production supervision: *Mary Carnis*
Cover design: *Lundgren Graphics Ltd.*
Manufacturing buyer: *Lorraine Fumoso*

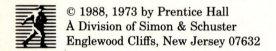

Printed in the United States of America

10 9 8 7 6 5

ISBN 0-13-038902-1 025

Prentice-Hall International (UK) Limited, *London*
Prentice-Hall of Australia Pty. Limited, *Sydney*
Prentice-Hall Canada Inc., *Toronto*
Prentice-Hall Hispanoamericana, S.A., *Mexico*
Prentice-Hall of India Private Limited, *New Delhi*
Prentice-Hall of Japan, Inc., *Tokyo*
Simon & Schuster Asia Pte. Ltd., *Singapore*
Editora Prentice-Hall do Brasil, Ltda., *Rio de Janeiro*

To my friend,
colleague, and wife,
Harriet

CONTENTS

**PART 2
PRODUCT
SERVICE—
PROCEDURE**

The appliances listed are among those most commonly found in households. The knowledge gained in working with these appliances can be used on other equipment not included here. To keep this book to a reasonable size, other household electromechanical equipment, such as furnaces, central air conditioners, hot tubs, attic fans, and deep-well pumps, are not included.

PREFACE

I hear and I forget
I see and I remember
I do and I understand

Author unknown

The information included in this revised text includes an added chapter on basic electronics and much updated product service information. It is yours for the taking—how much you get depends upon your efforts. We can only expose you to this accumulation of knowledge.

The contents of this book summarizes years of work by people who "grew up" with the appliance industry. Engineers, assemblers, inspectors, salesmen, servicemen, and customers have all contributed important and helpful facts.

As the title of Part I suggests, this book is intended to provide a basic electromechanical background as well as guidance in human relations and ethics—all of which are required by those of you who would succeed. Although educators and behavioral scientists often present information in the abstract, it has been found that many students learn more easily with "hands-on" experience. To this end, this book utilizes theory *with* the practical application of servicing appliances.

The materials presented in Part II are necessarily general, in order to present a broad overview of service methods. The use of specific makes and models to exemplify a technique is for our convenience in clarifying the description of "how to do it" and in no way reflects on a particular product's reliability. Rather, we especially thank these manufacturers for allowing us to use their material in the preparation of the book.

Technicians working for dealers will usually have manufacturers' literature for the products they sell, and we advise them to follow that information. This book will provide the necessary background to do that. The independent technician is not always so lucky. He will have good information on his specialty but may not have much on other products that customers might bring in. These product data will serve as a guide to aid him in using knowledge gained from experience with similar appliances.

The student who wants to expand his career beyond the service field will find that what is learned from this book will provide keys to open the doors of related fields.

The homeowner will find basic information on what different products are supposed to do and how they do it.

We suggest that you consider this volume a carefully worked master key that can open innumerable doors to self-satisfaction, independence, and service to your fellow man.

George Meyerink
Apache Junction, Arizona

Part I
Appliance Service—Fundamentals

1

The Occupation of Service

THE BROAD VIEW

Have you ever stopped to consider the growing dependence of people on machines? Who takes care of these machines?

The livelihood of people in homes, farms, stores, offices, and factories often depends on the machines that serve them. Anything that goes wrong with machines is at least disappointing to the operator. Someone has to keep the machinery of modern civilization running. This someone is not necessarily the owner of the machine. He may be too busy. He may not like to. He may not want to.

In an office, for example, even though good typists or computer operators can *use* their machines at high efficiency, *they may not have the knowledge or tools needed to service them.* The same is true in the home. Family members can use the array of appliances available to them but often do not know how to service them. In each example the machine user must rely on an outside source for help: the service techni-

cian. A homeowner may have service contracts on some appliances, and an office may subscribe

All set for a round of house calls. (Courtesy of Whirlpool-Tech Care)

to a preventive maintenance service which provides regular cleanup and examination of equipment to forestall a failure.

Having this preventive maintenance can save any office or factory untold dollars that would be lost if the machine failed and production was delayed. Yes, office work is production, too. Sending out bills, purchase orders for new material, sales information all must be done in the shortest possible time at the least cost.

There are many small companies that cannot afford a full-time maintenance man to service a machine breakdown or a control unit failure. *These plants are good prospects for a* **reliable** *servicing organization* of personnel trained in the installation and maintenance of industrial controls and production machines. These fields are becoming more specialized every day, and an expert in industrial control maintenance can be certain of an interesting, busy, and lucrative livelihood. The farmer who has a machine breakdown at harvest time just does not have time to repair it. He must call in outside help, even though he knows how to repair the machine. Thus the list of service applications grows because any machine made by people and used by people will, at sometime or other, require "fixing."

HOUSEHOLD APPLIANCES

The homeowner and his growing accumulation of appliances offer a tremendous market for service. His reasons for calling outside help are numerous. He may not know how to service the machine. For example, very few homeowners would know how to repair an automatic washer or a television set.

He may not have the special tools or parts required to fix the machine. In refrigeration, for instance, he would not have the gages or other special tools needed to open a system and check it (Fig. 1-1). He may not have time to do the job. *Example:* An electric range fails at breakfast time. Both adults work and must have the range fixed by evening for dinner. Or a homeowner is out of town and his family cannot wait until he returns to have a problem fixed, such as an oil burner failure in the wintertime. Or perhaps the owner just does not want to get involved—he would rather go fishing.

WHAT IS AN APPLIANCE?

For our purposes, we can define appliances as those machines or devices that contribute to

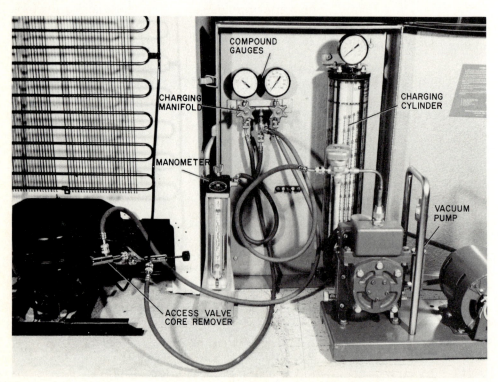

FIGURE 1-1. Testing a refrigeration system requires knowledge of the system and expensive, sophisticated test equipment.

man's comfort and health, and that aid in the care of his food and clothing. Although this distinction separates household appliances from business, office, commercial, and industrial machinery, which also need servicing at times, many of the practices and procedures used in household appliance service can be applied to these machines.

NATURE OF WORK

Appliance servicemen install and repair appliances—working in the customer's home or in a shop which may be part of a business establishment that deals in appliances, furniture, hardware, or other merchandise.

In order to repair an appliance, the serviceman must first determine what is wrong and why, then proceed to fix or replace the malfunctioning part. In doing so he will use hand tools, such as pliers, screwdrivers, and wrenches; power tools, such as a drill press, plus grinder, and lathe; and test equipment, such as voltmeter, temperature, and continuity tester.

In a small shop the serviceman may be called on to repair a wide variety of types and brands of appliances. In the larger shops, those having several servicemen, there is a tendency to specialization, with each man working on those machines with which he is most proficient.

RESPONSIBILITIES AND
CLASSIFICATION OF SERVICEMEN

The degree of responsibility depends on ability and knowledge. Obviously, a person having experience only in vacuum cleaner repair would not be sent to a home to repair a refrigerator. **A parts changer may not be a good diagnostician!**

We can classify servicemen into three sequential groups, each requiring additional knowledge, more responsibility, and paying more money. The benchman, shopman, and small appliance repairman are among the names used to describe a first level of achievement. The man in this group is able to repair

such small appliances as mixers, toasters, irons, and heaters that may be brought into the shop. He can change parts in the major appliances—working under direct supervision—and has little or no contact with the customers. His working conditions are normally indoors in a dry, adequately heated, and well-lighted shop. He can usually sit while working.

In the second group we have the installer, an outside serviceman who delivers, installs, and demonstrates the operation of an appliance to the customer. In general, there is no need for him to troubleshoot or replace parts, but he may work under a wide variety of conditions. He works with major appliances, such as washers, dryers, ranges, refrigerators, and air conditioners. He must know how they operate and what they need for best performance. Most of his work will be at the customer's home under a minimum of supervision. Presumably he must know the local building codes. He should be physically able and presentable.

Finally, the major appliance serviceman is the troubleshooter who makes house calls and repairs machines in the customer's home (Fig. 1-2). This job requires a wide range of knowledge and experience in order to handle problems as they arise. The serviceman in this category may make cost estimates and collect money for work done. He often plans his own schedule of work and does what he can to maintain good customer relations. He carries with him a large stock of parts and expensive test equipment for which he is responsible (Fig. 1-3). He, too, should be physically able, adaptable to changing working conditions, and well groomed.

PHYSICAL REQUIREMENTS
FOR SERVICEMEN

Appliance servicemen are active, doing a considerable amount of standing, bending, and looking. Small appliance servicemen lift only the smaller, lighter appliances and parts. The installer and major appliance servicemen have to move and handle the larger machines, machines that often weigh several hundred pounds. Help is provided when needed. Such

FIGURE 1-2. The service technician making house calls may be the only contact the housewife has with the company; thus he must make a good impression. (Courtesy of Sears, Roebuck & Co.)

men are subject to minor cuts and bruises, plus electrical shocks, and should note and follow the accepted safety procedures. See Chapter 14 on safety.

Those servicemen working outside the shop must be able to operate a motor vehicle.

PERSONALITY TRAITS AND APTITUDES DESIRED IN SERVICEMEN

Employers look for good health, neat appearance, average physical strength, good eyesight and dexterity, and a high mechanical aptitude with a sense of care for tools.

The ability to get along with people is most important, especially for those who have contact with customers.

The general aptitudes of importance are learning ability, numerical aptitude, reasoning ability, and form perception.

Desirable character traits are perseverance, patience, ingenuity, and accuracy. To understand how a machine works, to be alert to many things at once, and to recognize a malfunction are each needed for success in this field, *and—above all—integrity!*

THE SERVICE MARKET

Information obtained from the U.S. Department of Labor's *Occupational Outlook Handbook* indicates that there is a need for several thousand persons annually to replace those who

FIGURE 1-3. A well-organized selection of parts suitable for the calls to be made on any given day makes for an efficient operation.

retire or leave the appliance service field for other reasons. There is an additional need for technicians to care for the increase in appliance sales and usage. Servicing firms that hire technicians prefer that they be trained at a regular trade, vocational school, or college.

Training courses judged most helpful in preparing a technician for a job include: physics, basic electricity, electronics, mathematics, machine shop, English, and some introductory courses in the behavioral sciences, in addition to the course and laboratory work dealing directly with appliances. After having acquired a solid basic background, the technician should update his knowledge by attending manufacturers' product seminars.

Wage rates and income are difficult to discuss in a book because they change annually, and vary with skills and location. For example, service technicians in metropolitan areas usually earn more than technicians in rural areas. Also, the high demand for air-conditioning service technicians in warmer climates creates a better-paying market. In general, appliance service technicians' income is comparable to that of auto mechanics and TV and computer service technicians. You should be able to obtain data at your local library from the latest copy of the *Occupational Outlook Handbook* under the heading "Appliance Repairers." Also, the Association of Home Appliance Manufacturers (20 North Wacker Drive, Chicago, IL 60606) offers a booklet for [$0.75] a nominal fee entitled *Your Career as an Appliance Service Technician.*

ADVANCEMENT WITHIN THE SERVICE FIELD

Advancement within the appliance service field, as in any other work area, depends on many factors—ability, personality, knowledge, and willingness to accept responsibility. Employer policy and company size will determine if the serviceman can progress to any of the following positions.

Parts man: has responsibility of maintaining the inventory of parts needed for service. He issues them to the repairman as required or sells them "over the counter" to customers.

Dispatcher: receives service calls, routes the serviceman, establishes priority, and recommends what special tools and parts should be taken on the job.

Service manager: directs the operations of the service shop. He is responsible to both the customer and the shop owner to keep them satisfied with the product.

Field service technician (factory representative): has an in-depth knowledge of his company's product. He must be a skilled diagnostician and able to teach other servicemen. He must travel a great deal and many of his training classes may be held at night.

Technical writer: writes the instruction books and other service literature used by the repairman. He usually works from the factory with the engineers and engineering technicians who designed and tested the machine.

Such are the areas included in appliance service and what it takes to be a service technician. Let's look at the subject in more detail.

WHAT IT TAKES TO BE A SERVICE TECHNICIAN

The increasing complexity of modern electromechanical equipment has made it virtually impossible for the handyman to service appliances, business machines, controls, and similar devices that make our standard of living what it is today. *The service technician can no longer be just a part changer.*

The service technician must take a professional approach to the machine he is to service. In order to do so, he must be trained—not only to the techniques of diagnosis and repair but also in attitude—to the machine, its manufacturer, and its owner. It is his knowledge of the organization of the parts in the machine that gives him the confidence needed to do a good job.

A service technician who has a good sound electromechanical background can adjust to the servicing of any type of machine whether in the home, office, or factory, whether new or old, whether foreign or American.

There is much more to servicing than product knowledge. Many a good technician cannot hold a job or else he gets delegated to shopwork because he lacks sympathetic consideration of the customer's problem. The human or personal touch depends on you. Your attitude, integrity, outlook, grooming, and feeling will have much to do with your success in this field.

Empathy is a good word to describe this area. Put yourself in the other fellow's shoes. You have been called into a house because something has happened to an appliance. You answer as an authority. The customer has complete confidence in your ability or he would not have called you. It is up to you to maintain that confidence.

Looking the Part

Figure 1-4 shows a sharp-looking service technician at the mirror—be the part. Be neat, clean, and washed. You must not only be clean personally, but you must also be neat and clean in your work. When working in the home, keep your tools and parts in order. Clean up your work area after the job is finished. Wipe off the machine. A little polish does wonders for both the machine and customer relations. A few simple cleanup tools in your kit (a cloth, some polish, and a brush) will do much to make this gesture an easy and natural one.

Attitude

Your attitude, to the customer, the machine, your employer, and the manufacturer is important. To the customer you must appear confident, courteous, and cheerful. You must create a businesslike impression by keeping your specified appointment time. If for some reason you cannot keep the appointment, call the customer and establish a new time out of simple courtesy. A customer may have other things to do, or he may want to know well enough in advance so that he can make plans. If the product should

FIGURE 1-4. Say "good morning" to yourself in a mirror. Do you look like the service technician you would want coming to your home?

be unfamiliar to you, say so! Be honest with the customer. If you find that it is similar to something you have already serviced, you have a good starting point. If you cannot fix it, say so and explain what you intend to do. *You must always remember that you, too, are somebody's customer.* You know how you, as a customer, would like to be treated and you should know that other people want to be treated in the same manner.

You should respect the customer's privacy. Above all, do not gossip, either about other people or about the product. You are not the one to judge a customer's habits or whether he got a lemon. Even though you might think so—you just don't say it.

IT CAN BE FIXED

Actually, "lemons" are a state of mind. *Any machine that was built can be repaired* when approached with an intelligent attitude. A machine may have a defective part, but once that part is replaced, it should function just as well as any of its brothers. It is up to you to find that part and replace it. If the problem is a major one that will be costly, the customer must be consulted before the repair job or the servicing is done. It should be the customer's decision as to how much money is to be spent on a given machine. There are times when a machine can be repaired but it may not be economically feasible to do so, or the new models may be so much improved that this machine is obsolete and not worth repair money. However, the decision is the customer's. You can only present the facts as you see them. Don't try to sell—you may recommend, if asked.

TALKING TIME

When making a service call in a house, the question often arises as to how social or businesslike the serviceman should be. This is an area that has to be played pretty much by ear. You must bear in mind that your time is costing someone money. If your time is not productive, it is the customer who loses. Most people have a good sense of value, and there are many more housewives with mechanical aptitudes and understanding than you may realize. They are insulted, and rightly so, if your attitude implies that they "wouldn't understand." Moreover, how about their husbands, who will want to know what they are paying for? In such cases, you leave the defective parts and explain what has been done. If the housewive appears to have no interest, is busy, or has guests, she will appreciate your quick, efficient, and courteous service. *Having done all you can, make yourself scarce!*

Preservicing Discussion

Some discussion with the customer about the machine's normal operation, and what happened when the machine failed, will do much to save you time. In this discussion you must mentally overlay your knowledge of the machine's principles on the customer's description of the machine's performance. Then you can often pinpoint the basic fault to a specific operation or cycle. Your knowledge of the machine

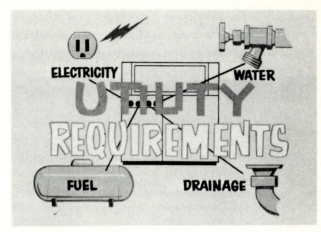

FIGURE 1-5. Check the installation. Something may be turned off.

will tell you which components were activated at the time. You should first check these parts individually. If there is no checklist for the machine being serviced, you would be wise to make your own, basing it on checklists for similar pieces of equipment that are outlined and discussed elsewhere.

On any service call it should be your practice to check the installation (Fig. 1-5), making certain that the machine is level, secure, well ventilated, and adequately supplied with electricity, water, air, gas, or whatever else it needs to function properly. Again, use the checklist provided for the machine in question. For an in-warranty service call, you should explain to the customer that the part failed, a connection came loose, or adjustments had to be made, as the case may be, and that there will be no charge for the call. Show regret for the inconvenience caused, and thank the customer for calling you promptly so that he or she can realize the proper performance of the machine.

Instructing the Customer

If you find the trouble to be the customer's fault—the machine was incorrectly used or proper operation was made impossible by an external factor, such as a clogged drain, inadequate water, or a plug loose in the wall receptacle—you must explain what you have found, how it was or could be corrected, and what the customer should do to prevent recurrence (Fig. 1-6). This is an educational program to the customer.

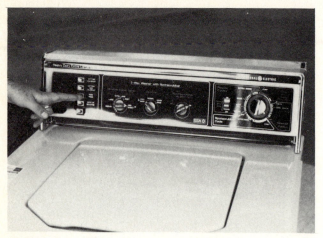

FIGURE 1-6. Demonstrate the correct operation of the machine and its controls.

Anything that you can do to make the machine trouble-free and capable of satisfactory operation will enhance your reputation as well as the product's reputation in the eyes of the customer. (Customers need to feel that they have invested wisely in their machine, whether you like that particular unit or not.)

As far as the customer is concerned, the two most important service calls you make are the first call on a specific appliance and the last one. The first, because the way you conduct yourself will determine whether you get called back for future service. This is a measure of customer confidence. The last, because here you must tell the customer that his appliance has had it. Again your attitude and choice of words will influence the customer's response and help him remain a customer.

In many instances, you are the only contact the customer has with the factory. You have an obligation to the manufacturer of the machine as well as to your employer (the dealer). The future of all depends on keeping the customer satisfied with the product, even though he may have to replace his appliance.

EFFECTIVE USE OF TIME

What you do for your boss affects both his livelihood and your own. The more successfully you can create an image of reliability in the mind of the customer, the more loyal he will be to the company you represent. This is part of good public relations and one of the intangibles that makes the difference between success and failure for many people. Your future may depend as much on your efforts in this area as on your knowledge of the product. A dishonest, unreliable serviceman soon develops a reputation that makes it difficult for him to live down and might make it impossible for him to find a job with anyone.

In order to succeed, you must be more than just a nice guy. You must know the product. You must recognize that occasionally a part will fail, be abused or damaged. You must learn the difference and tell the customer what you found and how to avoid a repeat of this same problem. You must not compare the performance of any particular appliance with a different make or model.

Each has different features. Each will produce different results. Each must operate on its own merits. The customer purchased this particular machine because, at the time he purchased it, it had the features he wanted for the price he was willing to pay. You must keep up to date with changes and improvements in your field so that you can freely discuss them when questioned.

SERVICE IS A NECESSARY OCCUPATION

Since few devices are completely trouble-free, every machine needs servicing at some time or other. It is not economically feasible to build electromechanical equipment so that it is completely expendable or completely failure proof. The constant discussion of price versus quality is a manufacturer's dilemma, and a serviceman's problem, because people (the customer) naturally want the best they can get at the lowest cost. Somewhere, in design and manufacture, a decision has been made as to the best compromise. You have no control over this, but you can control repeat service calls due to poor workmanship on your part. You can make your servicing easier if you will remember that *a machine is an organized collection of parts, so arranged as to perform a given job.* Each part has a specific function. Knowing this, you examine the whole for failure of the component, isolate the defect, then repair or replace as economy or need dictates.

Customers know that when they call a serviceman it is going to cost money for both parts and labor. Yet they are sometimes unprepared for the apparent size of the labor bill, and an explanation is in order. Why is the bill so high? (See Fig. 1-7.)

When a customer calls a shop for service, he expects the shop to have all the necessary knowledge, tools, and material to do the job. This calls for a big investment on the part of the shop owner. The shop must have enough income to meet its direct costs and overhead and still yield enough profit for the owner to keep him interested in staying in business.

All of this knowledge, as well as the tools and materials, is at the disposal of the appliance owner who called for a professional service technician. A professional is one who knows his job and does not "goof off."

NEVER STOP LEARNING

We have said that you must keep up to date in this rapidly changing and growing field. Here you will learn about the mechanics of appliances as they are being built today. We offer no assurance regarding the amount of additional information you will need to service a new model 5 or 10 years from now.

This textbook will provide a broad basic background that can be applied to any appliance now being used. It will be up to you to

FIGURE 1-7.

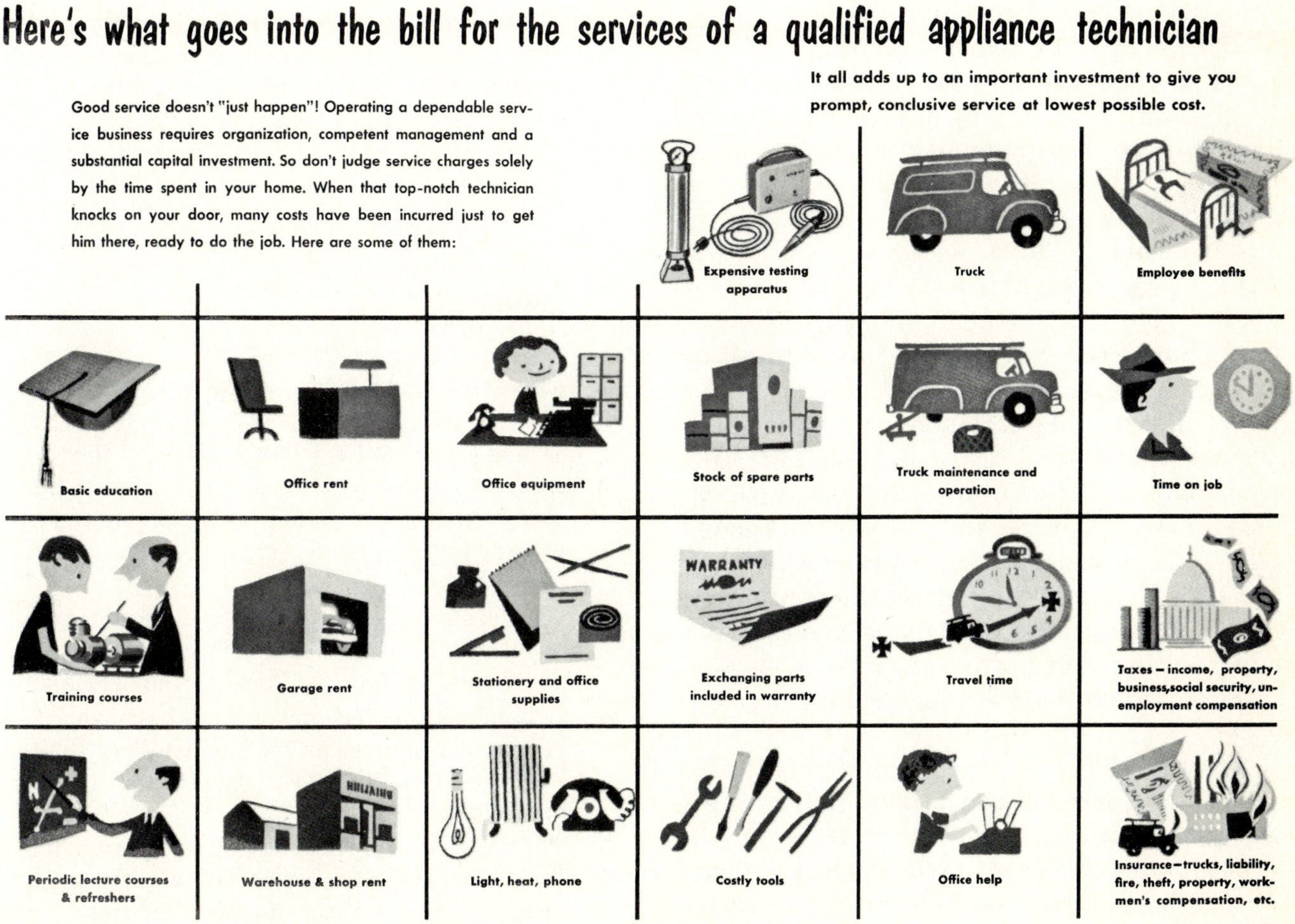

Here's what goes into the bill for the services of a qualified appliance technician

Good service doesn't "just happen"! Operating a dependable service business requires organization, competent management and a substantial capital investment. So don't judge service charges solely by the time spent in your home. When that top-notch technician knocks on your door, many costs have been incurred just to get him there, ready to do the job. Here are some of them:

It all adds up to an important investment to give you prompt, conclusive service at lowest possible cost.

Expensive testing apparatus	Truck	Employee benefits
Basic education	Office rent	Office equipment
Stock of spare parts	Truck maintenance and operation	Time on job
Training courses	Garage rent	Stationery and office supplies
Exchanging parts included in warranty	Travel time	Taxes — income, property, business, social security, unemployment compensation
Periodic lecture courses & refreshers	Warehouse & shop rent	Light, heat, phone
Costly tools	Office help	Insurance — trucks, liability, fire, theft, property, workmen's compensation, etc.

keep abreast of the new developments, changes, and improvements in the appliance field. There is a monthly publication serving the appliance industry.

It lists part sources, dates for service training sessions, as well as up-to-date information on what is happening in the service field.

Appliance Service News
5841 W. Montrose Avenue
Chicago, IL 60634

You must read current literature, attend training schools, and do everything in your power to keep faith with your customer, to provide him with the service he needs when he needs it.

The dividend to you that comes from "updating" your knowledge is that you will stay in business. You will always have job security as long as you want to work.

All that we have said about the people of the service field, their attitude, customer relations, the organization behind the serviceman, the costs of operation, and the need for a high degree of technical knowledge adds up to the fact that the service industry is a billion-dollar business that is still growing.

SUMMARY

An appliance is a machine that contributes to man's comfort and health and that helps in the care of his food and clothing. Any machine made by man will at some time or other fail. The owner may not be able to repair it because he does not have the time, the tools, or the knowledge. Thus he must rely on someone who does.

The serviceman (technician) must be physically able and must have high mechanical aptitude, learning ability, and rapport with people. There is a great need for qualified servicemen today.

Technically, the serviceman should have a broad electromechanical background and specific knowledge, in depth, about those appliances for which he has a service responsibility.

He should have a positive attitude to problems and a realization that it is the customer

who is paying his way. Consequently, he must make effective use of his time and keep abreast of the newest developments occurring in his field.

REVIEW QUESTIONS

1. What is an appliance?
2. Why is a serviceman needed?
3. What should an employer look for when hiring a serviceman?
4. List some job opportunities within the service field.
5. What does a customer expect from a serviceman?
6. How important are the installation and the usage of the appliance?
7. What (and why) are the two most important service calls?
8. List some of the costs that go into the price of a service call?
9. Why must you continue to learn more about appliances?

STUDY QUESTIONS

Service

1. A service technician who works outside the shop must (a) be aware of the safety rules, (b) be well groomed and polite, (c) be physically fit and able to drive a truck, (d) all of these.
2. Ability to get along with people is most important for (a) the benchman, (b) the engineer who designed the machine, (c) the serviceman, (d) all of them.
3. The technician who needs the largest variety of tools is usually (a) the serviceman, (b) the benchman, (c) the installer, (d) the salesman.
4. A handicapped technician would probably be best suited to serve as (a) the installer, (b) the home serviceman, (c) the shop owner, (d) the benchman.
5. The first decision any technician must make is (a) how to change a part, (b) why the appliance will not perform, (c) what is wrong, (d) which part to change.

6. The serviceman has responsibilities in four directions. Name them.
7. If a repair job promises to be expensive, what is the first thing a serviceman should do?
8. When making repairs in a home, a technician must keep the cost of his time in mind. Therefore, he should let the customer clean up when he has finished.   True or False
9. A technician must not only have a good knowledge of the machines he services but an adequate vocabulary and knowledge of human relations as well.   True or False
10. Once you have established yourself as a reputable technician, constant employment is all you need to keep up in your trade.   True or False

KEY TO REVIEW QUESTIONS

1. Page 2, paragraph 3
2. Page 2, paragraphs 3 and 4
3. Page 4, paragraphs 1, 2, 3, 4
4. Page 5, paragraphs 4, 5, 6, 7, 8
5. Prompt, honest, courteous, and knowledgeable service

6. Very! Correct installation and customer instruction will prevent many "in warranty" service calls.
7. Page 8, paragraph 2
8. See Figure 1-7.
9. Page 10, paragraphs 5, 8, 9

KEY TO STUDY QUESTIONS

Service

1. d
2. c
3. a
4. d
5. c
6. To customer, to employer, to manufacturer, to machine
7. Discuss it with machine's owner (customer).
8. False
9. True
10. False

2

Tools of the Trade

Ever since man became "the toolmaking animal," he has developed many devices to extend the range and usefulness of his five senses. Man's appliances, designed to improve his creature comfort, must be maintained with man's tools. Some of these are commonly used in many different ways and others are used only for a specific purpose.

Some tools serve as amplifiers of man's senses and are diagnostic, being used to locate the trouble. Such tools are called testers. Others, used to *turn, hammer, pry, hold, and cut,* as an extension of man's fingers and muscle, are called hand tools. Tools using a source of extra power in addition to man's muscle are called power tools, but they still do the same things that the hand tools do.

TOOLS THAT TURN

The tools used to handle nuts, bolts, screws, and such fasteners as must be turned to attach or remove are usually screwdrivers, nut drivers,

or wrenches and are available in many sizes and shapes.

Screwdrivers

The standard screwdriver is the flat-blade type (Fig. 2.1a). The blade is inserted into—and perpendicular to—a slot in the screw head (Fig. 2-1i) and the handle is turned to move the screw. (*Clockwise to tighten, counterclockwise to loosen is standard.*) The choice of a screwdriver size depends on the screw head. Always use the largest blade that comfortably fits the slot without extending beyond it, as shown in Fig. 2-1i. The blade end should be square and the sides parallel (Fig. 2-1h); otherwise it will tend to slip out of the slot, thus damaging the screw head and possibly the finish of the surface next to the screw.

A screw that has a damaged or clogged slot can often be removed by gently tapping the screwdriver blade into the slot to clear the rust or burr. *This is the only time a screwdriver handle should be hit.* Never should the screwdriver

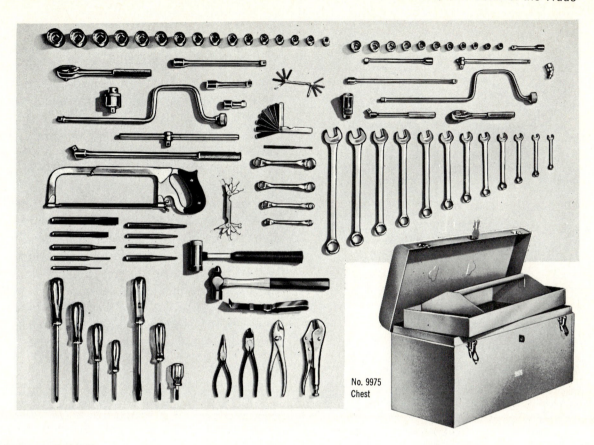

No. 9975
Chest

CONTENTS/NO. 9912 SET

NO.	DESCRIPTION
000 A	Feeler Gauge Set
000 D	Ignition Gauge Set
000 K	Spark Plug Gauge Set
41½	Center Punch, ½" x 5⅝"
47⁵⁄₁₆x⅛	Pin Punch, ⅛" x 5¼"
47⅜x³⁄₁₆	Pin Punch, ³⁄₁₆" x 6⅛"
47½x¼	Pin Punch, ¼" x 6¾"
50⁵⁄₁₆	Starting Punch, ⅛" x 5⁵⁄₃₂"
50⅜	Starting Punch, ³⁄₁₆" x 5²¹⁄₃₂"
50½	Starting Punch, ¼" x 6½"
86 A ⁷⁄₁₆	Cold Chisel, ½" x 6³⁄₁₆"
86 A ⅝	Cold Chisel, ¾" x 7³⁄₁₆"
207	Diagonal Cutting Plier, 7"
220	Chain Nose Plier, 6¾"
278	Slip Joint Plier, 8"
291	Lever Wrench Plier
352	Hacksaw
1210	Comb. Wrench, 12 pt. ⁵⁄₁₆"
1212	Comb. Wrench, 12 pt. ⅜"
1214	Comb. Wrench, 12 pt. ⁷⁄₁₆"
1216	Comb. Wrench, 12 pt. ½"
1218	Comb. Wrench, 12 pt. ⁹⁄₁₆"
1220	Comb. Wrench, 12 pt. ⅝"
1222	Comb. Wrench, 12 pt. ¹¹⁄₁₆"
1224	Comb. Wrench, 12 pt. ¾"
1226	Comb. Wrench, 12 pt. ¹³⁄₁₆"
1228	Comb. Wrench, 12 pt. ⅞"
1230	Comb. Wrench, 12 pt. ¹⁵⁄₁₆"
1232	Comb. Wrench, 12 pt. 1"
1312 P	Ball Pein Hammer, 12 ounce
1362	Plastic Tip Hammer, 16 ounce
2302	Ignition Point File, 5"

NO.	DESCRIPTION
2337	Carbon Scraper
5026 HPA	Spark Plug Socket
5210 H	Reg. Socket, 6 pt. ⁵⁄₁₆"
5212	Reg. Socket, 12 pt. ⅜"
5214	Reg. Socket, 12 pt. ⁷⁄₁₆"
5216	Reg. Socket, 12 pt. ½"
5218	Reg. Socket, 12 pt. ⁹⁄₁₆"
5219	Reg. Socket, 12 pt. ¹⁹⁄₃₂"
5220	Reg. Socket, 12 pt. ⅝"
5222	Reg. Socket, 12 pt. ¹¹⁄₁₆"
5224	Reg. Socket, 12 pt. ¾"
5226	Reg. Socket, 12 pt. ¹³⁄₁₆"
5228	Reg. Socket, 12 pt. ⅞"
5249	Ratchet, ⅜" square drive
5260	Extension Bar, 3½"
5261	Extension Bar, 7½"
5262	Extension Bar, 12"
5265	Hinge Handle
5270	Universal Joint
5280	Speed Handle
5285	Sliding Bar
5414	Reg. Socket, 12 pt. ⁷⁄₁₆"
5416	Reg. Socket, 12 pt. ½"
5418	Reg. Socket, 12 pt. ⁹⁄₁₆"
5419	Reg. Socket, 12 pt. ¹⁹⁄₃₂"
5420	Reg. Socket, 12 pt. ⅝"
5422	Reg. Socket, 12 pt. ¹¹⁄₁₆"
5424	Reg. Socket, 12 pt. ¾"
5425	Reg. Socket, 12 pt. ²⁵⁄₃₂"
5426	Reg. Socket, 12 pt. ¹³⁄₁₆"
5428	Reg. Socket, 12 pt. ⅞"
5430	Reg. Socket, 12 pt. ¹⁵⁄₁₆"

NO.	DESCRIPTION
5432	Reg. Socket, 12 pt. 1"
5434	Reg. Socket, 12 pt. 1¹⁄₁₆"
5436	Reg. Socket, 12 pt. 1⅛"
5438	Reg. Socket, 12 pt. 1³⁄₁₆"
5440	Reg. Socket, 12 pt. 1¼"
5447	Ratchet Adapter
5449	Ratchet, ½" square drive
5460	Extension Bar, 2½"
5461	Extension Bar, 5"
5463	Extension Bar, 10"
5468	Hinge Handle
5480	Speed Handle
5485	Sliding Handle
8160	Box Wrch., 12 pt. ⅜" x ⁷⁄₁₆"
8161	Box Wrch., 12 pt. ½" x ⁹⁄₁₆"
8162	Box Wrch., 12 pt. ⅝" x ¹¹⁄₁₆"
8163	Box Wrch., 12 pt. ¾" x ¹³⁄₁₆"
9604	Screwdriver, ¼" x 4"
9608	Screwdriver, ⅜" x 8"
9652	Screwdriver, ¼" x 1½"
9682	Phillips Screwdriver, #1
9684	Phillips Screwdriver, #2
9686	Phillips Screwdriver, #3
9688	Phillips Screwdriver, #4
9975	Metal Chest

This is a representative basic mechanics tool kit. Except for the ignition tools and carbon scraper, the appliance service technician could get along very well with this set.

Representative basic mechanics tool kit. Except for the ignition tools and carbon scraper, the appliance service technician could get along very well with this set.

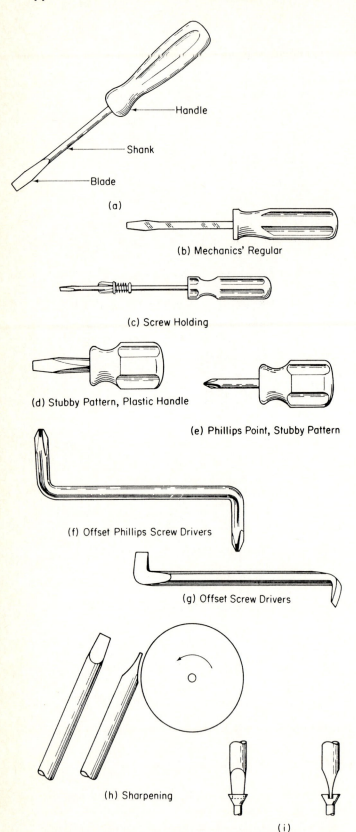

Handle

Shank

Blade

(a)

(b) Mechanics' Regular

(c) Screw Holding

(d) Stubby Pattern, Plastic Handle

(e) Phillips Point, Stubby Pattern

(f) Offset Phillips Screw Drivers

(g) Offset Screw Drivers

(h) Sharpening

(i)

FIGURE 2-1. Screwdrivers come in many shapes and sizes. Use the right one for the job. (Courtesy of Crescent and Proto)

be used as a pry bar or chisel; it simply is not built for that kind of work. A wrench may be used on a heavy-duty, square-shanked screwdriver to break loose an obstinate screw, but use some discretion as to how much force is applied.

A *Phillips*-type screwdriver (Fig. 2-1e) is needed for those screws that have two slots crossing at right angles in the center of the screw head. These screws are most often used with moldings and other decorative parts because there is less tendency for the screwdriver to slip and mar the finish. When using a Phillips screwdriver, you must exert somewhat more downward pressure to keep the screwdriver in the slots. *A worn Phillips screwdriver will tend to slip out of the slots and should be discarded, for it cannot be sharpened easily.* As with the flat-bladed screwdriver, use the correct size.

Nut Drivers

A nut driver (Fig. 2-2) is used, like a screwdriver, to turn nuts. Its handle is the same, but the working end is a socket designed to fit over a hexagonal nut. Each size nut requires a different driver. Common sizes, as measured across the flats, are $\frac{3}{16}$, $\frac{1}{4}$, $\frac{5}{16}$, $\frac{11}{32}$, $\frac{3}{8}$, $\frac{7}{16}$, and $\frac{1}{2}$ inch. These tools either fit or they do not; there is no sharpening or adjustment.

FIGURE 2-2. Hex-nut drivers have a standard handle color coding to match the nut size as measured across its flats. (Courtesy of Proto)

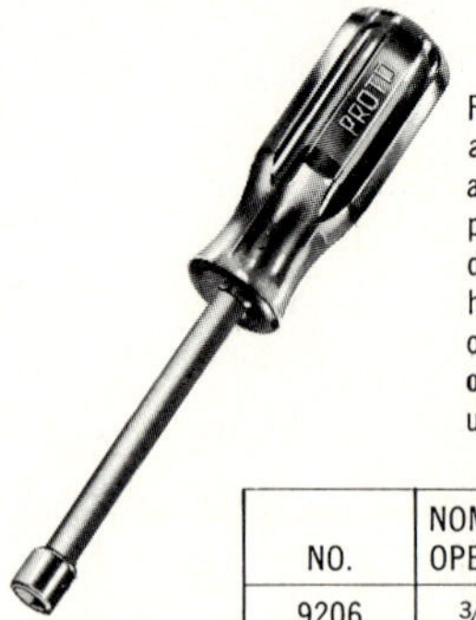

HEX NUT DRIVERS

For electronics, instruments, automotive, radio and ignition work. Fluted handles of fire-resistant plastic are size coded by colors. Chrome plated hollow steel shafts clear long bolts, reduce weight. Special **heat-treated** shafts give high wear resistance for use on hardened screws or bolts . . . long life. **Precision formed drive openings** fit snugly . . . hold tight for long usage.

NO.	NOMINAL OPENING	HANDLE COLOR	TOTAL LENGTH	WT., LBS.
9206	3/16"	Black	6⅛"	.14
9207	7/32"	Brown	6⅛"	.16
9208	1/4"	Red	6⅛"	.16
9209	9/32"	Orange	6⅛"	.16
9210	5/16"	Amber	6⅞"	.22
9211	11/32"	Green	6⅞"	.22
9212	3/8"	Blue	6⅞"	.22
9214	7/16"	Brown	7³⁄₁₆"	.30
9216	1/2"	Red	7³⁄₁₆"	.30
9218	9/16"	Orange	7³⁄₁₆"	.33
9220	5/8"	Amber	7³⁄₁₆"	.33

Wrenches

Wrenches are used to turn nuts. They differ from a screwdriver in the way they are handled. Nut drivers and screwdrivers are handled in the same line as the screw, whereas wrenches are handled at right angles to the screw. This arrangement permits much more leverage to be applied to a nut or bolt and allows more clearance for handle movement.

The various types of wrenches illustrated in Fig. 2-3 to 2-6 are the outgrowth of an assortment of needs. Your tool choice and usage will depend on your needs, but some fundamental rules for wrench care and usage must be followed for safe practice.

1. *Keep tools clean.* Dirt and grit will prevent them from seating firmly around a nut. Grease will make them slippery.
2. *Use the correct size.* The correct size wrench will fit securely on the nut with no danger of damaging the nut's corners.
3. *Place the tool in its correct position.* This means that the wrench must be square to the nut so as to maintain maximum contact. An angled or cocked position will slip.
4. *Always pull a wrench toward you.* If you should push a wrench and it slips, the chances are that you will skin your knuckles.
5. *Choose the correct type of wrench.* Match the wrench to the job, using socket or box wrenches for the heavier-duty jobs. Open-end wrenches for medium-duty work. Adjustable wrenches for light-duty and odd-sized nuts. A pipe wrench should only be used on pipes, *never on a nut,* for its teeth will damage the nut.
6. *Do not overload the wrench.* Do not hammer the handle or slip a pipe extension over it. There are special wrenches designed to take the punishment. Use them or an impact wrench to break loose a "frozen" bolt or nut.
7. *Position yourself correctly.* Your position when using a tool is most important. Assume a natural, well-balanced stance that will allow correct motion of the tool without endangering you should something slip.

TOOLS THAT HAMMER

A hammer is a tool to hit with. Its size, style, and material will vary with your needs (Fig. 2-7). Use of the wrong hammer for a given job may result in damage to it or something else. Obviously a sledgehammer would not be used to drive carpet tacks, nor should a soft-faced (plastic, rubber, wood) hammer be used to drive nails.

The appliance service technician will probably use a claw hammer for installation work and a ball peen hammer for most of his other work. Regardless of which he uses, the same rules of care and safety apply.

1. *Always be certain that the head is secure to the handle.*
2. *Hold the hammer near the end of the handle and swing so as to have the head strike its goal as squarely as possible.*
3. *Keep it clean; a greasy handle or head may cause it to slip.*
4. *Do not use the end of the handle for tamping or prying; it may split.*

TOOLS THAT PRY

The crowbar, ripping bar, and the claw hammer are the basic prying tools. The claw hammer is good for prying out nails, opening small boxes, and removing molding. Heavier-duty work requires the ripping bar; still heavier work is done with a crowbar.

TOOLS THAT HOLD

Among the best-known holding tools are the vise and plier family, but a screw-holding

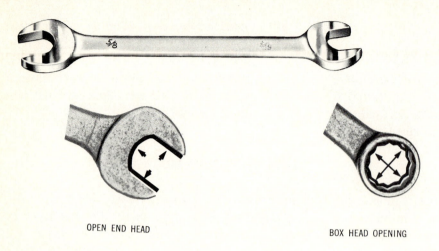

OPEN END HEAD

BOX HEAD OPENING

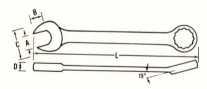

COMBINATION FLARE NUT WRENCHES

Designed to slip over tubing on to flare nut. Flare nut with open end. For use on tubing fittings—brake systems, compressors, fluid and gas lines, hydraulic units. Open end for light pressure or speed. Long handles for ample leverage. Thick heads won't damage soft metals. Forged from chrome plated special alloy steel.

NOMINAL OPENING SIZES (A)
¼″
⁵⁄₁₆″
¹¹⁄₃₂″
⅜″
⁷⁄₁₆″
½″
⁹⁄₁₆″
⅝″
¹¹⁄₁₆″
¾″
¹³⁄₁₆″
⅞″
¹⁵⁄₁₆″
1″
1¹⁄₁₆″
1⅛″
1¼″

COMBINATION BOX & OPEN END AND 15° WRENCHES

THE TYPE FOR BOTH SPEED AND SAFETY
Same opening size on each end. Use safe box end for breaking loose a nut, turn the wrench end for end, and then use speedy open end for running off nut. Reverse procedure for running on and cinching nut. Forged from special alloy steel.

OFFSET BOX WRENCHES

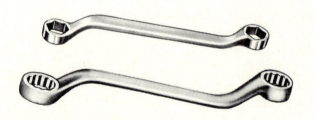

Reach into casting depressions, counterbored places and other tight spots where usual wrenches cannot be used. Heads are offset 45°, providing the needed height for many jobs. Thin, strong box walls allow maximum clearance. Forged from special alloy steel and chrome plated.

FIGURE 2-3. Wrenches are needed to suit the needs of fastening nuts to bolts. (Courtesy of Proto)

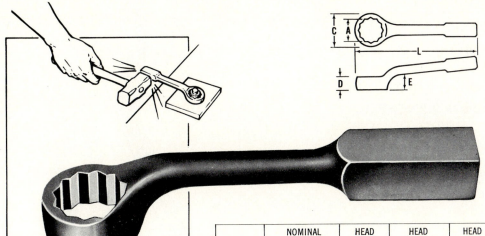

STRIKING WRENCHES, 12-POINT BOX TYPE

Striking wrenches have a square striking surface for use with a sledge or hammer. They are offset to allow maximum clearance. These wrenches are necessary for use on large bolts and nuts in restricted quarters. Like all of the heavy-duty Proto industrial wrenches, they are noted for their great strength, being forged from Proto-specified alloy steels. Although strong enough for extra heavy duty, they are streamlined to reduce unnecessary weight. Rust protection is by Parco-Lubrite finish.

NO.	NOMINAL OPENING SIZE (A)	HEAD O.D. (C)	HEAD THICK-NESS (D)	HEAD OFFSET (E)	TOTAL LENGTH (L)	WT., LBS.
2617SW	1¹⁄₁₆″	1.69″	.63″	.94″	8½″	1.70
2618SW	1⅛″	1.69″	.63″	.94″	8½″	1.60
2620SW	1¼″	1.88″	.81″	1.19″	10¾″	3.20
2621SW	1⁵⁄₁₆″	1.88″	.81″	1.19″	10¾″	3.00
2623SW	1⁷⁄₁₆″	2.19″	1.00″	1.50″	12″	4.70
2624SW	1½″	2.19″	1.00″	1.50″	12″	4.40
2626SW	1⅝″	2.41″	1 13″	1.50″	12¼″	4.65
2627SW	1¹¹⁄₁₆″	2.41″	1.13″	1.50″	12¼″	4.50
2629SW	1¹³⁄₁₆″	2.66″	1.19″	1.56″	13⁷⁄₁₆″	6.60
2630SW	1⅞″	2.84″	1.25″	1.66″	13⁷⁄₁₆″	6.80
2632SW	2″	2.84″	1.25″	1.66″	13⁷⁄₁₆″	6.60
2635SW	2³⁄₁₆″	3.38″	1.50″	2.00″	13½″	8.20
2638SW	2⅜″	3.38″	1.50″	2.00″	13½″	7.80
2641SW	2⁹⁄₁₆″	4.00″	1.50″	2.25″	13½″	8.70
2644SW	2¾″	4.00″	1.50″	2.25″	13½″	8.20
2647SW	2¹⁵⁄₁₆″	4.44″	1.88″	2.62″	16″	12.60
2650SW	3⅛″	4.44″	1.88″	2.62″	16″	12.00

½″ SQUARE DRIVE SOCKET WRENCH SETS

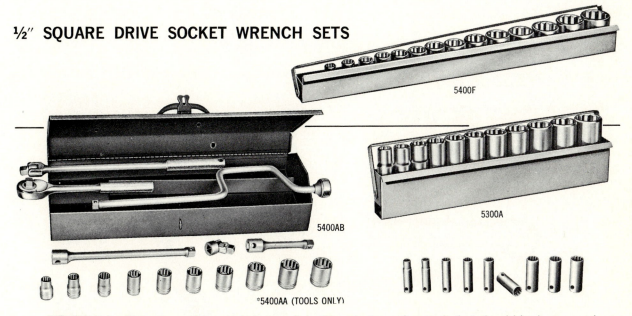

FIGURE 2-4. The square-drive socket wrench is the only type of wrench that should be hammered. The chart is included to indicate the size and weight of this type of tool. (Courtesy of Proto)

ADJUSTABLE WRENCHES

PIPE WRENCHES
CRESCENT

Heavy Duty

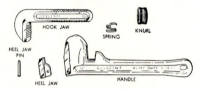

Normal Duty

PIPE WRENCH REPAIR PARTS
CRESCENT

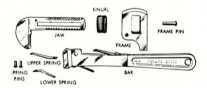

For Heavy Duty, WH Series

For Normal Duty, WS Series

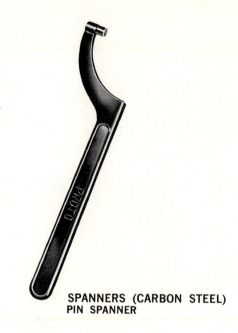

NO. 801 UNIVERSAL CHAIN WRENCH

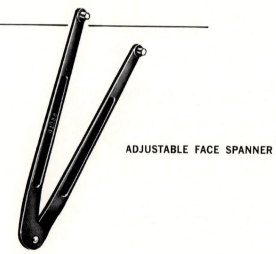

SPANNERS (CARBON STEEL)
PIN SPANNER

ADJUSTABLE FACE SPANNER

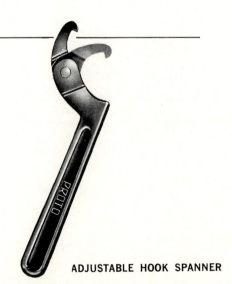

ADJUSTABLE HOOK SPANNER

FIGURE 2-5. Adjustable wrenches are available in these types in many sizes. (Courtesy of Proto)

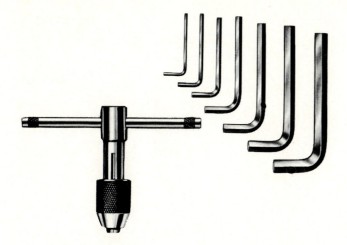

**ADJUSTABLE T-HANDLE
TAP WRENCHES**

Sliding T-handle facilitates usage, and allows use
of greater pressure. Used for holding taps, drills,
reamers, and other small tools which are turned
by hand. Chuck jaws and shell hardened for
normal use. Chuck head centered to allow use
on lathe centers or vertical drilling machines as
work starter.

FIGURE 2-6. Adjustable T-handle top wrenches and Allen wrenches. (Courtesy of Crescent)

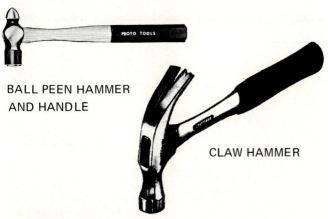

BALL PEEN HAMMER
AND HANDLE

CLAW HAMMER

FIGURE 2-7. The most commonly used hammer is the claw hammer. The ball peen hammer is used mainly by mechanics. (Courtesy of Proto and Sears)

screwdriver also holds and is of use in a tight spot (see Fig. 2-1c). A vise is not too portable; however, when mounted on a workbench, it becomes a third hand. When using a vise, consider its jaw faces and any damage they may do to the part being held. The rough jaws needed to prevent work from slipping can be covered with a piece of aluminum or other soft material to provide a smooth surface when needed. *Always try to hold the part in the center of the vise jaws so as not to put too much strain on the movable jaw and its sliding guide. Do not overtighten the jaws.* Doing so could damage the work or

the vise screw. And do not use the vise as an anvil; light tapping is harmless, but pounding could crack the vise body.

As shown in Fig. 2-8, pliers come in many sizes and shapes, each designed to fill a particular need. Basically, though, they hold or cut. Holding action is achieved by squeezing the handles together and maintaining the pressure until the part no longer needs holding. *Pliers can hold many things but are not very good on nuts because* those with rough jaws will damage the nut and those with smooth jaws will only grip on corners and slip off. Even the parallel-jawed style will not allow much pressure to be applied to the nut. Plier maintenance is not difficult—keep them clean and dry and use a drop of oil on the pivot pin occasionally.

TOOLS THAT CUT

Some combination pliers both cut and hold and are called *side cutters* (Fig. 2-8f and g) from the location of the cutting edge. Those designed for cutting only are called *diagonals* (Fig. 2-8c). The cutting action of pliers differs from scissors, shears, or tin snips (Fig. 2-9) in that the cutting edges of the pliers meet and squeeze apart the item being separated, whereas the cutting edges or blades on the others pass each other and shear an item apart.

A safety knife with a retractable blade is a handy cutting tool. The blade is protected from harm and doing damage while retracted, and when in use it cannot fold in to cut a finger.

The cold chisels (Fig. 2-10) used for cutting metal are made from a high-carbon steel that makes them tough and hard, but they could easily lose their temper (become soft) if overheated during sharpening. The chisel edge should always be kept sharp and free from nicks. When sharpening it, try to maintain the original angle by grinding only a little from each side at a time. Use very little pressure when holding the chisel on the wheel, and cool it frequently by dipping it in water. The chisel edge need not be square. In fact, it is better if it has a *slight* radius, higher in the center.

The other end of the chisel must also be dressed to prevent the upsetting, mushrooming, or rolling out of the metal being hit with the

**NOS. 201 & 202
THIN NOSE SLIP JOINT PLIERS**

(a)

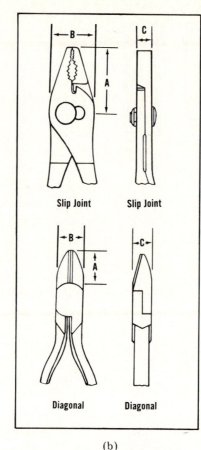

Slip Joint   Slip Joint

Diagonal   Diagonal

(b)

**NOS. 204, 205, 206, 207
DIAGONAL CUTTING PLIERS**

Thin, narrow jaws work easily in confined places. Useful for switchboard, telephone, ignition, electronic and electrical work. Induction-hardened cutting edge. High quality alloy steel.

(c)

CRESCENT UTILITY

Multiple Slip Joint

(d)

**NOS. 266 AND 268 LINEMAN'S
SIDE CUTTING PLIERS**

For utility linemen's work, wiring and electrical equipment installation and maintenance. Carefully matched induction-hardened cutting edges. Useful around filling stations, garages. High quality alloy steel.

(f)

**NO. 291R AND 292R LEVER-WRENCH
PLIERS WITH QUICK RELEASE**

Four tools in one—a powerful handvise, clamp, pipe wrench and plier. Lever action locks jaws in holding position, with one-handed pressure up to one ton. Quick release removes pressure instantly.

(e)

**(g) NOS. 226, 226G NEEDLE NOSE
SIDE CUTTING PLIERS**

The long nose, strong, extra sharp induction-hardened cutters make an ideal combination for electronic, telephone and other electrical work. A good long-reach plier for many types of jobs. High quality alloy.

(h) **TRUARC®*
RETAINING RING PLIERS**

FIGURE 2-8. Pliers. (Courtesy of Crescent & Proto)

hammer. *Always, when using a chisel, wear safety glasses to prevent chips from hitting the eyes.*

PUNCHES

There are many different-shaped punches (Fig. 2-10), each designed for a special job. The most useful to the service technician, however, is the center punch. Basically, it is a piece of small,

hardened steel rod ground to a point at one end. It is used to mark metal with an indent so that a drill can be started in a preselected spot without danger of its slipping off to some other place.

SAWS

A saw is also a tool that cuts. Saws can have many different teeth (cutting edges), each de-

TINNERS' SNIPS
CRESCENT AVIATION
No. V19L
No. V19R
No. V19S
Polished alloy steel blades; machine ground serrations. Streamlined shape. Self-locking nuts on bolts; flush mounted latch. Parkerized handles.
Nos.
V19L—Left hand, cuts right
V19R—Right hand, cuts left
V19S—Cuts straight
Standard Pattern

FIGURE 2-9. (Courtesy of Crescent)

FIGURE 2-10. (Courtesy of Crescent)

PIN PUNCHES
CRESCENT
CRESCENT TOOL CO. 488
SOLID PUNCHES
CRESCENT
CRESCENT TOOL CO. 489
CENTER PUNCHES
CRESCENT
CRESCENT TOOL CO. 486
COLD CHISELS
CRESCENT
CRESCENT TOOL CO. 160
STAR DRILLS
CRESCENT
CRESCENT TOOL CO. 161

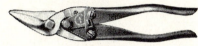

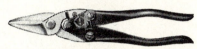

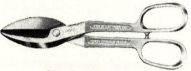

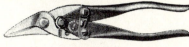

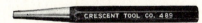

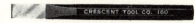

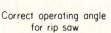

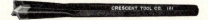

signed for a certain type of job and not really effective for a different use.

Saws for cutting wood come in two basic styles. One, which cuts in the directions of the wood grain, is called a ripsaw and can be recognized by its chisel-like teeth and their coarse spacing, approximately $5\frac{1}{2}$ tooth points to an inch. The other type for cutting wood across the grain is known as a crosscut saw and can be recognized by its beveled tooth shape and their close spacing, ranging from 8 to 12 points per inch. Saws for still finer work will have smaller teeth, up to 20 points per inch.

In use, a ripsaw is held at an angle of 60 degrees from the board being cut (Fig. 2-11), and the crosscut saw is held at an angle of about 45 degrees. In each case, use long easy stokes and *do not* force the cut. These natural cutting angles will provide enough force to feed the saw into the work comfortably.

A hacksaw (Fig. 2-12a) is used to cut metal and comes in two parts—the frame and the blade. The frame is usually adjustable to accept any length of blade from 8 to 12 inches long. The blade is a hardened steel band having sawteeth on its long side. The teeth are shaped and sharpened like those of a crosscut saw, and there can be as many as 32 points per inch. These blades cannot be sharpened and are discarded when worn.

In order to determine how coarse a blade to use, be guided by the material being cut and its thickness. The thinner the material, the more teeth the blade should have. *There should be at least two teeth in contact with the material being cut.* Very thin materials should be cut with shears.

FIGURE 2-11.

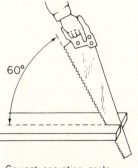

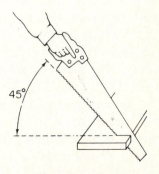

60°
Correct operating angle for rip saw
45°
Correct operating angle for cross cut saw

HACK SAW FRAMES
CRESCENT

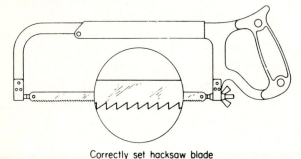

All steel frame, adjustable style. Handle is heavy gauge steel, pistol grip. Adjustable for 8 to 12 in. blades. Comes with 10 in. blade.

Correctly set hacksaw blade
(b)

FIGURE 2-12. [(a) Courtesy of Crescent]

The blade should be placed in the frame in such a way that the teeth point away from the handle (Fig. 2-12b), making the forward stroke the cutting stroke. Enough pressure must be applied to allow the teeth to bite into the metal. Rubbing the saw across the metal with little pressure will dull the teeth, as will too fast a cutting action; 40 to 50 strokes per minute would be about correct. On the return stroke, the blade should be lifted slightly so as not to rub the teeth on the metal.

FILES

Files are used to remove excess material and come in a wide variety of sizes, shapes, and styles. File teeth are like miniature chisels; as they are pushed across the part being filed, each tooth removes a small piece of material.

A file is hard and brittle and must not be hammered or used to pry. Its tang is soft, bends easily, and must be inserted into a handle to protect the user (Fig. 2-13b).

In use, it should be handled like a hacksaw with pressure on the forward stroke and relief on the return stroke. The teeth must be kept clean and any material buildup removed, either by shaking it out or brushing it out with a file card.

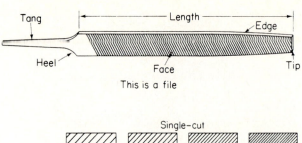

This is a file

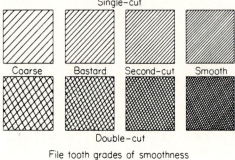

Single-cut

Coarse | Bastard | Second-cut | Smooth

Double-cut

File tooth grades of smoothness
(a)

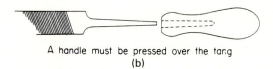

A handle must be pressed over the tang
(b)

FIGURE 2-13.

A file must be sharp to be effective. This point is particularly important when working with some of the tough copper and aluminum alloys. *These softer materials require the sharper and cleaner file.* When carrying a file in your toolbox, keep it covered and protected.

DRILLS

Drills are used for making round holes and come in two parts—the bit, which does the actual hole making, and the brace or drill, which turns the bit. Drill bits are available for wood, metal, or plastic and care must be exercised in making a choice. Size, of course, is the prime consideration, but the material being drilled must also be considered. A wood bit cannot be used on metal but a metal-cutting drill can be used on wood. There is a difference in how the drill bit is sharpened—to be most effective in hard or soft materials—and this factor must be considered if a great deal of drilling is to be done. Since many occasional drill users will merely use what is available, it would be wise to get a set of good high-speed steel drills (Fig.

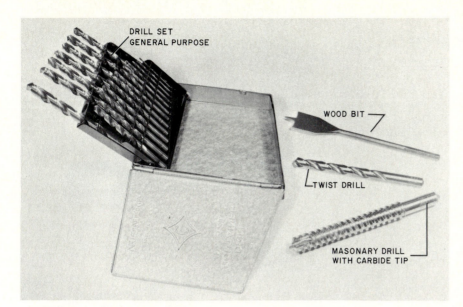

DRILL SET
GENERAL PURPOSE

WOOD BIT

TWIST DRILL

MASONARY DRILL
WITH CARBIDE TIP

FIGURE 2-14. General-purpose twist drill set will handle most of the technician's needs. The wood bit and masonry drill are used in installation work.

2-14) that are satisfactory for both wood and metal.

A special tungsten carbide-tipped drill is used for drilling holes in masonry. It is not one of the common tools a service technician would need unless he were doing a great deal of installation work.

Space does not permit discussion of all the hand tools and power tools available to the technicians, but the ones listed are the most common.

There is one important rule in all tool usage: Keep your tools clean, sharp, and protected.

POWER TOOLS

All the common hand tools mentioned are available as power tools. They do the same job in a faster and easier (for the user) manner.

As with hand tools, the best performance is obtained when the tools are kept clean and sharp and handled with care. Carelessness in tool handling can be dangerous.
For safety's sake:

1. Make certain that the power cord is kept away from the working end of the tool.
2. If the tool has a guard, be sure that it is in its proper place.
3. If an extension cord must be used, use one with large enough wire to ensure adequate voltage at the tool.
4. The part being worked on must be securely in its position and the tool must be firmly held. A slip in using a power tool can injure the user as well as damage the part.
5. All electrical power tools are provided with a grounding cord and plug. There is no problem when used in a grounded receptable, but when using an adapter, be sure that the adapter's grounding wire is properly grounded.

SPECIAL TOOLS

This discussion on tools would not be complete without mentioning special tools designed for a specific use.

The service technician should own tools listed in Table 2-1, and ought to own the test equipment. He will usually take better care of his personal tools. The average service technician is not expected to own special tools personally, but the shop for which he works is expected to have the tools needed to service the lines of appliances for which they have maintenance responsibility (Tables 2-2 and 2-3).

Generally the special tools are needed for "in-depth servicing." The part to be removed and replaced is so located or shaped that a standard tool will not fit. Or the parts have been

TABLE 2-1
Minimum recommended personal tool kit for
general appliance service

Tools
Phillips screwdriver set: Nos. 0, 1, 2, 3
Standard screwdriver set: $\frac{1}{8} \times 4$, $\frac{1}{4} \times 6$, $\frac{5}{16} \times 8$
Offset screwdriver
Stubby screwdriver
Allen wrench set: $\frac{1}{16}$ through $\frac{1}{4}$
Nut driver set: $\frac{3}{16}$ through $\frac{1}{2}$
Wrench set, open-end box: $\frac{3}{8}$ through $\frac{15}{16}$
Adjustable end wrench 10 in.
Screw starter
Vise grip wrench
Pliers, needle nose
Pliers, diagonals
Pliers, arc rib
Pliers, hose clamp
Pliers, terminal crimp
Drop cord extension

Test Equipment
Volt-ohmmeter—ac and dc
Temperature tester—50 to 1000°F (−45 to
535°C)

Miscellaneous
Floor cloth
Polish and cloths
Terminals
Tape
Fuses
Assortment—nuts, bolts, and screws
Tool chest

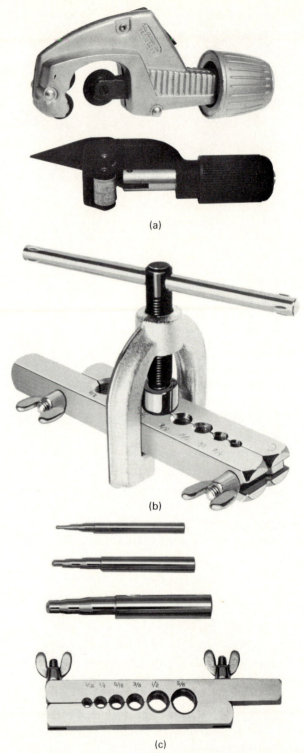

FIGURE 2-15. (a) Tubing cutters are used to cut copper, aluminum, and light steel tubing. The resulting cut is clean, neat, and square with the tubing. (b) The flaring tool is used to flare copper tubing for use with flare fittings. (c) A swedging tool is used to enlarge one end of a copper tube to fit over another of the original size. The tube to be swedged is firmly clamped in the holding bar, and the swedging tool is driven into the tube up to the shoulder of the next larger size.

assembled by expensive production tools or machines and the service tool is a simple adaptation of the machine.

Examples of special tools would be those required to remove and reinstall bearings and seals in washing machines, adjust switch contacts, impeller and clutch spacing (Fig. 2-16). Other examples are wrenches for removing special shapes of nuts, caps, and so on.

An ingenious technician can often adapt or convert a standard tool to do the job, but in most cases it will cost more than the special tool recommended by the manufacturer.

The same holds true for special gages and test equipment.

APPLIANCE TESTERS

In addition to tools for removing and replacing parts, the service technician needs something to test the appliance for performance. His most

TABLE 2-2
Minimum recommended shop tool complement for general appliance service[a]

Saber saw, variable speed
Oxyacetylane welder
$3\frac{1}{2}$–4-in. bench vise
6-in. Bench grinder–buffer
$\frac{1}{2}$-in. Drill press
Arbor press
Fire extinguisher (dry)
Heat lamps and holders
$\frac{1}{2}$-in. Portable drill
Recording thermometer
Combustion kit
Airflow meter
16-in. Manometer
Pliers, Tru-Arc
Pliers, wire stripper
Nut splitter
Bearing wheel puller
Hammer, No. 2 ball peen
Spray gun
Air compressor
Shop vacuum cleaner
Vacuum tester gage
Appliance test panel
Compound gauges kit
Capillary-pressure tool
Workbench
Swedging tool set

[a] It is assumed that both the personal and truck tool complement is available for shop use, or else there will need to be duplication. In addition to the tools and test equipment above, you will need miscellaneous parts and fittings for appliances most frequently serviced.

TABLE 2-3
Minimum recommended truck tool complement for general appliance service[a]

6–8 ft. Tape measure
Punch and chisel set
Hammer, claw
Socket set
Flare nut wrench set
Adjustable end wrench, 6 in.
Pipe wrench, 10 in.
Shears 10-in. compound
Saw keyhole hacksaw
File set, 4 pieces
Point dressing set
Soldering gun
$\frac{1}{4}$-in. Electric drill
Twist drill set
Wood bit set
Pry bar
Fire extinguisher (dry)
$\frac{3}{4}$-in. Star drill
14-in. Cold chisel
Propane torch set
Tubing cutter (Fig. 2-15a)
Flaring tool (Fig. 2-15b)
Tubing bend set
Service valve kit
Tubing brushes (2)
Appliance truck
Test cord, 110v

Specifically for Refrigeration
   Vacuum pump (0.1 μm, 2 ft³/min)
   Dial-A-charge cylinder
   Leak detector, propane

[a] In addition to the tools and test equipment above, you will need miscellaneous parts and fittings for appliances most frequently serviced. Normally, these tools belong to the employer.

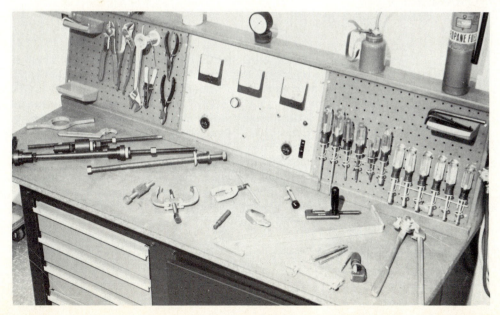

FIGURE 2-16. In addition to the basic hand tools, a number of special-purpose tools used in appliance servicing are shown on the bench.

important tool is the multimeter for electrical testing (Fig. 2-17). The next is a thermometer for temperature testing (Fig. 2-18).

After these two, the list lengthens to include many different types of testers, each needed in its special field. These are briefly mentioned and described in the accompanying illustrations.

Vacuum gage (Fig. 2-19)
Pressure gage
Leak detector (Figs. 2-20 and 2-21)
CO_2 indicator
Sling psychometer (Fig. 2-22)
Humidistat
Temperature recorder (Fig. 2-23)

FIGURE 2-17. The multimeter is the service technician's most important test instrument: (a) Triplett model 615 maintenance tester that tests for ac and dc volts, ac amperes, circuit resistance, temperature, and leakage current; (b) digital multimeter with test leads and battery charger.

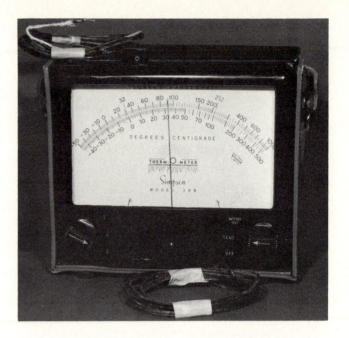

(a)

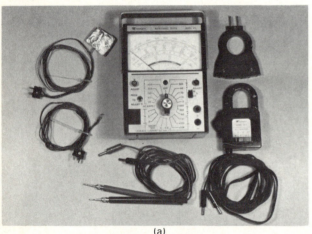

(a)

(b)

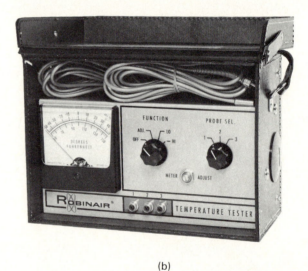

(b)

FIGURE 2-18. (a) (*Simpson Model 388*) This temperature-measuring instrument uses an iron–constantan thermocouple to include temperatures ranging from −50 to +1000°F (10 to 535°C), which allows for its usage on all appliances from freezers to self-cleaning ovens. (b) (*Robinair*) Low-temperature tester that ranges (in two scales) from −50 to +150°F (−45 to 65°C) and uses a thermister as a sensing unit. It is used principally in refrigeration servicing. This tester has provision for three inputs, thus allowing temperature reading of the freezer and refrigerator sections of a box and ambient or surrounding air simply by turning a switch.

Voltmeter—ammeter—wattmeter
Vacuum pumps
Refrigeration charging equipment
Air volume gage (Fig. 2-24)
Capacitor analyzer (Fig. 2-25)
Compressor test cord (Fig. 2-26)
Volt-amp-wattmeter (Fig. 2-27)
Clamp-on ammeter (Fig. 2-28)
Appliance cart (Fig. 2-29)

FIGURE 2-19. (*Robinair*) Double-port manifold with access valve is the proper name for this gage set used to determine the pressure of the refrigerant in a freezer, refrigerator, air conditioner, or dehumidifier. This is one of the service technician's most important tools.

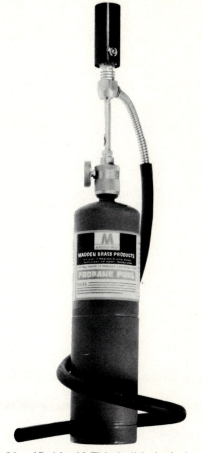

FIGURE 2-20. (*Robinair*) This halide leak detector uses propane gas for fuel and is used to locate a Freon gas leak. In operation, the torch is lighted and the snifter tube is moved along the system's piping. A leak is evidenced by a change in the flame color.

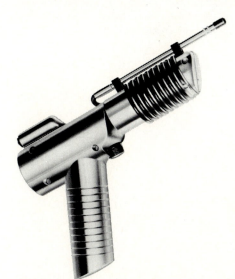

FIGURE 2-21. (*Robinair*) The electronic leak gun is an extremely sensitive refrigerant leak detector, capable of sensing a ½-ounce per year leak of the halogen gas contained in the Freon refrigerant. It is battery operated.

FIGURE 2-22. The sling psychrometer and the psychrometric slide rule are used to determine the relative humidity. The wick is saturated with water and placed over the bulb of one thermometer (wet bulb). The thermometers are swung through the air, allowing the water to evaporate from the wick, cooling the thermometer. The readings from each thermometer are matched on the slide rule and the arrow at the right-hand end of the scale will indicate the relative humidity.

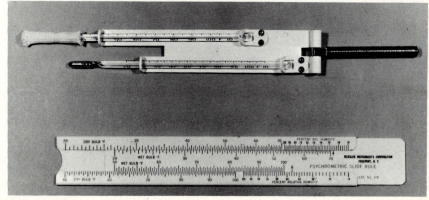

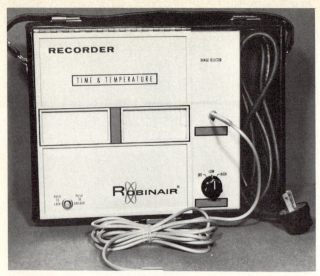

FIGURE 2-23. (*Robinair*) A time and temperature chart recorder provides a 24-hour record of temperature changes in a specific area: for example, in a refrigerator, freezer, or air-conditioned room.

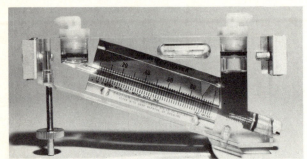

FIGURE 2-24. The air volume gage has a scale range of 0 to 1 inch and is used for checking the pressure drop across the coils in central air conditioning systems.

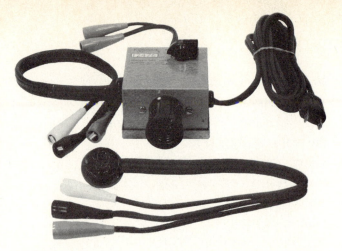

FIGURE 2-26. (*Robinair*) The compressor test cord is used to quickly test the compressor motor independently from the rest of the wiring.

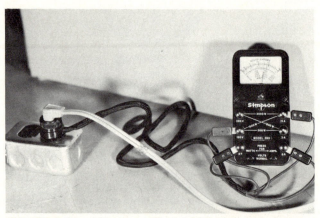

FIGURE 2-27. (*Simpson Model 390*) Volt-amp-wattmeter will measure ac/dc volts, ac amperes, and ac watts simply by shifting connections as indicated.

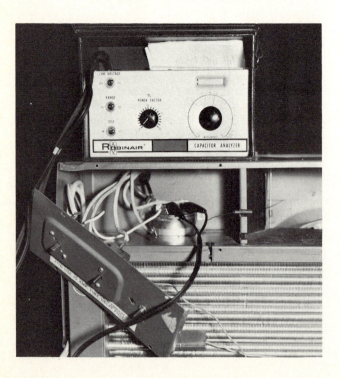

FIGURE 2-25. The capacitor analyzer measures both the capacity and power factor of any capacitor in common appliance usage. It is testing the value of a motor starting capacitor used in an air conditioner.

FIGURE 2-28. The clamp-on ammeter is used to measure the current flow (amperes) in an ac line without cutting into the wire itself. The transformer jaws are opened to encircle one conductor, then are closed. The meter will indicate the current flow in amperes.

FIGURE 2-29. This appliance cart is used to move appliances. It has a special belt track to aid in going up and down stairs.

STUDY QUESTIONS

Tools

1. Testing tools serve to amplify or magnify man's _____, and all others help to augment or increase man's _____ and _____.

2. Screwdrivers also make good narrow chisels. True or False

3. When a Phillips screwdriver becomes worn, it should be discarded.   True or False

4. List the two advantages of wrenches over nut drivers.

5. Wrenches are versatile tools that can be pushed as well as pulled and adjusted to any size nut. True or False.

6. An important fact to remember about the vise is _____.

7. There are three basic rules for handling and caring for all tools. List them.

8. Tools really do six jobs. After each tool below, write whether its main job is to turn, hold, cut, hammer, pry, or test.

   Knife   Thermometer
   Phillips screwdriver   Pliers
   Vise   Chisel

Drill   Hammer
Saber saw   Voltmeter
Crowbar   Ruler
Awl   Reamer

9. Sawtooth spacing should be closer than the material thickness being cut.   True or False

10. Since portable power tools are small, any size extension cord may be used.   True or False

11. If the receptacle for a power tool plug does not have a grounding terminal connection, it is all right to cut off the grounding pin on the cord's plug.   True or False

12. The most important piece of test equipment a service technician needs is a _____.

KEY

Tools

1. Senses, fingers, muscles
2. False
3. True
4. More leverage and right angle for handle clearance

5. False
6. Not to overtighten
7. Keep clean.
   Use right size and type for job.
   Use correct position.

8.

Tool	Main Job
Knife	Cut
Phillips screwdriver	Turn
Vise	Hold
Drill	Cut
Saber saw	Cut
Crowbar	Pry
Awl	Cut
Thermometer	Test
Pliers	Hold
Chisel	Cut
Hammer	Hammer
Voltmeter	Test
Ruler	Test
Reamer	Cut

9. True
10. False
11. False
12. Multitester or volt-ohmmeter

3

The Basic Approach

CAUSES OF FAILURE

Servicing is a process of observing and reasoning, in which you apply your general knowledge to the specific problem on hand. Many things can go wrong with any product. Some may not be within the machine itself; there may be several external factors influencing the machine's performance. These, too, must be checked. Basic product failures result from five general causes:

1. External factors
2. Physical damage
3. Poor workmanship
4. Part failure—deterioration
5. Obsolescence

Let's take a quick look at each of these items individually; all will be amplified later as they relate to specific products.

EXTERNAL FACTORS: ENVIRONMENT OR OPERATOR

The two major external factors contributing to failure are environment and operator. These should not be called a product failure because the product has not failed. Rather, *the appliance has not been supplied with the correct operating conditions.* The product may not have sufficient air, gas, electricity, water, or whatever else it needs to do its job properly. The serviceman must recognize this point and do everything he can to impress on the customer the need for a correct operating environment to enable the machine to perform properly.

The machine may not have been installed properly or it may have been moved and incorrectly repositioned. A fuse could have burned out (Fig. 3-1) or a strainer in a water line could have become clogged. Look the situation over very carefully before starting to service the appliance.

(a)

(b)

FIGURE 3-1.   Make certain that there is power to operate the machine. The problem may be a blown fuse (a) or a tripped circuit breaker (b).

The other external factor is operator knowledge. Many a service call has been made because *someone using the machine did not know how to use it* and blamed the machine for poor performance (Fig. 3-2).

Operator misuse of an appliance is a serious factor in service calls, for it will result in both poor performance and shortened life for the appliance. In either event, the customer is unhappy with the machine and he is willing to tell the world about the poor results he is getting.

PHYSICAL DAMAGE: ABUSE OR ACCIDENT

Physical damage to the machine might occur during transportation, whether from the factory to the dealer or from the dealer to the installation, or simply from being moved from one place to another within its normal area of usage (a vacuum cleaner). *Transportation damage* can

occur because of incorrect packaging. If a machine is loosely set in a carton or poorly padded in a truck, rub marks on the finish or dents in

FIGURE 3-2.   If you can get the customer *and the other appliance users* to read the manual, many nuisance service calls can be avoided.

the cabinet may develop. Or if the product is dropped, say from the tailgate of a truck to the ground or off a shipping dock, there may be damage to some of the parts within the machine. This is careless handling and can occur anywhere. These are "accidents."

Accidents can happen at home, too. Someone trips over the toaster cord, pulling it off the table, or a pipe springs a leak, flooding the oil burner motor.

The *user can abuse* the machine because he did not read the instructions and did not use it correctly. People usually tend to overload a machine, for example, trying to mow tall grass with a low-horsepower mower.

A machine can be physically damaged by an *unqualified serviceman who did not know the correct procedure for replacing a part.*

Each occasion poses a different service problem.

POOR WORKMANSHIP IN FACTORY, INSTALLATION, OR SERVICING

Early parts failure can occur in normal usage because of poor design. Yes, this happens occasionally when a designer does not take all factors of usage into consideration.

It may be that the tolerances were incorrect and the part never did fit properly, or being forced into position may have caused damage or misalignment of mating parts. A moving part that is assembled out of tolerance (or not quite the right size) may cause excessive wear (too tight) or noise (too loose). For example, if a bearing on a shaft is too loose, it would allow the shaft to vibrate and cause excessive wear, thereby shortening the life of the bearing. A bearing that is too tight would bind the shaft, thus causing that part to overheat and create still further bearing damage and possibly scoring the shaft itself.

Failures resulting from poor manufacturing and/or installation and service workmanship might well include such items as missing parts, like lock washers, loose parts that are not properly bolted together, or loose parts resulting from a poor welding operation that inspection let slip by. Subsequent handling and usage may cause a weld to separate. Some parts

may be incorrectly located, thus creating an interference with others, or a wrong component may have been used. These points will be corrected by the manufacturer when he is notified.

Again, as with part failure resulting from design, the manufacturer will do everything in his power to prevent such conditions from happening. Quality control is becoming more rigid all the time.

However, *the manufacturer cannot correct a fault unless he knows it exists. You must notify him when you find such a condition,* but be certain it is his fault and not a poor installation, a reworked machine, or an isolated instance.

PART FAILURE: (CHEMICAL AND PHYSICAL) DETERIORATION

Parts may fail in normal usage because they have not been properly maintained. It is probable that more power lawn mowers are relegated to the junk heap every year because of the lack of proper lubrication than any other one factor. With a power mower, lack of lubrication gets serious very rapidly; with other moving parts, it accelerates wear, causing premature failure.

Some wear is normal.

Part deterioration is serious because it often ends the life of the machine. It may result from *corrosion* or *material fatigue. Corrosion* is a chemical failure due to the action of air, moisture, detergents and other chemicals on the parts. Rust is one form of corrosion (Fig. 3-3). *Material fatigue* is a physical failure due to vibration, unbalanced operation, faulty installation, or misuse.

OBSOLESCENCE: DOLLARS VERSUS PERFORMANCE

Obsolescence is not generally planned into a product by the manufacturer; it usually results from an increased knowledge of the product as improvements in materials and processes are made. We know that our automobiles of 20 years ago, or even 10, are rather obsolete in terms of modern road conditions and the expectations that car owners have for their vehicles.

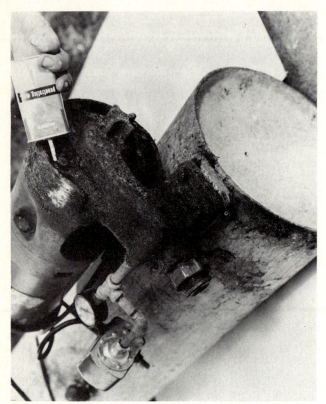

FIGURE 3-3. Rust is a great contributor to part failure. When nuts and bolts are rusted together and penetrating oil does not work, try (1) an impact wrench, (2) heat, (3) a nut cracker (there is such a tool), (4) a hammer and chisel. Is it worth the time and effort?

Certainly there are few of us who own a 20-year-old television set.

Deterioration and obsolescence sometimes *make it more practical for the customer to purchase a new machine rather than repair the old one.* This is a decision the customer must make; you can only present the facts. Much of his decision depends on which part has gone bad, the cost of repairs, and how much better the new model is. The customer must decide whether the new features are worth the extra money.

IS THERE A PROBLEM?

What steps should be considered when first approaching a machine for servicing? You have already talked to the customer, and from him have gained some information as to what to look for and where to look for it. *You must now make certain that there is a problem* (Fig. 3-4)

FIGURE 3-4. The freezer was okay. Only the indicator bulb had burned out. Check the simple and obvious first.

and that this is not a case of an inexperienced operator using the machine incorrectly or failing to turn on the correct switch.

MACHINE USAGE

Check adjustments of the machine to make certain that they are correct. Is it possible that an inexperienced operator has overloaded the machine? Was the appliance oversold? Could the customer have expected more from the machine than it was capable of delivering? In the appliance field, you must remember that you cannot compare dissimilar models even from the same manufacturer. It would be like saying that all dogs are good hunters. Each make and model has specific operating characteristics.

INSTALLATION

Check to make certain that any appliance using electricity has the right kind of electrical power available for it. For example, if the nameplate or data plate calls for 220 or 110 volts, single or three phase, that is what must be supplied. On simpler appliances, make certain that they are plugged in (Fig. 3-5); that power is available at the outlet, and that the wiring from the plug to the machine is able to carry the load. On a gas appliance, you must be sure that there is gas available and that the valve has not been shut off in the supply line. An oil-burning appliance must have fuel and the filters must be clean. An oil burner with a partially clogged filter can starve and not have sufficient fuel to operate properly. The same holds true for an appliance or machine that requires water. The water must be turned on, and there must be an adequate supply of the right kind. By right kind of water, we refer to the temperature of the water, its quantity and quality. Water that has a high iron or lime content will coat the inside of the tank and pipes quickly, thus decreasing the efficiency of immersion-type heater elements, for instance. These chemical impurities in water cause much damage to both household and industrial equipment.

OBSERVATION

Occasionally, a machine will be found in such condition that the service procedure is self-evident; that is, a machine that has been in a flood or fire will require a different service approach. There is no need to test this machine until it has been completely and thoroughly dried out and cleaned up. Actually, turning on a machine in this condition will do additional harm. Take care of the obvious before looking for more trouble.

FIGURE 3-5. Make certain that the appliance is plugged in and that the outlet is live.

FIGURE 3-6. Burned insulation has a very distinctive acrid odor. Hot dust and lint are different.

When you approach a machine that has simply failed in operation and you find that the external factors are all right, then **stop, look, listen, smell,** *and* **think.** This use of your basic senses can tell you much about the condition of the machine.

When you turn on the machine to try it, the *smell of burning insulation* is a warning to turn off or disconnect the machine at once (Fig. 3-6). You can usually track down the smell and discover where a part has burned out. You can see the charred area around the motor or that a piece of insulation has chafed and torn off, thus creating a short.

It is simple to remove a cover and examine the inside of a machine; then *listening* (Fig. 3-7) should tell if there are any *unusual squeaks, rattles, or bangs.* Unusual noises are generally due to something that is loose, vibrating, or moving out of place and hitting another part in the appliance. You will have to listen and look. You should look first for a foreign object. It could be anything: a bottle cap, a bobby pin, or a piece of scrap material that has worked itself down into the machine. Correcting such a malfunction might be a simple matter

FIGURE 3-7. If you know how a machine *should* sound, it is easy to recognize an unusual or unfamiliar sound.

load conditions. Belts should be checked for tension.

Looking carefully will often reveal a great deal about the condition of the machine (Fig. 3-8). Just plain looking will show whether

1. Parts are in their correct position.
2. Something is hanging out of place.
3. There is an unusually large amount of dirt or grime in a specified area.
4. There is a collection of dust, lint, or rags in some corner of the machine.

Mice have been known to build nests in appliances, and the nest material can get in the way of some vital part. An unusually large amount of grease or oil in any given area should be thoroughly checked. This could mean that there is a leak somewhere.

CLEANUP

A machine to be checked should be cleaned thoroughly both inside and outside. It will perform

FIGURE 3-8. A visual examination will reveal much about a machine's condition. *But* you must first know how a good machine looks.

of removing the foreign object. It may only be necessary to tighten some screws, adding lock washers to prevent a recurrence. There are also types of noises made by two parts rubbing against each other when they should not. These parts should be found and then separated, because rubbing of this nature will cause unnecessary wear and further damage eventually. Parts that should slide or rub are designed and lubricated for the job; others are not.

A bearing that is worn can be located by grasping the shaft and trying to move it. This is one way to diagnose damage if the shaft is badly worn. If the bearing is dry, it will make its own peculiar noise. Again, what is to be done in the way of repair depends on the condition. Sleeve bearings can be lubricated; ball bearings and needle bearings can be repacked if it is not too late. Otherwise they will have to be replaced.

A slipping belt will have its own particular noise, and it can create quite a few other problems in the machine. If it slips, some of the other parts that are being driven by the belt will operate at a slower speed than normal, particularly under a heavy load. However, they will usually operate at a normal speed under low better and it will be much easier to maintain. Many little things, like an oil leak or a water leak, are self-evident when you look for them, and it is much easier to find them in a clean machine.

DON'T JUMP TO CONCLUSIONS

Look before you leap!

First, look carefully and make an analysis before you start taking the machine apart to find the trouble (Figs. 3-9 and 3-10). Again, when you first approach any piece of equipment for servicing, stop, look, and listen; then analyze what you see before you start working.

Your quick and certain diagnosis of any trouble in a machine is entirely dependent on your knowledge of what the machine is supposed to do and what each part in that machine does. Always remember that any machine is simply a collection of miscellaneous parts, each of which has a certain thing to do.

SOLVE THE PROBLEM FIRST

All that you know, you have learned. That which you do best you have practiced the most, whether it be soldering or bowling. As you train your hands to do certain intricate jobs, you can train your mind to be more alert, sharper, or better able to solve problems.

And *appliance servicing is first a problem-solving job, then a repair job.* Your income depends on your efficiency in both areas.

EVERY PROBLEM REQUIRES
A DECISION

There are many different problem-solving techniques, but basically all use the following sequence in analysis to help you decide.

1. What is the apparent problem?
2. What is the real problem?
3. What are the related problems?
4. List all solutions:
    (a) Short range.

FIGURE 3-9. Do not assume. Check the wiring diagram and any other related information for the machine you are servicing.

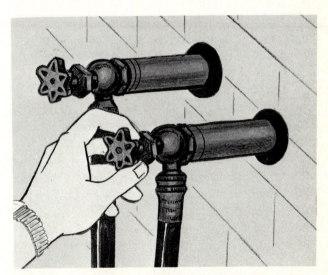

FIGURE 3-10. Always check the faucets, connections, and *hose screens* on a water-using appliance. These screens can clog with rust from the plumbing.

(b) Long range.
5. ACT!

For example, on a dishwasher service call the complaint is that the water does not drain out; this is the apparent problem.

The real problem may be a restricted drain line, jammed pump, defective motor, timer, or wiring. Your check-out shows the drain line to be clogged with a piece of bone that has caught some other food particles—finally plugging the line. Thus the real problem was a clogged drain line.

The related problem is "How did the drain line get clogged?" Was the strainer incorrectly placed? Were the dishes incompletely rinsed prior to loading the machine?

The short-range solution is to clear the drain line and make the machine operative.

The long-range solution is to instruct the customer on correct dish rinsing, loading, and strainer placement.

SUMMARY

The five general causes of product failure are external factors, physical damage, poor workmanship, part failure, and obsolescence.

Each contributes to the decision as to whether the machine needs repair or cannot be repaired. The service technician must first determine whether there is a problem, next isolate it, then repair it.

He observes the installation, discusses the usage, and examines the machine, plotting his knowledge of the machine against what he discovers.

Once a problem has been determined, he must check for the real, apparent, and related problems and then list all solutions, finally using the ones that satisfy both the immediate failure and its cause.

REVIEW QUESTIONS

1. List five general causes of product failure.
2. Name two external factors that may cause product failure.
3. What can cause parts to fail?
4. What problems can arise from an incorrectly installed appliance?
5. List the four observations to be made before commencing to service the machine.
6. Describe the steps in a problem-solving procedure.

STUDY QUESTIONS

1. The two major external factors that might cause machine failure are (a) climate and age, (b) operator and installation, (c) product failure and environment, (d) operator and environment.
2. A factory-caused failure might be (a) poor installation, (b) incorrect tolerances within the appliance, (c) ignorant operator, (d) transportation.
3. If the fault in an appliance is factory caused, (a) wait until you hear from the manufacturer to repair the part, (b) communicate immediately with manufacturer, (c) be ingenious and fix up something to keep your customer happy, (d) tell your customer how to get in touch with factory.
4. Two frequent reasons for encouraging a new purchase rather than the repair of an appliance are _____ and _____.
5. The first step in learning that an appliance needs repairing is (a) find the broken part, (b) know what the machine is supposed to do, (c) check all adjustments, (d) be sure that machine is plugged in.
6. Five basic actions are involved in approaching a service call. Name them.
7. Using one's nose may tell that (a) a bearing is worn, (b) there is no power for the machine, (c) insulation is burned off somewhere, (d) a leak has developed.
8. By using one's sense of touch to approach a repair job, one can learn (a) there is no power at the source, (b) parts are too worn to function well, (c) a part is getting too hot, (d) all of these.
9. By using one's eyes one can see whether (a) there is an unauthorized leak, (b) there is enough oil in the proper places, (c) the machine is plugged in, (d) bearings are worn.
10. Before an appliance is serviced, what must happen?

KEY TO REVIEW QUESTIONS

1. Physical damage, part failure, external factors, obsolescence, and poor workmanship
2. No power supply, inexperienced operator
3. Foreign object in machine, friction, poor or incorrect maintenance

4. Wrong power, unstable, unlevel
5. Look, listen, smell, think
6. Make analysis, know what the machine *should* do

KEY TO STUDY QUESTIONS

1. d
2. b
3. b and c
4. Obsolescence and deterioration (cost of repair)
5. b
6. Stop, look, listen, smell, think
7. c
8. b and c
9. a and c
10. Its problem must be diagnosed.

4

Installation

The correct installation of an appliance is an important step in ensuring its proper functioning. It should be one of the first factors for a service technician to check when he makes a service call.

GENERAL INSTALLATION NEEDS ▬▬▬

The technician should check the appliance and the space around it for enough clearance to ensure adequate ventilation. The appliance should also be level, secure, and have proper utility connections.

This may sound like an oversimplification, but *many appliances are installed by untrained personnel.* Frequently, such appliances do not work right, thereby leaving the customer with a poor impression of the product. And if she is "angry" enough, the result will cost the company an "in-warranty" service call.

1. *All appliances should be level.* Check the working surface with a spirit level (Fig. 4-1).
2. *All appliances should be securely placed on the floor.* Most appliances have leveling screws attached to their legs. (These screws can be adjusted so that all the feet make contact with the floor. Further adjustment can then be made to make sure the working surface is level.) *Note:* The floor itself should be secure and free from abnormal spring.
3. *Location should be chosen to:*
   (a) Provide an *adequate work area* for the user.
   (b) Provide *accessibility* to the machine's parts for the service technician.
   (c) Provide adequate *ventilation* to prevent working parts from overheating.
   (d) Allow for *cleaning* dust, lint, cobwebs, etc. from the back and inside of the machine.
   (e) Allow easy *connection to an adequate source of electricity, gas, water, oil, drain, chimney,* or whatever else the machine needs to operate as it should.

FIGURE 4-1. Always level the appliance on its working surface, which for this built-in oven is a shelf.

UTILITY CONNECTIONS

All utility connections must be made in accordance with the local plumbing, electrical, and building codes. Unless licensed to do so, *do not get involved in installing an electrical outlet, gas, or water line, or connection to a sewer!* These jobs must be done by those who are legally liable.

You can connect to already-existing utility outlets, but when you do, make certain that

1. The electrical outlet supplies the correct voltage (Fig. 4-2).
2. The electrical circuit can handle the new load.
3. The voltage drop under load is not excessive. (The limit is 10 percent.)
4. The gas supply is the right kind for the appliance burner.
5. The gas line is clear of moisture, rust, and other contaminants.
6. The gas line is large enough to supply the volume of gas needed.
7. The water pressure is within the operating limits of the appliance.
8. The flow rate is high enough to supply all the water needed within a reasonable length of time (Fig. 4-3).

(a)

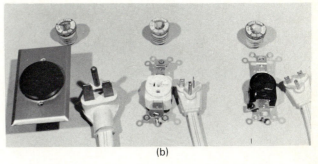

(b)

FIGURE 4-2. (a) Standard fuse, receptacle, and cord used with air conditioners rated at less than 15 amperes at 120 volts; (b) fuses, receptacles, and cords used with 220-volt air conditioners rated as follows: 30 amperes (left); 20 amperes (center); and 15 amperes (right). Note that these cords cannot be connected to a standard 110-volt receptacle.

FIGURE 4-3. An easy way of measuring water flow rate is timing the filling of a two-quart bottle or carton. Flow rate is stated in gallons per minute:

$$\text{GPM} = \frac{\text{quantity}}{\text{time}} = \frac{\frac{1}{2}\text{ gallon}}{\text{seconds}/60} = \frac{30}{\text{seconds}}$$

Thus a time reading of 30 seconds = 1 GPM; 15 seconds = 2 GPM; 10 seconds = 3 GPM.

9. The water temperature is high enough (or low enough) to allow the machine to do its job.

10. The plumbing drain has a trap between the appliance and the stack.

11. The drain has sufficient pitch to allow flow away from the appliance.

12. The chimney flue is not clogged or leaky.

This is a brief summary of basic appliance-installation needs. Each type of appliance has specific installation requirements that must be met in order to obtain proper appliance performance. Having ascertained that the environment is satisfactory one should now install the appliance.

FIGURE 4-4. An appliance cart makes moving appliances a great deal safer and easier.

DELIVERY AND PREINSTALLATION

An appliance may be delivered to a customer either crated or uncrated. The particular condition has been agreed to by the customer and the salesman. Some people want to receive their appliances in factory-sealed cartons because then they know that the merchandise is new and no one has tampered with it. Others prefer buying the floor sample, for that is the one they saw demonstrated and they know that it works.

In either case, the installer who delivers the machine must handle it with care to ensure that the customer gets the machine in the condition he bought it.

THE UNCRATED MACHINE

When delivering an uncrated machine, make certain before loading it on the truck that it works, is cleaned and polished, and that all the parts and literature are with it. Wrap the machine carefully in furniture pads and strap it securely in the truck.

Loading and unloading should be done carefully to avoid damaging the finish, denting the cabinet, or bending a leg. Use an appliance cart to move it into the house (Fig. 4-4).

Do not remove the protective pad until the machine has been placed in its final location. It should then be checked for any damage.

THE CRATED MACHINE

Before loading a crated machine on the delivery truck, *check the carton to make certain that it is the make and model the customer ordered.* Also, *inspect the carton for any signs of damage.* Corners are particularly vulnerable.

At delivery time, as with the uncrated machine, transport it upright and securely strapped into the truck.

Do not remove the carton until the machine has been delivered to its final location. In removing the carton, follow the manufacturer's instructions to avoid damaging the cabinet.

Many of the major appliance cartons are removed by cutting the carton at the base and lifting it up and over the machine. The base is later removed from under the machine by unscrewing the shipping bolts or blocks and lifting the machine off the base when setting it in its working position.

As soon as the machine has been uncrated, inspect it for external damage. Then remove the various tapes, ties, blocks, and other shipping retainers used to keep the parts where they belong during transit.

Next, remove instructions, owner's manual, detergent samples, and the installation kit. These items are usually packed inside the machine. Inspect the machine carefully for missing parts and any visible damage.

Now you can go about the business of installing the machine.

A final step—show the customer how to use the machine.

WORKMANSHIP

Not mentioned *but implied in all installations is the quality of workmanship.* The appearance and neatness of the installation, in addition to a properly operating machine, will help ensure customer satisfaction.

An important part of this workmanship is your ability to *explain to and show the customer how to use the machine.* Then leave all the instructions, warrantees, and so on with the customer.

One last word on general information, *read the manufacturer's instructions* on unpacking, preparing for installation, and installation of the particular model involved. The preceding information can guide you through *most* installation jobs but *should not* supercede the manufacturer's word for his machine.

STUDY QUESTIONS

Installation—General

1. True or False
   (a) All appliances should be leveled.
   (b) All four feet should be on the floor.
   (c) The amount of available work area is not important.
   (d) An appliance should be well ventilated.
   (e) Since the back of a machine cannot be seen, dust can be allowed to accumulate.
   (f) If an outlet is not handy, it is all right to use an extension cord.
   (g) A 20 percent voltage drop is allowable.
   (h) A gas burner adjusted for natural gas will work on any gas.
   (i) the water-flow rate should be reasonably fast.
   (j) Appliances do not need traps between their outlets and the main stack.

2. What can be considered as good workmanship?
3. How should an appliance be delivered to a customer?
4. What is the last thing an installer must do with an appliance?

SPECIFIC REQUIREMENTS

The following installation requirements listed for specific appliances are excerpted from a number of different manufacturers' installation manuals. Slight variations will occur from model to model, and part of what is included here may not have direct bearing on a specific machine. The fundamental information, however, is the same for all of each given type of appliance. Installation instructions are included for those appliances in common use that have definite needs. Each will be handled individually.

CENTRAL VACUUM SYSTEMS

The built-in (central) vacuum cleaning system has a permanently installed power unit connected with tubing or pipe to wall outlets. The wall outlets are designed to accept a vacuum cleaner hose and to self-seal when not in use. The power unit collects the dust and dirt.

Location of Power Unit

The power unit should be located on the main floor or basement level, preferably in the garage or utility room.

The outside exhaust should be situated away from living and traffic areas to prevent the exhaust noise and air from disturbing the occupants of the building.

The power unit must be firmly mounted on the wall, high enough to allow the dust collector to be emptied easily (Fig. 4-5).

Other considerations are

1. 120-volt ac, 60-hertz power source
2. Connection of the $1\frac{1}{4}$-inch suction line (dust-conveying system) to the intake
3. Connection of the exhaust tube to an outside vent

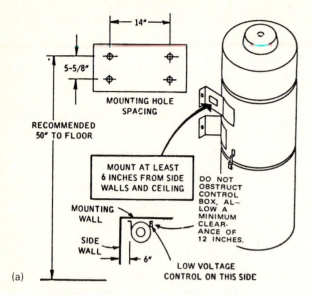

FIGURE 4-5. (a) When mounting the power unit, make certain that the wall is secure and at least one set of mounting screws are in a stud. Observe the need for clearance around the unit. (b) If the wall is not too secure, mount a piece of plywood on it to span three studs. Then mount the cleaner on the board. [(a) Courtesy of Sears Roebuck & Co.]

4. For a wet system, adequate water supply and drain facilities

Location of Inlet Valves

A central vacuum system uses a 25-foot flexible cleaning hose. Thus the inlet valves should be located in such a way as to allow cleaning accessibility to all parts of the house.

In addition, the built-in dust transmission line ($1\frac{3}{4}$-inch OD plastic pipe) should be located in a way that offers the least number of bends between the inlet valves and the power unit.

Figures 4-6 to 4-8 illustrate a typical installation.

Electrical Controls

There are two recommended methods of controlling the electrical circuit.

1. A single, manually operated wall switch for turning the power unit on and off.
2. A low-voltage system wired to each wall valve, allowing the user to turn the power unit on or off at each location. Figure 4-9 is a schematic wiring diagram for this type of installation.

All electrical wiring must be done according to the local electrical code.

STUDY QUESTIONS

Installation—Central Vacuum Systems

1. What does the power unit do?
2. Where should the power unit be located?
3. Should the power unit exhaust into the garage?
4. Where should the inlet valves be located?
5. What size and kind of tubing should be used in the dust transmission line?

DISHWASHERS

Location

A built-in or undercounter dishwasher is usually located under the kitchen counter, near the sink in a space that would otherwise have been used for a storage cabinet.

Most of the current machines will fit into a $24\frac{1}{2}$-inch-wide space and require 35 inches free height and 25 inches depth. Figure 4-10 is typical of dishwasher dimensions.

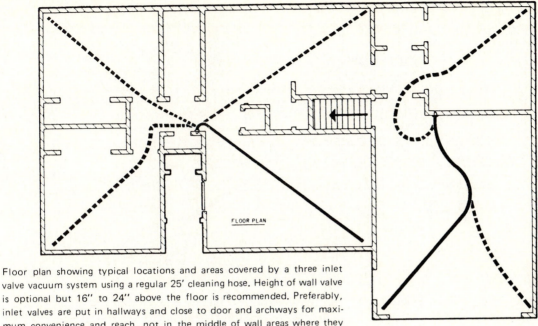

Floor plan showing typical locations and areas covered by a three inlet valve vacuum system using a regular 25' cleaning hose. Height of wall valve is optional but 16" to 24" above the floor is recommended. Preferably, inlet valves are put in hallways and close to door and archways for maximum convenience and reach, not in the middle of wall areas where they could be obstructed by large pieces of furniture and equipment.

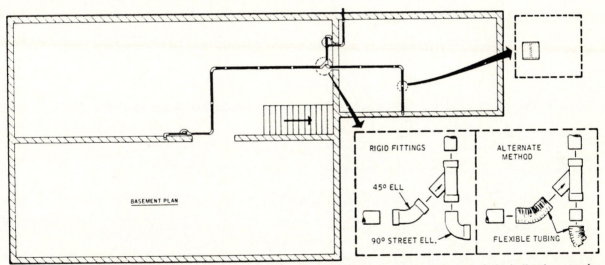

Typical plan and elevation views of a three valve vacuum cleaning system installed in a ranch style one-story home, with basement. One wall valve in regular living area, one in garage, and one utility valve in basement. Tubing conveying system and power unit installed in basement with tubing runs at ceiling height.

The conveying system schematic is basically the same for any one-story design. If there is no basement, mount the power unit in main floor utility room or in carport or garage. The tubing system may be installed over-head with vertical runs down to the inlet valves or it may be installed in an under-floor crawl space or in the slab with vertical runs up to the main floor inlet valves, similar to basement installations.

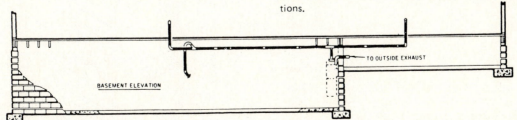

FIGURE 4-6. Typical installation in ranch-style home with basement. (Courtesy of Sears Roebuck & Co.)

FABRICATION PROCEDURES AND SCHEMATIC LAYOUT OF CONVEYING SYSTEM

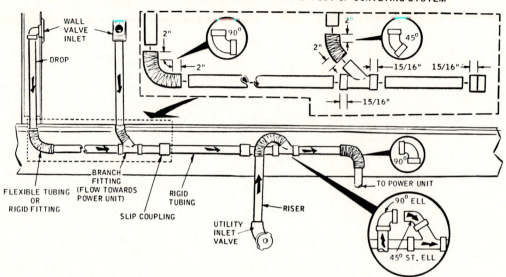

FIGURE 4-7. The preferred method of joining rigid tubing and fittings is with pliobond applied to the outside of the tubing being inserted into the fitting. Twist the fitting about one-fourth of a turn as you press it over the tube. This will spread the pliobond more evenly. **Caution:** Do not get *any* pliobond inside the tube. Dust will stick to it and plug the line. All fittings and risers must set so they feed *down* into the transmission line. (Courtesy of Sears Roebuck & Co.)

FIGURE 4-8. Exploded view of a wall valve assembly. (Courtesy of Sears Roebuck & Co.)

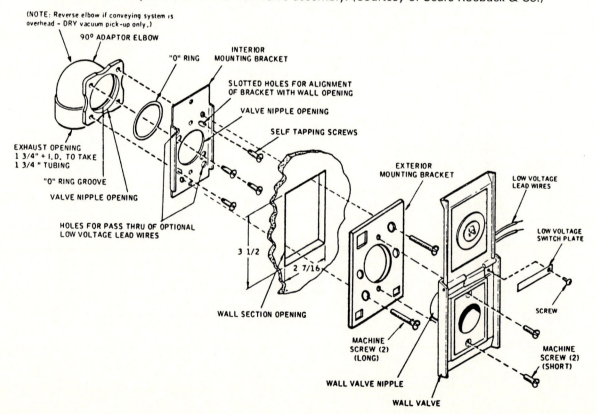

	or	
LENGTH		*WIRE SIZE*
50 ft		22 GA
100 ft		20 GA
200 ft		18 GA

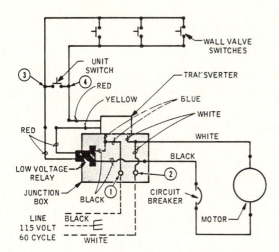

CONNECTIONS 1, 2, 3, & 4 TO BE MADE BY INSTALLER. ALL OTHER CONNECTIONS ARE MADE AT THE FACTORY.

FIGURE 4-9. The wall valve switch wiring is low-voltage dc operating the relay that switches the motor. If the length of wiring is not over 50 feet, No. 22 wire may be used. If the length is up to 100 feet, use No. 20 wire.

FIGURE 4-10. Dimensions for the typical undercounter dishwasher. Play it safe and leave at least ½ inch of extra space.

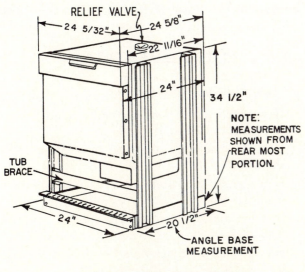

FRONT OF ACCESS PANEL FLUSH WITH DOOR, 4" ABOVE FLOOR

Utilities

The dishwasher requires electricity, hot water, and a drain. The location and routing of these utilities are important for ease of installation and service (Fig. 4-11).

Electricity. Generally, the electrical wiring should be a pair of No. 12 conductors with a ground wire extending 3 to 3½ feet from the rear into the dishwasher space. It should enter from near the lower left corner of the rear wall (Fig. 4-11) so that it can be dressed to the left side of the dishwasher when it is installed. The junction box for this wire is usually located at the front of the machine, directly in back of the bottom access panel (Fig. 4-12).

Water. Only a hot-water line need be connected to the dishwasher. Usually ⅜-inch copper tubing is used with a flare nut connection to the water inlet valve. Check the local plumbing code in case it specifies something different. As shown in Fig. 4-11, the tubing should enter from the lower rear, but not too close to the corners. This should clear the motor to allow dressing the tubing to the left front side of the machine. When making the connection, leave some form of a loop or bend in the tubing

FIGURE 4-11. If all plumbing and wiring service is kept within the rear window area outlined, the space will be clear for most machines.

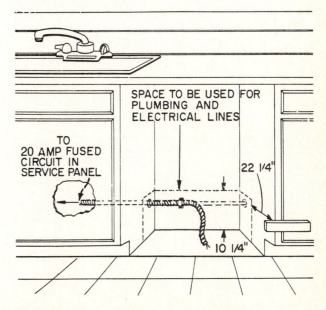

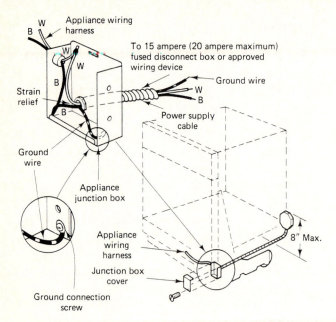

FIGURE 4-12. Make certain that the power supply grounding wire is securely attached to a screw in the metal frame of the dishwasher.

so as not to strain the fitting while the machine is in operation.

Drain. The drain hose is generally mounted on the right side of the machine and includes an antisyphon loop. Additional hose is added to this loop to allow connection to the house drain, which may be in a disposer or as shown in Fig. 4-13.

FIGURE 4-13. The dishwasher drain may enter the disposer or a "Y" drain under the sink. In *all* installations, it must connect before the trap.

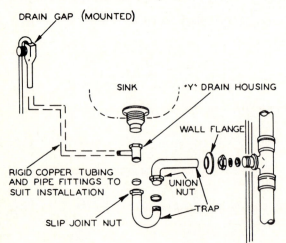

If it is necessary to connect the hose to the house drain system, a licensed plumber or other qualified person should make the connection and bring out a trapped standpipe that will accept the dishwasher drain hose.

Fastening

Once the dishwasher has been set in its final position, it must be leveled and secured there. Mounting holes in the feet permit securing it to the floor.

When setting the machine in its place, *be sure that it rests on the finished floor.* That is, if linoleum is to be the finished floor, the machine should be placed on the linoleum so that if it has to be pulled out of its cavity, it can be done on the same level surface.

If the linoleum is installed later, the machine will have to be pried up and over the linoleum edge for removal, which could lead to complications.

The rear screws are awkward to reach and so are sometimes omitted. The result will cause the dishwasher to tip forward when a dishrack full of dishes is pulled out of the machine. *Do not omit the rear screws.*

Testing and Final Inspection

The machine should always be tested for operation through a complete cycle. It should be checked for leakage, drainage, vibration, and excessive noise.

STUDY QUESTIONS

Installation—Dishwashers

1. How much space does an undercounter dishwasher usually require?
2. How far should the electrical wire extend from the wall?
3. How should the water line be connected to the dishwasher?
4. Is it permissible to connect the drain line to a disposer?
5. How should the machine be secured in its cavity?

DISPOSERS

Mounting the Flange

A disposer can be installed in any sink having a 3½- to 4-inch drain opening. Apply a generous amount of permanently pliable "plumber's putty" all around the underside of the disposer's sink flange. Then set it firmly in position in the sink drain opening. Mount the pressure plate and support plate, as shown in Fig. 4-14, and secure the flange to the sink. *Make certain that the flange is square to the sink;* otherwise there may be a leak and the disposer will not hang straight.

This flange-mounting procedure is common to many models of disposers, but be sure to check the instructions for any unit you may have to install because it might have a different flange arrangement.

Hanging the Disposer

The disposer should mount on the flange as shown in Fig. 4-15. Any manual switches attached to the disposer should face the front for accessibility. If there are no manual switches, the drain outlet should be located so as to allow for convenient plumbing connections.

After "catching" the mounting screws or nuts, tighten them *evenly* to ensure the secure seating of the disposer to the flange gasket and correct hanging from the sink. *Do not over-tighten!*

Connecting to Plumbing

Next, connect the tailpiece from the disposer outlet to the drain trap that connects to the stack. Figures 4-16 and 4-17 illustrate one of several alternative procedures for doing so.

FIGURE 4-14. Sink flange (b) fits into the sink (A) drain hole. Slip gasket (C), pressure plate (D), and mounting ring (E) over the sink flange and secure them by installing snap ring (F) in the sink flange groove. Turn the three screws (G) through the holes in E until they contact ring D. Set the disposer in position under the flange, then rotate the mounting ring and screws to align with the disposer mounting tabs. Tighten all three screws, making certain that the assembly is level.

FIGURE 4-15. When mounting the disposer on the flange, make certain that the splash guard or chamber seal is in position. There should be no metal-to-metal contact between flange and disposer.

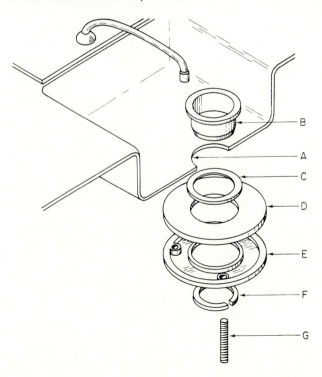

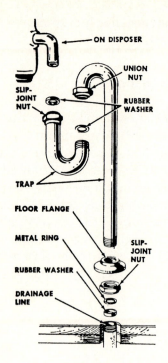

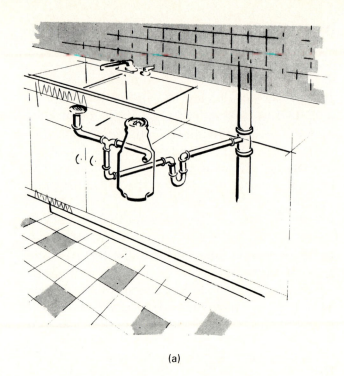

(a)

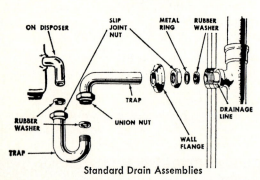

Standard Drain Assemblies

FIGURE 4-16. The standard drain assembly parts can be purchased at most hardware stores. These sketches show the two most common methods of connecting a disposer to a stack, and includes a trap.

Electrical Requirements

The electrical needs are for a 120-volt ac line of No. 12 wire with ground protected with a 20-ampere slow-blow fuse or breaker. The wiring of the disposer can be either direct to the cable or through a flexible cord to an outlet (Fig. 4-17b).

In any event, *the wire should have enough slack so that vibration will not damage the wire.* Yet it must be secure from damage that could result from retrieving stored items under the sink.

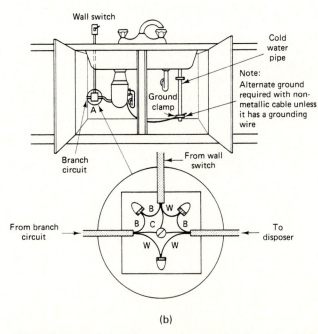

(b)

FIGURE 4-17. (a) Always make certain that the drain trap is between the disposer and the stack. (b) Wiring for a continuous-feed disposer requires the wall switch, connected as shown, to control the disposer. The batch feed type of disposer *does not* require a wall switch. Its control is activated by the cover switch.

Testing the Disposer

After the installation is complete, test the disposer by turning on the cold water at about a 3-gallon per minute flow rate, and check under the sink for leaks. If all is well, start the disposer and recheck. *Never run the disposer unless water is flowing through it; otherwise, the seal will be damaged.*

STUDY QUESTIONS

Installation—Disposers

1. Where is a disposer installed?
2. Can the disposer outlet connect to the sink drain trap?
3. What are the electrical needs of a disposer?

DRYERS

Dryers are available in both electric and gas-heated models. Some common needs should be studied, however, before discussing the specific heating supply.

Preparing for Installation

1. Remove dryer from carton, usually by cutting around the bottom and lifting the carton up and over the machine.
2. Remove shipping blocks from base and all literature and miscellaneous parts from drum.
3. Give the operating instructions and literature to the customer to read while you are completing the installation.
4. Remove the tape that holds the drum to the cabinet and make certain the drum revolves without interference. Remove any other tape that may have been used for shipping protection.
5. Many dryers have self-leveling legs in the rear and adjustable ones in the front. Adjust the screw-type leveling legs to ensure that the dryer will be level. Check it with a spirit level, both side to side and back to front, after the machine has been moved to its permanent position.

Exhaust System Requirements

All dryers require an exhaust system to carry the warm, moist air out of the dryer to the outside of the house. All duct installations should be made through the building wall or a window.

Never exhaust into a chimney! Standard 4-inch aluminum pipe, elbows, and backdraft eliminator can be used for this duct. The length of duct with a backdraft eliminator should not exceed 20 feet. If any elbows are used, subtract 2 feet for each elbow. Thus the installation shown in Fig. 4-18, should not exceed 16 feet of duct.

FIGURE 4-18. The exhaust duct can be run in any direction, but in no case should a duct with two elbows be longer than 16 feet.

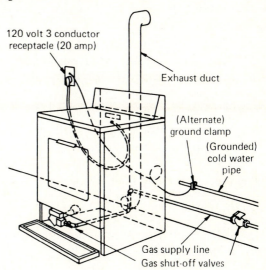

120 volt 3 conductor receptacle (20 amp)
Exhaust duct
(Alternate) ground clamp
(Grounded) cold water pipe
Gas supply line
Gas shut-off valves

Typical gas installation

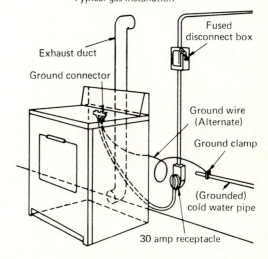

Exhaust duct
Ground connector
Fused disconnect box
Ground wire (Alternate)
Ground clamp
(Grounded) cold water pipe
30 amp receptacle

Typical electric installation

Four-inch flexible duct may be used, if approved by local codes, but it should not exceed 8 feet in length. Grounding of the dryer must be in accordance with local codes.

Electrical Requirements for the Electric Dryer

Most electric dryers are designed for 240-volt, 60-hertz, three-wire operation, with ground. Check the nameplate to make certain that the dryer matches the power available. *The power wiring from the dryer to the power distribution panel must be a minimum of No. 10 three-wire with ground cable protected with a 30-ampere dual breaker or two 30-ampere slow-blow fuses.*

Do not use an extension cord. Local codes may permit the use of a flexible-type power supply cord and plug that can connect to a specially wired outlet.

Generally, the dryer connection is made directly to the line from the main service panel without an intervening attachment plug and socket, as shown in Fig. 4-19. Leave about 3 or 4 feet of slack in the line to allow moving the dryer far enough out from the wall for cleaning or servicing.

Electrical Needs for the Gas Dryer

In addition to gas, the gas-heated dryer requires electricity for the drum motor and other controls. A nominal 120-volt, 60-hertz ac, 15-

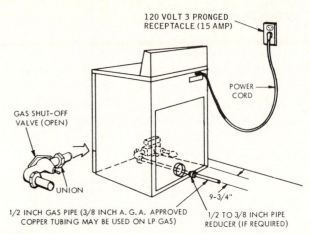

FIGURE 4-20. A gas dryer gets its electrical power from a standard 120-volt outlet.

ampere fused electrical supply is required (Fig. 4-20). It is recommended that a separate circuit be used, but if this is not possible, make certain no other appliances are in use on this same circuit while the dryer is being used. A 20-ampere circuit wired with No. 12 wire with ground would be much better.

Gas Supply

Check the dryer nameplate to make certain it is equipped with the correct burner for the type of gas in the home. (Sometimes this information will be found on the burner data label on the burner base.) If nameplate and burner do not match, do not connect. Check with the salesman who sold the machine.

The gas supply line should be a $\frac{1}{2}$-inch pipe if not over 20 feet from the meter. For a longer run, use larger pipe. (A $\frac{3}{8}$-inch copper tubing may be used for LP gas.)

The supply line should have a shutoff valve readily accessible and near the dryer.

Adequate air supply to aid ventilation around the dryer must be provided, and all work and materials used must meet local code requirements.

Connecting to the Gas Supply

After connecting to the gas supply line, open the shutoff valves and test for leaks in the gas line, using a soap solution. *Motion of the soap bubbles will indicate a leak.* These must be cor-

FIGURE 4-19. An electric dryer is usually directly connected to the line. A heavy-duty three-terminal connection block is provided.

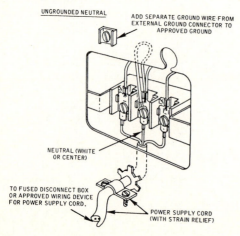

rected before proceeding. **Never test gas leaks with a flame!**

Plug in the power cord and purge the air from the gas line as follows:

1. *Automatic ignition models.* Turn the machine on by setting the dial to a 15-minute *heat* position. If the burner does not ignite within one minute, shut off the machine, allow to stand 10 minutes, then repeat procedure.
2. *Models with manual ignition.* There are several varieties of manual ignition pilots, so the safest thing to do is to refer to the "Light Gas Pilot" instructions on the machine.

   Usually it involves holding a lighted match over the pilot orifice while pushing a red lever. Then hold the lever for a minute before releasing it. I have found the wooden kitchen-type matches good for this purpose. Some burners have a ZIP tube, which is lighted, and this carries the flame to the pilot.

Final Test

Always run a test cycle to make certain that the dryer functions correctly in all phases of its operation.

STUDY QUESTIONS

Installation—Dryers

1. Does a gas dryer need electricity?
2. What size wire should be used to connect an electric dryer to the fuse panel?
3. What size fuse should be used?
4. How long can an exhaust duct be?
5. What is the best method to use for locating leaks in a gas line?

FREEZERS AND REFRIGERATORS

The modern refrigerator-freezer normally offers no installation problem, for it is usually a re-

placement for an older one that had occupied the space previously.

However, there are a few precautions.

1. *Make certain that there is ventilation* space *around* the box to allow enough airflow over the condenser. Models *without* forced-air condenser cooling require a 3-inch clearance on the sides and top of the machine. Models with forced-air circulation over the condenser need only enough space for installation and moving.
2. *The floor under the unit must be strong enough to support the weight of the machine and its load.* A large box might easily weigh 500 pounds or more. If the floor sags or is "springy" it would be wise to place a ¾-inch plywood sheet on the floor under the box; then level the box on that. The size of this sheet should be as much larger than the refrigerator floor space needs as is practical. Even though a machine has been leveled on a sagging floor, the added weight of the food load could cause enough sag to rack the cabinet and cause a poor door seal.
3. Make certain that the refrigerator is *level* and that the door swings freely and easily without bumping into another cabinet when only partway open. This can ultimately lead to a dented door or sprung hinges that will cause a poor door seal.
4. For best results, the refrigerator *should not be located next to a stove or other heating device.* Nor should it be exposed to direct sunlight.
5. The electrical service cord should *plug directly into a wall outlet* of the correct voltage. Ideally, this should be an independent circuit, fused with a 20-ampere slow-blow fuse or a breaker. Voltage tests should be made when the compressor starts and should be within 10 percent of the refrigerator nameplate rating.

6. Check the door seal by closing the door on a piece of thin paper; then pull the paper out. *A correct gasket seal will cause a slight drag on the paper.* This check should be made on the hinge side as well as near the upper and lower outer corners.

7. Check the door switch, fan, and light operation. The thermostat should shut the compressor off when in the OFF position, and the compressor should start in the ON position. Normal operation should be midway on the dial, and the installer should set it there.

8. Before leaving the job, explain the thermostat setting to the customer, requesting that she adjust it to suit her needs *after* it has run for several hours. Also, explain any new or different features this model has that the previous refrigerator did not have: an automatic ice cube maker, for example. This feature requires a connection to a cold-water line. It is usually done with copper tubing and a saddle valve as shown in Fig. 4-21.

STUDY QUESTIONS

Installation—Freezers and Refrigerators

1. How much ventilation clearance should be left around a freezer?
2. What size extension cord should be used with a freezer, if one is needed?
3. How would you check a door gasket for fit?
4. How is the water line of an ice cube maker usually connected to the supply pipe?

INCINERATORS

Location

Choose a location that is convenient for the user and *as close as possible to a flue connection.* This location must be properly ventilated so as to *provide sufficient air for combustion* and space for proper clearance from combustibles.

Clearances should be not less than 48 inches at top and front, 12 inches at sides and rear, and 18 inches from flue pipe.

Reduced clearances can be obtained by the use of proper fireproof insulation and wall-

FIGURE 4-21. (a) A saddle valve is used to tap a water line from an existing pipe to supply water to a humidifier, ice-cube maker, or some other low-volume water user. Installation is made by drilling a hole (usually $\frac{3}{16}$ inch) in the pipe and clamping the valve over it. The neoprene washer serves as a leakproof seal. The strainer is included to prevent foreign objects in the piping from getting into the fill valve of the appliance being used. (b) A completed installation places the saddle valve in a convenient location so that it can be shut off if the need arises. The valve should never be mounted on the underside of the pipe!

(a)

(b)

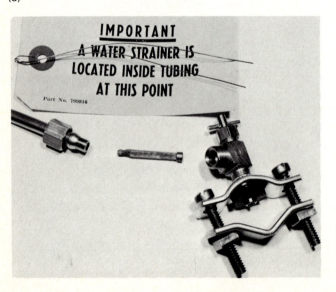

board—for example, 4 inches from a brick wall. However, check with your local building inspector before using them, and get his approval.

Connecting to the Chimney Flue

Venting (running the flue) through a combustible wall requires a metal thimble 12 inches larger in diameter than the flue pipe and filled with noncombustible insulation, as shown in Fig. 4-22.

The vent pipe from incinerator to chimney flue should be 6 inches in diameter, galvanized stovepipe, not less than 24 gage in thickness. It should be pitched upward *at least ¼ inch* per foot of length toward the chimney.

This pipe should be as short and direct as possible, with not more than two 90-degree elbows. A long horizontal run must be firmly supported at intervals of not more than 6 feet.

A long run, or one exposed to cooling drafts, should be insulated, as shown in Fig. 4-23, to prevent chilling of flue gases, which would result in improper combustion.

It is recommended that *no* pipe damper be installed in the flue pipe. If excessively high

FIGURE 4-22. This Figure shows minimum installation clearances from the incinerator to a combustible wall. Sheet metal and rockwool protection must be used when the unit is in a tight spot or if the flue passes through a partition.

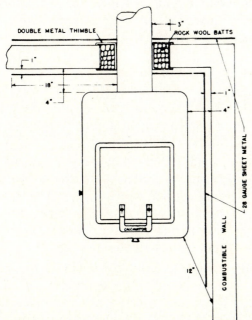

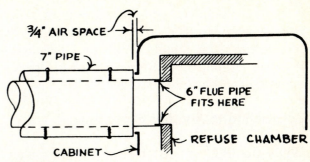

FIGURE 4-23. A simple way of insulating the flue pipe is to run a larger pipe over it and space this pipe away from it with metal screws.

draft is noted, a barometric draft control, preset for a 0.03-inch water column, should be used. It would be best to check with the local building inspector for a listing of approved types.

When assembling the flue pipe, all joints should be made with the male end pointed in the direction of flue-gas travel, and they should be fastened with sheet-metal screws.

Entering the Chimney

The point where the flue pipe enters the chimney flue should be a few inches above the bottom to prevent stoppage by an accumulation of fly ash and erratic draft, due to eddy currents (see fig. 4-24).

FIGURE 4-24. (a) In cases where flue extends to the basement floor, the draft can usually be improved by filling the base of the chimney with sand to within 12 inches of the smoke pipe and relocating the cleanout door. (b) If the appliance vent enters the chimney a short distance from the bottom, cut off the end of the pipe at a 45-degree angle. This will prevent an eddy at the bottom of the chimney.

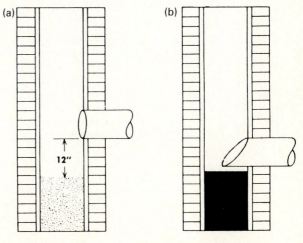

If another vent pipe, possibly from a furnace or water heater, enters the same flue, the incinerator's flue should be about 1 foot below it. Also, it must not enter a chimney directly opposite another flue.

Connecting to Another Stove Pipe

When the flue connection from the incinerator must be made to another pipe, it must be a 30-degree Y connection, flowing to the chimney as in Fig. 4-25.

The T connection shown in Fig. 4-25 is incorrect, for it will cause turbulence and restrict the free flow of gases.

Setting in Position

When the incinerator is set in its final position, it must be leveled and stable on the floor, which we are assuming to be concrete. Before installing on a wooden floor, check with your local

FIGURE 4-25. Connections to another pipe must always be a Y pointing in the direction of normal airflow to the chimney.

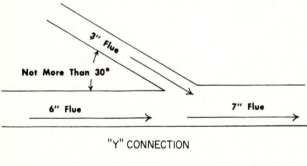

"Y" CONNECTION

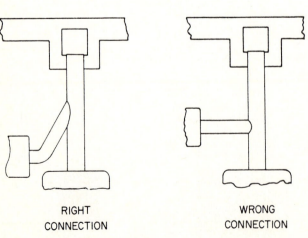

RIGHT CONNECTION   WRONG CONNECTION

building inspector. He should tell you the types of floor protection he would approve.

So far all that has been said holds true for either a gas-fired or an electric incinerator.

Electrical Connections

No electrical connections are required for a gas-fired incinerator. An electric incinerator operates on either 115 or 240 volts ac, depending on its design. Make certain that the voltage called for on the nameplate is available at the outlet. Usually, a 15-ampere-fused circuit, wired with No. 14 wire, is large enough to handle the unit.

Gas Piping

The gas piping must be at least $\frac{3}{8}$-inch rigid wrought iron or steel (a $\frac{1}{2}$-inch on longer runs). It should be firmly supported at suitable intervals and pitched down to avoid traps where condensation can collect.

Gas piping should be protected against freezing temperatures or sudden changes in temperature.

An accessible shutoff valve must be provided near the incinerator, as shown in Fig. 4-26.

After all connections are made, the gas can be turned on and the system tested for leaks by the soap bubble method. *Do not smoke on the job or use an open flame near gas line during installation and testing!*

Checking Out the Burner

1. Make certain that all gas-line supply valves are open and that gas is available to the burner.
2. Read the pilot-lighting procedure instruction plate located near the timer and control valve. This procedure may have to be repeated several times on the initial lighting, depending on the gas supply in the new lines.
3. Once the pilot is on, the main burner may be tested by depressing the timer knob and rotating it clockwise about $\frac{1}{4}$ turn (15 minutes) and releasing it.

4. After the burner has warmed up for about 10 minutes, the air shutter should be adjusted. Loosen its locking nut (Fig. 4-27) and turn the shutter until a soft blue flame is obtained. *A flame that "blows" or tends to lift off from the burner indicates too much air, and a flame that burns yellow indicates not enough air.*

Smoke Test

Before leaving an installation, either gas or electric, the incinerator should be checked for smoke leaks and draft.

A draft check can be made after the unit has warmed up by crumpling four sheets of newspaper and placing them against the burner shield. The newspaper should ignite in a few minutes, and the products of combustion should go up the flue. When the top lid is opened, air should be sucked down into the unit. Be sure that the ash drawer is closed at all times during operation.

Any joints that show smoke leaks should be sealed with a furnace cement applied from within the unit. *Note:* It may be necessary to remove a top or some trim in order to gain access to the leaky area.

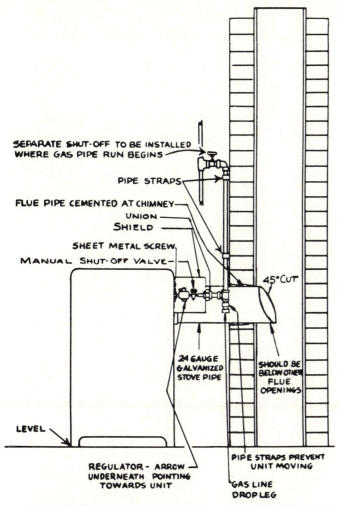

FIGURE 4-26. All these factors must be considered in a complete installation as well as any others that your local building code requires.

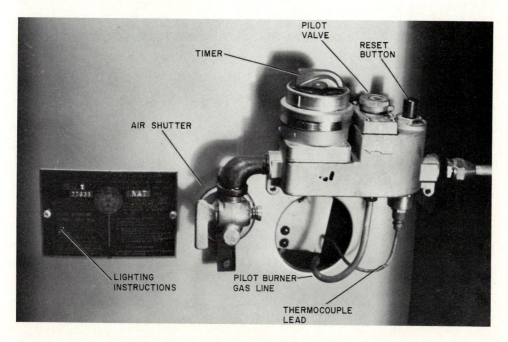

FIGURE 4-27. The incinerator burner control valve is located on the side of the incinerator and is readily accessible for adjustments.

Before leaving an installation, be sure that the customer has all the literature that came with the incinerator and that she knows how to operate the device.

STUDY QUESTIONS

Installation—Incinerators

1. What size flue pipe is run from an incinerator to the chimney?
2. When venting an incinerator into the same chimney as the furnace, should the incinerator vent be over or under the furnace pipe?
3. Are electrical connections made to a gas incinerator?
4. Where will you find the pilot-lighting procedure instructions?
5. What size wire should be used for an electric incinerator?

CHECKING THE CHIMNEY

Gas-burning units, such as incinerators, furnaces, water heater with input of over 5000 Btu per hour, and others as noted on their rating plates, will require venting to the outside of the house. Generally, a chimney is used.

In the interest of both safety and proper product performance, the chimney should be inspected before connecting to it.

Outside Chimney Inspection

This step can be done quickly before entering the house. Look for the following characteristics of a satisfactory chimney:

1. *Brick, stone, or other masonry construction* must be capable of withstanding high temperatures and thick enough to insulate surrounding structure from chimney heat.
2. *It must have sufficient flue area* to handle all the appliances feeding into it. Nominally this should be twice the cross-sectional area of the appliance

TABLE 4-1
Area of various diameter flue pipes[a]

Pipe Diameter (in.)	Area (in.^{2})
3	7.1
4	12.6
5	19.7
6	28.3
7	38.5
8	50.4
9	63.8
10	78.5
11	95.5
12	113.0

[a] An incinerator with a 7-in. flue (38.5 in.^{2}) should connect to a chimney flue of not less than 77-in.^{2} area or it should be 10 in. in diameter.

flue. Table 4-1 relates flue-pipe diameter to cross-sectional area.

3. The chimney should extend at least *2 feet above any part of the building* or trees within 10 feet of it and at least 3 feet above the roof from which it emerges. Figure 4-28 illustrates these conditions.

FIGURE 4-28.

The chimney height will be determined by surrounding roof and trees. Chimney should extend at least 24-in. above high point of roof with no obstructions from nearby roofs, trees.

Trim or remove any nearby trees that would interfere with the chimney draft.

(a)　　　　　　(b)　　　　　　(c)　　　　　　(d)

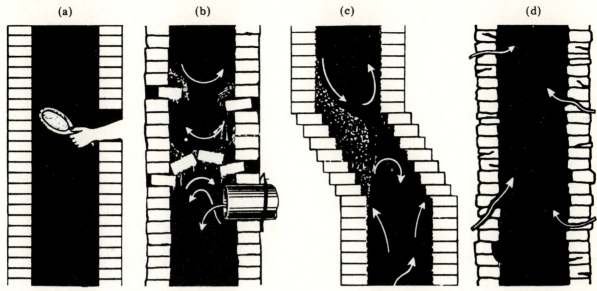

FIGURE 4-29. If the chimney is clear and firm (a), it is all right to use but if it shows loose bricks or mortar (b), or is otherwise obstructed (c) or leaky (d), recommend to the customer that the chimney be repaired before you install an incinerator.

Inside Chimney Inspection

1. To prevent smoke and odor in the house, the chimney should be checked for proper draft. A wad of newspaper can be crumpled, placed in the chimney, and ignited. If the draft is proper, flame and smoke will be drawn up the flue.

2. The easiest way to check for obstructions in the chimney is to open the cleanout door, or remove a furnace pipe, and shine a light up the chimney and examine it with a mirror, as shown in Fig. 4-29, that illustrates conditions that may be found.

A defective chimney should be repaired by a qualified mason before connecting any other appliance to it.

STUDY QUESTIONS

Installation—Checking the Chimney

1. How high should a chimney be above a roof?
2. How do you check a chimney for proper draft?

RANGES AND OVENS

Location

Ranges and ovens are usually designed to fit into a standard kitchen cabinet and counter arrangement. But not all have the same dimensions, so it is best to check the available space against the dimensions of the appliance before trying to move it into its final position.

The floor must be solid and reasonably even. Leveling feet will take care of some unevenness.

Venting

Normally, ranges and ovens do not require special venting duct work or flues. However, some special units may include an exhaust fan that must be vented out of doors.

If the range is on an outside wall, this position should offer no problem. Simply cut the required hole through the wall and install a hood and backdraft eliminator.

For a unit that requires duct work of any length to reach the outside wall, make certain that it is metal, securely joined at each section, and having a minimum of elbows. Its size will depend on the exhaust-fan capacity, but it should be at least 10 inches by 3 inches. Here again you must be guided by the manufacturer's installation manual.

Electrical Requirements

The Underwriters' Laboratory, in their requirements for range approval, request that the range have a long enough cable to allow connection to a power outlet that is accessible without moving the range (Fig. 4-30). Most electric ranges require a "220"-volt, three-wire with ground electrical supply. This may be 208 volts or 240 volts. *Check the nameplate for rating on the range, and measure the available voltage to determine whether or not they are compatible.* An electric range operating on the wrong voltage will not be efficient.

The wire size must be at least No. 8 three-wire with ground cable protected with a 40-ampere fuse or breaker.

A gas-heated range requires a 120-volt, 15-ampere circuit to operate its lights and controls.

Gas Supply

The gas-supply line from the meter to the range location should be installed by someone who is qualified or licensed under the local building code. There will be a shutoff valve and a suitable threaded fitting at the end of the line to which the range can be connected.

A flexible pipe can be used where the code permits. Otherwise the connection must be made with rigid pipe and a union. *The shutoff valve and union must be accessible without moving the range from its working position.* All connection must be tested for leaks with a soap solution.

Testing the Range

After all connections are made, the range must be tested for operation.

The gas range first requires lighting the pilot, then adjusting the primary air supply at each burner (Figs. 4-31 and 4-32).

The electric range requires turning all heating elements ON to make certain that they will heat properly.

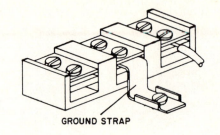

(a)   GROUND STRAP

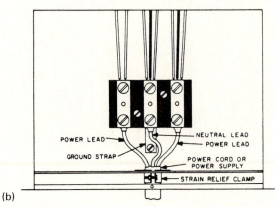

POWER LEAD

NEUTRAL LEAD
POWER LEAD

GROUND STRAP

POWER CORD OR
POWER SUPPLY

STRAIN RELIEF CLAMP

(b)

FIGURE 4-30. (a) An electric range terminal block usually includes provision for grounding the neutral lead to the range frame. (b) The electric supply line should connect as shown.

In any event, check the manufacturer's manual for the particular model, and be sure to follow all special instructions included.

Special Installations

The built-in oven and the drop-in countertop range require equal care in checking the physical size, gas, and power supply.

When installing a self-cleaning oven, observe all the clearance and insulation rules set by the oven manufacturer and your local building code. These units run pretty hot during the cleaning cycle.

STUDY QUESTIONS

Installation—Ranges and Ovens

1. Do ovens require special venting ductwork?
2. What size wire and fuse should be used for an electric range?
3. Does a gas range require an electrical connection?
4. Is it important to level a built-in oven?

ORIFICE HOOD

HI-LO VALVE STEM

MANIFOLD MOUNTING

MINIMUM FLAME ADJUSTMENT SCREW

CORRECT FLAME LENGTHS¹ FOR BURNERS² AND PILOTS, USING VARIOUS GAS TYPES

		L. P. GAS		NATURAL & MOST CITY GAS		MANUFACTURED GAS	
		BY-PASS	HIGH	BY-PASS	HIGH	BY-PASS	HIGH
OVEN BURNERS		$\frac{1}{8}$"	$\frac{9}{16}$"—30" $\frac{1}{2}$"—40"	$\frac{1}{8}$"	$\frac{9}{16}$"	$\frac{1}{8}$"	$\frac{5}{16}$"—30" $\frac{1}{4}$"—40"
BAR-B-KEWER BURNERS		LOW	HIGH	LOW	HIGH	LOW	HIGH
		$\frac{1}{8}$"	$\frac{7}{16}$"	$\frac{1}{8}$"	$\frac{7}{16}$"	$\frac{1}{8}$"	$\frac{3}{16}$"
SURFACE BURNERS	**SINGLE & GRID-ALL**	LOW	HIGH (Reg. / Thermo-Trol³)	LOW	HIGH (Reg. / Thermo-Trol³)	LOW	HIGH (Reg. / Thermo-Trol³)
		$\frac{1}{8}$"	Reg. $\frac{7}{16}$" Thermo-Trol $\frac{9}{16}$"	$\frac{1}{8}$"	Reg. $\frac{1}{2}$" Thermo-Trol $\frac{11}{16}$"	$\frac{1}{8}$"	Reg. $\frac{5}{16}$" Thermo-Trol $\frac{1}{2}$"
	DUAL	INNER (Keep Warm / Simmer)	OUTER	INNER (Keep Warm / Simmer)	OUTER	INNER (Keep Warm / Simmer)	OUTER
		Keep Warm $\frac{1}{8}$" Simmer $\frac{3}{8}$"	$\frac{3}{8}$"	Keep Warm $\frac{1}{8}$" Simmer $\frac{3}{8}$"	$\frac{5}{16}$"	Keep Warm $\frac{1}{8}$" Simmer $\frac{3}{16}$"	$\frac{1}{4}$"
MINI-PILOT		A BALL OF FLAME AROUND THE TIP OF THE MINI-PILOT ABOUT $\frac{1}{4}$" IN DIAMETER.					
YELLOW FLAME PILOT		A SHORT YELLOW FLAME ABOUT $\frac{1}{2}$" LONG.					
QUIK-LITE PILOT		ABOUT THE SIZE OF AN ORDINARY BOOK MATCH HEAD.					
TOWER PILOT		A MINIMUM STABLE FLAME ON THE SLOTTED CARRY-UP PORT.					

¹All flame lengths measured from outside edge of port wall to end of primary cone.
²On double row ported burners, dimensions given are for largest (and longest) port flame.
³There is no low adjustment for the Thermo-Trol burner.

FIGURE 4-31. The orifice hood is adjusted to provide the correct flame length. The hood adjustment controls the amount of gas flow when the valve is completely open.

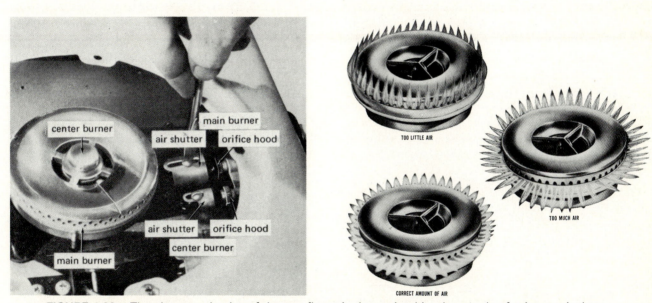

center burner — main burner — air shutter — orifice hood — air shutter — orifice hood — center burner — main burner

TOO LITTLE AIR

TOO MUCH AIR

CORRECT AMOUNT OF AIR

FIGURE 4-32. The shape and color of the gas flame is determined by the supply of primary air that is controlled by the air shutter. It should be a clean, blue flame.

WASHERS

The installation needs for an automatic washer are as follows:

1. *A firm, level floor.* A weak, springy, or slanting floor may cause excessive vibration and allow the machine to "walk," or move out of place.

2. The washer should be leveled as close to the floor as possible, making certain that all four feet are in secure contact with the floor. Then use the jam nuts to lock the leveling screws in position

because vibration during use may cause them to shift.

3. *A tub or 1½-inch-diameter standpipe at least 34 inches above the base of the washer* to receive the drain water (Fig. 4-33). If this drain is too low, the water will syphon out of the tub, even when you do not want it to. The drain outlet must *not* be over 6 feet above the washer base or else the pump cannot lift all the wastewater out of the machine.

   The drain hose must not be securely sealed into the standpipe because a siphon action may prematurely drain water from the tub or sewer water may flow back into the washer. There must be an air gap between the hose and standpipe.

4. An adequate supply of both hot and cold water (Table 4-2a). The water should be clean—free from rust and sediment. The hot water should be 140°F (60°C) in the machine.

   Warm water is obtained by mixing water from the *hot-* and the *cold*-water supply lines. Thus as the *cold*-water temperature changes from summer to winter, it may be necessary to adjust the *cold*-water flow rate at the faucet to adjust for a correct *warm*-water temperature [100 to 110°F (38 to 43°C)]. Both hot and cold water should have a flow rate of 3 gallons per minute and a pressure range between 20 and 100 psi.

5. A 117 ± 10%-volt, 60-hertz electrical power supply. No less than No. 12 wire with ground fused with a 20-ampere fuse. There should be no other appliances in use on this circuit when the washer is being used. Table 4-2b shows the average power consumed for the various wash cycles.

   Voltage tests should be made while the washer is operating, preferably at the beginning of the spin cycle.

6. For personal safety, the washer must be properly grounded with a wire connecting from the back of the washer to the nearest metal cold-water pipe. Also, all new machines are equipped

FIGURE 4-33. Typical installation of a suds-return model. For a non-suds-return model, there would be no suds-return hose assembly. For either model you may use a stand-pipe drain assembly (see the dashed line).

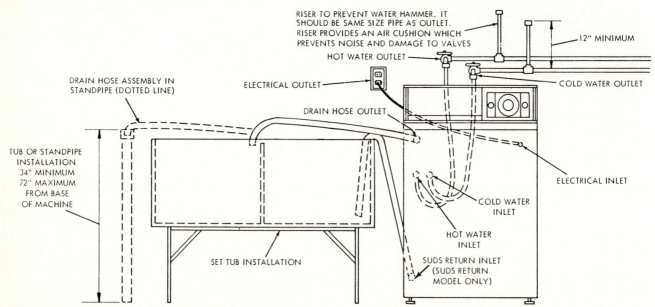

This illustration shows a typical installation of a suds return model. For a non-suds return model there would be no suds return hose assembly. For either model you may use a standpipe drain assembly (see dotted line).

TABLE 4-2
Average power and water usage for washing machines[a]

a. Water consumption

Water Level (in.)	Total Water (gallons)
3	13–15
$5\frac{1}{2}$	22–24
8	31–33
10	38–40
12	46–50

b. Power consumption[a] *for 10 to 12 lb load*

Cycle	Motor	Watts
Wash	$\frac{1}{3}$	450–500
Spin start	$\frac{1}{3}$	750–800
Wash	$\frac{1}{2}$	550–600
Spin start	$\frac{1}{2}$	650–700

[a] If the voltage drops more than 10% below normal line voltage at the spin start it is an indication of inadequate wiring to the machine.

with a grounding-type three-prong plug, which must be inserted into a matching electrical receptacle (Fig. 4-34). *Do not cut off the grounding terminal to make connection* to an older-type ungrounded receptacle. Either use an adapter or have a qualified electrician change the outlet.

FIGURE 4-34. All washers have a three-pronged grounding plug on the power supply cord. This is fine for an outlet having a ground receptacle. If the receptacle is not a grounding type, use an adapter as shown.

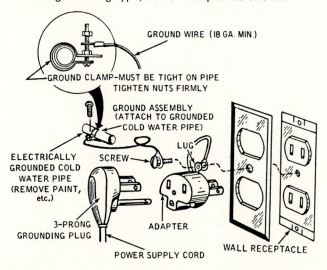

GROUND WIRE (18 GA. MIN.)

GROUND CLAMP—MUST BE TIGHT ON PIPE TIGHTEN NUTS FIRMLY

GROUND ASSEMBLY (ATTACH TO GROUNDED COLD WATER PIPE)

ELECTRICALLY GROUNDED COLD WATER PIPE (REMOVE PAINT, etc.)

SCREW

LUG

3-PRONG GROUNDING PLUG

ADAPTER

POWER SUPPLY CORD

WALL RECEPTACLE

STUDY QUESTIONS

Installation—Washers

1. Is it important to level an automatic washer?
2. How high should the drain be above the floor?
3. What is the maximum height the drain outlet can be above the washer top?
4. Should the drain hose be securely sealed into the drainpipe?
5. What is the smallest-size wiring and fuse that can be used with an automatic washer?
6. Is the grounding terminal on the washer cord very important?

WINDOW AIR CONDITIONERS

Location

Obviously, the first thing the installer must do is learn where the air conditioner is to be installed. It would be embarrassing to install a unit in the wrong room.

If you have a choice of windows, try to install the air conditioner in a north window or any window not in direct sunlight. Sun shining on an air conditioner will decrease its cooling ability.

Btu Capacity

Having determined where the air conditioner is to be installed and how much area the customer expects to cool, the alert technician will check the cooling capacity of the air conditioner. A reasonable estimate of room needs can be made using the floor-area cooling guide (Table 4-3), which indicates the Btu capacity required to cool a room having an 8-foot ceiling and the square footage of the floor area shown. Be certain to choose the correct column on the chart for the room in question. You *must* consider the room exposure and the type of space over the ceiling.

The room may have an excessive heat load, such as many windows, undoored openings to other rooms, high ceilings, and heat-producing

TABLE 4-3
Cooling guide[a]

Approx. Btu Capacity Required	Space Above Room Being Cooled								
	Occupied Room			Attic			Insulated Flat Roof		
	Area Being Cooled has Exposed Walls Facing:								
	North or East	South	West	North or East	South	West	North or East	South	West
4,000	160	90	45	120	60	30	120	60	30
6,000	325	205	150	270	175	120	240	160	115
8,000	490	330	255	420	285	220	355	250	195
10,000	670	450	360	580	400	325	475	345	280
12,000	885	580	465	735	520	420	600	445	360
14,000	1200	720	580	900	640	525	730	540	445
16,000	—	880	690	1095	760	630	865	645	535
18,000	—	1090	820	—	885	745	1005	745	625
20,000	—	—	970	—	1030	855	1185	855	720
22,000	—	—	1185	—	1195	975	—	985	820
24,000	—	—	—	—	—	1120	—	1125	930
26,000	—	—	—	—	—	—	—	—	1060

[a] This cooling guide lists the main factors to consider when making a "quick" estimate of cooling needs for average rooms. Extra windows, higher ceilings, open hallways, etc., will always increase the cooling load.

TABLE 4-4
Typical electrical requirements for air conditioners[a]

Btu Count Capacity	Volts	Wattage	Main Wire Size	Fuse Size	Amperes	
					Run	Start
6,000	115	1100	14	15	9.5	36
8,000	115	1350	12	20	12	50
10,000	115	1400	12	20	12	50
12,000	220	2500	12	20	12	50
14,000	220	2650	12	20	12	52
16,000	220	2650	12	20	12	52
18,000	220	2900	12	20	13.5	64
20,000	220	3350	12	20	15	61
22,000	220	3700	10	30	18.0	73
24,000	220	3750	10	30	19	76
26,000	220	4200	10	30	19.5	75
28,000	220	4000	10	30	19	79
30,000	220	4100	8	40	20	76
32,000	220	4800	8	40	23	110

[a] At 95°F dry bulb and 75°F wet bulb conditions.

equipment. If that is the case, be certain that the air conditioner has ample extra capacity.

Physical Size

Before commencing installation, make certain that the air conditioner will fit in the chosen window space. This aspect may become a prob-lem in some of the older homes or houses having other than double-hung window design.

It is possible that some carpentry work will have to be done to *reinforce,* enlarge, or otherwise change the window frame before installation. Do not forget the weight of the machine and its need for external mounting brackets.

Electrical Requirements

An air conditioner requires one of the following power supplies:

> 115 volts ac, 60 hertz single phase
> 208 volts ac, 60 hertz single phase
> 220 volts ac, 60 hertz single phase
> 230 volts ac, 60 hertz single phase

In many other countries, the same voltages are used but the frequency is 50 hertz.

The available voltage *must* match the required voltage as listed on the nameplate of the appliance.

The wattage or amount of power used by the air conditioner will depend on its size. Table 4-4 offers an approximate correlation between Btu capacity, wattage, and ampere needs. The outlet into which the unit is to be connected *must* be able to provide this power when needed.

The easiest test is to plug in the unit and measure the voltage when the unit starts and is running. The no-load voltage must be within ± 10 percent of the nameplate-rated voltage and must not drop more than 10 percent below rated voltage while in operation. *Any abnormal voltage drop is reason for checking the wiring from the outlet to the meter.* Look for a loose connection or undersized wire. Low voltage at the meter should be reported to the power company.

Continuous operation of an air conditioner at low voltage will overheat the motor and shorten its life.

Operating an air conditioner from an extension cord is not recommended. However, if one must, then use the shortest possible heavy-duty type designed for air-conditioner usage. These cords use No. 12 wire and have molded ends to match the 115-volt receptacles and connecting plugs on the air conditioner.

Preparing the Window

After checking to make certain that the window sill is clean, level, and secure, mark a center line on it in order to center the air conditioner mounting bracket, cabinet, or frame in the window. The mounting device will vary from model to model, but it must always be level and secure. An example of a manufacturer's instruction is shown in Fig. 4-35. Figure 4-36 shows an alternative bracing method.

STUDY QUESTIONS

Installation—Window Air Conditioners

1. If you have a choice of windows in which to install the air conditioner, which should you use?
2. Is the condition of the window frame important in an air-conditioner installation? Why?
3. Are external mounting brackets used to support an air conditioner?
4. If so, are they above or below the air conditioner?
5. How much tolerance is allowed in the voltage supplied to an air conditioner?
6. Is this voltage measured under no-load or full-load conditions?
7. Does a window air conditioner need to be leveled?

KEY

Installation—General

1. (a) True, (b) True, (c) False, (d) True, (e) False, (f) False, (g) False, (h) False, (i) True, (j) False
2. Appearance and neatness of installation
3. Well protected
4. Show the customer how to use it.

Installation—Central Vacuum Systems

1. Collects dust and dirt
2. On a basement, garage, or utility room wall
3. No
4. On room walls so as to be within 25 feet of any area in the house
5. $1\frac{3}{4}$-inch-o.d. plastic pipe

Installation—Dishwashers

1. $24 \times 34 \times 34\frac{1}{2}$ high
2. 3 to $3\frac{1}{2}$ feet
3. Flare fitting

MARK CENTER LINE ON WINDOW SILL.

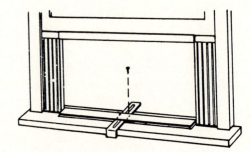

INSTALL SILL CLAMP IN LIEU OF SILL FASTEN-
ING SCREWS FOR STONE SILLS OR OTHER SUCH
MATERIALS.

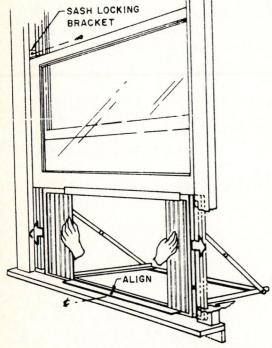

PLACE UNIT MOUNTING ASSEMBLY IN WINDOW
ALIGNING CENTER HOLE IN LOWER FRAME
WITH CENTER LINE MARK ON WINDOW SILL.
LOWER WINDOW TO TOP OF FRAME. EXTEND
SIDES OF MOUNTING ASSEMBLY TO FULL WIDTH
OF WINDOW. AS SHOWN BY ARROWS. INSTALL
SASH LOCKING BRACKETS AFTER SEATING
WINDOW FIRMLY ON ASSEMBLY FRAME.

INSTALL TOP RAIL AND
SEAL ASSEMBLY ON TOP
OF UNIT CASE IN THE
HOLES PROVIDED FOR
FLUSH MOUNTING ONLY.
SEE "STEP 12" FOR
OTHER THAN FLUSH
MOUNT.

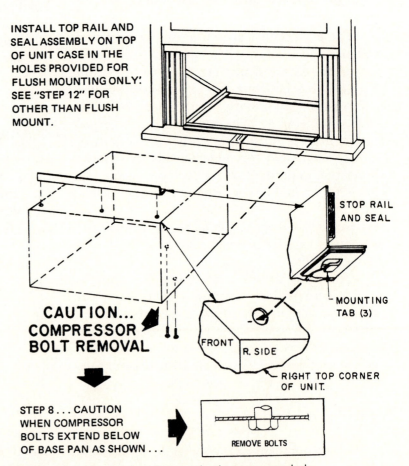

CAUTION...
COMPRESSOR
BOLT REMOVAL

STEP 8...CAUTION
WHEN COMPRESSOR
BOLTS EXTEND BELOW
OF BASE PAN AS SHOWN...

REMOVE BOLTS

FIGURE 4-35. During installation include such weather-stripping and sealing methods as are needed
to provide a reasonably good weather seal between the air conditioner and the window frame.

ADDITIONAL CABINET SUPPORT

FIGURE 4-36. Support brackets run from the window frame down to the cabinet mounting rails. Please note that supports can also be located under the unit and attached to the wall.

4. Yes
5. By four leg bolts

Installation—Disposers

1. In any sink having a 3½- to 4-inch drain opening
2. Yes
3. A 120-volt ac circuit having a 20-ampere fuse

Installation—Dryers

1. Yes
2. No. 10 three-wire cable
3. 30 amperes
4. 20 feet
5. Soap bubbles

Installation—Freezers and Refrigerators

1. 3 inches
2. No. 12 two-wire with ground cable
3. Use a thin piece of paper between the gasket and the box and feel the pull when dragging it out.
4. With a saddle valve

Installation—Incinerators

1. 6 inches
2. Under
3. No
4. On a plate next to the pilot valve
5. No. 12 with ground

Installation—Checking the Chimney

1. 3 feet
2. Place a crumpled wad of newspaper in the chimney and light it. The smoke and flame should be drawn up the flue.

Installation—Ranges and Ovens

1. No
2. No. 8 three-conductor cable, 40-ampere
3. Yes
4. Yes

Installation—Washers

1. Yes
2. 34 inches
3. 6 feet
4. No
5. No. 12 with ground wire, 20-ampere
6. Yes

Installation—Window Air Conditioners

1. On the north side of the house
2. Yes, it must be secure.
3. Yes
4. Either
5. ±10 percent
6. Loaded
7. Yes

5

The Machine:
A Box of Parts

The mystery of servicing even a highly complex machine may be easily solved. Just remember that any machine or appliance is simply a collection of miscellaneous parts. These parts are coordinated to perform in a specific order. However, *you must first know what these parts are, what they do, how they do it, when they do it, and where they are located in a specific machine* (Fig. 5-1). Then you must know their individual relationship to the whole.

THE MACHINE

Let's look at the broad picture. Details come later. For convenience, we will divide a machine into six areas of part usage:

1. Frame
2. Power source
3. Controls
4. Workers
5. Linkage and wiring
6. Fasteners

These parts are within the machine. Since we do not have perpetual motion, the machine must have a source of fuel or energy in order to operate. This fuel may be electricity, gasoline, oil, coal, air, water, or any number of other chemical compounds that can be used or consumed by the machine to produce the type of energy you need.

Now back to the machine. We will give some examples of parts that fit into these various categories. Bear in mind that some parts may belong in more than one category, depending on their usage.

FRAME

The frame of a machine is that area on which the working parts, controls, and power source are mounted so as to hold them in the desired relationship to each other (Fig. 5-2). This is what *makes the machine an organized collection of parts*. Of course, surrounding the frame, in most instances, is a shell, cabinet, or housing.

PARTS LIST FOR MODEL SP1 - PP09 (Z125) AND SP2 (Z126) PRESTO ELECTRIC SHOE POLISHER

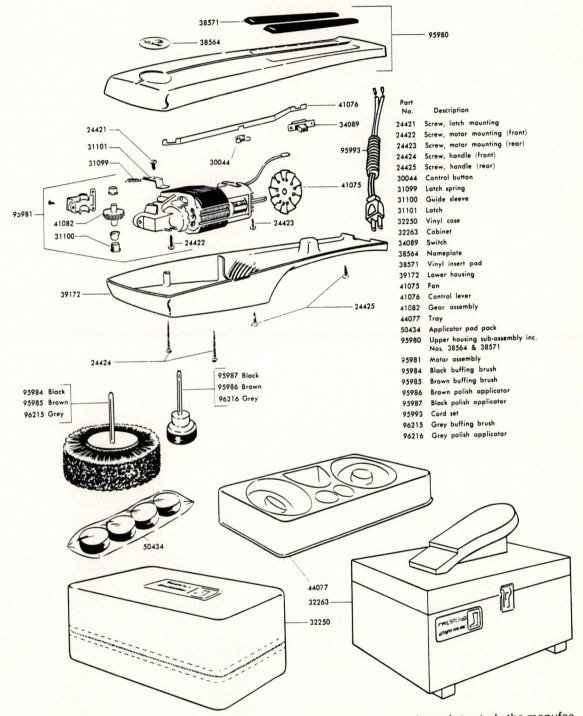

Part No.	Description
24421	Screw, latch mounting
24422	Screw, motor mounting (front)
24423	Screw, motor mounting (rear)
24424	Screw, handle (front)
24425	Screw, handle (rear)
30044	Control button
31099	Latch spring
31100	Guide sleeve
31101	Latch
32250	Vinyl case
32263	Cabinet
34089	Switch
38564	Nameplate
38571	Vinyl insert pad
39172	Lower housing
41075	Fan
41076	Control lever
41082	Gear assembly
44077	Tray
50434	Applicator pad pack
95980	Upper housing sub-assembly inc. Nos. 38564 & 38571
95981	Motor assembly
95984	Black buffing brush
95985	Brown buffing brush
95986	Brown polish applicator
95987	Black polish applicator
95993	Cord set
96215	Grey buffing brush
96216	Grey polish applicator

FIGURE 5-1. A good way to start learning about parts, their names, and uses is to study the manufacturer's parts' lists and product drawings such as those illustrated. (Courtesy of National Presto Industries)

FIGURE 5-2. Many appliances have parts mounted on subframes to allow servicing a related group as a complete assembly. In this instance the compressor and condenser assembly can be moved from under the refrigerator to a more convenient spot for servicing.

This covering may be plastic, sheet metal, wood, or almost anything else that will provide protection for the parts and improve the overall appearance. Also, in some cases, the shell or covering of the machine will support the parts, as well as protect people from getting hurt by moving parts or by some action of the machine. In most cases where the shell is of sheet metal, it has a protective coating. Wood cabinets may have a paint, varnish, or other finish to protect the wood and also to improve the appearance.

POWER SOURCE

The prime *power source* for most appliances is the electrical outlet to which the machine is connected. However, in a motor-driven appliance (vacuum cleaner, washer, blender, etc.) the electrical power is converted into mechanical energy by the motor. *Thus the motor becomes the power source for the mechanical actions* within the appliance. This is an instance where one part fits in more than one category, a mechanical power source and an electricity user.

Other power sources could be an air or hydraulic motor as in a compressor, turbine, or even a windmill or old-fashioned waterwheel.

And let's not forget the gasoline engine in your car or power mower.

CONTROLS

Controls are those devices that *control the power or action of a machine* (Fig. 5-3). Examples could be the wall switches that turn lights off and on in your home, or that turn on the radio. The volume knob on your television set is a control. The accelerator on your car is a control in that it regulates the amount of fuel that goes to the engine. The brake pedal is another control, which affects the deceleration or stopping of a car. A timer, as used on a washer, is a control; it determines which parts do what in the machine. The thermostat on a toaster controls toast quality. A relay is a control, as is a voltage regulator, which controls the voltage and maintains it at a preset level.

Controls may be manual—that is, they are worked by the machine operator—or they may be automatic—preset to perform in a specific manner by design and unalterable by the operator. Frequently, we find a semiautomatic control in which the operator sets the control and the appliance will operate at that setting. (An example would be the temperature control in a refrigerator or the timer in a washer.) Electronic controls are discussed more fully in Chapter 6.

FIGURE 5-3. This humidistat, using human hair as a sensing element, is a control. The changes in hair length that result from changes in humidity will actuate a switch to control either a humidifier or a dehumidifier.

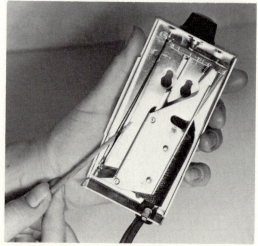

WORKERS

Workers are those devices that *convert the available energy to the mechanical work* that is needed. For example, if the available energy is compressed air, then the worker could be an air motor as in a power screwdriver or a piston. The electricity, as a power source, will run motors, solenoids, relays, vacuum tubes, and many other objects or components that are all parts of more complex machines. A solid background in electricity is important to the service technician because the electrical circuitry has undergone the greatest changes in appliance control design.

LINKAGES AND WIRING

Linkages and wiring are channels that *transport energy from one part of a machine to another*. They are connectives, joining a power source to a working part. Insulated copper conductors—the wiring in an appliance—carry the electricity to the parts that control and use it (Fig. 5-4). In an electric dryer, wires connect the timer and thermostats to the heater. Mechanical linkages come in many forms—gears, connecting rods, and belts, to name a few. The automatic washer is a good example of an appliance that uses several types of mechanical linkage to transmit power from place to place. The motor, with its pulley, drives a belt (Fig. 5-5), which in turn drives other pulleys on the pump and transmission. Within the transmission are gears and connecting rods to translate the rotary motion from the pulley to the reciprocating motion used in washing.

FASTENERS

Fasteners or affixing connectives, are the nuts and bolts that hold the parts together and at-

FIGURE 5-4. The wiring harness is a collection of wires that connects machine controls to the electrical working parts and the power source. The wires are bundled together and routed through the machine in such a manner as to avoid damage to the insulation.

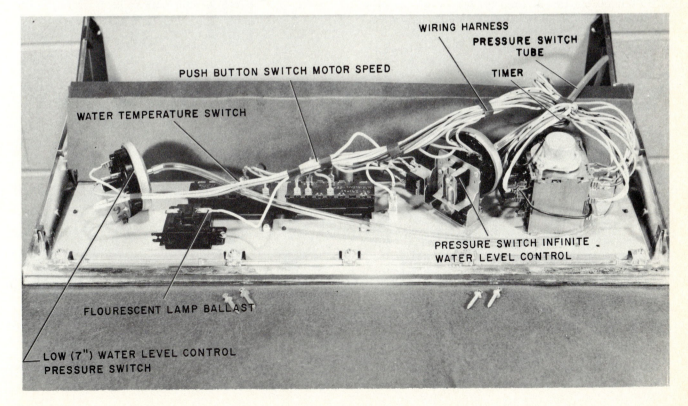

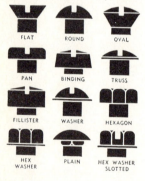

FIGURE 5-5. In many motor-driven appliances, the belt is a most important mechanical linkage. It should be tested on every service call for *tension* and *wear.*

tach them to the frame. These items come in a wide variety of sizes and shapes (Fig. 5-6). Some are permanent; others are removable. Essentially, we are talking about a part or parts used *to hold two or more parts to each other.* Wires are also connected to each other by special fasteners, as shown in Fig. 5-7.

Soldering, brazing, welding, and riveting as a means of connecting two parts together

FIGURE 5-6. All service shops should have a collection of fasteners that can be used to replace those lost or damaged in service. *Do not omit any when reassembling the machine.*

These are the more common screw head types.

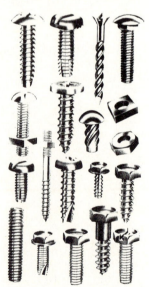

An assortment of sheet metal, self-tapping, and standard screws.

constitute a separate subject and will be discussed independently.

Actually, the major difference in choice between using nuts and bolts or the welding or riveting methods for fastening two parts together is based on whether or not the parts need to be separated for service. Sometimes this serves as a clue in determining where to take something apart.

FIGURE 5-7. (a) This preinsulated splice cap consists of a copper sleeve covered with a nylon insulator. In use it is slipped over the ends of the stripped wire and then crimped with special pliers. The nylon is translucent, to allow visual inspection of the connection. (b) This type of wire connector is twisted on the stripped ends of wire to make a secure insulated connection that can be removed and reused (all without tools). (Courtesy of Buchanan Electrical Products)

(a)

(b)

NYLON INSULATOR   UNIQUE "WEDGE EDGE" SPRING   FINGER GRIP FINS

Each part in a machine has its job to do. It was included for a definite reason. Everything from the finish of the paint job on the outside to the motor that drives the mechanism is interrelated and must do its share of the work. A failure of any one part will, to some degree, affect the operation of other parts and the machine as a whole. It is your duty as a service technician to interpret the conditions of the parts when you check out a machine, and then determine which parts need replacing, which part can be repaired, and which part simply needs cleaning to restore the machine to its normal operation.

There are scores of different parts. Each is the result of hundreds of man-hours of thought, design, manufacturing, and field experience by many people of different companies. As parts are used, they are refined and improved. Generally speaking, you will find that most of the components used in modern machinery are reasonably failure-proof. *You will also find that many of the same parts appear in different machines.* The reason is that a part such as a thermostat—used to control the temperature in an oven—will control the temperature regardless of the brand name on the oven or range. There are only a few manufacturers of thermostats; this is their specialty. So if you learn about one, you will be able to apply that knowledge to many. As we continue, you will be able to see the similarity between parts that perform the same function. You will see *that parts made by one or two manufacturers are used in the appliances made by several manufacturers.* This will simplify your study of servicing. *Learn these basic parts and their functions well.*

STUDY QUESTIONS

Machine

1. The six areas of a machine are _____ .
2. All machines need _____ to run.
3. The shell and its protective coating are part of the _____ of the machine.
4. Electricity could be considered a _____ _____ for a machine.

5. The accelerator of a car is one of its _____ .
6. Welds and rivets are _____ .

SOME OF THE MORE COMMONLY USED PARTS

The easiest way to learn about parts is to get them from a friendly dealer who is "scrapping out" some appliances that may be too costly to repair. Some of the parts may be good and others defective; it should not matter to you which you get. *Study their construction while you study the parts descriptions* that follow.

SWITCHES

Fundamentally, *a switch is a mechanical device for completing or interrupting an electric circuit.* It includes a pair of mating contacts, one fixed and one movable. The material, size, and shape of the contacts have much to do with their life. The contact pressure when the switch is closed and the speed with which the contacts separate when the switch is opened are the other major construction factors governing switch life.

The switch manufacturer will rate a switch for the type of service and the maximum current allowable for that design, but he has no control over its usage. Appliance design engineers usually choose the best available switch for the application, and they pay attention to its protection when used in a product.

When possible, the service technician should replace a damaged or defective switch with a duplicate of the original. ***Do not*** experiment with a lower-rated or less-expensive switch. All switches have their voltage and current ratings marked on the case or mounting bracket.

Construction

The switch contact points are commonly made from a good electrical-conducting, high-corrosion-resistant metal, typically silver or platinum. The mechanism that opens and closes the contacts is usually a copper or brass alloy, which allows good electrical conduction from the terminals to the contacts and offers high resistance to atmospheric corrosion. The moving part is spring-loaded in such a manner as to ensure a fast making and breaking action and enough distance from ON to OFF position to discourage any electrical arcing between the contacts.

These parts are housed in an insulating case designed to keep dust, moisture, and other foreign matter out of the mechanism and *to protect the user from electrical shock.*

Figure 5-8 shows various typical switch constructions. Figure 5-9 describes switch terminology.

In usage, two major problems can occur in a switch:

1. *It will not make contact (OPEN).*
2. *It will not open contact (SHORTED).*

Both conditions can be tested with an ohmmeter by testing from terminal to terminal. An open switch will read infinity (∞) in either switched position, and a shorted switch will read zero (0) (or very close to it) ohms in either switched position.

A defective switch is not serviceable and should be replaced.

FIGURE 5-8. (a) Rocker switch (DPDT) with ¼-inch male spade terminals. The basic switch mechanisms are the same as those used in toggle switches. (b) Toggle switch (DPST): simple on–off type shows dentent action that provides the snap action and holds the toggle mechanism in position. (c) Snap-action switch (shown here as DPST). When the button or plunger is pressed, the movable contact travels from the normally closed (NC) position to the normally open (NO) position. (d) Miniature snap-action switch (SPDT) is a light operating force unit used in timers, door interlocks, thermostats, record players, vending machines, etc. (Courtesy of McGill)

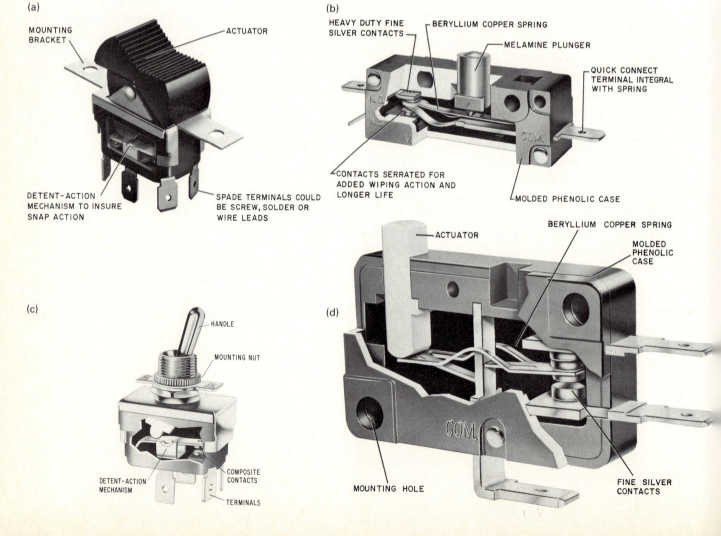

(a)
MOUNTING BRACKET
ACTUATOR
DETENT–ACTION MECHANISM TO INSURE SNAP ACTION
SPADE TERMINALS COULD BE SCREW, SOLDER OR WIRE LEADS

(b)
HEAVY DUTY FINE SILVER CONTACTS
BERYLLIUM COPPER SPRING
MELAMINE PLUNGER
QUICK CONNECT TERMINAL INTEGRAL WITH SPRING
N.C.
N.O.
COM.
CONTACTS SERRATED FOR ADDED WIPING ACTION AND LONGER LIFE
MOLDED PHENOLIC CASE

(c)
HANDLE
MOUNTING NUT
DETENT–ACTION MECHANISM
COMPOSITE CONTACTS
TERMINALS

(d)
ACTUATOR
BERYLLIUM COPPER SPRING
MOLDED PHENOLIC CASE
COM.
MOUNTING HOLE
FINE SILVER CONTACTS

GENERAL TERMINOLOGY

Noninductive Load—Consists of such applications as irons, toasters, ranges, etc. Tungsten loads such as incandescent lamps fall in this category except that there is a greater change in resistance from cold to hot condition. The inrush current on a cold lamp is as much as ten times the rated current when the filament lamp is hot.

Inductive Load—Consists of motors, transformers, solenoids and other devices having an electromagnetic circuit.

Stalled Rotor Test—As required by U.L. is an overload test on the switch consisting of 50 cycles of operation making and breaking the lock-rotor current at the rate of 10 cycles per minute.

Tungsten-Filament Lamp Ratings — T rating applied specifically to a switch rated DC and AC tungsten filament lamp applications.

L rating applies specifically to a switch rated for AC tungsten filament lamp application only.

Horsepower Rating—

HP Rating	Full Load Amps 125V	250V	Lock-Rotor Amps 125V	250V
¼	5.8A	2.9A	34.8A	17.4A
½	9.8A	4.9A	58.8A	29.4A
¾	13.8A	6.9A	82.8A	41.4A
1	16. A	8. A	96.0A	48.0A
1½	20. A	10. A	120.0A	60.0A
2	24. A	12. A	144.0A	72.0A

All switches listed in this catalog exceed the rating shown. The operational life of a switch depends on such diverse factors as duty cycle, mode of operation and environmental conditions.

Bounce — Rapid rebounding of contact after closing.

Break—An opening or interruption of a circuit. Simultaneous interruption of a circuit in two different places is described as double break.

Clearance—Air space, usually ¹⁄₁₆" minimum, between live metal parts of opposite polarity or to ground.

Detent—A catch or holding device.

Dielectric Strength—The maximum potential gradient that an insulating material can withstand without rupture.

Dead Break—An unreliable contact made near the trip point, at low contact pressure. The circuit is interrupted but the switch does not "snap over".

Double Throw—A switch which alternately completes a circuit at each of its two extreme positions. It is both normally open and normally closed.

Power Factor—In AC loads, a measure of inductive or capacitive characteristic of a load. The ratio of real power to apparent power. Underwriters' Laboratories require a 75%

power factor for testing switches. A 100% power factor is a pure resistive load.

Repeatability—The ability of a switch to repeat its operating point.

Silver, Fine—Fine silver is 99.95 pure. It has the highest electrical and heat conductivity of any metal and is widely used as contact material.

Contact Size—Large contacts are desirable for heavy loads because they dissipate the heat from an arc. Silver is a very good contact material because of its excellent conductivity and low electrical resistance. This allows the contact to carry the heat away from the point where the arc is burning and dissipate it over some area. The thickness of a contact has much less effect on its ability to carry heavy loads or to dissipate heat than does its surface area.

Contact Transfer—Whenever an arc is present between two materials, a portion of the contact surface tends to be vaporized, transported in the arc and deposited on other contact. The direction of this transfer depends on the magnitude and polarity of current flow. In an AC circuit the net buildup of material is zero since the transfer over a great number of cycles will be the same in both directions.

Contact Gap — The effective open gap distance between stationary and movable contacts.

CIRCUITRY ABBREVIATIONS

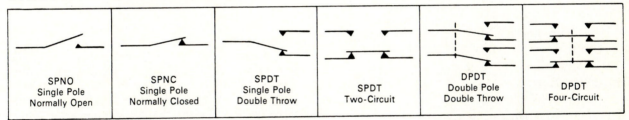

| SPNO Single Pole Normally Open | SPNC Single Pole Normally Closed | SPDT Single Pole Double Throw | SPDT Two-Circuit | DPDT Double Pole Double Throw | DPDT Four-Circuit |

SNAP-ACTION SWITCH TERMINOLOGY

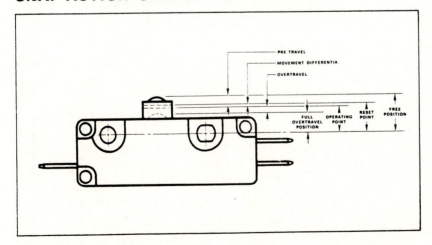

Operating Force—That straight-line force in the designated direction applied to the actuator to cause the switch contacts to snap to the open contact position.

Release Force—The value to which the

force on the actuator must be reduced to allow the contacts to snap from the operated contact position to the normal contact position.

Force Differential — The difference between operating force and release force.

Free Position—The initial position of the actuator when there is no external force (other than gravity) applied on the actuator and the switch is in the specified position.

Operating Position—The position of the actuator at which the contacts snap to the operated contact position.

Releasing Position—That position of the actuator at which the contacts snap from the operated contact position to the normal contact position.

Pretravel—The distance or angle through which the actuator moves from the actuator free position to the actuator operating position.

Overtravel—The movement of the actuator beyond the operating position.

Overtravel Limit Position—That position of the actuator beyond which further overtravel would cause damage to the switch or actuator.

Total Travel—The distance or angle through which the actuator moves when traveling from the free position to the overtravel limit position.

Movement Differential—The distance or angle from the operating position to the release position.

FIGURE 5-9. (Courtesy of McGill)

Switches

1. It does not matter what size switch is used for a motor as long as it fits the mounting hole. True or False

2. A switch that reads ∞ ohms in one position and 50 ohms in the other position is good.   True or False

PRESSURE SWITCHES

Construction

The pressure switch is a single-pole, double-throw (SPDT) switch actuated by a diaphragm responsive to pressure changes. The switch mechanism and the pressure calibration spring that tends to maintain the low-pressure contact are on one side of this diaphragm (Fig. 5-10a). Generally the other side is an airtight chamber having an access hole for the control air. In some designs the diaphragm is in direct contact with the fluid whose level or pressure is to be controlled (Fig. 5-10b).

In dishwashers and clothes washers this switch operates as the water-level control.

Operating Range

Some switches are set for a definite water level and should not be changed. For example, Maytag automatic washer Models A300 and A500 have a normal setting that will range from the second row of holes in the tub to slightly above the top row of holes. Kenmore and Whirlpool models operate at about 11 inches, as do Philco washers. The adjustable pressure switch will adjust to a low level of about $3\frac{1}{2}$ to 4 inches lower.

Westinghouse washers, being front loaders, must operate at a lower level. Their regular setting allows from $5\frac{7}{8}$ to $7\frac{1}{8}$ inches of water fill.

Installation

For proper operation, the *pressure-switch diaphragms must have ready access to the fluid being measured.* In the case of the open diaphragm units, this means no foreign matter should rest on the diaphragm; and for the en-

(a)

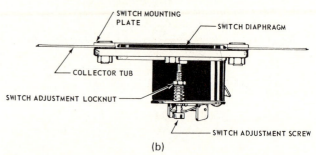

(b)

FIGURE 5-10.   (a) This switch has been cut open to show the diaphragm and switching mechanism. When installed in a washer, the knob shaft will project through the console panel, and its knob will be the water-level control. (b) In this type of pressure switch, the switch diaphragm is in direct contact with the water and is located at the bottom of the tub. The outer edge of the diaphragm, clamped between the tub and the switch mounting ring, serves as a seal to prevent water leakage.

closed diaphragm units, it means that the air tube must be clear of any obstructions and sealed against air leakage.

Electrically this is usually a SPDT switch and can be tested as such. Pressure for a test can be provided by blowing into the pressure tube.

Pressure switches are usually not serviceable. It is essential to replace the entire unit.

Pressure Switches

1. What happens if the tube bringing air to the switch becomes clogged?

2. What happens if the tube bringing air to the switch develops a leak?

THERMOSTATS

A thermostat is a switch that is activated by a change in temperature. All the design needs, uses, and construction precautions that affect good switching operation must be observed in the thermostat's switching mechanism. The aspect that makes a thermostat different is the heat-sensitive part that must open and close the switch.

Heat Sensors

A number of heat-sensing methods can be harnessed to control a switch:

1. Thermocouple
2. Thermistor
3. Bimetal
4. Hydraulic expansion

Some, such as the thermocouple and the thermistor, are used for temperature measurement and are accurate in their response to temperature changes. However, they may require costly parts and amplifying circuitry to be used as thermostats. The bimetal and the expansion types are most commonly used in appliances and are explained below.

Bimetal Element. The bimetal element consists of strips of two different metals securely bonded together so that any change in temperature will affect both metals. Since each different metal has a different coefficient of expansion, one of the metals will expand more rapidly than the other, thus causing the element to bend in the direction of least expansion (Fig. 5-11a). This motion can be harnessed to activate a switch.

Expansion Element. A heated liquid will expand and exert pressure on its container. Continued heating will convert that liquid to a gas producing excellent expansion and pressure capabilities. The liquid can be captive in a tube attached to a leakproof bellows that will move outward as the liquid expands or inward as the liquid cools off. This motion can be harnessed to operate a switch (Fig. 5-11b).

Motion and Pressure

In both instances, the motion is slow but positive, exerting pressure on a switch. The slow make and break at switch contacts operated directly from either a bimetal or expansion type of thermal element usually cause short contact life. A number of switch mechanisms are designed to be triggered by the motion of the thermal element, then to snap either on or off. One type of snap action is built into a bimetal element by "dishing" the element. The bimetal is shaped like a saucer with the more expansive metal on the inside curve. As the temperature is increased, the inside metal expands until it overcomes the restraining action of the outside metal. At a critical point the disk will suddenly reverse to allow the contacts to open or close quickly (Fig. 5-11).

All other thermal elements require a mechanical assist to provide the fast, positive make and break action needed for best switch life. *A properly operating thermal element must move with a change in temperature as long as it is free to do so. A restrained thermal element will exert pressure with a temperature change.* Failure of the element to either move or exert pressure would indicate a defective element.

Direction of Motion

A bimetal strip will always move in a direction toward the side having the least expansive material when the temperature is increased. *The hydraulic thermostat's bellows will always expand when the temperature is increased* (Fig. 5-12). Obviously, in both cases, the motion is reversed when the temperature is decreased. Thus the motion of the thermal element is dependent on temperature change, and its direction of motion depends on the direction of the temperature change.

The switching mechanism design determines whether the contacts will make or break on a rise or fall of temperature (Fig. 5-13). The

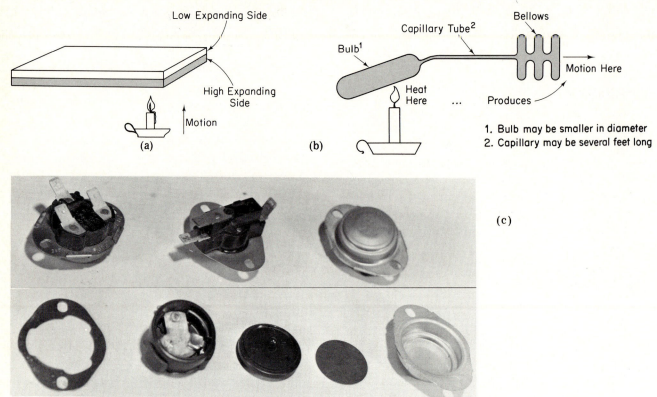

FIGURE 5-11.   The disc-type thermostat provides a snap action when it switches and is used in many applications requiring action or a predetermined temperature. It is not adjustable.

thermal element really does not care; it simply responds to temperature changes.

Testing

But as a service technician you must know what the switch is supposed to do before you can test it. You must know whether the switch contacts are open or closed at a particular temperature,

FIGURE 5-12.   Refrigerator temperature control in which the switch is activated by the changing pressure of the gas in the capillary tube.

and you must know whether they open or closed with an increase or decrease in temperature. For instance, a room thermostat that controls a heater would normally be off at 80°F (27°C), but a room thermostat that controls an air conditioner would be on at 80°F (27°C) (Fig. 5-14).

Usage

Bimetal-type controls are generally used where the wiring, switch, and control can be packaged or contained in one place (Fig. 5-15). The hydraulic type is used where the sensing element must be located at a point other than the location of the switch and wiring section of the control. Such an installation is exemplified in the oven control of a range. The sensing bulb is located in the oven and the switching end of the control is located on the range's control panel.

Very little can go wrong with the thermal element of a bimetal thermostat. But *be careful in handling the thermal element of the hydraulic thermostat.* Avoid sharp bends, kinks, and

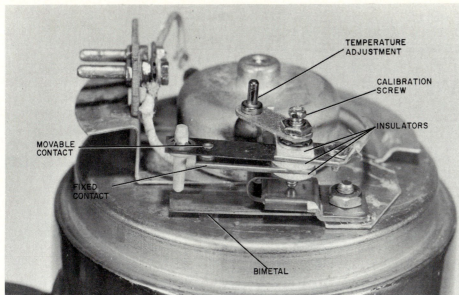

FIGURE 5-13. Bimetal thermostat used to control the temperature of a coffeemaker. As the bimetal warms, its free end will bend away from the coffeemaker to move the insulator and its contact arms away from the fixed contact.

excessive handling or bending of the capillary tube. Vibration, flexing, chafing, or rubbing will damage a capillary tube and allow it to leak. The bulb must be securely mounted but not so tightly as to crush it. Tubing must be routed away from sharp edges and protected from usage interference.

Servicing

Thermostats are switches that will be either open or closed at a given temperature and so can be tested for continuity at that temperature. A change in temperature should activate the thermal element, which in turn should close

FIGURE 5-14. Central heating and cooling wall-mounted control. This type of temperature control is the one with which we are most familiar. (Courtesy of Honeywell)

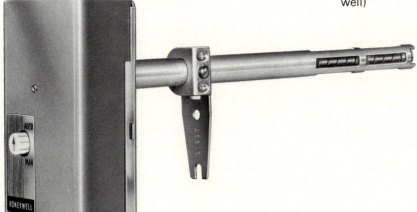

FIGURE 5-15. This type of thermostatic control is most often found in the flue pipe from an oil burner. The active element is a spiral-wound bimetal element. (Courtesy of Honeywell)

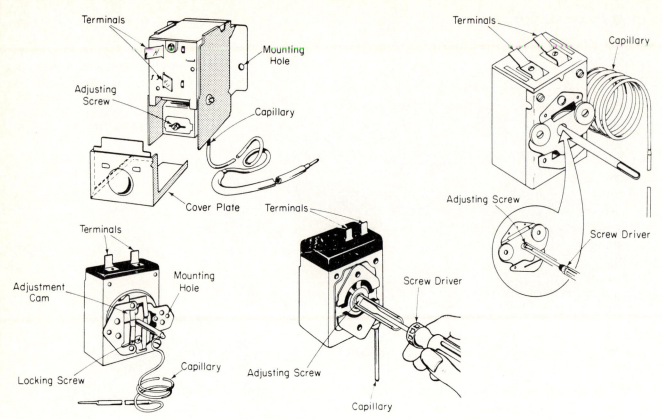

FIGURE 5-16. Adjustable thermostats: Ranco (upper left); Robertshaw (lower left and center); Wilicolator (right).

or open the switch contacts. Failure to obtain a change in continuity with a suitable change in temperature would indicate a defective unit.

Control defects may occur in either the switch contact or the linkage of thermal power elements. Normally thermostats are not serviceable and should be replaced as complete units.

Many adjustable thermostats, such as those used in ovens, may have to be recalibrated to have the actual oven temperature match the numbers on the control. As a rule, this step is done at 300°F (150°C), and a good thermometer is placed in the oven for temperature reference. Figure 5-16 shows the recalibration procedure for several typical thermostats.

STUDY QUESTIONS

Thermostats

1. What activates a thermostat?

2. Name the two types of thermostats most commonly used in appliances.

3. Which switch depends on a liquid to activate it?

4. Which switch requires a bellows as a component?

5. Thermostats require _____ and _____ to operate a switch.

6. If a thermal element is free, a change of temperature will cause it to do what?

7. Increase of temperature will cause the bellows of a hydraulic thermostat to _____.

8. As an appliance technician, is it important to know the details of thermostat switches?

9. Would an oven thermostat be ON or OFF if the dial is set for 350°F (175°C) and the thermometer reads 375°F (190°C)?

10. What is that part of the switch called that reacts to heat?

11. In handling an expansion-type thermostat, what particular care should be taken?

12. If a thermostat proves to be faulty, what is the suggested service procedure?

COLD AIRFLOW CONTROL

Figure 5-17 illustrates the use of a hydraulic thermostat to actuate the mechanical opening and closing of an air damper directly. The damper controls the flow of subzero air entering the food section.

Note: A small heater is wrapped around the body of the thermostat; then the assembly is insulated. This step is to ensure that only the capillary tube is the controlling factor.

Figure 5-17a, with the capillary exposed to room temperature, shows the damper open to allow cold airflow. Figure 5-17b, with the capillary inserted into the cup of ice cubes at 32°F (0°C), shows the damper closed to the cold airflow.

This sytem of cold airflow control operates with a modulating action in that the damper can be partially open or closed. The damper position will vary with the degree of expansion in the tube.

TIMERS

The timer in an appliance has often been called the "brain" of the machine, for it controls the sequence of operation.

A mechanical or clock timer generally consists of three basic components assembled into one unit.

1. The motor
2. The escapement
3. The cam switchbox

Electronic timers are discussed in Chapter 6.

The Motor

The motor is an electric-clock-type motor geared down to a small pinion gear that drives the escapement. These motors develop considerable torque for their 2- to 3-watt input power. Output speed range could be from $\frac{1}{30}$ rpm to 2 or 3 rpm, depending on the usage.

Usually the motor is mounted on the timer by two easily accessible screws, and it can be removed for service replacement. Obviously, if the timer motor is defective, only the motor should be replaced, not the entire timer.

When replacing a motor, use an exact replacement, or else the entire machine's time cycle will be altered.

The Escapement

The purpose of the escapement is to rotate the cam shaft in a series of timed pulses. *This conversion of the slow steady speed of the timer*

FIGURE 5-17.

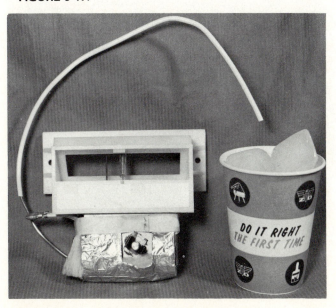

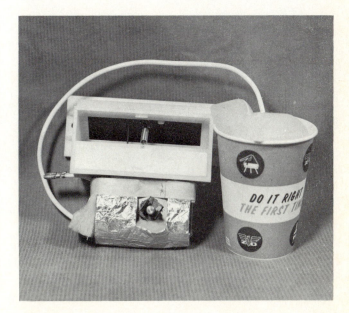

motor to a series of steps provides the quick make and break (snap) action needed at the cam-drive switch contacts (described later).

Most escapement mechanisms consist of a power spring stretched by an offset cam driven by the motor. At a given point, the spring arm slips its cam and forces the switching cam to move forward rapidly a few degrees.

Also included in the escapement mechanism is the ratchet assembly that allows the timer to be advanced quickly by turning the knob manually. *It also prevents the timer from being turned backward.*

The escapement mechanism is *not* generally serviceable in the field, and a problem here normally requires timer replacement.

The Cam Switchbox

The cams in a timer open and close the electrical switch contacts, thereby controlling the sequence of operations. Figures 5-18 and 5-19 show their construction and operation.

This has been a quick, broad review of timer operation. There are many deviations.

For example,

1. Some single-cam timers as used in dryers do not have an escapement (Fig. 5-20).
2. Some timers include the main ON–OFF power switch in a PUSH or PULL motion of the timer dial, and there is no standardization for that action. If a push does not start the machine, try a pull.
3. The defrost timer used on "frost-free" models is essentially a 12-hour clock that turns the refrigerator off for 20 minutes or so twice a day (Fig. 5-21).

STUDY QUESTIONS

Timers

1. What are the three basic parts of a timer?
2. Can a timer be advanced manually?
3. Can a timer be backed up manually?
4. What does a timer control?

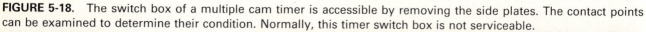

FIGURE 5-18. The switch box of a multiple cam timer is accessible by removing the side plates. The contact points can be examined to determine their condition. Normally, this timer switch box is not serviceable.

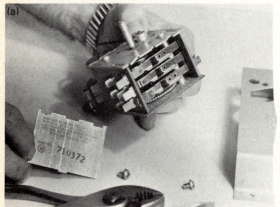

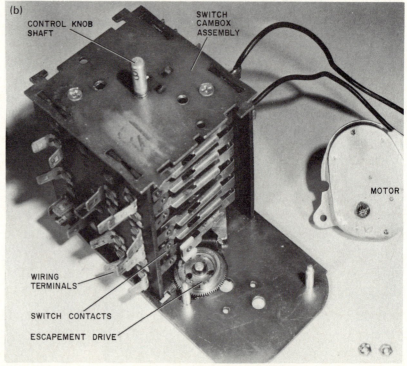

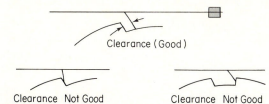

Clearance (Good)

Clearance Not Good   Clearance Not Good

FIGURE 5-19. Position of a cam follower on a cam is "clock" adjustment. If the follower rides part of the way up the cam, poor contact pressure will result in damaged contacts.

FIGURE 5-20. Cam and switch mechanism of a typical timer used as a dryer control. As shown, both the motor and heater elements are ON. As the cam rotates clockwise the follower will drop to the first ledge, turning off the heater. As rotation continues, the follower will drop to the bottom level, turning off the motor.

FIGURE 5-21. The defrost timer on a refrigerator is set to defrost the unit at regular intervals.

SOLENOIDS

A solenoid is an electromagnet having a movable ferromagnetic (iron) core (Fig. 5-22). The mechanical motion of the core being drawn into the coil does the work. The electromagnetic pull-

ing action of all solenoids is basically the same as described in Table 5-1. The data shown in the following chart lead to a number of conclusions on solenoid operation:

1. *Pull is proportional to voltage.* A higher voltage produces more pull.
2. *The farther out the armature, the more current needed and the weaker the pull.*
3. *An intermittent-duty solenoid requires more current and develops more pull.* It also heats up more and, if used for continuous duty, would overheat and burn out.

Therefore:

Low voltage may weaken the solenoid's pull so that it will not close under normal load. This would cause the coil to overheat and burn out.

The armature may jam before closing, or an overload may prevent the armature from closing. This would cause the coil to overheat and burn out.

It is work done by the moving core that makes the solenoid useful in appliances.

Figure 5-23 illustrates the range of usefulness of a solenoid.

FIGURE 5-22. Typical solenoid used to operate valves, open shutters, shift belts, and many other mechanical duties that are remote from the control center.

TABLE 5-1

Operating characteristics of a typical solenoid such as shown in Fig. 5-22

| Open Distance (In.) | Constant-Duty Design | | | | Intermitten-Duty Design, 120 Volts | |
| | 90 Volts | | 120 Volts | | | |
	Pull (Pounds)	Current (Amperes)	Pull (Pounds)	Current (Amperes)	Pull (Pounds)	Current (Amperes)
1	4.5	4.9	6.4	6.5	6.7	7.9
$\frac{3}{4}$	4.6	4.1	7.6	5.5	10.2	7.2
$\frac{1}{2}$	5.3	3.1	9.0	4.2	11.7	5.9
$\frac{1}{4}$	4.6	2.1	9.2	2.8	12.2	3.8
0	9.2	0.3	12.0	0.4	16.0	1.0

Servicing

Electrically, two things may happen to a coil: it may short or it may go open. In any case, it will have to be replaced. Some construction will allow the coil and its housing to be replaced; other designs require replacement of the entire assembly.

Mechanically, the armature may get jammed and stick in an ON or OFF position. If the jam can be cleared, do so; if not, replace the entire solenoid assembly.

STUDY QUESTIONS

Solenoid

1. What moves in a solenoid?
2. Will an armature that is stuck open damage a solenoid?

RELAYS

A relay is an electromagnet having a fixed core and a movable ferromagnetic (iron) armature. In most applications, the armature is held away (or open) from the core by a return spring and is pulled in (activated) when the coil is energized.

The armature usually moves a set of switch contacts to open or close electric circuits. Thus a relay can be used to switch large currents and voltages on command from a low-voltage or a low-current source.

Relays are also used to energize and deenergize the starting windings in appliance motors. These are called starting relays.

Switching Relays

Figure 5-24 should allow you to recognize and

FIGURE 5-23. This assortment of solenoids includes one that translates its straight-line pull to rotary motion and another that *pushes.*

FIGURE 5-24. Switching-type relays of the type used in some appliances as well as other electrical control circuits.

determine the function of the general-purpose switching relay.

Motor-Starting Relays

There are several types of motor-starting relays, and although their appearance, construction, and method of operation differ, they do the same job. They energize the motor-starting winding until the motor is almost up to speed; then they open the circuit to deenergize the start winding, thus allowing the motor to operate on its run winding only. (See Split-Phase Start Motors—page 94.)

Magnetic Current Relays

Figure 5-25a and b includes representative samples of the magnetic current relay. Figure 5-25c is a schematic diagram showing its electrical connections. It shows the coil to be in series with the run winding and the contacts in series with the start winding. Normal position is for the start switch contacts to be open. When current is supplied to the motor, the amperage draw of the run winding is great enough to develop a strong magnetic field in the relay coil, thereby activating the armature so that it closes the start winding contacts.

As the motor nears its running speed, its amperage draw decreases, thus weakening the magnetic pull to the relay armature. When this pull becomes too weak, the armature will release, thus opening the start winding contacts.

Most current relays depend on *gravity to release* the *armature* from the magnetic field and must be *installed* with the *TOP* up as labeled!

Magnetic Potential or Voltage Relays

As shown in Figs. 5-26 and 5-27, the potential relay responds to the voltage appearing across

FIGURE 5-25. Magnetic current relays. (a) Relay has built-in overcurrent protection. If the current draw becomes too great, a resistance wire will heat a bimetal strip, causing a normally closed contact to open the circuit. As with all current relays, the correct relay size *must* be used with a given motor size. (b) Typical starting relays that plug onto the terminals of a compressor. (c) Magnetic-current type of start relay and motor connnections for the relay of part (b).

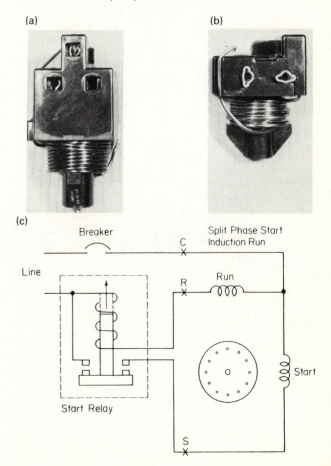

FIGURE 5-26. Typical application of the relay shown in Figure 5-25b. The relay terminals connect to the start and run winding terminals and the third connection (upper terminal) is the common from the overload breaker.

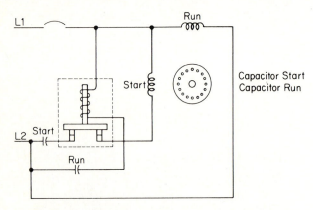

Potential Relay Disconnects Start Capacitor from Circuit

FIGURE 5-27.

FIGURE 5-28. This solid-state relay can replace several sizes of current relays.

the start winding, and its normal position is with the starting contacts *closed.*

This relay operates on the principle that during periods of high-amperage draw there is a corresponding drop in voltage. The high-amperage draw occurs during the starting period from 0 rpm to about three-fourths of normal running speed. At this point the voltage increase develops enough magnetic pull in the relay coil to open the switch.

The major advantage of the potential relay over the current relay is that its switch contacts open at a period of low-current draw, thereby decreasing contact arcing. The current relay must match the motor current with which it is being used. Therefore, there will be a different electrically sized relay for each different horsepower-rated compressor.

Thanks to developments in solid-state electronics, an improved starting relay has been developed that will replace several sizes of current relays (Fig. 5-28). In addition, this relay has no moving parts or contacts.

Hot-Wire Relays

As shown in Fig. 5-29 the hot-wire relay connects the line to both the start and run winding through a series resistance. As current flows through the resistance wire, it expands, allowing a mechanical linkage to open the start winding circuit. If the current flow is correct, it will remain in that position and the motor will run normally. If there is excessive current flow, the wire will expand farther until the winding contact has been broken. This relay includes overload protection and can be mounted in any position.

Bimetal Relays

The bimetal relay, shown in Fig. 5-30, consists of two bimetal switches, each in series with a winding and a small resistance heater in series with the line. The starting current causes the heater to open the start bimetal element and allows the run bimetal to remain as is. An overload will cause the run bimetal to warp open, shutting off the motor. This relay can be mounted in any position.

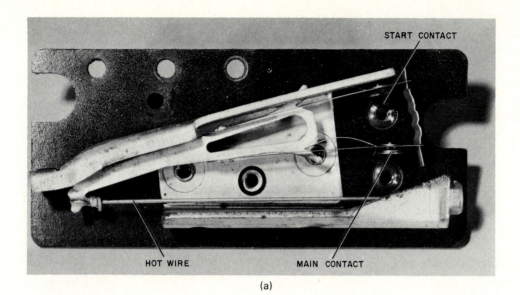

START CONTACT

HOT WIRE   MAIN CONTACT

(a)

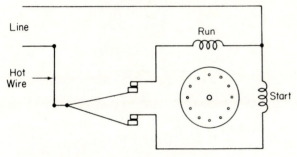

Line

Hot
Wire

Run

Start

Position 1 — Both Contacts Closed at Start
Position 2 — Start Contacts Open at Run
Position 3 — Both Contacts Open at Overload

(b)

FIGURE 5-29. (a) Hot-wire starting relay and overload protector; (b) schematic diagram. The relay is not serviceable; when it fails, it must be replaced. Make certain to choose one that is an exact replacement or that matches the compressor motor rating.

FIGURE 5-30. (a) Bimetal-type motor starting relay and protector; (b) schematic diagram.

(a)

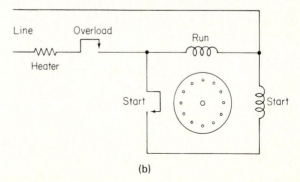

Line   Overload   Run

Heater

Start   Start

(b)

The most common failures in starting relays are:

1. Open coil
2. Burned contacts
3. Faulty wiring
4. Incorrect mounting
5. Wrong part

Each of these relays is designed to operate with motors in a definite size range, and a correct replacement must be used.

If the complaint is "does not start," and

voltage is available at the relay's L1 terminal, and the compressor will start on a test cord, replace the relay.

STUDY QUESTIONS

Relays

1. What moves in a relay?
2. Is it inside or outside the coil?
3. What causes a magnetic-current relay to operate?
4. What causes a potential relay to operate?
5. Does a hot-wire relay provide overload protection?
6. Can a starting relay be serviced?

WATER INLET VALVES

Pilot-operated water valves come in all sizes and shapes, but all operate on the same principle. *Water pressure seals the valve when the pilot hole is closed and lifts the diaphragm to open the valve when the pilot hole is opened.*

Operation

Refer to the cross-sectional drawing of the water inlet valve in the closed position (Fig. 5-31a). The water pressure at B and C has equalized because the bleed hole opening permits water to flow from B to C. Pressure at A is normal air pressure. The valve remains closed because the higher water-line pressure at C is greater than the atmospheric pressure at A. When the solenoid coil is energized by the timer, the plunger (armature) is withdrawn from the diaphragm opening. Water in compartment C will then escape through the diaphragm opening, thus causing the pressure in compartment C to fall to almost atmospheric pressure. The higher water-line pressure, at B, forces the diaphragm away from its seat, thus opening the valve. Water then flows through the screen, under the diaphragm, and out the valve, as indicated in the cross-sectional view showing the valve open (Fig. 5-31).

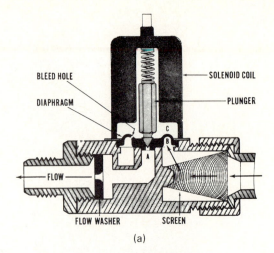

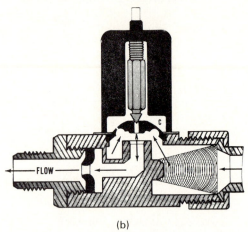

FIGURE 5-31.   (a) Valve in closed position; (b) valve in open position; (c) valve diaphragm showing bleed holes.

When the solenoid releases the plunger, the hole in the center of the diaphragm is sealed. Water then flows through the bleed hole (Fig. 5-31c) in the diaphragm until the pressure in compartment C rises to water-line pressure. Since the area of the diaphragm at water-line pressure in compartment C is greater than the area at water-line pressure in compartment B, the diaphragm is forced closed. The pressure at A is always lower than full-line pressure because water passing through the valve is escaping to open air.

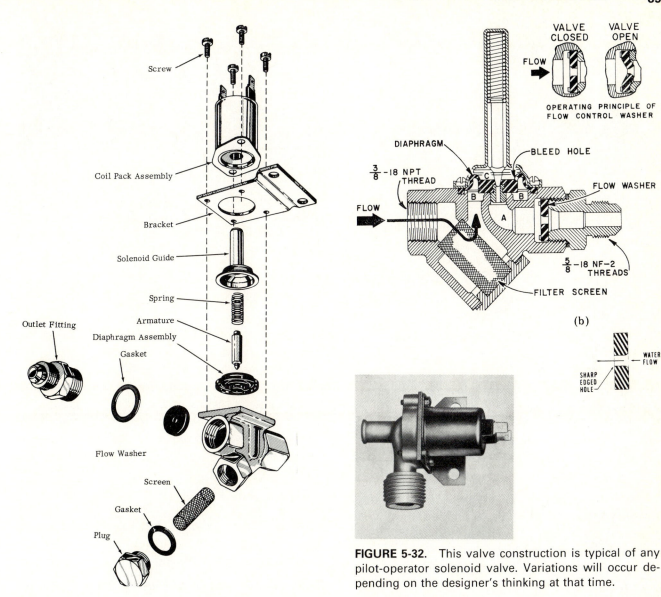

FIGURE 5-32. This valve construction is typical of any pilot-operator solenoid valve. Variations will occur depending on the designer's thinking at that time.

Construction

The construction of this type of valve is shown in the accompanying exploded views (Fig. 5-32a). The flow-control washer and its retaining ring are only included in valves that are used in machines in which the timer controls the duration of fill time. The total amount of water used is equal to the fill time multiplied by the flow rate. For example, a fill time of 3 minutes at a flow rate of 2 gallons per minute will provide 6 gallons of water in the machine within the time allowed by the timer. It requires a coordinated design of both the timer and the flow control washer to provide the correct amount of water in the machine. The purpose of this washer is to maintain a constant flow rate within a wide pressure range from 20 to 80 psi, the size of the hole determining the flow rate (Fig. 5-33). The flow-control washer is flexible, allowing an increase in water pressure to bend it and thus making the hole smaller (Fig. 5-32b). The balance between hole size and pressure will maintain a constant flow of water.

Servicing

Most valves can be disassembled by removing the four screws that attach the solenoid, valve bracket, and valve body to each other (Fig. 5-32).

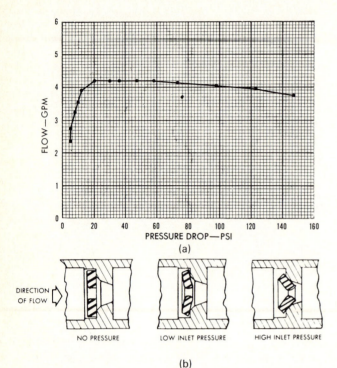

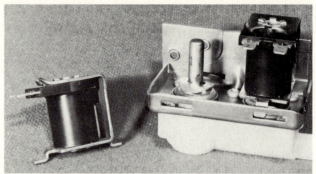

FIGURE 5-34. The construction of this valve allows the solenoid coil to be removed without a screwdriver. Depress the locking tab and turn the coil counterclockwise. Since the coil resistances vary from 200 to 500 ohms, check any questionable reading against a good coil from a valve of the same model number before condemning it.

FIGURE 5-33. (a) Characteristic operating curve for a 4-GPM flow control washer. (b) Effect of pressure on the shape of the flow control washer, showing how the opening squeezes to a smaller size with an increase in pressure.

The solenoid guide, spring, armature, and diaphragm can then be disassembled for inspection. If the diaphragm bleed holes are obstructed, they must be opened with a pin or small needle. *When replacing diaphragms, locate those with a single bleed hole so that the hole is toward the outlet side of the valve.* Diaphragms with two bleed holes should be so located that the holes are in line with the flow of water through the valve.

In hard-water areas the solenoid guide, valve body, and other parts within the valve may become encrusted with lime. These parts can be cleaned by soaking them in vinegar; boiling will hasten the process.

When reassembling, make certain that all parts are positioned as shown in the sectional view of the valve.

On every service call, always check the screen in the valve. It takes little rust, scale, or other foreign matter in the water supply to clog this fine mesh screen. *Do not omit the screen under any circumstances.* To do so will invite clogging of the bleed holes, which in turn

will prevent the valve from closing. The result is an overfill or flooding condition.

The solenoid (coil) that operates the valve can be tested for continuity to determine whether it will work or not (Fig. 5-34). An open or otherwise defective coil can be replaced. *As a rule,* only one or two screws hold it in position and *they can be removed without disturbing the rest of the valve.*

STUDY QUESTIONS

Water Inlet Valves

1. What principle of physics is the key to water inlet valve operation?
2. What determines the flow rate of water through a valve?
3. Does it matter where you put the holes in a diaphragm when replacing it?
4. What would you use to clean the bleed holes in a diaphragm?
5. Is there any way encrusted lime can be removed from components? If so, how?
6. What usually operates a water valve?

ELECTRIC MOTORS

Electric motors used in appliances must develop enough rotational torque to move a load without overheating or otherwise endangering the appliance or its user.

TABLE 5-2
Motor-type selection chart

Motor Type	Shunt	Universal	Split Phase	Capacitor Start	Capacitor Run	Shaded Pole
Current supply	Dc	Ac, Dc	Ac	Ac	Ac	Ac
Duty	Continuous	Intermittent	Continuous	Continuous	Continuous	Continuous
Starting torque	High	High	Normal	High	Low	Low
Current	Normal	Normal	Normal	Normal	Low	Low
Speed	Adjustable	Adjustable	Fixed	Fixed	Limited Adjustable	Limited Adjustable
Rotation[a]	R.R.	O	R.O.	R.O.	R.R.	O

[a] R.R., Reversible at rest or run by switch; R.O., reversible at rest only by switch; O, reversible by design change.

Consequently, the product design engineers have had to consider many factors regarding motor type, power, speed, and ventilation, as well as final shape and mounting needs.

Motor Types

Of the many types of motors available, appliance usage has centered around the few fractional horsepower, single-phase motors described in Table 5-2.

Universal Motors

The series universal motor is the only ac-dc motor designed for common usage in the appliance field. *It operates at about the same speed and output power on either direct or alternating current of approximately the same voltage.*

Figures 5-35 and 5-36 show the series universal motor to consist of a wound armature connected, in series, to the field coil through a system of brushes and a commutator.

This type of motor is self-starting, has a high starting torque, and a high speed at no load. But its speed is sensitive to input voltage and load. As the load is increased, the speed will decrease; and as the input voltage is decreased, the speed will decrease.

This sensitivity to voltage leads to several methods of speed control:

1. Varying the strength of the magnetic field by tapping the field coil (Fig. 5-36b).

2. Using an automatic switch that controls the field current as a governor which automatically connects a re-

FIGURE 5-35. This universal series motor is the type used in hand mixers. The worm drive to the beater gears serves as a speed reducer.

COOLING FAN   WORM DRIVE   FIELD   ARMATURE   BRUSH HOLDER   BRUSH   COMMUTATOR

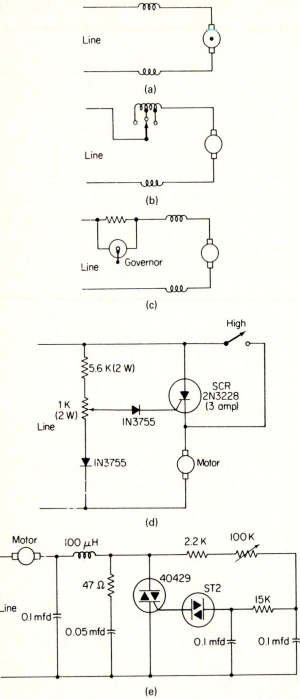

FIGURE 5-36. (a) Basic wiring, one speed. (b) Speed-control field winding is tapped. (c) A series resistance will decrease speed and can be governor controlled. (d) This circuit will provide half-wave speed control with regulation. Closing the "high" switch allows the motor to operate at its maximum speed with no control. (e) This circuit will provide full-wave motor speed control with regulation. The 1-microfarad capacitor and the 100-millihenry coil provide a reduction in radio interference. The 47-ohm resistor and the 0.05-microfarad capacitor protect the triac from transient voltages.

sistor inserted in series with the field winding decreases the current, as in Fig. 5-36c.

3. Using an SCR or triac and its associated phase control circuitry, as in Figs. 5-36d and e.

General usage for this type of motor is in vacuum cleaners, fans, mixers, blenders, floor polishers, and small power tools where the load is more or less fixed; and the high speed can be utilized to produce a high power output.

The service reference guide (Table 5-3) will list the most common troubles and their causes.

AC Induction Motors

The remainder of the motors listed are all *single-phase ac only, induction-type motors, and as such require some arrangement to make them start.*

In order to understand the starting of an ac motor, you must know that a rotating or moving magnetic field is needed to pull the armature from rest up to its operating speed.

This is easily done by supplying three-phase power to the motor. Three-phase power will result in a sequence of magnetic pulls within the motor that produce a rotating field, which the armature will try to follow.

However, on single-phase power, this sequencing must be done by delaying the magnetic field buildup on part of each north and south polar area (Fig. 5-37).

Slip

Normal operating speed will always be somewhat less than the synchronous speed, depending on the "slip," which is the difference in speed between the rotating electrical field speed and the armature speed.

Within the power-operating range of the motor, the slippage is somewhat self-regulating. As the load is increased, the slippage increases, thereby making the speed differential between the electrical field and the armature

TABLE 5-3
Reference guide to probable causes of motor troubles

Motor Type:	Ac Single Phase					Brush Type (Universal, Series, Shunt, or Compound)
	Split-Split Phase	*Capacitor Start*	*Permanent-Split Capacitor*	*Shaded Pole*	*Ac Polyphase (Two- or Three-Phase)*	
Trouble	Probable Causes[a]					
Will not start	1,2,3,5	1,2,3,4,5	1,2,4,7,17	1,2,7,16,17	1,2,9	1,2,12,13
Will not always start, even with no load, but will run in either direction when started manually	3,5	3,4,5	4,9		9	
Starts, but heats rapidly	6,8	6,8	4,8	8	8	8
Starts but runs too hot	8	8	4,8	8	8	8
Will not start, but will run in either direction when started manually—overheats	3,5,8	3,4,5,8	4,8,9		8,9	
Sluggish—sparks severely at the brushes						10,11,12,13,14
Abnormally high speed—sparks severely at the brushes						15
Reduction in power—motor gets too hot	8,16,17	8,16,17	8,16,17	8,16,17	8,16,17	13,16,17
Motor blows fuse, or will not stop when switch is turned to off position	8,18	8,18	8,18	8,18	8,18	18,19
Jerky operation—severe vibration						10,11,12,13,19

Source: Bodine Electric Company.
[a] Probable causes:
1. Open in connection to line.
2. Open circuit in motor winding.
3. Contacts of centrifugal switch not closed.
4. Defective capacitor.
5. Starting winding open.
6. Centrifugal starting switch not opening.
7. Motor overloaded.
8. Winding short circuited or grounded.
9. One or more windings open.
10. High mica between commutator bars.
11. Dirty commutator or commutator is out of round.
12. Worn brushes and/or annealed brush springs.
13. Open circuit or short circuit in the armature winding.
14. Oil-soaked brushes.
15. Open circuit in the shunt winding.
16. Sticky or tight bearings.
17. Interference between stationary and rotating members.
18. Grounded near switch end of winding.
19. Shorted or grounded armature winding.

greater. Then the motor draws more current from the line, thus increasing the magnetic pull. The increase in torque, resulting from the increase in magnetism, will cause the armature to try and "catch up" to the field speed.

Shaded Pole Motors

The shaded-pole motor has one winding in its stator and a portion of each pole face is "shaded" by a copper band to provide the rotating magnetic field necessary for starting (Fig. 5-38).

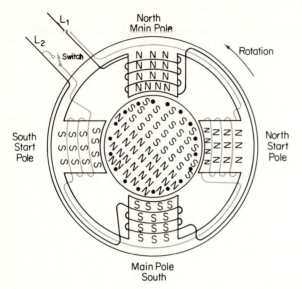

FIGURE 5-37. Because the start winding and the main winding have different impedances, each will peak their magnetic field at a different time. This shifting of magnetic field strength with a polar area attracts the armature to follow it, making it turn.

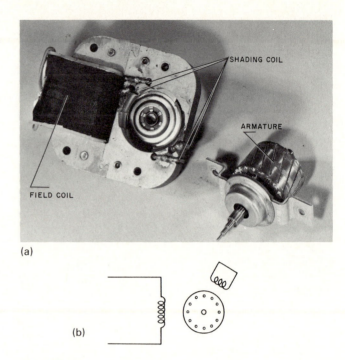

FIGURE 5-38. (a) Shaded-pole motor; (b) schematic.

Once started, the shaded-pole motor runs as an induction motor, almost at synchronous speed. The direction of rotation is determined by the position of the shaded pole in relation to the coil winding. (See Table 5-2) for other characteristics.) Typical usage is for small fans, record players, and clocks.

Split-Phase Start Motors

The split-phase start motor uses a main (run) winding and *a start winding having different inductive reactances from the main winding. This difference "splits" the single phase of the power supply to allow the start winding magnetic field to build to a peak after the main winding magnetic field does.* This slight time difference in the magnetic field buildup provides the rotating magnetic field needed to start the armature turning (Fig. 5-39a).

Once the armature has reached about 75 percent of its rated speed, the start winding is no longer needed and is automatically disconnected by either the centrifugal switch or the starting relay (Fig. 5-39b). (See notes on relays page 85.)

The motor can be reversed by reversing the

connections to the start windings, but it cannot be adopted to a variable speed control.

Multiple-Speed, Split-Phase Motors

In the multiple-speed, split-phase motor, several run windings are wound around different numbers of poles to produce the various synchronous speeds. A switch selects the desired run winding.

These motors are used in fans, dryers, washers, and many other appliances.

Capacitor Start Motors

The capacitor start motor is basically the same as the split-phase start motor except for the addition of a capacitor in series with the start winding (Fig. 5-40). *The addition of a capacitor to the start winding allows its magnetic field to build to a peak sooner.* This result increases the time differential between the start and run winding magnetic fields and produces more starting torque with less input power.

It has the same usage as the split-phase motor but is more efficient in starting, and thus preferable for stop and start applications.

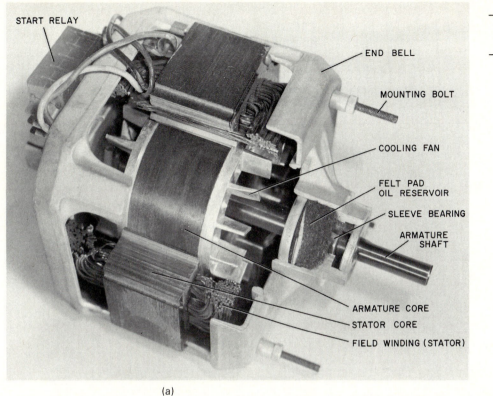

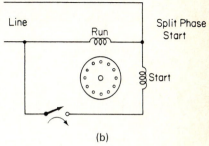

(b)

(a)

FIGURE 5-39. (a) Split-phase start motor typical of those used to power washers and dryers. Note the start switch mounted on the outside of the motor frame, allowing it to be changed without entering the motor. (b) Schematic.

Permanent-Split Capacitor Motors

The permanent-split capacitor motor has a capacitor in series with the auxiliary winding, which is designed to be left in the circuit while running.

Since both windings are in the circuit at all times, it does not need a starting relay or centrifugal switch (Fig. 5-41). When the two windings are made identical, the phase shift occurs when the capacitor position makes the motor reversible by shifting the capacitor connection from one winding to another.

Like the shaded-pole motor, its speed can be varied by changing the available voltage.

FIGURE 5-41.

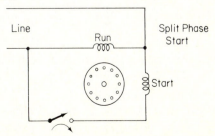

FIGURE 5-40. The capacitor start motor has a capacitor (about 100-microfarad) connected in series with the start winding to provide better starting action than the split-phase motor.

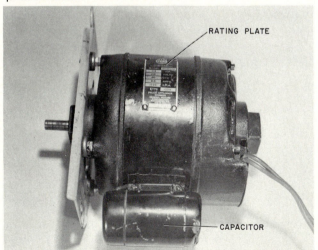

STUDY QUESTIONS

Electric Motors

True or False

1. A universal motor will operate on either ac or dc.
2. The speed of a universal motor can be controlled.
3. An induction motor will run on only one winding.
4. "Slip" occurs when the supply voltage is too low.
5. A shaded-pole motor is self-starting.
6. A shaded-pole motor is reversable.
7. A split-phase motor can be reversed by reversing its starting winding.
8. The starting winding is disconnected from the shaded-pole motor when it reaches full speed.
9. The speed of a split-phase motor can be controlled by a transistorized circuit.
10. The capacitor-start motor has a higher starting torque than the split-phase motor.
11. A permanent-split capacitor motor is reversable.
12. A permanent-split capacitor motor can be speed controlled.

HEATING ELEMENTS

Electric-heating elements used in appliances are possible because the resistance of a conductor to the flow of electricity through it produces heat.

Most elements are made with Nichrome wire as the heating conductor. Nichrome is a nickel-chromium alloy having both a high resistance to current flow and high strength at high temperatures (Table 5-4). A type of nichrome wire used in many appliance heating elements is an alloy of: 60 percent nickel, 16 percent chromium, and 24 percent iron. It has good heat resisting properties up to 1700°F (925°C).

The wire's resistance and operating temperature will depend on its particular alloy percentage and its *diameter*.

A heating element's heat output is measured in Btu, which can be calculated from its

TABLE 5-4
Some typical data on nichrome wire

Size B & S	Ohms Per Foot at 68°F (20°C)	Amperes Needed for Stated Temperatures	
		500°F (260°C)	1500°F (815°C)
15	0.2078	5.8	18.6
20	0.659	2.4	7.9
25	2.107	.96	3.4

wattage input, for 3412 Btu per hour are produced by 1 kilowatt-hour, and from Watt's and Ohm's laws,

$$\text{Watts} = \text{volts} \times \text{amperes}$$

$$\text{Resistance} = \frac{\text{volts}}{\text{amperes}}$$

or

$$W = I^2R \quad \text{or} \quad \frac{E^2}{R}$$

In practice, the resistance can be measured by an ohmmeter and the voltage by a voltmeter. Since this is strictly a resistive function, these formulas are good for both ac and dc.

Figure 5-42 should serve to acquaint you with the great variety of heaters in common usage so that you will recognize them in the field.

Selecting a Replacement Heater

1. Check the mounting ring or flange for size and type.
2. Select the element needed.
   (a) *Range:* 6 or 8 inches in diameter, single coil or two coil.
   (b) *Oven:* bake or broil, plug-in, hinged, permanent mounting, size.
   (c) *Water heater:* length must be less than tank diameter.
3. Check the voltage.
   (a) 115 to 120 volts select element rated at 120 volts.

(a)

(b)

(c)

(d)

FIGURE 5-42. (a) Heating element and mounting flange as used in an electric water heater. This is an immersion-type element having a 75-watt per square inch watt density. This type of element depends on the water to conduct its heat away rapidly. If operated in air, it overheats and burns out. (b) The electric range heater element with an adapt-o-kit is used to replace many other elements. (c) Electric range heater element: monotube type with trim ring adapter. (d) A defrost heater consists of a resistance wire in a plastic insulating sheath attached to an aluminium foil reflector. This type of heater is used with frost-free refrigerators to prevent water accumulation around a door from freezing. [(b) and (c) Courtesy of Chromalox]

(b) 208/230/240 select element rated at 240 volts.

4. Check the required wattage.

(a) Match wattage of original element. (For faster heating, use the next size larger.)

(b) For 208 volts use 240-volt element having a one-third higher wattage. Otherwise, the lower voltage will produce a lower wattage, which will not matter in many instances.

Example:

Range Element	6 Inches	8 Inches
240 volts	1500 watts	2600 watts
208 volts	1125 watts	1950 watts

STUDY QUESTION

Heating Elements

1. Can a heating element designed for use in water be safely used in air?

KEY

Machines

1. Frame, power source, controls, workers, linkage and wiring, fasteners
2. Fuel or energy
3. Frame
4. Power source
5. Controls
6. Fasteners

Switches

1. False
2. False

Pressure Switches

1. The appliance will flood (no water level control); switch will not work.
2. Same as number 1.

Thermostats

1. *Change* of temperature
2. Bimetal and expansion
3. Expansion switch
4. The expansion switch
5. Motion and pressure
6. Move
7. Expand
8. Yes
9. OFF
10. Sensing element

11. Do not bend, kink, or stab the thermal element capillary.
12. Replace complete unit.

Timers

1. Motor, escapement, cam switch
2. Yes
3. No
4. The sequence of operation of a machine

Solenoids

1. The core
2. Yes

Relays

1. The armature
2. Outside
3. Current flow
4. A drop in voltage
5. Yes
6. No

Water Inlet Valves

1. Pressure differential
2. The size of the hole and pressure
3. Yes, should align with water flow
4. A pin or needle
5. Yes, by soaking or boiling in vinegar
6. A solenoid

Electric Motors

1. True
2. True
3. True
4. False
5. True
6. False
7. True
8. False
9. False
10. True
11. True
12. True

Heating Elements

1. No

6

Basic Electricity and Electronics

A basic knowledge of electricity is necessary to service an appliance properly. This does not mean that you must be an electrician or an electrical engineer in order to be an appliance technician, but having a knowledge of what electricity is, what it can do, and how it is controlled will make you more efficient when servicing electrical equipment of every kind. Although this chapter cannot cover all there is to know about electricity, it will provide enough basics to get you by.

HISTORY

Man has known about electricity for centuries. In fact, more than 5000 years ago man knew about a stone that had power to attract iron. But it was not until much later that the Chinese took advantage of the magnetic attraction of this stone to use it as a compass to help them find their way across the Gobi Desert. About a 1000 years or so ago someone noticed that the "north-seeking" end of one magnet sought the

"south-seeking" end of another magnet (Fig. 6-1) but that the *like* ends *repelled* each other. In the meantime the Greeks had discovered that when they rubbed amber with felt or wool, it attracted light substances, such as straw, leaves, or paper (Fig. 6-2). About the fourth century B.C., other Greeks evolved the theory that all matter was composed of tiny particles called atoms.

This is about where electricity stood until 1600, when Sir William Gilbert, an English physician and physicist, gathered all available data and wrote a series of books on the subject of electricity. He gave the subject its name—borrowed from the Greek word *electron*, meaning amber.

In about 1660 a German physicist named Otto von Guericke built the first static generator and proved that electricity could be *generated*. He also discovered that electricity could be made to travel to the end of a thread and that all charged bodies had properties similar to those of magnets. *Like charges repelled each other* and *unlike charges attracted each other*

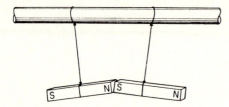

FIGURE 6-1. Unlike poles attract.

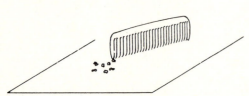

FIGURE 6-2. A charged nonconductor will attract small particles that have an opposite charge.

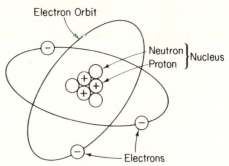

FIGURE 6-4. The atom is structured as a nucleus of protons (+) with an equal number of electrons (−) in orbit around it.

(Fig. 6-3). These discoveries lead to a number of experiments by other people, and additional interest and theories about electricity were developed.

In 1750 Benjamin Franklin's famous kite experiment proved that lightning is a form of electricity. He also established a theory about the nature of electricity, stating that it existed in two different forms, *positive* and *negative*. It was this theory that became the foundation of most experiments that have been carried on ever since. It has taken thousands of years for man to get to where he is in his use and knowledge of electricity—even though it has been witnessed since the beginning of time.

THE ELECTRON THEORY

The *electron* theory is the key to the understanding of electricity. Simply stated, the electron theory says that all matter is made of *atoms,* and atoms are made of smaller particles called *electrons* and *protons* (Fig. 6-4), each of which has an electrical charge. *Electrons have*

FIGURE 6-3. Like charges repel each other; unlike charges attract each other.

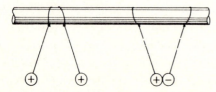

a negative charge and protons have a positive charge.

In order to understand better how these electrons behave, we should know that matter is any physical object that takes up space and has weight—all solids, liquids, and gases are matter. Thus, for example, ice, water, and steam are examples of matter in all three forms, even though they are the same *compound,* H_2O (which is the scientists' shorthand for a *molecule* of water that consists of two parts of hydrogen and one part of oxygen). The molecule is the smallest particle into which a compound can be divided and still retain its identity. The *atom* (which in this case would be the hydrogen and the oxygen atoms) is the smallest particle into which an *element* can be divided and still retain its *identity*.

Nature has provided 92 basic elements (Fig. 6-5) (man has made more), each of which differs from the other in the number of electrons and protons it contains. For example, hydrogen, which is the simplest, consists of one electron revolving around one proton, and uranium (Fig. 6-6), which is the most complex, has 92 electrons in orbit around the core of 92 protons. Between these two are the remaining elements, each differing in the number of electrons in *orbit* around the core or nucleus and the number of protons in the nucleus. The number of protons must equal the number of electrons.

It might be helpful if we consider the atom as we would our solar system with the sun being the nucleus and the planets being the electrons in orbit around it. When we talk about electrons in orbit, we must note that not all the electrons are in the same orbit. They travel in different

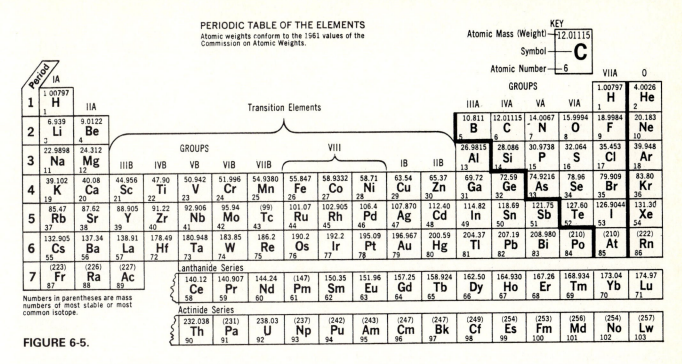

PERIODIC TABLE OF THE ELEMENTS
Atomic weights conform to the 1961 values of the
Commission on Atomic Weights.

FIGURE 6-5.

Numbers in parentheses are mass
numbers of most stable or most
common isotope.

rings of orbit. The hydrogen atom, having only one electron, has just one ring. However, copper has 29 electrons in four rings.

In all atoms, the nucleus, which is composed of protons and *neutrons,* weighs much more than the electrons. The neutrons are not charged, but they have the ability to hold the protons together in the nucleus. We may consider the gravitational pull of this mass, as well as the attraction of the negative electrons to the positive core, as being a means of maintaining the orbit around the nucleus. Since these

FIGURE 6-6. Uranium is the heaviest natural element, with 92 electrons spinning around the nucleus in seven different orbits, each orbit a specified distance from the nucleus.

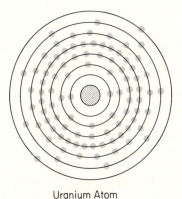

Uranium Atom

electrons orbit at varying distances, the ones that are closest are more strongly attracted to the nucleus. These are called the bound electrons.

Electrons in the outer ring are not so tightly bound and are called the free electrons. The latter are the ones of greatest interest to us because they allow the motion of electricity. Free electrons can be knocked out of orbit from their nucleus to the outer orbit of another nucleus, and so on.

Valence

The outer ring of electrons is often referred to as the valence ring. The valence of an element is defined as the *capacity* of its atom to combine with other atoms to form a *molecule.* It is the tendency of every atom to complete its outer ring by having the maximum allowable electrons in its outer ring. This number is eight electrons. If the number of electrons in the valence ring is less than four, the electrons are held to the core rather loosely, making the material a good *conductor.* These electrons can move easily from atom to atom.

If the number of electrons in the valence ring is greater than four, the electrons are held more firmly to the core and the material is classified as an *insulator,* for these electrons do not move freely from atom to atom.

Ions

Normally, atoms are *neutral* in that the + charges in the nucleus are balanced by the − charges in the orbiting electrons. However, when an electron is added to or taken away from an atom, an unbalanced condition exists and the atom is then called an *ion* (Fig. 6-7).

An atom that loses an electron is called a positive ion and an atom that gains an electron is called a negative ion. The process of adding and subtracting electrons from atoms is called *ionization.*

Conductors

Obviously, any material that allows electrical current to flow is a conductor. Specifically, a conductor must have free or loosely bound electrons to permit current flow. The metals, silver, copper, and aluminum are the best conductors, but we must also include *ionized* gases and liquids as conductors. A fluorescent light is possible because its gas atoms ionize. Both the electroplating process and the storage battery owe their success to ionized liquids.

Insulators

An *insulator* is an electrical nonconductor, which is a material lacking in free electrons

or ions—that is, its structure consists of atoms having tightly bound electrons. Such materials as glass, ceramics, rubber, plastics, paper, mica, and dry wood or fabric are classified as insulators. So are oils, shellacs, and certain gases.

THE "FLOW" OF ELECTRICITY

In order to do work, electricity must move. The flow of electrons through a conductor is called *current* and is measured in *amperes.* The force (pressure) that drives the current through the conductor is called the *electromotive force* (voltage) and is measured in *volts.* Opposition to the flow of current through a conductor is called *resistance* and is measured in *ohms* (Fig. 6-8). The rate at which electrical energy is used is called the *power* and is measured in *watts.* The amount of work done by electricity is found by multiplying the power by the time that the electricity was used and is called a watt-hour. (1000 watt-hours is called a kilowatt-hour, and your electric bill is calculated from the number of kilowatt-hours used.)

The only way that electricity can move is through a complete conducting path from the negative terminal of the source back to the positive terminal of that same source.

As an electric current flows through a conductor, two things happen:

1. Heat is generated within the conductor.
2. A magnetic field is generated around the conductor.

The amount of heat generated depends on the resistance of the conductor and the amount of current flowing in it.

The amount of magnetism, or strength of the magnetic field, is determined by the way

FIGURE 6-7. When sulfuric acid is placed in water, it breaks up into H^+ ions and SO_4^- ions. The ions make the solution conductive, with each charge being attracted to the opposite electrode.

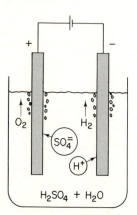

FIGURE 6-8. A voltage *E* does the work of pushing *I* through *R* and in so doing uses some of its energy and thus becomes smaller. *I* loses nothing, and *R* never wears out.

$$E = IR$$

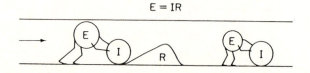

the conductor is coiled and the amount of current flowing in it.

Both of these properties of current flow are used by man and will be discussed in more detail.

Theory of Flow

We have briefly discussed electricity but have not explained its sources or characteristics. It is a form of energy inherent in all matter, requiring some other external force to liberate it. If we consider electricity to be the flow of free electrons from atom to atom in a conductor, and call this flow the current, we can explain much about the action of electricity.

In order to have a flow of electrons, there must be a source of electrons at one end of the conductor and a *deficiency* at the other, creating a *potential* difference between the ends of the conductor. Electrons will flow from the area where there are too many to the area where there are not enough. This potential difference in charge pressure is the electromotive force or *voltage*. The ability of the conductor to handle this flow of current depends on its size (diameter and length) and composition. There is always some opposition to the electron flow in a conductor. This opposition is called resistance. It causes a loss of energy resulting from interelectronic collisions that usually produce heat.

SOURCES OF ELECTRICITY

There are a number of ways to make electrons move, and these ways can be called sources of electricity (Table 6-1).

Friction

Electricity caused by friction is called static electricity and is produced by rubbing certain insulators called dielectrics together. Rubbing a hard rubber rod with a piece of fur or wool will transfer the free electrons from the fur to the rod, thus imparting a negative charge on the rod. In the same manner, a positive charge is left on a glass rod when it is rubbed with a piece of silk.

Since the rods and the fabric are insulators, there will be no current flow within the material. Being motionless, this stored energy is called *static* electricity. Although there is no flow of current, charged bodies are surrounded by an *electrostatic* field in which lines of force extend from the body and affect other nearby materials, diminishing in strength as the distance is increased. Two like charges (positive and positive) or (negative and negative) will repel each other. Two unlike charges (positive and negative) will attract each other. Touching a charged rod to a conductor will allow the electrons to reorganize to a normal neutral state, and the charge is gone.

Heat

When two dissimilar conducting metals are joined together in a complete circuit, the metal with the fewer number of free electrons will attract some from the other metal and there will be an electron flow that will increase as the temperature goes up (Fig. 6-9). This junction is called a *thermocouple*. A number of thermocouples connected together in series is called a *thermopile*. Thermocouples, frequently iron

TABLE 6-1
Sources of electricity

Source	Energy	Example
Rod and fabric	Friction—rubbing (static)	Air purifier
Thermocouple	Heat	Safety valve
Photoelectric cell	Light	Exposure meter
Piezoelectric	Pressure	Record player pickup
Chemical	Chemical change	Battery
Electromagnetic	Mechanical—rotating	Generator
Radioactivity	Electron decomposition	

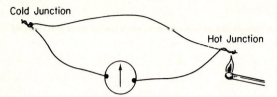

FIGURE 6-9. Voltage will increase at a fixed rate of millivolts per 1 degree rise in temperature.

and *constantan* are used for temperature measurement and safety controls in gas-burning appliances (see Table 6-2).

Light

Light can also create the electron activity in some kinds of materials. When two dissimilar conducting materials are in contact, there will be some electron migration. If one of these materials is selenium and the other is iron, electron flow will increase as the light intensity on the selenium gets brighter.

This effect is used in photographic exposure meters and photoelectric controls, such as numerical counters, door openers, and streetlight switches.

Pressure

When certain crystals, such as quartz, rochelle salt, and tourmaline, are subjected to pressure,

there will be a displacement of charges that free electrons to produce a potential difference (voltage) between the crystal faces. This generated voltage only occurs when there are squeezing and motion. Conversely, applying a potential difference across the crystal faces will cause it to change dimension. This is known as the *piezoelectric* effect. Practical uses for crystals are in phonograph pickups, sonar, and frequency controls.

Chemical Reaction

One of the better-known sources of electricity is the battery. Two common types are the "storage" type used in automobiles or the "dry" type used in flashlights and portable radios.

To understand battery action better, we must know that when two *dissimilar* metals (electrodes) are *immersed* in an ionized solution (electrolyte), there will be electron movement with an excess of electrons collecting on the more negative of the metals. Thus when the external circuit is completed, there will be a flow of electrons (current) at some pressure (voltage) determined by the position of the metals in the *electrochemical* series of metals. The more negative metal gradually dissolves as it gives up its electrons to produce this flow of

TABLE 6-2
Temperature–millivolt relations for thermocouples

°F Above Cold-Junction Temperature	Copper– Constantan (mV)	Iron– Constantan (mV)	Chromel– Alumel (mV)	Platinum– 10% Rhodium (mV)
50	1.11	1.44	1.08	0.148
100	2.23	2.88	2.20	0.313
150	3.40	4.36	3.34	0.493
200	4.65	5.84	4.50	0.687
250	5.93	7.36	5.65	0.892
300	7.23	8.87	6.77	1.108
350	8.60	10.40	7.88	1.333
400	9.99	11.24	8.99	1.565
450	11.40	13.47	10.11	1.804
500	12.87	15.01	11.24	2.048
600	16.03	18.08	13.53	2.549
700	—	21.15	15.86	3.067
800	—	24.23	18.20	3.597
900	—	27.33	20.56	4.136
1000	—	30.46	22.93	4.686

current. This is why dry batteries are thrown away when they no longer produce voltage.

A cell in which one *electrode* is consumed is known as a *primary* cell because it cannot be recharged.

A cell that can be recharged is known as a *secondary* cell or storage cell because, during discharge, one electrode combines with the electrolyte, thereby changing its chemical composition. Then it can be charged by passing direct current electricity through it in the opposite direction to restore the electrodes to their original condition.

A battery consists of two or more cells connected to each other.

Magnetism

Since this source of electricity is the most commonly used one, and many books have been written on how the process occurs, we will simply summarize the action.

Natural magnetism occurs in certain iron ores. Man can make permanent magnets from alloys of iron, nickle, or cobalt.

Electromagnets are made by winding a number of turns of insulated wire around an iron core, which may be solid for direct current and must be laminated for alternating current (Fig. 6-10). As electricity flows through the wire, a magnetic field will be induced in the iron.

An electric current will be generated in a coil of wire as it is *moved* through a magnetic field (Fig. 6-11). The direction of current flow in this coil depends on the direction of the motion. Obviously, as the coil of wire rotates in the magnetic field, the direction of current flow will reverse each revolution. Thus this gener-

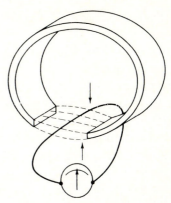

FIGURE 6-11. There must be motion between the wire and magnetic field to generate electricity.

ated electricity is alternating current and remains so if it is connected through slip rings. If it is connected through a commutator, it leaves the generator as direct current. Machines operating on this principle are called dynamos, generators, or alternators, depending on their specific use and output. Such machines convert mechanical energy into electrical energy in the same basic manner and are the most common source of the electricity we use.

Radioactivity

Radioactivity is an additional source of electricity that is being considered and studied by scientists, but it is not yet ready for our consideration.

Summary

All of these sources of electricity have one thing in common: *there must be some active form of energy* to stimulate the electron flow through a conductor.

In each case, some form of energy is being consumed to supply the electrons we use as electricity. *We do not get something for nothing.* The engineers' concern is for improving the efficiency of this transfer of energy into electricity.

As technicians, our concern is to keep machines that use electricity operating at their most efficient level in order to convert electrical energy into the type of energy the customer needs. Consequently, we must know the basic properties and laws of current flow in electrical circuits.

FIGURE 6-10. Two uses for electromagnetism.

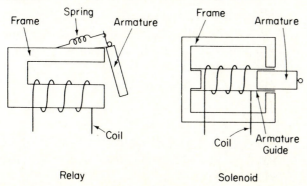

Relay

Solenoid

ALTERNATING AND DIRECT CURRENT

Electricity produced by a generator can be either alternating current (ac) or direct current (dc), depending on the design of the generator.

Alternating current is a flow of electricity that changes its polarity periodically (Fig. 6-12). When this change occurs 60 times a second, it is said to have a frequency of 60 hertz (modern name for cycles per second).

Direct current is a flow of electricity that always maintains the same polarity.

LOADING THE GENERATOR

The generator output is of no value until it can be led to a load that will use its electrical energy. This load may have resistive, capacitive, and inductive properties or a combination of all three. Each does its own thing, which will depend on whether it is suppled with ac or dc.

In practice, a "load" can be any electrical appliance in the home. Electric ranges, water heaters and light bulbs are examples of resistive loads. Electric motors and doorbells are examples of inductive loads. Except for fluorescent lights, no appliance is primarily a capacitive load, but *capacitors* fill an important part in the operation of many appliances. A radio or TV uses all three properties as well as both ac and dc.

The amount of energy used by an appliance is what you pay the power company for at a rate of so many cents per kilowatt hour. A kilowatt-hour is 1000 watts × 1 hour. It could be

100 watts per hour for 10 hours or 5000 watts for 12 minutes. Any combination of power (watts) × time.

How do these properties of resistance, inductance , and capacitance relate to ac and dc to determine how much power is used? Let's look at a few laws.

OHM'S LAW

The basic law governing the flow of electricity is called *Ohm's law,* which expresses the relationship between the current, voltage, and resistance in a closed circuit. It applies to *linear* circuits, not to semiconductors, electrolytes, or ionized gases.

The basic formulas involving Ohm's law are included in the "circle of formulas" (Fig. 6-13). In order to use this "circle of formulas" for finding volts, amperes, ohms, or watts, you must have any two of the others. The quantity to be found is located in a quadrant of the inner circle. Use the two known quantities as indicated in the outer circle in its quandrant. *Example: E = IR.*

DEFINITION OF TERMS

Coulomb (Q). A quantity of electricity equal to 6.28×10^{18} electrons.

Ampere (I). Also intensity of electron flow. An ampere is the flow of 1 coulomb per second past a fixed point ($I = Q/t$).

Volt (E). Also known as electromotive force or potential difference. A volt is the electrical pressure required to force 1 ampere through a resistance of 1 ohm.

Ohm (R). Resistance to electron flow. An ohm is the unit of electrical resistance that will limit the current to 1 ampere when 1 volt is applied.

Ohm's Law. The current in amperes in a circuit is equal to the applied voltage divided by the resistance ($I = E/R$).

Watt (P). A watt of power is the work done by 1 volt of electrical pressure to move 1 ampere ($W = EI$).

FIGURE 6-12. The same voltage as it appears in ac and dc. This is 1 hertz of ac.

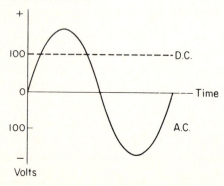

```
I = Current (Intensity)
E = Volts   (Electro-motive Force)
R = Resistance
W = Watts  (IXE)
```

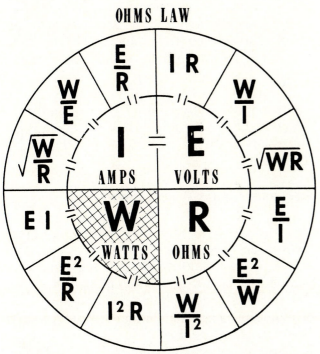

FIGURE 6-13. To use this "circle of formulas" for finding volts, amperes, ohms, or watts, you must have any two of the others. The quantity to be found is located in a quadrant of the inner circle. Use the two known quantities as indicated in the outer circle in its quadrant.

ELECTRICAL CIRCUITS

An electrical circuit is a complete path through which an electric current can flow.

It must have a supply voltage (EMF).

It must have an electrical conducting path (wire).

It must have resistance to current flow (load).

It should have a control element (switch).

The two basic ways in which circuit parts can be connected to each other are *series* and *parallel*.

Series Circuits

In a series circuit (Fig. 6-14a), the following are true:

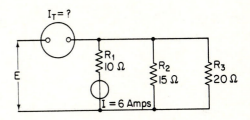

$R_T = R_1 + R_2 + R_3$

$R_T = 10 + 15 + 25 = 50$ Ohms

$I = \dfrac{E}{R} = \dfrac{100}{50} = 2$ Amperes

$E = IR = 2 \times 15 = 30$ Volts

(a)

$E = IR_{R_1} = 6 \times 10 = \underline{60\ \text{Volts}}$

$I_2 = \dfrac{E}{R_2} = \dfrac{60}{15} = 4$ Amps

$I_3 = \dfrac{E}{R_3} = \dfrac{60}{20} = 3$ Amps

$I_T = I_1 + I_2 + I_3$

$I_T = 6 + 4 + 3 = \underline{13\ \text{Amperes}}$

(b)

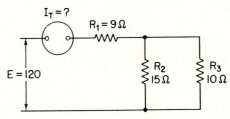

$R_T = R_1 + \dfrac{1}{\dfrac{1}{R_2} + \dfrac{1}{R_3}}$ $I_T = \dfrac{E}{R} = \dfrac{120}{15} = 8$ Amperes

$= 9 + \dfrac{1}{\dfrac{1}{15} + \dfrac{1}{10}}$

$= 9 + \dfrac{1}{\dfrac{5}{30}}$

$= 9 + 6$

$= \underline{15\ \text{Ohms}}$

(c)

FIGURE 6-14. (a) Series circuit; (b) parallel circuit; (c) series–parallel circuit.

1. The current in any part of the circuit is equal to the current in any other part of the circuit.

$$I_1 = I_2 = I_N$$

2. The total resistance is the sum of all the resistances in the circuit.

$$R_T = R_1 + R_N$$

3. The sum of the total voltage drops equals the supply voltage.

$$E_T = E_1 + E_2 + E_N$$

Parallel Circuits

In a parallel circuit (Fig. 6-14b), the following are true:

1. The current in any branch may vary, depending on its actual resistance, and the sum of the currents equals the total current.

$$I_T = I_1 + I_2 + I_N$$

2. The total resistance is the reciprocal of the sum of the reciprocals of the branch circuits.

$$R_T = \frac{1}{1/R_1 + 1/R_2 + 1/R_N}$$

3. The voltage across a parallel circuit is the same and is equal to the supply voltage.

$$E = V \text{ at any parallel part}$$

Series–Parallel Circuits

A series–parallel circuit includes *both* series and parallel current flow (Fig. 6-14c) in its branch circuits. The simple circuits can be solved by applying Ohm's law to each branch and replacing groups of series or parallel resistances by equivalent single resistances to reduce the circuit to a single equivalent resistance eventually (Fig. 6-14c).

CIRCUIT DIAGRAMS

All appliance-wiring diagrams are adaptations of these basic circuits. Instead of resistors, there will be symbols for motors, solenoids, lights, heaters, and such parts that use electricity. Instead of a battery as a source of EMF, there may be a power cord to connect to a power source.

Normally you will not have to go into a mathematical analysis of a circuit to find what is wrong. But you will have to read the diagram and relate it to the conditions existing in the machine.

Appliance-Circuit-Wiring Diagrams

One of the biggest stumbling blocks appliance technicians meet is the variety of wiring diagrams. Each manufacturer has his own ideas on how a circuit diagram should be presented. The result is that each one has different types of diagrams even for the same product—and especially from product to product. Sometimes a pictorial schematic diagram is used (Fig. 6-15). A pictorial diagram consists of a phantom view of the appliance with sketches or pictures of the parts and with heavy black lines running from part to part, showing the connecting wires. This type of presentation is suitable for an appliance or product with a simple circuit. However, as circuitry gets more complex, and as more and more electrical parts are being added to an appliance, the pictorial diagram becomes extremely impractical to read.

Other manufacturers present a literal schematic diagram that uses symbols for the various parts. The parts are located on a piece of paper in much the same relative position as they are used on the appliance (Fig. 6-16). The wires are connected from part to part in the same patterns as used in the product. Consequently, parts are scattered all over the paper and wires go in various directions. As a result, it is often extremely difficult to trace the wiring from point to point. It is similar to the op art we have today—the eye is not sure which way is out.

A third method groups parts in one area and then runs all the wiring to it from a control center. Let's call this a cable schematic (Fig. 6-17). Some will run the basic wiring across the page, and some will run the pattern up and down the page.

The waterfall schematic (Fig. 6-18) illustrates a diagram in which all the wiring to the working components falls vertically from the timer or control unit.

FIGURE 6-15. Diagram of the parts, showing the routing of the interconnecting wires. (Courtesy of Admiral)

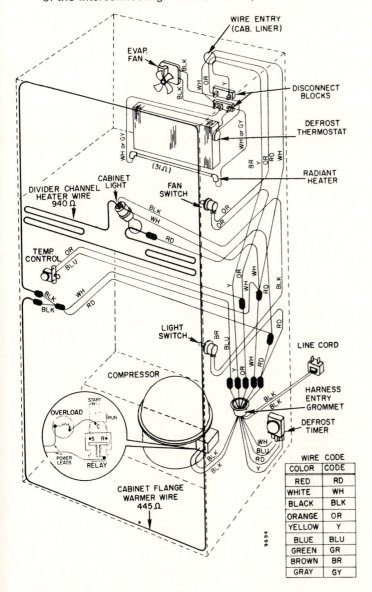

Schematic Symbols

In spite of the variations of diagram presentations, we have one thing going for us: manufacturers are now using the American National Standard symbols No. 32.2-1967 (Fig. 6-19). Like the basic vocabulary learned in elementary school, *these symbols must be learned.*

There is no need to go into all the various appliance diagrams anymore than there is a need to know all the words in a language before you read a book. *But* a working knowledge of the basic patterns is mandatory!

Ladder Diagrams

We suggest that the ladder type of wiring diagram is the most effective. Strictly a diagram, it uses symbols for the parts, and running wires from these components to their control units, as simply as possible (Figs. 6-20 and 6-21). Wiring is arranged as follows: The common leads from the power line, which we shall say are the black wire and the white wire in this case, are run vertically up and down the paper much as the side rails of a ladder are located. Then each individual circuit is placed crosswise as the rungs of a ladder would be. Illustration shows the electrical parts, their controls, and how the wires connect the controls to the working parts. The timer is enclosed in a dotted line.

Circuits

Consider an individual circuit as a complete circuit for a working component and its controls (Fig. 6-22). The working part is the part that converts electricity to some useful function. It could be a light, a solenoid, a motor, a heater element, and so on. The controlling unit or switch turns the working component on or off.

For our purposes, controls will be referred to as *functional* or *safety* controls. A functional control is one that *deliberately* turns the working unit on or off. The safety control *automatically* turns the working component off in case of malfunction. These are usually referred to as fuses, circuit breakers, or overload protectors. Whether or not automatic reset action is

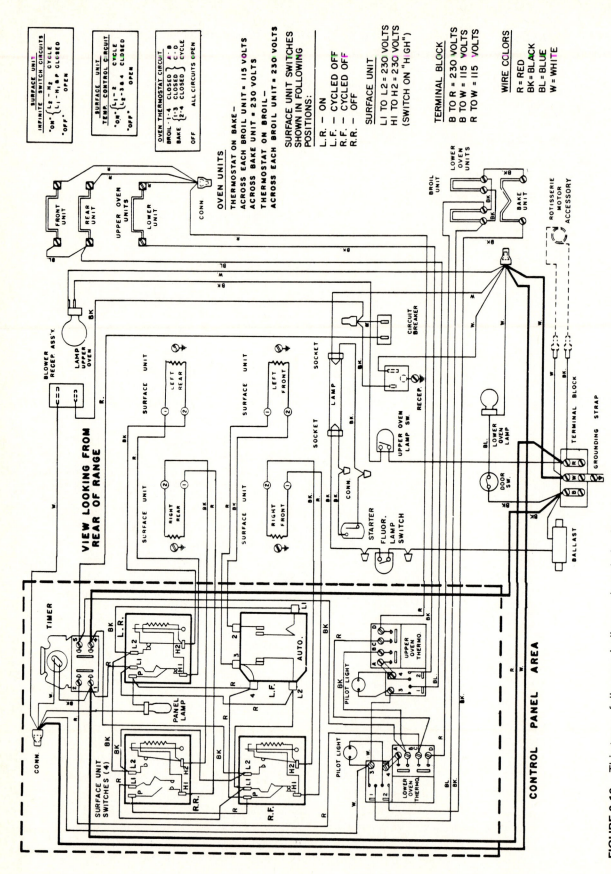

FIGURE 6-16. This type of diagram indicates the relative location of parts as well as the interconnecting wires.

110

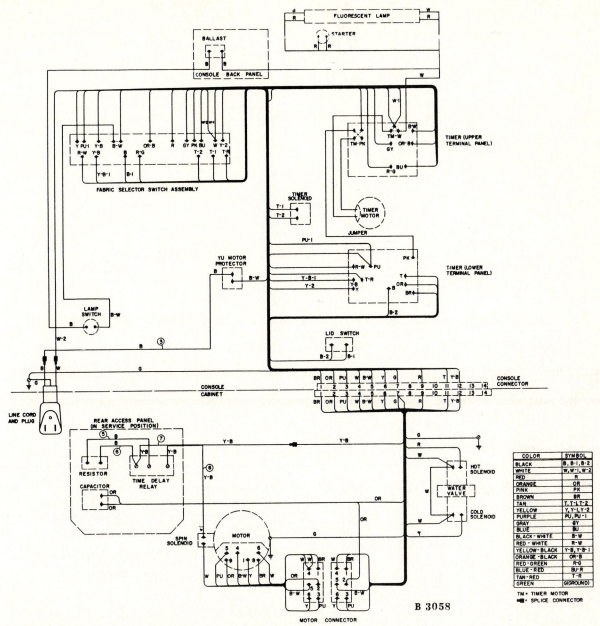

FIGURE 6-17. A cable, highway, or trunkline type of connection diagram shows the individual wires merged into cables instead of being shown separately.

involved is the designer's option. Thus there are three essential elements in each circuit: the part that works, the part that controls, and the wire that connects. That is all there is to an individual circuit.

Since an appliance is a collection of circuits, understanding circuitry and the interrelation of several components is an important tool for the appliance technician. By working with a ladder diagram, one can readily see the function and control of any given circuit.

Cycle Charts

A cycle-sequence chart tells when a circuit is active, plus at what time in the cycle which valve opened, or heater turned on, or motor started. The information is conveyed by manufacturers in many different ways. From Fig. 6-23, note that some are much easier to interpret than others. The easiest to interpret is a bar chart showing the particular action occurring on a side and the specific increment of time

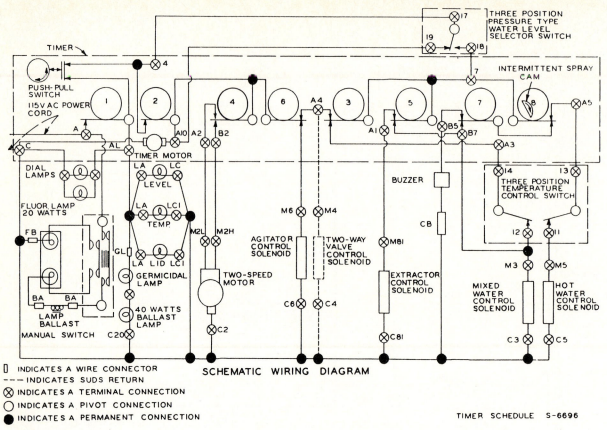

SCHEMATIC WIRING DIAGRAM

▯ INDICATES A WIRE CONNECTOR
--- INDICATES SUDS RETURN
⊗ INDICATES A TERMINAL CONNECTION
◯ INDICATES A PIVOT CONNECTION
● INDICATES A PERMANENT CONNECTION

TIMER SCHEDULE S-6696

FIGURE 6-18. The timer is shown in detail and the parts controlled as semi-symbols. All wiring is clearly separated.

FIGURE 6-19. Selected graphic symbols for electrical diagrams.

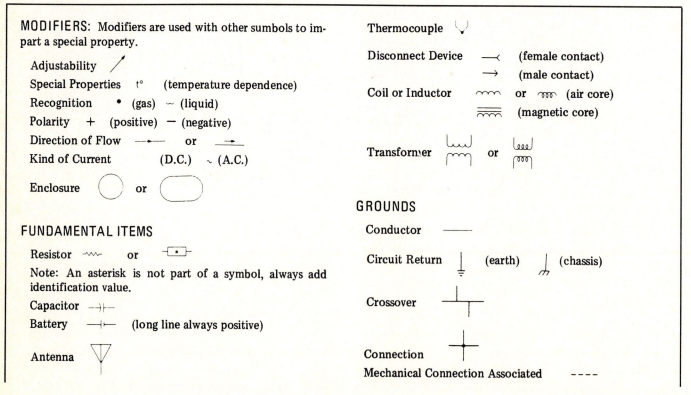

MODIFIERS: Modifiers are used with other sumbols to impart a special property.

Adjustability /

Special Properties t° (temperature dependence)

Recognition • (gas) ~ (liquid)

Polarity + (positive) − (negative)

Direction of Flow —▸ or ⟶

Kind of Current (D.C.) ~ (A.C.)

Enclosure ◯ or ▭

FUNDAMENTAL ITEMS

Resistor ⌇ or ▭

Note: An asterisk is not part of a symbol, always add identification value.

Capacitor ⊣⊢

Battery ⊣⊢ (long line always positive)

Antenna ▽

Thermocouple ⋎

Disconnect Device —◁ (female contact)
⟶ (male contact)

Coil or Inductor ⌒⌒⌒ or ⌒⌒⌒ (air core)
≡ (magnetic core)

Transformer ⌣⌣⌣ or ⌣⌣⌣

GROUNDS

Conductor ——

Circuit Return ⏚ (earth) ⎏ (chassis)

Crossover ╁

Connection ╂

Mechanical Connection Associated ----

112

FIGURE 6-19. (Continued)

SWITCHES

Closed Contact Normal

Normally Open Contact

Relay Coil

Pushbutton   (N.O., make)   (N.C., break) or   (two circuit)

Multiposition   or

Limit   (N.O.)   or   (N.C.)

Circuit Breaker

Safety Interlock   (circuit opening)

Time Delay   or   (normally open with time delay closing)

Flow- Actuated   (N.O., closes on increased flow)

or   (N.C., opens on increased flow)

Liquid-Level-Actuated   (N.O.) or   (N.C.)

Pressure- or Vacuum-Actuated   (N.O.) or

  (N.C.)

Temperature-Actuated   (N.O.) or   (N.C.)

Thermostat   (opens on temperature rise)

  (closes on temperature rise)

Humidistat   (opens on humidity rise)

Centrifugal

Governor or Speed Regulator

Fuse   or

SEMICONDUCTORS

Diode   or

Transistor   (PNP)   (NPN)

Unijunction Transistor (UJT)

Field Effect Transistor (FET)

Thermistor

Photosensitive Cell   (resistor)   (solar cell)

Thyristor (SCR)

Bi-Directional Thyristor (trial)

Magnetron Tube

Bell   Buzzer

Flourescent Lamp

Gas Filled (neon) Lamp

Incandescent Lamp

Germicidal Lamp

Induction Motor

Series Motor

Shaded Pole Motor

Applications on diagrams may show induction motors as:

Single speed

Two speed

Three speed

Refrigeration compressor

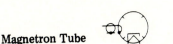

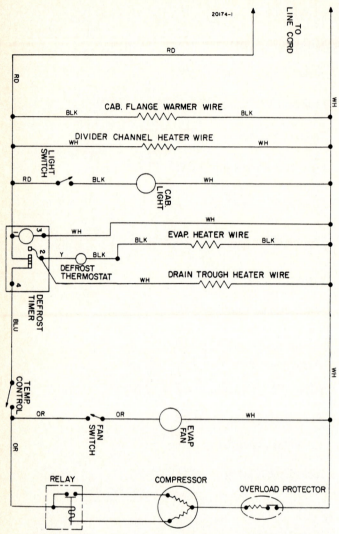

FIGURE 6-20. The ladder diagram is becoming more popular in appliance diagramming. This is the ladder version of the pictorial shown in Fig. 6-15. (Courtesy of Admiral)

FIGURE 6-21. Ladder diagram for a typical dishwasher.

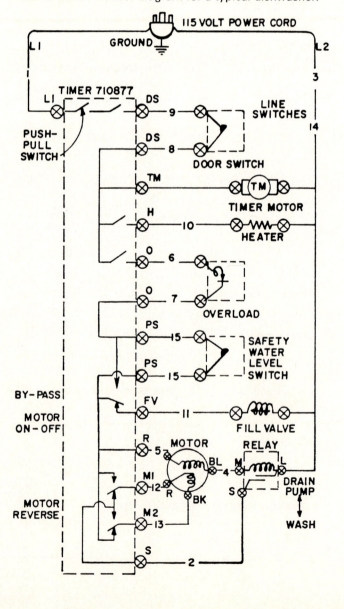

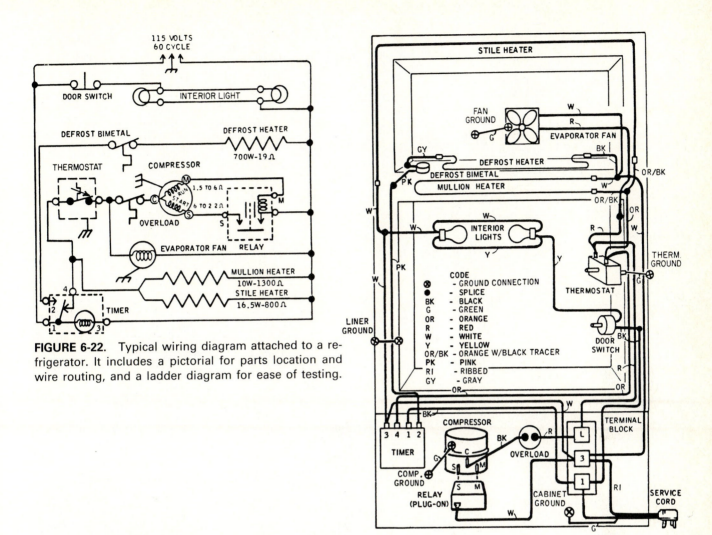

FIGURE 6-22. Typical wiring diagram attached to a refrigerator. It includes a pictorial for parts location and wire routing, and a ladder diagram for ease of testing.

CODE
⊗ - GROUND CONNECTION
● - SPLICE
BK - BLACK
G - GREEN
OR - ORANGE
R - RED
W - WHITE
Y - YELLOW
OR/BK - ORANGE W/BLACK TRACER
PK - PINK
RI - RIBBED
GY - GRAY

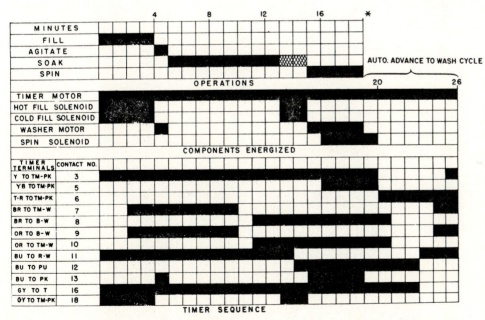

FIGURE 6-23. A cycle sequence chart is usually included with the diagram for appliances that use a timer to control their operation. Illustrated are several examples.

■ ─ 60 SECONDS. 6° ADVANCEMENT OF DIAL

▨ OVERFLOW WHILE SOAKING

✱ 20TH ADVANCEMENT REQUIRED TO PERMIT WASHER MOTOR TO STOP BEFORE SPIN SOLENOID IS DEENERGIZED.

A 1073

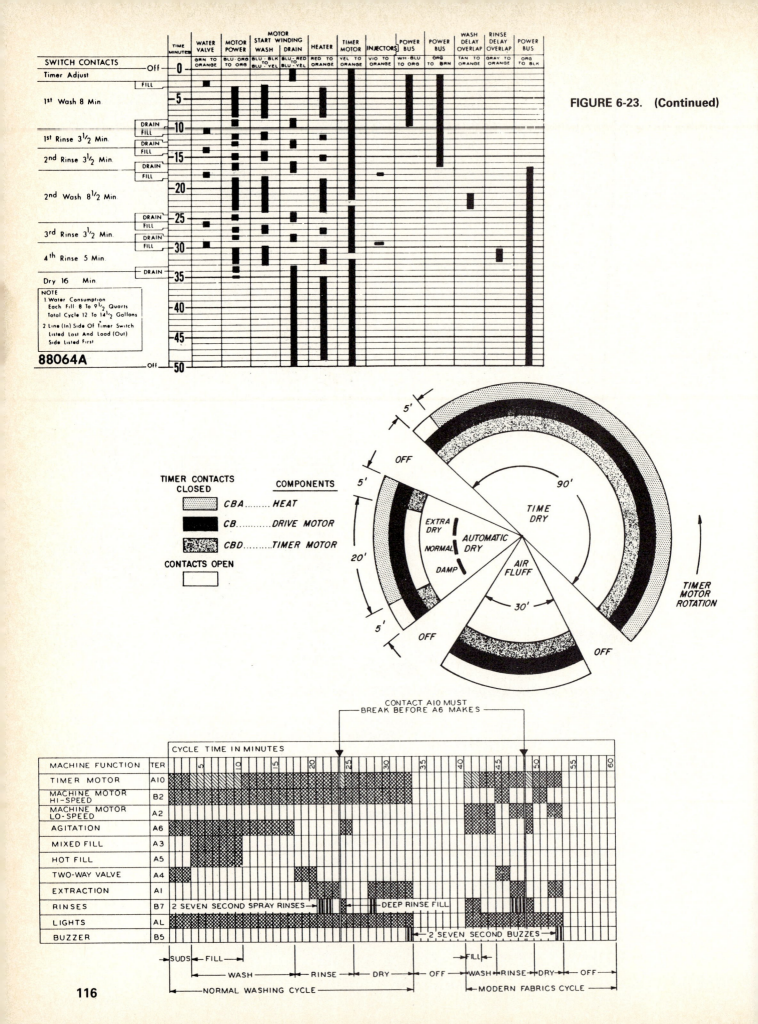

FIGURE 6-23. (Continued)

88064A

116

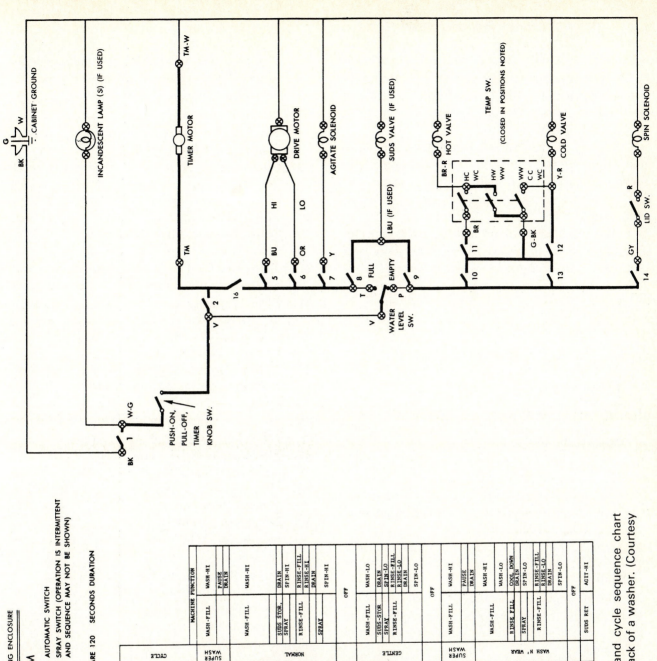

FIGURE 6-24. Typical wiring diagram and cycle sequence chart as they would appear mounted on the back of a washer. (Courtesy of Whirlpool)

across the top, with the blacked-out areas indicating the times when the working components are in operation. This type of cycle chart can be used with any appliance that has an automatic-timing device.

Once you know how to interpret the schematic-wiring diagram and the operation chart for an appliance you will find that servicing the appliance is much simpler and more accurate (Fig. 6-24).

You must know what is supposed to happen and when it is supposed to happen before you can tell whether there is anything wrong!

STUDY QUESTIONS

Electricity

1. The south-seeking end of one magnet will _____the south-seeking end of another magnet.

2. About 1660, von Guericke proved that _____could be _____.

3. The American whose theories laid the foundation of most electrical experimentation was _____.

4. The electron theory states that atoms are the smallest particles in all matter.   True or False

5. Matter is _____.

6. What one compound can be an example of matter in its three states? What are the three kinds?

7. Which is smaller, a molecule or an atom?

8. The smallest particle that carries a positive charge is the _____, whereas a negative charge is carried by _____.

9. Electricity uses mostly _____ electrons.

10. Electrons usually come from the nucleus. True or False

11. An atom with an unbalanced condition, because of an added or missing electron, is called an _____.

12. Ionization is the process of adding or subtracting _____ from _____.

13. What are some good conductors *besides* metals?

14. List at least seven good electrical insulators.

15. The force that drives current through a conductor is called amperes.   True or False

16. Resistance is measured in _____.

17. Power is measured in _____.

18. The rate at which electricity flows is called power.   True or False

19. One could say that the electricity was on but that there were not enough amperes to allow the appliance to function.   True or False

20. In spite of adequate voltage, an appliance could be inoperable if it had too many ohms.   True or False

21. Electricity does not have to return to its original source to keep moving.   True or False

22. Electricity is a source of _____. It can be found in all _____, but it _____an _____ force to _____it.

23. A deficiency of electrons at both ends of the conductor causes current to flow.   True or False

24. The more ohms a conductor has, the hotter it will get.   True or False

25. Static electricity is when _____ remains_____.

26. One source of electricity is called static. True or False

27. In the rod and fabric experiment, one can discharge the rod by touching it to a pipe.   True or False

28. Street lights might use a _____ as the source that actuates their switches.

29. Batteries are considered a _____ source of electricity.

30. A primary cell can be recharged.   True or False

31. The most commonly used source of electricity is _____.

32. Alternators, dynamos, and generators change electrical energy into mechanical energy. True or False

33. Name the six main sources of electrical energy.

34. Ohm's law can be applied to transistors. True or False

35. A quantity of electricity equal to a specific number of electrons is _____.

36. The pressure needed to force one ampere past one ohm is a _____.

37. Name the three requirements for an electrical circuit.

38. In a series circuit, the total resistance is the reciprocal of the sum of the reciprocals of the branch circuits.   True or False

39. In a parallel circuit, the branch voltage equals the supply voltage.   True or False

40. A safety control turns off the working parts automatically.   True or False

41. Name three safety controls.

42. The two essential elements in a circuit are the working parts and the controls.   True or False

43. What equipment will test any electrical circuit?

44. When using an ohmmeter, what must you be sure of?

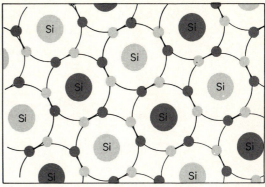

Covalent Bonding

FIGURE 6-25. Covalent bonding. Each silicon atom has four electrons in its outer ring. The electrons interlock the rings to each other.

ELECTRONIC CONTROLS

With the increasing use of electronic clocks, timers, sensing elements, switches, and so on, it becomes important for the technician to know how electronic controls do their job. Electronic controls are possible because of the developments in the manufacture and application of solid-state devices such as diodes and transistors to integrated-circuit design. The following information on valence bonding and semiconductors is included to help explain how these solid-state devices work.

When the valence ring of an atom contains just four electrons, the atom can become either a good insulator or a good conductor, depending on its purity and composition. These materials are classified as semiconductors.

The peculiar properties of semiconductors have made transistors possible. One of the materials in wide use in *transistors* is *silicon,* which has four electrons in its outer ring. Pure silicon in *crystalline* form has very tight bonding between electrons resulting from electrons in the outer rings of the silicon atoms, joining to fill up and share their outer rings. This is known as *covalent bonding* (Fig. 6-25). As shown in the illustration, the *covalent*-bonded silicon shows eight electrons in the outer ring, thus making the outer ring complete. Crystalline silicon is an extremely good insulator. However, when certain other materials are added to the silicon crystal, the resultant structure is no longer a good insulator and becomes a semiconductor.

P AND N MATERIALS

The *doping* process—that is, the process of adding *impurities* to silicon—when done under carefully controlled conditions, affects the silicon crystal in one of two ways, depending on whether the doping material has three or five electrons in its outer ring. If an element such as phosphorus, which has five electrons in the outer ring, combines with silicon, covalent bonding occurs, but one electron is left over (Fig. 6-26a). The result is a well-bonded material with free electrons floating around within it. Having an extra electron makes a negative-type, or N-type, material. If an element such as boron, which has only three electrons in the outer ring, is added to the silicon crystal, we have a condition in which the bonded outer ring has only seven electrons, leaving a hole in this ring (Fig. 6-26b). Thus we have a situation where the atom is shy one electron. The material with the hole or lack of an electron is called a P-, or positive-type material.

CURRENT CARRIERS

If we regard the excess electrons as negative-current carriers, we must regard the holes as being positive-current carriers. That is, the hole

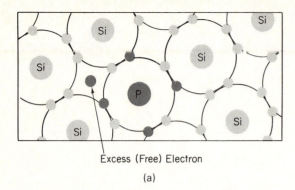

Excess (Free) Electron

(a)

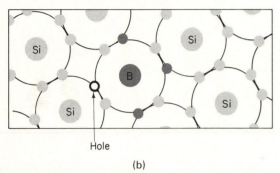

Hole

(b)

FIGURE 6-26. (a) The excess electron is free to move within the structure until it finds an opening into which it can move. This is N material. (b) A shortage of electrons leaves a hole in the outer ring, the hole being ready to receive any free electrons coming its way. This is P material.

can effectively move from atom to atom just as an electron can move from atom to atom. However, as long as no external *voltage* is applied to the material, there will be no flow of electricity even though there is electron or hole *mobility*.

The moment a battery (or another source of voltage) and a current-limiting resistor are correctly connected to this semiconductor material there will be a flow of current. In the N type of semiconductor this flow consists of the movement of the free electrons through the material. In the P type of material (that is, a material having a deficiency of electrons or an excess of holes) current will also flow as a movement of holes. We might say that the current flow consists of the free electrons moving into holes.

In order to better understand the action of current flow in a P-type material, we must remember that like charges repel and unlike charges attract. The positive battery voltage will attract electrons to the positive battery ter-

minal and the negative terminal will repel electrons. This flow of electrons from negative to positive occurs because electrons from *adjacent* atoms will move into the nearest hole in the positive (+) direction, and when they do, each leaves a hole. Thus there is a flow of electrons from negative to positive as they move from hole to hole. At the same time these holes then *migrate* from positive to negative—that is, the opposite direction to the electron flow.

We must understand that there is mobility within the material for both the electrons and the holes. It is a continuous process up to the edge of the material. Of course, at the edge of the material the positive lead removes the excess electrons and the negative current-carrying lead supplies electrons to enter the available holes.

DIODES

There are two semiconductor materials, a P type and an N type. When joined together, the two materials form a diode, which has the property of allowing electricity to flow through it in one direction only. This process occurs when the negative battery *terminal* is connected to the N type of material and the positive battery terminal is connected to the P-type material. It is known as the *forward-bias connection* (Fig. 6-27a). When the connections are reversed—that is, with the positive terminal connected to the N material and the negative terminal to the P material—we have a *reverse-bias condition* and the diode will not allow current to flow through it (Fig. 6-27b). With no battery connection to the diode, a state of balance exists and nothing happens.

The ability of a diode to allow current flow in one direction and block it from the other makes it a very useful device. When connected to an alternating-current source it will allow only half of the wave to pass through it in the forward direction, thus rectifying the ac to pulses of dc (Fig. 6-27a), or it can block a dc flow coming from the reverse direction (Fig. 6-27b).

Electron Flow

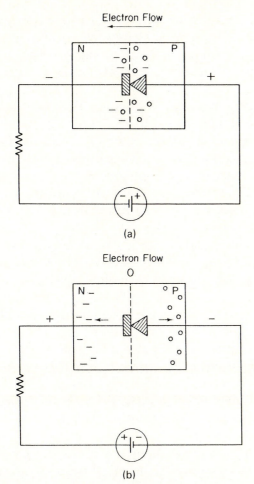

(a)

(b)

FIGURE 6-27. (a) *Forward*-bias condition for a diode will allow current to flow. It requires positive connection (+) to the P material and negative connection (−) to the N material. (b) *Reverse* bias will not allow current to flow through a diode. Positive connection to N material attracts electrons from the boundary layer. Negative connection to P material attracts holes from the boundary so that there is no interchange.

TRANSISTORS

A transistor is a three-layer semiconductor device having a very thin middle layer of P or N material. Sometimes P is between two layers of N; sometimes the reverse. When the two outer layers are a P-type material and the center layer is an N-type material the device is called a PNP transistor (Fig. 6-28a). When the device has a P-type material in the center and N-type material in the two outer layers, it is called an NPN transistor (Fig. 6-28b).

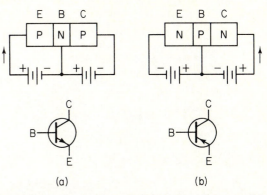

(a)   (b)

FIGURE 6-28. (a) Forward-bias PNP transistor; (b) forward-bias NPN transistor. Note the position of the arrow in the emitter lead of the symbols and the battery polarities.

As with diodes, a transistor should be connected only in a forward-bias connection. Also, *do not* connect a PNP transistor in place of an NPN transistor; *it won't work!* A transistor is capable of matching low-resistance circuits to high-resistance circuits. It is this *trans*fer of *resist*ance that gives the transistor its name. A transistor amplifies by power gain.

OTHER SOLID-STATE DEVICES

P and N materials may be varied in composition and may be combined in different ways to produce solid-state devices having a wide range of applications.

INTEGRATED CIRCUITS

Integrated circuits may consist of a number of transistors, diodes, resistors, and capacitors manufactured as a single device to operate as a complete circuit. Figure 6-29 shows a collection of several types.

CIRCUIT BOARDS AND MODULES

The parts used in electronic control circuits are usually mounted on a printed circuit board or sealed in a module. For service purposes these boards and/or modules are removable.

For in-the-home service these modules can

FIGURE 6-29. Integrated circuits (ICs) and a socket (upper left). The lower right IC is set in a conductive foam pad to prevent damage due to static charges.

be removed and replaced. The defective unit can then be returned to the manufacturer or repaired in the shop, depending on company policy and the shop's capabilities. After repair it can be used as a future replacement.

Always handle circuit-board assemblies carefully because they are fragile. Also, identify them with a tag indicating the machine make and model number in which they are used.

THIS DOES NOT MEAN THAT ALL PRODUCT FAILURES ARE DUE TO CIRCUIT-BOARD FAILURE!

Generally, the circuit board can be considered as failed when it does not provide the proper control voltage for a given function at the right time. The technician must be certain that the part requiring this voltage will operate when supplied with the correct voltage. *Do not blame the circuit board until you know that the rest of the machine is in working condition!*

CONTROL FAILURES

Electronic controls are sensitive to voltage. Normally, the control voltages are supplied by a low-voltage direct-current power supply built into the appliance. These voltages are distributed to the various circuits, such as the touch module, display assembly, clock, or microcom-

puter, through an interconnecting cable. All or some of these circuits may be on one or more individual panels. Some of these voltages will appear as pulses to activate an SCR, triac, or relay.

Any unusual voltage in this system can create a false signal to an operating part of the appliance. This will result in a malfunction of the machine. Sometimes simply turning off or disconnecting the appliance from its outlet for a while (15 to 20 seconds) will clear the system and the appliance will function properly when it is restored.

The unusual voltage is usually generated outside the appliance and enters through its power line. It may be called a *voltage surge, spike,* or a *transient.* Transients are caused when a motor or some other inductive load is turned off. The closer the transient-causing device is to the electron control, the more likely its spike will cause trouble. For this reason it is recommended that each electronically controlled appliance be installed on its own circuit from the power distribution panel.

If the transient peak voltage is high enough, it could damage one of the components in the electronic circuit. Lightning striking a nearby power line could do this much damage. It is also recommended that all electronically controlled appliances be turned off during a thunderstorm. It would be wise to disconnect these appliances during an area-wide power outage *before the power is restored.*

Damage could also occur in a home situation where a dishwasher with electronic controls and a disposer are on the same circuit. It sometimes happens when the disposer is turned on and off while the dishwasher is running.

UNUSUAL SERVICE PROBLEMS

With increasing use of electronic control circuits in appliances such as microwave ovens, washing machines, dishwashers, and dryers, some strange failures are being reported. These are not normal part failures due to use of the appliance. They are caused by an unusual happening in the 120-volt, 60-hertz household wiring system.

The National Electrical Code® standard (normal) house wiring for base or wall outlets (duplex receptacles) into which we plug our appliances should be connected according to specific rules (Fig. 6-30). The black wire (hot) must connect to a brass screw, the white wire (neutral) must connect to a white-plated screw, and the green or bare wire (equipment ground) must connect to a green screw.

All of these wires connect to the distribution panelboard. The black wire connects to the fuse or circuit breaker. The white wire connects to a ground terminal. The bare or green wire connects to the same ground as the white wire. All wiring from the wall outlet to the panel *must be continuous.* The ground terminal at the panel must connect to an earth ground, usually the cold-water pipe as it enters the house but before it connects to the water meter. Normally, several outlets are connected to the circuit coming from a specific fuse or breaker. So much for what is supposed to be.

Technicians have found black and white wires reversed and equipment ground wires not connected back to the panel board. Lack of valid equipment ground leaves the appliance open to static charge pickup, which can amount to a high enough voltage to allow leakage to sensitive electronic parts, causing them to fail or malfunction. The reversal of the hot (black) and neutral (white) wires can present a serious problem to an electronic circuit which has one point grounded to the appliance frame.

Even when all the internal house wiring is correct, it and the wiring entering the house can pick up unexpected random voltages. These can come from lightning striking the power line or a transient voltage, or surge, produced when a nearby machine is turned off. These transient voltages will pass into the electronic control system and act like a control signal, causing the machine to do something for which it was not programmed, such as stopping or skipping a cycle. A strong transient voltage can break down the insulation in a part, causing it to fail.

Figure 6-31 is a typical schematic diagram for an electronically controlled washer.

STUDY QUESTIONS

Electronics

1. What elements are used in transistors?
2. What is the property or electron placement that makes a good semiconductor?
3. Electrons or "holes" will move from ring to ring without external suggestions such as voltage. True or False
4. Electrons and "holes" flow in the same direction.   True or False
5. When a P-type material and an N-type material are joined, they form a _____.

FIGURE 6-30. Correct wiring to this 120-volt outlet as shown will provide the correct polarity for the appliance cord, which also must be correctly connected *in* the appliance.

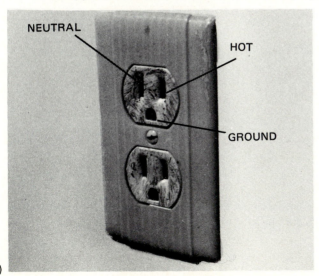

(a)

(b)

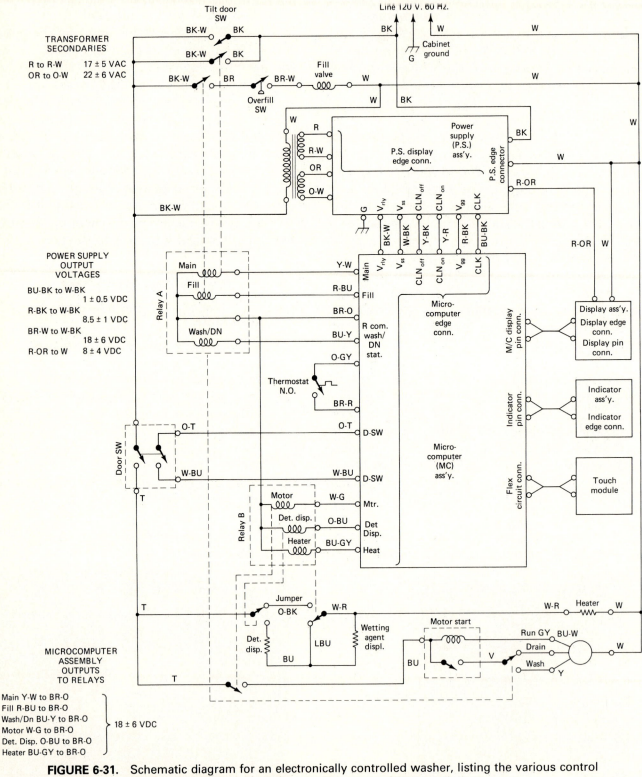

FIGURE 6-31. Schematic diagram for an electronically controlled washer, listing the various control voltages as well as the interrelated circuit boards. (Courtesy of Whirlpool)

6. Diodes allow current flow in one direction only. True or False

7. Some crystalline silicon is a good semiconductor and some is a good insulator.   True or False

8. What is the "doping" process?

9. What is the general result?

10. P-type material has a (an) _____ of electrons.

11. Unlike charges _____.

12. Within current carriers either electrons move or "holes" move, not both.   True or False

13. A reverse-biased NPN transistor would have a negative influence on its immediate circuit. True or False

14. A transistor reduces power gain.   True or False

15. Circuit boards and modules should be treated like any other part of the machine.   True or False

16. Control voltages are usually supplied by dc power.   True or False

17. Unusual voltages may shorten a machine's cycle.   True or False

18. Why is it important for electronically controlled appliances to be on their own circuit?

KEY

Electricity

1. Repel
2. Electricity could be generated.
3. Ben Franklin
4. False
5. Any physical object that takes up space and has weight
6. Water—solid/liquid/gas
7. Atom
8. Proton/electrons
9. Free
10. False
11. Ion
12. Electrons/atoms

13. Ionized gases and liquids
14. Glass, ceramics, plastics, drywood, rubber, fabrics, oils, shellacs
15. False
16. Ohms
17. Watts
18. True
19. True
20. True
21. False
22. Energy/matter/requires,needs/external/move, liberate, free
23. False
24. False
25. Stored energy/motionless
26. False
27. True
28. Photocell
29. Chemical
30. False
31. Magnetism
32. False
33. Heat, friction, light, pressure, chemical, magnetic
34. False
35. Coulomb
36. Volt
37. Voltage; conductor (wire, path); resistance, (load)
38. False
39. True
40. True
41. Fuses, circuit breakers, overload protectors
42. False
43. Volt-ohmeter
44. Disconnection from power line

KEY

Electronics

1. Silicon and germanium
2. A valence of four electrons in outer ring

3. False
4. False
5. Diode
6. True
7. False
8. Adding impurities to silicon
9. Makes P or N material
10. Deficiency

11. Attract
12. False
13. True
14. False
15. False
16. True
17. True
18. To minimize transients

7

Lubrication

FRICTION

Any appliance with moving parts has need for lubrication to reduce the friction between these parts, whether door hinges or agitator shafts on washing machines.

Friction is defined as the resistance that tends to oppose motion. Sometimes friction is desirable, as in the fastening action of a nail or the decelerating action of car brakes. However, our concern with friction is how to reduce it to improve the operating efficiency of our machines.

Excessive friction resulting from poorly aligned, inadequately lubricated, or contaminated bearings will create a power loss, evidenced by an increase of temperature in the bearing area.

The power used to overcome the extra friction is dissipated as heat, and if the loss is great enough to generate excessive temperatures, damaged bearings will result.

DESIGN

In the original design of an appliance, engineers have included some basic considerations regarding its bearing and lubrication needs.

1. In addition to "Normal" operation, the system should function under extremes of load and ambient conditions.
2. Ideally, the bearing system should be sealed with a lifetime supply of lubricant so that dust and moisture cannot enter the system and the lubricant cannot leave the system.
3. Where this cannot be done, easily accessible lubrication points should be provided, including adequate instruction for relubrication.

In actual practice, many things can happen to upset the design considerations. In manufacturing, accumulation of tolerances could create either abnormally tight or loose fitting of bearing to shaft, or some degree of misalignment. Most manufacturers do not let these marginal machines get into the field but occasionally some slip by. Soon the service technician has a customer complaining about abnormal noise in the case of a loose bearing or slow speed and heating in the case of tight bearings.

Careless transportation, handling, and installation can also affect the machine's operation. *Anything that could warp the frame can misalign the bearings.* The use of excessive amounts of detergent or cleaning fluids can absorb the oil from a bearing. (So can dust.) Manufacturers are doing what they can to seal off bearing areas to prevent this loss of lubricant.

BEARING TYPES

Each of the different bearing types requires its specific lubrication for most effective operation. This and the usage to which the bearing is to be put should serve as a guide for proper relubricating of a bearing.

Sleeve Bearings

The most commonly used bearing is the porous sintered type. These are prelubricated by soaking in oil for 24 hours or by a vacuum process. Typically, they contain 15 to 25 percent oil by volume. In operation, the oil is instantly available, flowing freely throughout the bearing and thus aiding in cooling (Fig. 7-1). The large oil capacity of these bearings makes lifetime lubrication a practical reality. Where necessary, additional oil can be stored in a reservoir and conducted to the bearing by a wick (Fig. 7-2). Dust protection is important to this bearing because the *dust can bleed away the oil just as easily as it can be drawn from a reservoir to the bearing.* Oil used for this type of bearing must be nongumming, nonoxidizing, and noncorrosive to the steel shaft and bronze bearing. The oil typically has SAE 20 to 30 characteristics.

Needle Bearings

Needle bearings are usually self-contained in an outer shell packed with either oil or grease,

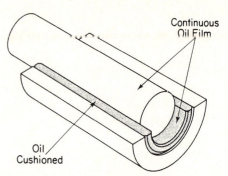

FIGURE 7-1. Porous sintered metal sleeve bearings are available to fit any shaft from ⅛ to 4 inches in diameter in a number of wall thicknesses.

depending on the application. The moving shaft is the inner race on which the bearings ride and should be hardened to at least 58 Rockwell C to promote maximum bearing life (Fig. 7-3).

Both the bronze and needle bearings require a smooth, accurately ground shaft, which must be clean and rust free. In disassembling a unit having these bearings, make certain that all rust, burrs, and other foreign matter have been cleaned off the shaft before pulling it through the bearing. If the shaft does not show wear, it can be reassembled after cleanup and relubrication. *A worn bearing or shaft must be replaced.*

Ball or Roller Bearings

These bearings ride between an inner and outer race (Fig. 7-4) and are usually lubricated with a lifetime supply of grease when assembled. The entire bearing assembly is a *light* press fit over the shaft and into the bearing retainer. A loose fit will allow the bearing raceways to move and wear their mating parts. *The only motion should be that of the bearings within their raceways.*

Since the bearing load is carried by metal-to-metal rolling contact, the lubricant serves to dissipate heat from the point of metal contact and to seal out moisture, dirt, and other contaminents.

Whenever gears or sliding parts are used in an enclosed housing, an oil lubricant is preferred. These lubricants are usually heavy-bodied oils with additives to impart the specific properties needed (SAE 50). For assemblies that may be exposed or that cannot contain an oil, a grease must be used.

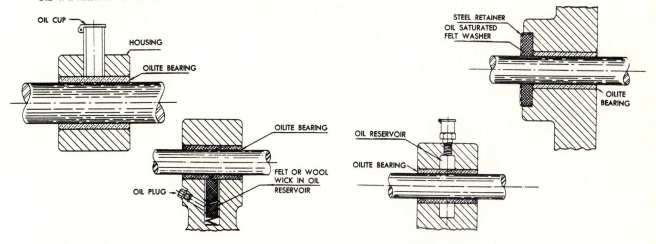

OIL—IT IS NECESSARY TO FEED OIL ONLY TO ROD OR END OF BEARING.

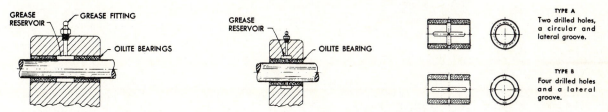

GREASE—GREASE IS FED TO A RESERVOIR OR DIRECTLY TO THE SHAFT THROUGH HOLES AND GROOVES IN BEARING

FIGURE 7-2. Bearings that need relubrication are equipped with oil cups or grease fittings. Learn to look for them! (Courtesy of Ampex Division, Chrysler Corp.)

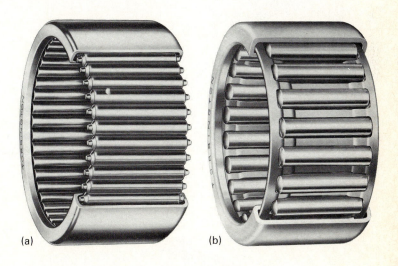

FIGURE 7-3. (a) The hardened shell forms the outer raceway for the bearing. The shaft on which the rollers operate acts as the inner race and must have a hardened surface. The retainer that serves as a roller spacer and guide only contacts the rollers at their ends. (b) In the drawn cup needle bearing construction the hardened outer shell, when backed by a proper housing, forms the outer raceway for the rollers. Rollers are mechanically retained. (Courtesy of Torrington)

Plastic Bearings

Du Pont has accomplished a great deal with some of their plastic materials, such as nylon (Fig. 7-5a), Teflon, and Delrin for light-duty self-lubricating bearings (Fig. 7-5b). *These materials are resistant to corrosion, detergent, and water* and are found in increasing frequency in smaller appliances (Fig. 7-5c).

LUBRICANT PROPERTIES

Without going into a great deal of highly technical detail to describe the various lubricants, we can only provide a broad outline of their purposes and properties.

The four basic functions of a lubricant are:

1. *Reduce friction and wear.*

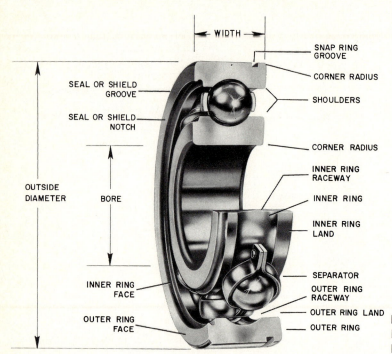

WIDTH

SNAP RING GROOVE

CORNER RADIUS

SHOULDERS

SEAL OR SHIELD GROOVE

SEAL OR SHIELD NOTCH

CORNER RADIUS

INNER RING RACEWAY

INNER RING

INNER RING LAND

OUTSIDE DIAMETER

BORE

SEPARATOR

OUTER RING RACEWAY

OUTER RING LAND

INNER RING FACE

OUTER RING FACE

OUTER RING

FIGURE 7-4. Typical ball bearing. (Courtesy of New Departure Hyatt, bearings division of General Motors Corp.)

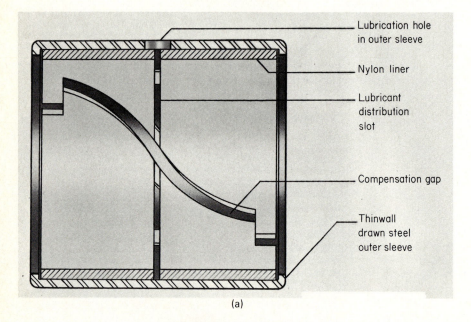

Lubrication hole in outer sleeve

Nylon liner

Lubricant distribution slot

Compensation gap

Thinwall drawn steel outer sleeve

(a)

FIGURE 7-5. (a) Nylon bearing unit; (b) another view of the nylon bearing as supplied for replacement use; (c) nylon bearing liners—simply insert between shaft and housing. (Courtesy of Thomson Industries)

(b)

(c)

2. *Conduct heat away from the working surfaces—"cool it."*

3. *Seal out contamination.*

4. *Prevent rust and corrosion.*

Some oils are light and run or pour easily; others are heavier and pour more slowly. This difference in pourability is referred to as *viscosity*. An oil that is too light for the job will quickly break down; an oil that is too heavy will develop high temperatures and create a power loss because its film cannot penetrate into the friction area.

An oil that is stiff in cold weather would not properly lubicate a bearing that has been standing outside in the winter. An oil that is too thin could seep out of the bearing in extremely hot weather. In either case, the resultant lack of lubrication will shorten the bearing life.

Gear case oils are usually heavy and well laced with such additives as antifoam, anti-wear, extreme pressure, corrosion, and oxidation inhibitors.

Refrigeration compressor oils are in a special class; they must be wax free, dehydrated and flow at low temperatures. They must be compatible with both the refrigerant and the materials used in the compressor. *Never use anything but a refrigeration oil in a sealed system* and keep the can closed to prevent moisture and other contaminants from mixing into the oil.

Much work has been done by lubrication engineers to provide the correct lubricant for each application. Product design engineers work closely with them to provide the best lubrication for the machine—they have a "lubricated for life" goal. So when you have to relubricate, try to use the recommended oil or grease and use the best you can get (Table 7-1). This is one of the little things that can bring big results.

A word of warning: Overlubrication of ball or roller bearings is harmful. It causes excessive churning, friction, and heat, which will "break down" the lubricant and ultimately damage the bearing. The space surrounding these bearings should never be more than one-third to one-half full of grease.

Solid or sintered sleeve-type bearings are

TABLE 7-1
Lubrication general recommendations

Use	Typical Lubricant
In refrigeration sealed system	Texaco Capella D Ansul 150, Sun Oil Co. Suniso 4G
Ball and needle bearings	Any good wheel bearing grease. Manufacturer's use—Shell Alvania No. 2, Texaco Regal Starfak 2, Cities Service Tellus 41, Esso Andok B, Mobilplex EP No. 2, Chevron BRB
Worm and other enclosed gearing	Cities Service Trojan M-2 grease SAE-90 or equal automotive gear and axle lubricant
Automatic washer gearbox	SAE-50
Sintered sleeve bearings	SAE-20 or 30, depending on operating temperatures
Small high-speed bearings	Light spindle or turbine oil
Sliding members	Light silicon/lithium grease

not generally injured by overlubrication. The danger here is the damage resulting from an oil leak on food or clothes.

STUDY QUESTIONS

Lubrication

1. Friction is the _____ that tends to _____ motion.

2. Excessive friction might be suspected if a bearing area seems to be too hot.   True or False

3. What are ambient conditions?

4. A bearing system seals out _____ and _____ and seals in _____.

5. Since bearings are inside of most machines, they do not cause trouble until the machine has been well used.   True or False

6. Dust protection is not important with the porous sintered-type bearing.   True or False

7. Three factors that must be included in considering oil for the porous sintered bearing are _____.

8. Should needle bearings be replaced on a shaft that is worn in the bearing area as long as there is plenty of lubricant?

9. The objects of the lubricant in ball bearings are _____ and _____.

10. Why are the plastic bearings an advantage?
11. The four basic functions of a lubricant are
    (a) reduce _____ and _____
    (b) conduct _____
    (c) seal _____
    (d) prevent _____ and _____
12. A highly viscous oil is best to use in cold weather.   True or False
13. Any good-quality oil may be used in servicing refrigerators.   True or False
14. What is one of the goals of lubrication that product design engineers aim for?
15. Is it possible to overlubricate any types of bearings?

BEARING FAILURE

A bearing that is failing will generally be noisy and allow the shaft excessive radial motion ("bearing slop"). When this warning occurs, it will be useless to try to clean, lubricate, or otherwise salvage the bearing because it will continue to get worse. Eventually complete failure may show as a frozen or "seized" bearing, which becomes stuck to the shaft, preventing it from turning freely, or as a worn or "burned out" bearing, which allows the rotating member to hit a stationary part.

Occasionally someone gets lucky and salvages a seized bronze bearing by careful shaft cleanup and relubrication, but often the work involved is not justified when compared to the parts cost. And *there is no assurance as to how long the salvaged bearing will last.*

The decision to replace bearings in appliance motors, gearboxes, and other areas must include

1. Relative costs of a new assembly
2. General condition of the parts
3. Bearing costs
4. Time required to do the job
5. Availability of the proper tools and working conditions

In so far as possible, the replacement bearing should be ordered from the appliance manufacturer by its part number. Sometimes this step is not practical and a replacement bearing will have to be used. The best source for standard replacement bearings will be the *local sales representative* for one of the *major bearing companies.* He will usually be listed under *Bearings* in the yellow pages of your telephone book.

Bring him the bearing and the mating parts to give him a chance to check measurements.

BEARING REPLACEMENT

A bearing that has been damaged, worn, or that otherwise no longer serves to maintain shaft position must be replaced. But first check the other mating parts—the shaft and the housing—for their condition, because unless these parts are good the new bearing will not fit correctly and soon it, too, will be "burned out."

Replacement Procedure for Ball Bearings

Bearing Removal from Shaft:

1. Make certain that shaft is clean of dirt, rust, and burrs before trying to slide a bearing over it.
2. Use a wheel or bearing puller to remove bearing. Try to catch the hooks in close to the shaft *on the inner race* (Figs. 7-6 and 7-7).
3. Examine shaft for wear in bearing area.

Bearing Installation on Shaft:

1. Make certain that the shaft is clean.
2. Check the shaft outside diameter and the bearing inside diameter to ensure correct fit. Normally only a ball or roller bearing is pressed on the shaft. *The bearing ID may be from 0.0001 to 0.0005 inch less than the shaft OD. It should never be larger. "A light press fit (Fig. 7-8a)."*
3. Use a piece of pipe that just clears the shaft and rests *squarely* on the bearing's inner race (Fig. 7-8b). Avoid pressure on the outer race or the bearing

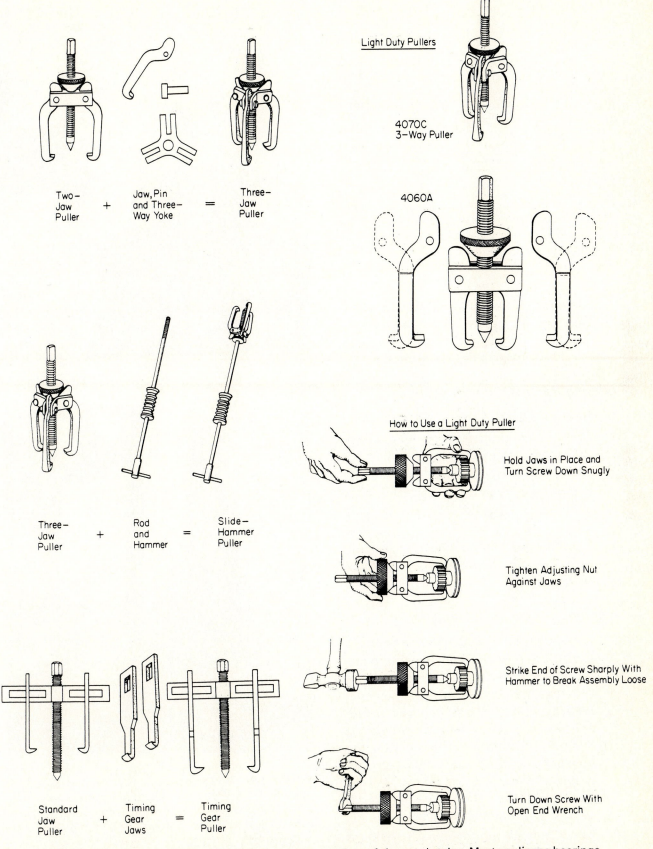

Light Duty Pullers

4070C
3-Way Puller

4060A

Two-
Jaw
Puller   +   Jaw, Pin
and Three-
Way Yoke   =   Three-
Jaw
Puller

Three-
Jaw
Puller   +   Rod
and
Hammer   =   Slide-
Hammer
Puller

Standard
Jaw
Puller   +   Timing
Gear
Jaws   =   Timing
Gear
Puller

How to Use a Light Duty Puller

Hold Jaws in Place and
Turn Screw Down Snugly

Tighten Adjusting Nut
Against Jaws

Strike End of Screw Sharply With
Hammer to Break Assembly Loose

Turn Down Screw With
Open End Wrench

FIGURE 7-6. Wheel pullers are available in a great variety of sizes and styles. Most appliance bearings can be removed with the two-jaw, light-duty puller shown here. (Courtesy of Proto)

FIGURE 7-7. If no bearing puller is available, use two screwdrivers. Place them opposite each other with their ends against the *inner race* and gently pry the bearing off the shaft.

(a)

cage. Either use an arbor press or carefully tap the pipe with a hammer.

Caution: This pipe must be clean and free of chips and burrs.

Bearing Removal from Housing:

1. Remove all retainers and oil wicking.
2. Clean surrounding area to remove grease, dirt, burrs, and so forth.
3. Drive out defective bearing by using a pipe, shaft, or other metal slightly smaller in diameter than the bearing OD.
4. Examine housing for wear in bearing area. *A worn housing will have to be replaced.*

Bearing Installation in Housing:

1. Make certain that housing is clean.
2. Check bearing OD and housing bearing seat ID. Usually needle and sleeve bearings are pressed into the housing. *Bearing should be a light press fit into the housing. Thus its OD should be 0.0001 to 0.0008 inch larger than the housing ID.* Hard materials, such as cast iron, require less of an *interference fit* than do the softer materials, such as zinc or aluminum castings.
3. Press bearing into housing by using a

(b)

FIGURE 7-8. (a) The bearing raceways must be a light-press fit. Do not force! A coating of light oil on the shaft will do much to ease the job. (b) When pressing a bearing over the end of a long shaft, use a piece of pipe or other hollow tube.

rod or pipe slightly smaller in diameter than the bearing OD. It must be started squarely and pressed evenly into the housing seat. For this operation, an arbor press will do a much better job than a hammer.

4. Any oil wicking that had been removed must now be reinstalled and loaded with oil. Needle bearings also should be oiled. Use a light SAE 10 oil for the smaller, higher-speed bearings. A heavier oil if operating temperature is apt to be high.

It has been assumed that all this work was done in a clean, dust-free area and that no metal chips from the tools have been deposited on the bearings.

Cleanliness is the key to successful bearing replacement. Any grit, dust, or metal chips that get into a bearing will quickly score it or jam a needle or ball. If the part has to be set aside, cover it with a clean cloth.

When reassembling the shaft to the housing, there should be a slight (0.0002 to 0.0005 inch) clearance between the shaft and a needle or sleeve bearing, enough to allow the shaft to spin freely with no wobble.

A ball or roller bearing must have a light press fit between both its races and the shaft or housing. When pressing these bearings in place, the pressure must always be on the race being positioned. *Pressure on the ball cage or the other race will damage the bearing.*

Heating to Relieve Pressure

Applying a gentle heat to either the bearing or the housing as needed will ease the bearing fit.

Metals normally expand when heated and so do holes in them. Thus heating the bearing's inner race will enlarge its bore, thereby often allowing it to fit a shaft with no external pressure. When it cools, the hole will close in to grip the shaft firmly.

The same holds true for the fit of bearing into a housing, only here the housing is heated.

A light flame from a torch or the radiant heat from a light bulb is all that is necessary. Warning: **Do not exceed 300°F or heat parts for a long period of time!**

STUDY QUESTIONS

Bearings

1. If a bearing becomes noisy, it should be lubricated more frequently.   True or False

2. Bearing failure is indicated when (a) parts will not move easily, (b) parts move easier than usual, (c) stationary part is hit by revolving member, (d) both (a) and (c), (e) none of these.

3. Sometimes steel bearings can be saved by careful cleanup.   True or False

4. When buying replacement bearings it is as important to bring the dealer the mating parts as it is the bearing.   True or False

5. When installing a bearing on the shaft the ID of the shaft and the OD of the bearing must be carefully checked.   True or False

6. When removing a bearing from the housing, do not remove the oil wicking and retainer. True or False

7. Bearing removal is one service approach when the cleanup should be done first.   True or False

8. There is an allowable interference fit of about 0.008 inch between bearing and housing. True or False

9. Zinc and aluminum rquire more tolerance than iron.   True or False

10. Installation of a new bearing should be done with a(n) _____ rather than a hammer.

11. After a new bearing is installed (a) the project is finished, (b) oil wicking should be added, (c) neither of these, (d) not mentioned.

12. The most important factor in bearing replacement is _____.

13. When fitting a new bearing into place, what is one of the most convenient aids?

14. What two sources of the aid in Question 13 are suggested?

15. What two factors are cautioned *against* in using the aid mentioned in Question 13?

16. What is the danger signal of worn-out bearings?

17. When deciding whether or not a bearing should be replaced, what five factors should be considered?

18. One of the main facts to keep in mind in all bearing replacement techniques is _____.

19. An important tool to have available for bearing service is the _____.

20. Once the bearing is set correctly over its position, it helps to hammer it in soundly. True or False

Replacement Procedure for Sleeve Bearings

A sleeve bearing is normally a free fit on the shaft and press fit into the housing. Thus the shaft will slide out of the bearing and the bearing will have to be pressed out of the housing. Avoid hammering the bearing directly or using a drift, punch, or screwdriver that puts abnormal pressure on one edge and none on another. It may damage the housing as well as the bearing.

Examine both the shaft and housing for scoring, rust, burrs, or other visible signs of damage or wear.

A worn shaft or damaged housing requires much more expense to repair or replace than a simple bearing replacement. At this point, consideration of repair costs and general condition of the appliance must be weighed against the costs of a new appliance, and the customer must be given the opportunity to make that decision.

Bearing Removal. Before removing a sleeve bearing from its housing, make certain that all retainers, hole plugs, and oil wicking are removed.

The best method of removal is to press the bearing out of the housing with an arbor press, using a short pipe or rod slightly smaller in diameter than the bearing OD. Use it between the bearing and the ram. Sometimes it may be necessary to heat the housing, *gently and carefully.*

Bearing Installation. The bearing should be pressed into its housing in much the same manner as was used to remove it.

Note especially,

1. The bearing must be positioned so as to align oil holes in the bearing with the oil reservoir in the housing.
2. The bearing must be square with the housing so that it enters evenly.
3. The hub must rest firmly on the table of the press.
4. Apply pressure steadily and gently (a little oil on the bearing and hub will help). Avoid excessive pressure. It means something is not fitting as it should. Reexamine the situation and correct the trouble before proceeding.

Sometimes it may be easier to use the shrinkage method of press fit to lock a bearing in its housing. This can be done by cooling the bearing, heating the housing, or both.

The bearing can be shrunk by packing it in dry ice or by setting it in a deep freeze. The housing can be heated in an oven set for 200°F (93°C). Heating should be carefully done to avoid warping the housing or damaging any attached parts that might be sensitive to heat.

RUNNING CLEARANCE

The running clearance in sleeve-type bearings is the amount of space between the shaft and the ID of the bearing. *This space accommodates the lubricant,* permitting a protective oil film to form between the moving parts. It also allows for some expansion due to temperature rise when in operation.

For appliances, pumps, motors, and other general machine practices, use this table of recommended running clearances.

Shaft Diameter Under:	$\frac{1}{2}$ inch	$\frac{3}{4}$ inch	1 inch	$1\frac{1}{2}$ inch
Minimum clearance	0.0008	0.001	0.0015	0.002

TABLE 7-2
Recommended allowances for press fit

Housing bore	$\frac{1}{2}$-inch	$\frac{3}{4}$-inch	1-inch	$1\frac{1}{2}$-inch	2-inch
Steel, cast iron	0.00025	0.0004	0.0005	0.00075	0.001
Heavy section:	0.0005	0.00075	0.00075	0.001	0.00125
aluminum or die cast					
Light section:	Line	0.00025	0.00040	0.0005	0.00075
aluminum or die cast					

PRESS FIT PRACTICES

The purpose of a press fit, shrink fit, or interference fit is to lock the bearing securely into the housing. This step is done by making the bearing OD larger than the housing or hub bore.

The dimensions given in Table 7-2 are the amount the diameter of the housing bore should be under the outside diameter of the bearing.

The normal "close-in" of bronze sleeve bearings is 80 percent of the amount of press and will vary, depending on the relative strength of the bearing and housing materials, expansion rates, and the bearing and bore tolerances.

STUDY QUESTIONS

Sleeve Bearings

1. When removing an old bearing it will have to be pressed off the shaft, but it should slide out of its housing.   True or False

2. A bearing replacement is much less costly than a shaft or housing replacement.   True or False

3. As soon as an old sleeve bearing is removed, be sure to remove wickings from its housing. True or False

4. The only thing to remember when pressing a bearing into its housing is to keep the pressure even.   True or False

5. Is there any method besides pressing to install a bearing in its housing? If so, what is it?

6. The space between the shaft and the ID of a bearing is called the ⸻.

SHAFT SEALS

When a liquid or gas has to be held in a container having a projecting, rotating shaft, a seal must be used. Examples of this usage are

1. The crankshaft extension from a refrigeration compressor
2. The impeller shaft extension from a water pump or disposer
3. The agitator shaft from the gearbox of a washing machine

Locating a Leaky Seal

The seal should be suspected in any case of leakage from a crankcase, gear housing or a pump. The easiest way to test for this situation is to clean the area thoroughly and watch for drippings or spray at the point where the shaft exits from the housing. This will indicate leakage out of the container.

In the case of a washing machine, a seal is necessary to keep water and detergents out of the gearbox (transmission). In other words, we must prevent water from leaking in. Oil and water dripping out of a relief hole in the gear housing would indicate a leak of this nature.

In no case will a seal repair itself. The only satisfactory solution is to install a new seal. A new bearing may also be needed, since the leakage may have caused excessive bearing wear.

Seal Types

Two basic types of seals are used with rotating shafts:

1. Lip seal
2. Face seal

Lip Seals. The lip seal is pressed into the housing with the seal's inner diameter being a flexible lip (Fig. 7-9), smaller in diameter than the shaft. When the shaft is inserted through the seal, it will make a pressure contact all around the shaft (Fig. 7-10). Take care to *install the seal so that the liquid pressure is on the underside of the lip to* cause a tighter contact to the shaft as pressure increases.

The part of the shaft that mates to the lip of the seal *must* be clean and glass *smooth.* Such a finish can best be obtained by polishing rather than by sanding or grinding. Polishing motions should be *around* the shaft, not along the length of it. An engineer would specify a 10 to 20 rms finish and that's *pretty smooth!*

Steel and stainless steel shafts are best for use with lip seals and they should be at least a 30 Rockwell C hard; otherwise the seal will wear a groove in it.

The seal will be effective with a little shaft eccentricity or runout, but it should not exceed 0.010 inch for the average motor speed of 1725 rpm. At higher speeds, the runout should be less.

High-operating temperature will shorten

FIGURE 7-9. Construction of a lip-type shaft seal. Note how the garter spring keeps the sealing element in contact with the shaft. (Courtesy of Chicago Rawhide Mfg. Co.)

GARTER
SPRING

SEALING
ELEMENT

← WIDTH →

the life of a seal. Generally, 200°F (93°C) is considered the safe maximum, although there are special materials that will be safe at higher temperatures.

The safest thing is to replace a seal with the correct part as listed in the parts list for the specific machine (Fig. 7-11).

Face Seals. The face seal is usually a two-part unit; one half is pressed into the housing and the other half is pressed onto the shaft. The two mating surfaces then meet to form the seal (Fig. 7-12).

As with the lip seal, all working and fitting areas must be *clean.* Prelubrication of the shaft and the housing seat will ease installation of the seal. *Make certain that the seal seats securely and squarely on both shaft and housing. Any misalignment of the seal faces could allow leakage.*

The seal faces, typically ceramic on one half and bronze or carbon on the other half, must be absolutely clean of any foreign matter, including fingerprints.

Seating

Face seal halves are a light interference fit into the housing and over the shaft. Both seats must be clean and burr-free. A loose fit of seal to seat will allow liquid or gas leakage around it. *Sometimes a sealing compound is used to aid in seating the seal.* A fit that is too tight may damage the seal or distort the seat. It is best to use an exact replacement as listed in the parts list for the particular machine.

It is also important to be careful in removing a defective seal so as not to damage either the housing or shaft seats.

Leakage During Storage

Sometimes, after an appliance has not been run for a long time, the seals may leak even though they were good when used last. An automobile air conditioner's loss of Freon gas after a long winter's idleness is an example of this situation. The air conditioner manufacturers recommend operating the air conditioner at least once a

SEAL INSTALLATION DETAILS

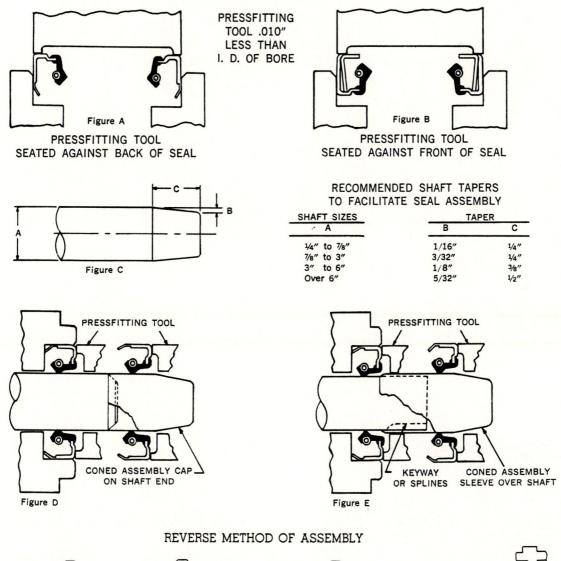

PRESSFITTING TOOL .010" LESS THAN I. D. OF BORE

Figure A
PRESSFITTING TOOL SEATED AGAINST BACK OF SEAL

Figure B
PRESSFITTING TOOL SEATED AGAINST FRONT OF SEAL

Figure C

RECOMMENDED SHAFT TAPERS TO FACILITATE SEAL ASSEMBLY

SHAFT SIZES	TAPER	
A	B	C
¼" to ⅞"	1/16"	¼"
⅞" to 3"	3/32"	¼"
3" to 6"	1/8"	3/8"
Over 6"	5/32"	½"

Figure D
PRESSFITTING TOOL
CONED ASSEMBLY CAP ON SHAFT END

Figure E
PRESSFITTING TOOL
KEYWAY OR SPLINES   CONED ASSEMBLY SLEEVE OVER SHAFT

REVERSE METHOD OF ASSEMBLY

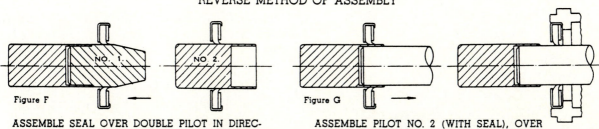

Figure F
NO. 1   NO. 2
ASSEMBLE SEAL OVER DOUBLE PILOT IN DIRECTION SHOWN TO GET WIPING EDGE OF PACKING IN PROPER POSITION. REMOVE PILOT NO. 1.

Figure G
ASSEMBLE PILOT NO. 2 (WITH SEAL), OVER SHAFT, THEN PRESS SEAL INTO HOUSING AND REMOVE PILOT NO. 2.

ASSEMBLY BULLET AND THIMBLE USED WHEN REVERSE METHOD OF ASSEMBLY IS NECESSARY FOR THE C/R TYPE G OIL SEAL.

FIGURE 7-10. Shaft seals must be installed carefully to avoid cutting, distorting, or otherwise damaging the lip. The outside mounting ring must be square to the shaft and securely sealed. Sometimes a sealant is used. (Courtesy of Chicago Rawhide Mfg. Co.)

TYPE SELECTION CHART

BONDED DESIGNS – RECOMMENDED TYPES FOR NEW APPLICATIONS

CROSS-SECTION	CLASS	TYPE	SHAFT SURFACE SPEED	SHAFT RUNOUT
 CRS HMS	Bonded single lip spring loaded standard duty	CRS HMS	0-1000 fpm 1000-2000 fpm 2000-3600 fpm	.020 TIR (0-1000 RPM) .015 TIR (1000-2500 RPM) .010 TIR (2500-4500 RPM)
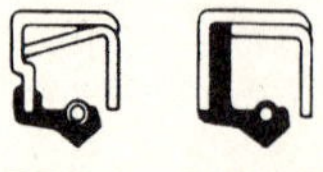 CRSH HMSH	Bonded single lip spring loaded heavy duty	CRSH HMSH	0-1000 fpm 1000-2000 fpm 2000-3600 fpm	.020 TIR (0-1000 RPM) .015 TIR (1000-2500 RPM) .010 TIR (2500-4500 RPM)
 CRSA HMSA	Bonded double lip standard duty	CRSA HMSA	0-1000 fpm 1000-2000 fpm 2000-3600 fpm	.020 TIR (0-1000 RPM) .015 TIR (1000-2500 RPM) .010 TIR (2500-4500 RPM)
 CRSHA	Bonded double lip heavy duty	CRSHA	0-1000 fpm 1000-2000 fpm 2000-3500 fpm	.020 TIR (0-1000 RPM) .015 TIR (1000-2500 RPM) .010 TIR (2500-4500 RPM)
 HM	Bonded single lip non-spring loaded	HM	0-500 fpm 500-1000 fpm 1000-2000 fpm	.005 TIR .005 TIR .005 TIR

FIGURE 7-11. Lip seals are available with many lip designs. Each will do a certain type of job better than a different style would. (Courtesy of Chicago Rawhide Mfg. Co.)

month during its season of nonuse to help prevent leakage.

Shaft Seal Lubrication

Either lip seals or face seals are used on rotating shafts having a considerable amount of

FIGURE 7-12. General construction of a face seal. (Courtesy of Garlock, Inc.)

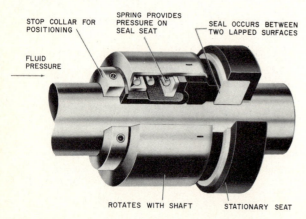

STOP COLLAR FOR POSITIONING

SPRING PROVIDES PRESSURE ON SEAL SEAT

SEAL OCCURS BETWEEN TWO LAPPED SURFACES

FLUID PRESSURE

ROTATES WITH SHAFT   STATIONARY SEAT

speed. Therefore, *unless properly lubricated and cooled, there would be early seal failure* due to friction heat.

O-RING SEALS

Shafts having a slow occasional rotating or sliding motion can be sealed with an O ring. O-rings are also excellent as a static seal (gasket) between parts.

O-ring seals fit into a groove that may be in either the shaft or the bearing and that project into the other area. The result is a sealing pressure between shaft and bearing. The O-ring is shaped like a round ring "doughnut" to very exact tolerances. These seals are made from a number of different compounds, depending on the usage. Size range from $\frac{1}{32}$ inch ID to 2 feet or more ID. Some compounds used are neoprene, butyl, silicon, Teflon, and polyethylene.

O-rings used as static seals require no lubrication, but when used as dynamic seals they

DESIGN HINTS FOR EFFECTIVE O-RING SEALING

BEVEL FOR INSTALLATION

O-rings installed on shafts should be protected from damage during assembly by providing a chamfer at the bore entry. Likewise when the O-ring is installed in the housing bore the leading edge of the shaft should be chamfered.

CHAMBER ANGLE
15° TO 30°

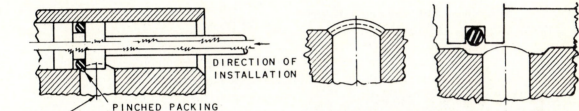

FREE RING IN GROOVE

CROSS-DRILLED PORTS

DIRECTION OF INSTALLATION

PINCHED PACKING

CROSS-DRILLED PORT

Either chamfer hole junction with bore or undercut bore as illustrated at right. Undercut is recommended because O-ring is free when passing port.

LUBRICATION

Dynamic O-ring sealing is subject to rapid wear if run dry. This problem is not present with sealing fluids having good lubricating value. Pneumatic applications or commonly used inert gases such as oxygen, acetylene, carbon dioxide, etc., require means for lubricating the O-ring.

Lubrication for dry running seals can be accomplished in several ways. One of the most satisfactory is to install a felt washer adjacent to the O-ring and provide an oil hole. Good results are also obtained by installing two felt washers, one on either side the O-ring, saturated in lubricating oil.

FIGURE 7-13. O-rings are precision sealing members and must be treated as such. When installing an O-ring, review the hints and obey them. (Courtesy of National Seal Division, Federal-Mogul Corp.)

are lubricated by the fluids being sealed (oil, water, etc.). Dry running applications can be lubricated by installing an oil socket felt washer next to the O-ring and providing an oil hole for relubrication.

Failure

Obviously, as with any other seal (Fig. 7-13), O-ring failure will be evidenced by a leak.

Once having failed, the O-ring cannot be repaired; it must be replaced.

When removing an O-ring, examine it, its

groove, and the mating surface for cause of failure. This failure must be corrected before installing the new seal. *Some causes of O-ring failure are listed* below:

1. O-ring pinched during installation
2. O-ring damaged by sharp edges during installation
3. Excessive temperature rise
4. Excessive pressure rise
5. Corrosion/abrasion due to contaminents in fluid being sealed

6. Metal finish of groove or bearing not good enough
7. Running dry

Installation

When installing a new O-ring, make certain that the cause of failure has been corrected and that your handling of the O-ring has not damaged it or allowed it to collect chips or dust prior to assembly. Use the factory-recommended replacement, if possible; otherwise, follow the recommendations of the nearest O-ring distributor. He will be listed in the Yellow Pages under *Industrial Equipment and Supplies, Bearings,* or *Rubber Products.*

STUDY QUESTIONS

Seals

1. Before diagnosing for a leaky seal, clean the area thoroughly.   True or False
2. In some cases, given a little time, a damaged seal may fix itself.   True or False
3. Where a seal mates to a shaft, the latter must be _____ and _____.
4. Which type of seal is usually in two parts?
5. Replacing seals does not call for the same care in cleanliness that replacing bearings does. True or False

KEY

Lubrication

1. Resistance, retard
2. True
3. Surrounding air
4. Dust and moisture, lubricant
5. False
6. False
7. Nongumming, nonoxidizing, noncorrosive
8. No
9. Dissipate heat and seal out moisture and contaminents
10. They are resistant to corrosion, detergent, and water

11. Friction and wear, heat away, out contaminents, rust and corrosion
12. False
13. False
14. Lubricate for life
15. Yes

Bearings

1. False
2. d
3. False
4. True
5. False
6. False
7. True
8. False
9. False
10. Arbor press
11. b
12. Cleanliness
13. Heat
14. Small torch flame or light bulb
15. High temperature and length of time
16. Noise
17. Relative costs of new assembly
    Condition of parts
    Costs of new
    Time required to do job
    Availability of tools
18. Cleanliness
19. Arbor press
20. False

Sleeve Bearings

1. False
2. True
3. False
4. False
5. Yes, shrinkage
6. Running clearance

Seals

1. True
2. False
3. Clean, glass smooth
4. Face seal
5. False

8

Refrigeration

HISTORY

It was not until 1683, when a Dutchman named Anton van Leeuwenhoek invented the microscope and discovered microbes, that scientists began to learn that there were many microorganisms, some essential to our health and others dangerous.

Continuing investigations revealed that these microscopic organisms flourished in a warm, moist environment and became dormant (to some extent) when the temperature dropped to 50°F (10°C) or colder. Also, they learned that cold did not kill them but simply checked their growth activity.

Originally the idea of cooling food to preserve it spread slowly, but, once accepted, ice delivery became a worldwide business. The Clipper ships carrying ice harvested in the New England lakes to the southern states and, farther, to the ports of London, Cairo, and Bombay returned with their famed loads of spices and silks.

Previously, food that had to be preserved was dried, smoked, spiced, salted, or pickled—all of which, in the process of preservation, would alter the taste of the food. Refrigeration preserved fresh foods safely for a reasonable period of time.

Actually, the Greeks and the Romans used their slaves to bring snow down from the mountains to cool their drinks. The ancient Chinese used ice to make their summer drinks more tasty. They were the first to cut and preserve ice for use during the summer months. It was a luxury item for cooling and comfort, never considered for food preservation. Thus for centuries man knew of and used refrigeration, but he was unaware of its most valid use.

As the demand for ice increased, man tried to produce it artifically. In the late eighteenth century some laboratory models were developed, but the first commercially successful ice-making machine was not patented until 1834. Sales of ice-making machinery were limited to breweries, meat-packing plants, and other commercial ventures because the public was afraid that artificial ice was unhealthy. It took the

American Civil War and a warm winter in 1890 to demonstrate the value of machine made ice for use in the home icebox.

It was not until electricity entered many houses and small electrical motors were developed that the ice-making machine could be considered for household use. The early machines were often installed in the old family icebox with the cooling unit located in the ice compartment and the rest of the machine in the basement or under the back porch. This cooling unit, installed to replace the cake of ice, did the refrigerating and ice became a by-product.

HEAT REMOVAL

As we have seen, ice is used to cool man's foods, to make them more palatable, and to preserve them (Fig. 8-1). Also, ice is used to cool air to make man more comfortable.

This work is now being done by refrigeration, which can be defined as the process of *removing heat* from where it is not wanted.

It is all done mechanically. Heat is absorbed by a set of tubes containing a boiling refrigerant, which carries the heat from the area that needs cooling to another area where the heat will do no harm (Fig. 8-2).

To understand *how* this happens, we should review a few fundamentals of heat flow, refrigerants, and their associated machinery.

HEAT LAWS

In order to understand refrigeration, we must know a few basic laws about *heat*. If you will

FIGURE 8-1.   Ice has long been used to cool man's food and body.

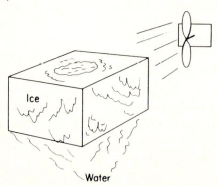

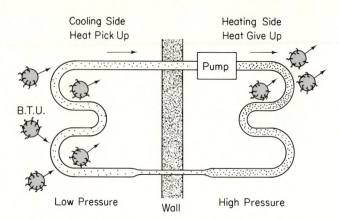

FIGURE 8-2.

accept the following facts to be true, we can more easily explain how a refrigeration system does its job.

1. *Refrigeration* is the removal of heat.
2. *Heat* is the energy in a substance resulting from its molecular motion.
3. *Temperature* is an indication of the degree of intensity of heat and will vary as molecular motion varies. It is measured with a thermometer calibrated in degrees, usually Fahrenheit in this country and Celsius in most of the rest of the world. Both scales use the freezing point of water and the boiling point of water at sea level as standards.

   The Fahrenheit scale sets the freezing point of water at 32°F and the boiling point of water 180 divisions farther up the scale at 212°F.

   The Celsius scale sets the freezing point of water at 0°C and the boiling point of water 100 divisions farther up the scale at 100°C.

   To convert from one scale to another:

   $$F = \tfrac{9}{5}C + 32°$$

   $$C = \tfrac{5}{9}F - 32°$$

   Our discussions on temperature will use the Fahrenheit scale.

4. The measurement of temperature has no relation to the *quantity* of heat. That is, an ice cube has the same temperature as a 50-pound cake of ice (Fig. 8-3).

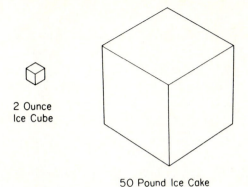

FIGURE 8-3. Both are at the same temperature, but the larger one will last longer.

FIGURE 8-5. Heat flows from warm to cold, from any direction, by any means—conduction, convection, radiation, etc.

5. The amount of heat energy in a substance is called a British thermal unit (Btu), which is defined as the amount of heat required to raise the temperature of 1 pound of water 1 degree Fahrenheit (Fig. 8-4). For example, to raise 5 gallons of water from 60°F to 110°F will require 2075 Btu. *Five gallons × 8.3 pounds per gallon × (110 − 60) = 2075.*

6. Heat *always* travels from the higher temperature to the lower temperature. Thus heat flow is from the warmer object to the colder one regardless of relative size (see Fig. 8-5).

7. The greater the temperature difference, the faster the heat flow.

8. Heat is always in motion and can travel in three ways:
   (a) *Conduction* is the flow of heat through a substance. There must be physical contact between the hot and the cold objects. Example: If you touch a hot soldering iron, your hand will conduct heat from the iron, and your hand will feel hot. If you hold an ice cube in your hand, the ice cube will conduct heat from your hand, thus causing the hand to feel cold and the ice to melt.
   (b) *Convection* is the flow of heat by means of a fluid medium, either liquid or gas. A hot-air furnace is a good example. The air is heated in the furnace and then this warm air is discharged into the room to be heated.
   (c) *Radiation* is the transfer of heat by waves similar to light or radio waves. Standing in front of a fireplace would illustrate this point. The parts of your body facing the fireplace get warm, but the rest remains cool. The sun radiates its heat energy to the earth.

9. A change of state occurs when enough heat is applied to a solid to cause it to liquify (melt) or to a liquid to cause it to vaporize (boil). This also works in the other direction. Removing heat from a vapor will condense it to a liquid and removing heat from a liquid will freeze it into a solid. Our most common example of matter in its three states is ice, water, and steam representing the solid, liquid, and gas states of the same compound, H_2O.

FIGURE 8-4. Adding heat raises the temperature.

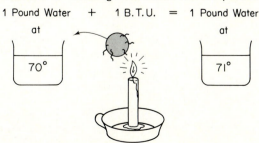

1 Pound Water at 70° + 1 B.T.U. = 1 Pound Water at 71°

Any upward change of state will require additional heat (Btu) at its critical temperature to make the conversion (Fig. 8-6). The additional Btu will not change the temperature but will be *used to increase the molecular energy of the substance*. We can say that the extra Btu are stored in the material and will be returned, in full, when the process is reversed and the matter returns to its original form.

Water is a good example. It boils (change of state from liquid to gas) at 212°F (100°C). Regardless of the amount of heat applied to boiling water, its temperature will not exceed 212°F (100°C). The extra heat absorbed by the water changes it to steam and leaves the kettle *with* the vapor, taking with it 970.4 Btu of heat energy for every pound of water that was converted to steam. When the steam condenses back to water, the same molecular energy (Fig. 8-7), is converted into heat energy—it gives up heat.

10. *Latent heat* is the heat energy (Btu) required for a change of state and

FIGURE 8-6. Heat used for a change of state does not increase temperature.

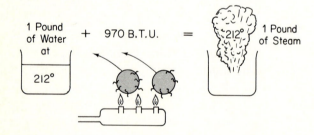

FIGURE 8-7. Adding or subtracting heat does not change the amount of substance.

cannot be measured with a thermometer. Depending on the change of state, it is called:
(a) Latent heat of fusion when the change is from solid to liquid, or liquid to solid.
(b) Latent heat of vaporization when the change is from liquid to gas, or gas to liquid.
(c) Latent heat of sublimation when the change is from solid to gas, or gas to solid. The latent heat of sublimation is equal to the sum of the latent heat of fusion and the latent heat of evaporation.

11. Sensible heat is the heat energy required to change temperature and is measured with a thermometer.

12. The temperature at which a liquid boils depends on the pressure being exerted on it. As the temperature increases, the molecular activity increases, and the pressure of the molecules trying to escape from the liquid increases. When this *vapor pressure* reaches the external pressure, boiling occurs. Or as *pressure* on a gas is *increased* to compress the gas, its molecular energy is confined and the excess becomes heat energy, which increases its temperature.

Conversely, as the external pressure on the liquid decreases, the boiling temperature will decrease.

From this we get the "perfect gas law," which describes the relationships of the three basic factors controlling gas behavior: pressure, temperature, and volume.

$$\frac{P_1 \times V_1}{T_1} = \frac{P_2 \times V_2}{T_2}$$

13. Refrigeration is a practical application of these pressure–temperature relationships and heat laws (Figs. 8-8 and 8-9, and Table 8-1).

14. A refrigerant is a fluid that picks up heat by *evaporating* at a low tempera-

TABLE 8-1

Effect of pressure decrease on boiling point of water

Reading (in. Vacuum)	Pressure (in. Mercury)	Boiling Temperature of Water [°F (°C)]
0	29.9	212 (100)
10	19.9	192 (89)
20	9.9	162 (72)
25	4.9	133 (56)
27	2.9	114 (46)
28	1.9	101 (38)
29	0.9	79 (26)
29.5	0.4	53 (12)
29.7	0.2	35 (0.2)
29.8	0.1	20 (−6.7)

FIGURE 8-8. These gages, used to measure refrigerant pressure, tell much about the condition of a refrigeration system. Look carefully at the vacuum end of the compound gage when you study Table 8-1.

FIGURE 8-9. This manometer reading is 0.5 inch mercury, which is equal to a 29.4-inch vacuum. In this vacuum, water will boil at 59°F (15.0°C). Remember these figures; they are important when you have to clear out a refrigerant system before changing the gas.

ture and pressure and gives up heat by condensing at a higher temperature and pressure.

REFRIGERANTS

Theoretically, any gas that can be alternately liquified and vaporized *within* mechanical equipment can serve as a refrigerant. But some characteristics make one gas more desirable than others.

A satisfactory refrigerant should:

1. Absorb heat readily at the temperature required by the load.
2. Be capable of being used over and over again within the same system.
3. During normal operation, the working pressures should remain within a reasonable range.
4. Be stable, nonpoisonous, nonexplosive, nontoxic, nonflammable, and noncorrosive.
5. Leaks should be easy to detect and locate.
6. Be compatible with lubricating oil, but not absorb moisture (Fig. 8-10).
7. Have a constant pressure–temperature relationship.

Many different refrigerants are available, but the two most commonly used in household refrigeration and air conditioning are R12 and R22. The R stands for refrigerant, and the iden-

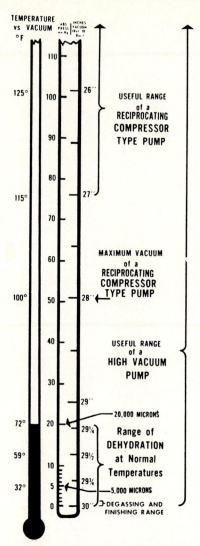

TEMPERATURE
vs VACUUM
°F

ABS. PRESS	INCHES VACUUM

USEFUL RANGE
of a
RECIPROCATING
COMPRESSOR
TYPE PUMP

MAXIMUM VACUUM
of a
RECIPROCATING
COMPRESSOR
TYPE PUMP

USEFUL RANGE
of a
HIGH VACUUM
PUMP

20,000 MICRONS

Range of
DEHYDRATION
at Normal
Temperatures

5,000 MICRONS

DEGASSING AND
FINISHING RANGE

FIGURE 8-10. We place much emphasis on the boiling of water in a vacuum, a process called dehydration. Any moisture in a refrigeration system will cause trouble— as freeze-up, corrosion, or both. You must dehydrate the system *completely* before putting in fresh gas.

tifying number has been standardized by the American Society of Heating, Refrigeration, and Air Conditioning Engineers to indicate a particular formulation.

Both compounds are classified in the National Refrigeration Safety Code Catalog as group one. They are among the safest refrigerants, for they are not a fire hazard and are among the least toxic.

R12 and R22 are both Freon gases formulated especially for refrigeration usage. They have many more desirable characteristics than some of the earlier refrigerants, such as ammonia, sulfur dioxide, and methyl chloride.

PRESSURE–TEMPERATURE RELATIONSHIP

We have said that heat is absorbed by a boiling liquid and that the boiling point of a liquid is controlled by the pressure on it.

In a sealed refrigeration system, this factor makes the pressure–temperature relationship of a liquid important to a refrigeration service technician. The graph (Fig. 8-11) showing the pressure–temperature curves for R12 and R22 are the saturated vapor curves of these refrigerants. Any point on the curve represents a balanced condition in which the evaporation of liquid is balanced by the condensation of vapor. It is the point where any increase in pressure will cause the vapor to condense and any decrease in pressure will allow the liquid to vaporize.

If left alone, a refrigerant in a sealed system will tend to equalize its pressure and balance in a liquid-saturated vapor state at some point on its curve. Refrigerant vapor cannot exist at a temperature lower than its saturation temperature.

As the refrigerant goes through its change of state, there must be an exchange of its *latent* heat energy. For R12 this is 68.2 Btu per pound and for R22 it is 93.21 Btu per pound. (Latent heat is vapor heat minus the liquid heat.)

MEASUREMENT

Since the Freons have a constant pressure–temperature relationship, it is possible to learn

FIGURE 8-11.

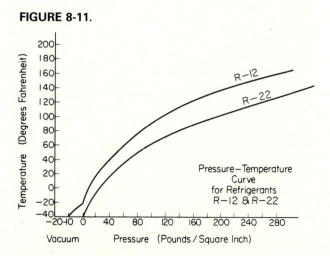

Temperature (Degrees Fahrenheit)

R-12
R-22

Pressure–Temperature
Curve
for Refrigerants
R-12 & R-22

Vacuum   Pressure   (Pounds / Square Inch)

much about their condition in a sealed system by simply installing gages to measure the pressures at strategic locations.

The two locations that will tell us the most are at the input and the output of the compressor, as in Fig. 8-12.

The input, or low-pressure side, to a compressor is measured with a combination gage (Fig. 8-8) having a range of 30 inches of vacuum to 250 psi pressure.

The output, or high-pressure side, of a compressor is measured with a pressure gage (Fig. 8-8) having a range of 0 to 500 psi.

THE REFRIGERATION CYCLE

We have said much about heat laws, refrigerant gases, and temperature–pressure relationships. Now let us put this together into a mechanical package and see how it works.

It is said that *mechanical* refrigeration is accomplished by continuously circulating, evaporating, and condensing a fixed supply of refrigerant in a closed system. Evaporation occurring at low *pressure* absorbs heat, thus causing low *temperature*. Condensation occurring at

high *pressure* gives off the heat it absorbed while evaporating, at a high *temperature*. Thus the flowing gas conducts the heat from one place to another.

The essential elements that compose the refrigeration system are

1. Evaporator or the cooling unit
2. Pump or compressor
3. Condenser
4. Liquid metering device (restriction)
5. Refrigerant

These elements are connected to each other as shown in Fig. 8-13. Operation occurs in the following sequence:

1. The suction pump reduces the pressure on the liquid refrigerant in the evaporator until it boils, forming a vapor. As the pump continues to remove the vapor and lower the pressure, the temperature within the evaporator decreases until a pressure level is reached, then the incoming liquid bal-

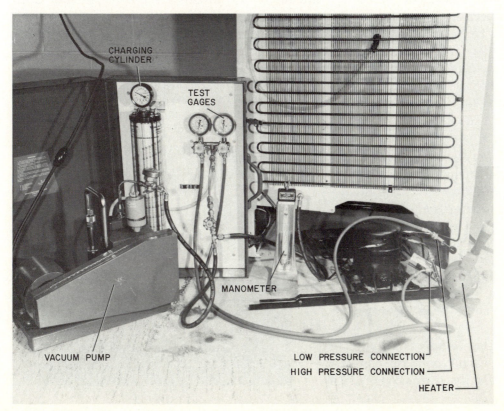

FIGURE   8-12. Complete setup needed to properly evacuate and recharge a sealed refrigeration system.

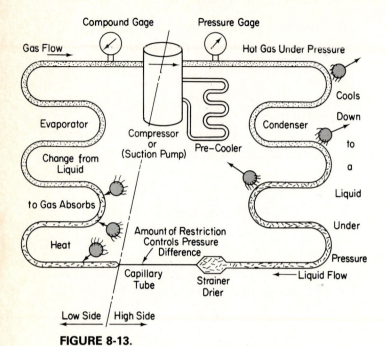

Compound Gage   Pressure Gage

Gas Flow →

Hot Gas Under Pressure →

Evaporator

Cools

Down

Condenser

to

Compressor
or
(Suction Pump)

Pre-Cooler

a

Change from
Liquid

Liquid

to Gas Absorbs

Under

Amount of Restriction
Controls Pressure
Difference

Pressure

Heat

← Liquid Flow

Capillary
Tube

Strainer
Drier

Low Side / High Side

FIGURE 8-13.

ances the vapor being withdrawn and a stable temperature results.

2. The suction pump, which had withdrawn the vapor, compresses it in the condenser to a pressure that will make its temperature above the room air temperature. The heat in the vapor will flow from the condenser to the cooler room temperature, causing the refrigerant vapor to condense to a liquid.

   The operating temperature of the condenser will be as many degrees above room temperature as is necessary to dissipate the heat absorbed in the evaporator and compressor.

3. The liquid metering device is a restriction to the flow of refrigerant from the condenser to the evaporator. The degree of restriction determines the rate of flow of refrigerant into the evaporator.

Thus the combination of parts and a refrigerant are united in a balanced system to provide refrigeration at some predetermined temperature.

This resulting operating temperature will be in a different range for each type of refrigerant-using machine. Specifically, water coolers,

dehumidifiers, air conditioners, refrigerators, and freezers all operate on this basic cycle. Only the operating *range* is different. For example, there is no need for pressure to be too low in a water cooler because ice is not required, just *cold* water (about 40°F (4°C).

THE MECHANICAL PARTS

A refrigerator's parts have been mentioned in the theory discussions but have not been described. A service technician should recognize all the parts he is going to work with, so Figs. 8-14 to 8-17 should serve as an introduction to the parts. We hope you also have an opportunity to see, feel, handle, and test these parts in real life.

THE ELECTRICAL SYSTEM

In its simplest form, the only electricity-using part in a refrigerator is the compressor motor. Its only control was once a thermostatically operated switch, but as time went on, the refrigerator was improved to make it more versatile and more convenient for the housewife. Each improvement added more electrical parts and controls. These components are shown in Fig. 8-18.

FIGURE 8-14. This compressor is the pump that moves the refrigerant. This is a typical unit that will be found in most domestic refrigerating and air-conditioning units. It is a sealed unit and, if defective, must be replaced.

FIGURE 8-15. For those who want to know what is inside the black box—this is an ''opened'' compressor showing the rotary pump. The electric motor is under the pump. The entire motor-pump assembly is spring mounted to absorb vibration and decrease noise.

FIGURE 8-16. Shown here are the compressor of Fig. 8-15 plus the other parts that normally appear *outside* the cabinet. These are all high-side parts (except for the suction line).

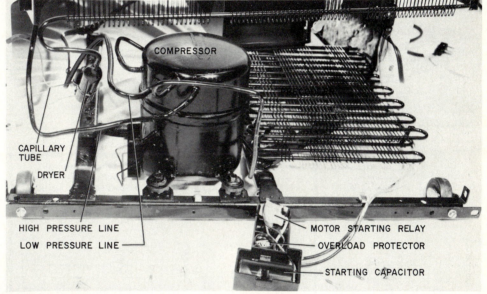

FIGURE 8-17. The evaporator and its associated parts are usually located behind a panel on the rear wall of the freezer compartment, as shown. Some models have the evaporator located under the freezer compartment or above the refrigerator compartment.

REFRIGERATION SYSTEM

- **SIMPLE SYSTEM**
  Single capillary tube
  No solenoid valves
  Single evaporator (1) cools both compartments (2) (3)
- Freezer thermostat (4) (air sensing) controls compressor (5) and condenser fan (6).
- Timer (7) stops refrigeration and starts defrost.

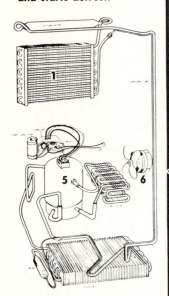

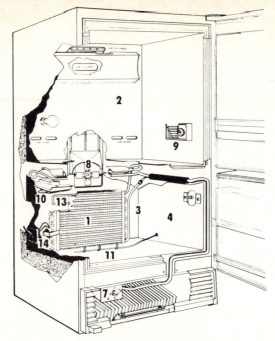

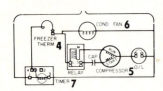

DEFROST SYSTEM

- Timer initiates defrost (each 12 hours). Refrigerator turned off 20 minutes.
- Shroud (12) and liner drain (11) heaters on for 20 min.
- Evaporator and header drain heaters (10) (13) on, until thermo disc (14) (on evaporator) warms to 55° or 20 minutes, maximum.
- Thermo disc resets at 22°.
- Defrost water drains from drain pan and evaporator into bottom of liner through drain hole (15) to unit compartment pan (16).

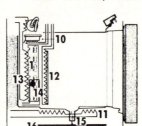

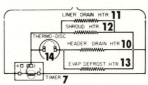

AIR SYSTEM

- Single two-speed fan (8) — delivers cold air to both comparments (2) (3).
- Speed of fan is controlled by refrigerator thermostat (9) (air sensing):
- Fan runs at high speed when refrigerator thermostat calls for cooling.
- Fan runs at low speed continuously when refrigerator is satisfied, except
- Fan is off when freezer door is open or during defrost.

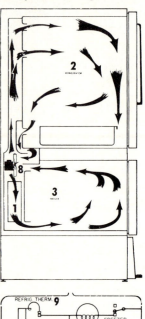

COMPONENT TESTING
Check Voltage at Outlet — Unplug to Make Electrical Checks

REFRIGERATION SYSTEM

1. Compressor does not run.
   A. Check freezer thermostat for proper setting and continuity.
   B. Check compressor with starting cord.
   C. Check overload for continuity.
   D. Check relay — replace with known good relay.
   E. Check timer — timer motor must run — remove wires, check continuity, terminals 1 and 4.

2. Compressor runs, insufficient cooling.
   A. Check condenser fan motor by direct connection.
   B. Clean condenser.
   C. Partial restriction — frost at restriction; check for moisture by applying heat.
   D. Complete restriction (hot compressor, cool condenser, no frost).
   E. Leak or low charge — see Service Manual.

AIR SYSTEM

1. Fan motor — make direct connection to check.
2. Timer — motor must run — remove wires, check continuity, terminals 1 and 4.
3. Refrigerator thermostat — check proper setting and continuity.
4. Speed Resistor — O.K. if fan runs on low speed (500 ohms).
5. Freezer door switch — check continuity, closed door position.

DEFROST SYSTEM

1. Timer — motor must run — set to 2 o'clock (audible click); check continuity terminals 1 and 2.
2. Heaters — remove breaker trim at crossrail — remove wires at terminal board, check continuity of like colors.
3. Thermo disc — must contact evaporator firmly — check at terminal board (brown wires), opens 55° and resets 22° (approx.).

FIGURE 8-18. Typical frostless refrigerator-freezer showing the fundamental relationship between the electrical, refrigeration, and air systems. Except for minor variations in part locations, the principles shown will hold for most of the self-defrosting (automatic) refrigerators. (Courtesy of Coldspot)

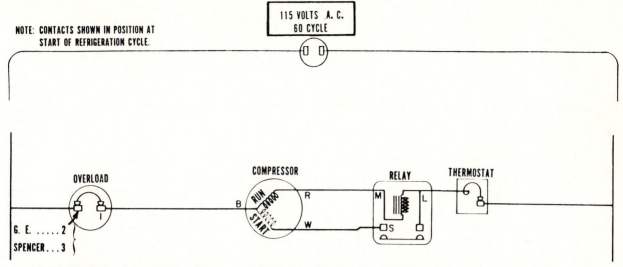

FIGURE 8-19. This is the minimum electrical circuit used in a refrigeration system. It consists of the compressor motor and its basic controls. (Courtesy of Coldspot)

No discussion on an electrical system is complete without wiring diagrams. Figures 8-19 and 8-20 illustrate the basic growth in electrical parts usage and how they relate to each other.

THE CABINET

Since refrigeration is the removal of heat from one place to another, we must consider the space we are trying to keep cool—normally called "the box."

Basically, the refrigerator cabinet consists of an inner and outer shell, separated by thermal insulation. The outer shell, usually steel, is shaped and painted to appeal to the owner's aesthetic sense. The inner shell, also usually steel, is porcelain coated to provide easy cleaning and a sanitary appearance. Both shells should be continuous and all fastening and utility holes should be plugged to prevent moisture from entering the insulation (Fig. 8-21).

Door construction is similar to the box construction. The seal between the door and the box is made by a rubber gasket. Most seals in current usage have a magnetic strip buried in the rubber. This strip serves as a means of holding the door closed and ensures better sealing contact between gasket and cabinet.

The sealing gaskets are mounted on the door rather than on the box, and the magnetic strip extends around *three* sides, *leaving the hinge side more compressible.*

The interior of the cabinet will contain the evaporator (cooling unit), shelves, drawers, ice cube trays, and such items as the designer felt the customer needed. Some of these parts may be made of glass, plastic, aluminum, or plated steel. If any are damaged, they will have to be replaced. Some cannot be repaired; in other instances, the cost of repair may be greater than the cost of replacement.

The distribution of air within the cabinet is of great importance to ensure proper cooling in all parts of the cabinet. The "boxes" without air-circulating fans tend to have less-uniform cold distribution than the units having fans. For those boxes having air duct work and fans, make certain that the duct remains open and that the air exits are clear enough to allow free air circulation.

Figure 8-22 illustrates some of the various air-circulation systems in use.

OTHER REFRIGERANT USERS

We have said that there are many other appliances besides a refrigerator that use the basic refrigeration system. Operation of and service on these machines are included elsewhere in this book.

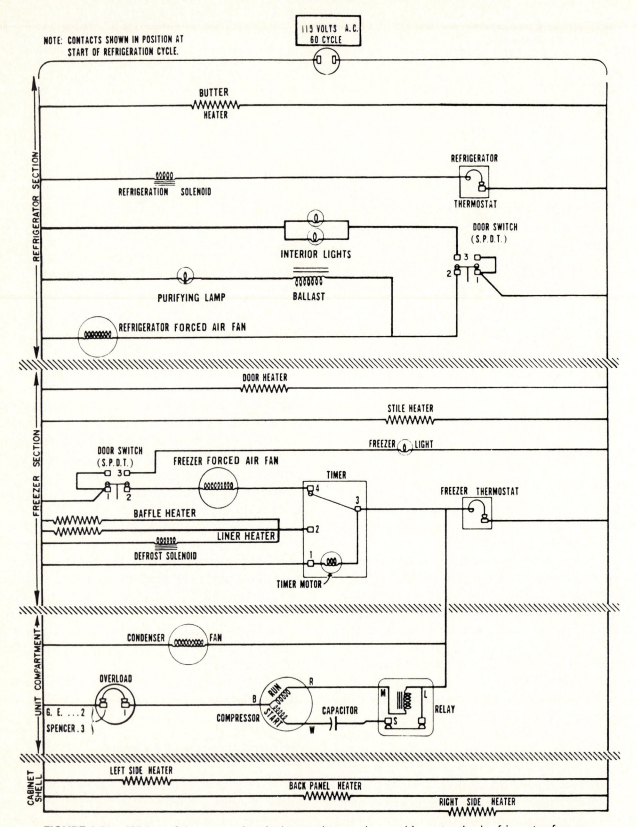

FIGURE 8-20.   Wiring of the many electrical parts that can be used in a standard refrigerator-freezer.

FIGURE 8-21. In addition to the thermal insulation used to fill the space between the cabinet and the liner, there is a series of plastic strips around the opening to the cabinet. These are called "breaker strips" because they break the conducted flow of heat from the outside to the inside that would occur with a metal connecting strip. These are removable to allow access to the insulation, heaters, evaporator, and such other parts as may be located in the insulation areas.

FIGURE 8-22. Airflow patterns typical of the forced-air-cooled refrigerators in current use. The service technician should be aware that anything which interferes with the airflow will decrease cooling efficiency. (Courtesy of Whirlpool)

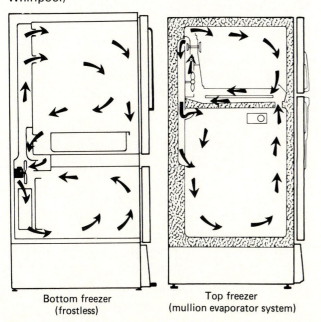

Bottom freezer (frostless)

Top freezer (mullion evaporator system)

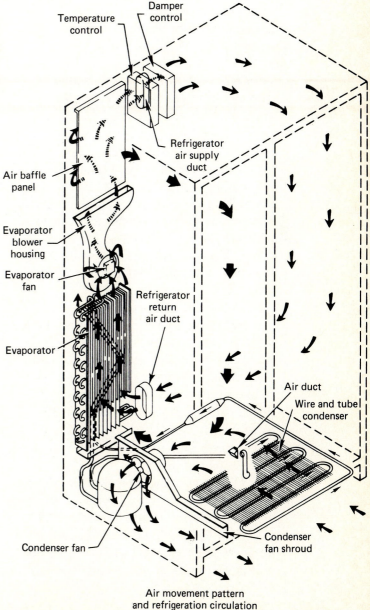

Air movement pattern and refrigeration circulation

TROUBLE DIAGNOSIS

Each refrigeration unit design will have certain operating conditions that make it different from any other design. These differences will result in individual service problems. The best place to get information on performance characteristics and service aids is from the manufacturer's service manuals for each specific model.

However, there is enough similarity in the refrigeration family to allow discussion of some general diagnostic procedures.

1. Always check the installation and operating environment. Make certain that the machine is secure, has power and adequate air circulation around the condenser. An overheated condenser will not allow the vapor refrigerant to condense to a liquid and thus permits hot gas to enter the evaporator and causes inefficient refrigeration.

2. Does the compressor run? If not, and there is voltage to the machine, the easiest test is to connect a test cord and wattmeter directly to the compressor motor terminals. If it starts and operates properly, the trouble is obviously in one of the motor controls: thermostat, overload capacitor, or starting relay. These parts can be tested individually and the defective part replaced.

   If the compressor does not start, the trouble must be in the motor-compressor (sealed unit) assembly, which will have to be replaced. Replacement should only be considered if the rest of the appliance is in good enough condition to warrant the expense. This must be the customer's decision.

3. The compressor runs, but there is little or no refrigeration.

   Obviously, the liquid refrigerant is not getting into the evaporator in enough volume to be effective. The cause could be a leak, a restriction, or an inefficient compressor. Test gages will be required to isolate this trouble.

The compound gage connects to the suction line at the compressor and the pressure gage connects between the compressor output and the condenser.
   (a) Little or no pressure on both gages would indicate a leak.
   (b) Abnormally high head pressure and a low suction reading would indicate a restriction. Generally this will be in the capillary tube.
   (c) Less than normal pressure differential between the two gages would indicate an inefficient compressor.

4. Other problems may be
   (a) Defective fans.
   (b) Clogged drain.
   (c) Worn door hinges.
   (d) Leaky door gasket.
   (e) Defective door switch.
   (f) Open heater element.
   (g) Failed timer.
   (h) Failed thermostat (other than the operating one).

   Many of these problems are self-evident on careful inspection. Although they affect the quality of operation, they do not directly affect the basic refrigeration system.

ENTERING THE SEALED SYSTEM

Having determined that the problem is in the sealed system, you are now ready to cut into it to test and service. But before you do, list in your mind the only three reasons for entering the system.

1. A restriction to the flow of refrigerant:
   (a) Partial
   (b) Complete
2. An incorrect refrigerant charge:
   (a) Overcharge
   (b) Undercharge
3. Compressor failure:
   (a) Stuck—will not run.
   (b) Defective motor windings
   (c) Inefficient compressor

After the trouble has been located, the repair must be made. Most parts replacement procedures are self-evident. The manufacturer's service manuals give detailed replacement methods and their instructions should be followed.

The big jobs in refrigeration service are compressor replacement and resealing a system after repairing a leak.

The following sealed system service sequence is offered as a proven method for resealing a system or replacing a compressor.

Sealed System Service Sequence

1. Remove unit from cabinet as far as is necessary to gain access to all tube connections at compressor. Care must be taken to avoid sharp bends or kinks in tubing that may result in restriction.
2. Install processing valves on unit (Fig. 8-23). Install high-side access valve on line between compressor and condenser and low-side access valve on line between compressor and evaporator.

   Caution: When installing access valves on a system charged with refrigerant, locate saddle nuts a safe distance from existing brazed joints.
3. Connect gage manifold test set to access valves as shown in Fig. 8-24. (Close both manifold valves.) Gage readings should provide indication of basic problem. Refer to Fig. 8-25 for summary of gage reading meanings.
4. For restriction in system,
   (a) Locate restriction.
   (b) Proceed to step 6.
5. For leak in system,
   (a) Add small test charge to system as needed.
      (1) Connect supply cylinder with outlet valve up to charging line on manifold set (Fig. 8-24).
      (2) Set refrigerant supply cylinder with outlet valve *up* and open manifold valves to release some gas into unit.

FIGURE 8-23. Process valves may be either soldered to the line (as these are) or may be special saddle valves, which clamp on the line. In any case, make certain that there is no leak at the valve.

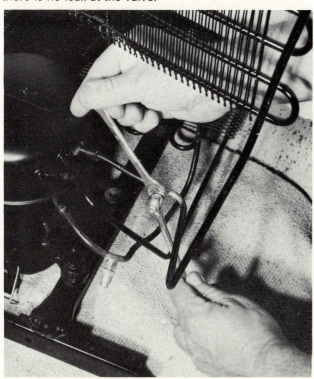

FIGURE 8-24. Connection of evacuating and charging system to refrigerator. With this hookup, all pressure tests, system purging, drawing vacuum, and recharging can be accomplished without disconnecting any part.

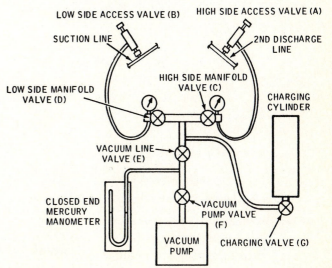

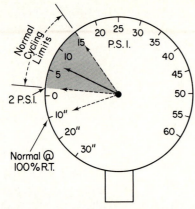

Normal Operation—A slight leak, slight restriction or slightly low
capacity compressor will read in the same range. However, the On time
for the compressor will have increased.

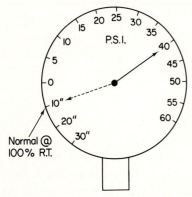

An inefficient or low capacity compressor will run constantly, never
cooling down the unit enough to turn it off. The gage readings will be
higher than normal.

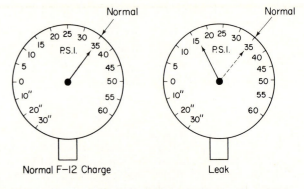

Normal F–12 Charge   Leak

The equalized pressure inside a system occurs when the unit has been
allowed to stand idle for an extended period of time bringing all parts of
the system to room temperature. A lower than normal equalized
pressure would indicate a loss of refrigerant, most likely due to a leak.

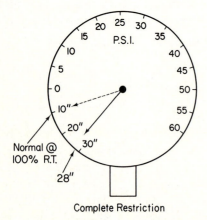

Complete Restriction

A restriction in the system will limit the flow of refrigerant to the
evaporator, the compressor may run a 100% of the time and the gage
will read lower than the normal 100% running time reading.

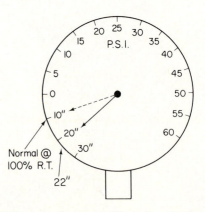

A leak will show as a shortage in refrigerant causing the compressor to
be On more than normal, possibly 100%, and the gage reading will be
lower than normal.

FIGURE 8-25. Gage readings.

(b) Locate leak.

(c) Close valves to supply cylinder.

(d) Proceed to step 6.

6. Discharge system.
   (a) Remove charging line from supply cylinder and connect to purging hose.
   (b) Open both manifold valves to discharge gas out of doors.

7. Cut drier out of system (do not use torch).

8. Repair restriction. Proceed to step 15.

9. Repair high-side leak. If there had been some pressure in the system and little chance of moisture having been drawn into the system, proceed to step number 15.

10. Repair low-side leak. Usually, a low-side leak allows an excessive amount of moisture to enter the system, thus contaminating the oil, which must be changed.

11. Change oil in the compressor.
    (a) Unbraze lines from compressor and remove compressor from mounting frame.
    (b) Pour oil out of high-side stub into container and accurately measure the amount removed.
    (c) Accurately measure an equal amount of clean, dry refrigeration oil and inject it into the compressor through the same stub.
    Note: A clean plastic squeeze bottle with nozzle (Fig. 8-26) will do an excellent job.

12. If a new compressor is being installed, check to make certain it contains enough oil.
    (a) If you are not certain, follow steps 11(b) and (c).
    (b) If the compressor is completely dry, inject the required amount of oil. If you do not know how much it needs, try 10 to 11 ounces.

13. Blow out condenser and precooler with refrigerant gas. If compressor motor was burned out and the oil looks a dirty brown, flush the con-

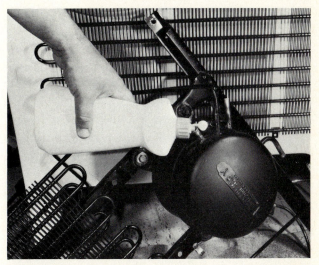

FIGURE 8-26. This plastic squeeze bottle originally held a detergent. Make certain that the bottle is completely dry and clean before filling it with oil. Also be sure to keep the cap closed tightly so that moisture and dust will be kept out of the oil.

denser and precooler with liquid refrigerant.

14. Reinstall compressor.
    (a) Mount on frame.
    (b) Rebraze lines.

15. Install service dryer.
    (a) Thoroughly clean outside of cap tube (capillary tube).
    (b) Score it with a file; then break off end to get clean, unrestricted hole.
    (c) Insert the cap tube into dryer to the screen; then withdraw $\frac{1}{4}$ inch to make certain it is not bottomed.
    (d) Crimp cap tube in dryer and silver solder dryer to system.

16. Connect unit to evacuation and charging system as shown in Fig. 8-24.
    (a) Remove mercury keeper from manometer tube.
    (b) Be sure charging valve is closed.
    (c) Open high- and low-side manifold valves, vacuum line valve, and vacuum pump valve.

17. Using heat gun, apply heat to compressor, evaporator, etc., to save time in evacuation. As an alternate method, heat motor windings by using starting cord to wire 250-watt

heat lamp in series with run winding as shown in Fig. 8-27.

18. Start vacuum pump and note manometer. Reading should increase steadily and should read 28 inches of mercury within 2 to 3 minutes. (If it doesn't, check for and correct leaks.)

19. Continue pump down until manometer stabilizes at 29.8 inches or more. Then close vacuum pump valve and stop the vacuum pump. (If 29.8-inch reading is not reached in 10 to 15 minutes, check for leaks. Also check vacuum pump for capacity.)

20. Read manometer.
    (a) If reading holds steady for 30 seconds, system is tight and adequately evacuated. Proceed to step 21.
    (b) If reading drops, there is a leak in the system or system is not sufficiently evacuated. Make sure all hose connections and valve seals are tight. Start vacuum pump and open vacuum pump valve. Apply heat to evaporator, condenser, and compressor to drive out moisture. When manometer stabilizes at 29.8 inches, close vacuum pump valve, stop vacuum pump, and again read manometer. If reading holds steady for 30 seconds, proceed to step 21.

21. Prepare system for charging.
    (a) Be sure that there is enough

FIGURE 8-27. Motor windings can be heated by screwing a 250-watt heat lamp into the fuse socket on the test cord, then connecting the leads to the compressor.

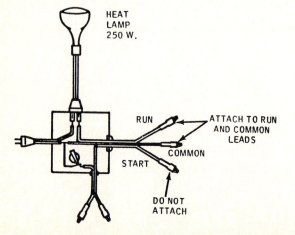

refrigerant in charging cylinder to charge the unit.
    (b) Set scale for proper refrigerant and gage pressure.
    (c) Close vacuum line valve and low-side manifold valve.
    (d) High- and low-side access valves and high-side manifold valve should be open.

22. Charge unit through high side by cracking charging valve intermittently while watching gage. Close valve when proper charge is indicated. If specific charge will not flow into unit through high side, it must be pumped in.
    (a) Close high-side access valve and charging valve.
    (b) Open low-side manifold valve and see that low-side access valve is open.
    (c) Start unit compressor.
    (d) Crack charging valve intermittently, allowing refrigerant to enter system in small spurts to avoid liquid refrigerant slugging compressor. Watch gage and close charging valve when proper charge is indicated (charging tolerance is $\pm\frac{1}{2}$ ounce).

23. Clear refrigerant from lines.
    (a) Let unit run for about 1 minute to draw refrigerant from lines; then close low-side access valve.
    (b) Stop unit compressor.

24. Remove lines and seal access valves immediately with sealing caps.

25. Clean off all flux with wet cloth and recheck system for leaks.

26. Reinstall unit in cabinet.

STUDY QUESTIONS

Refrigeration

1. Bacterial dormancy usually takes place at about _____ degrees.

2. Refrigeration does what to foods?

3. What is often an important by-product of refrigeration?

4. Define refrigeration.
5. Heat is the result of _____ motion in a _____ .
6. How is difference in heat measured?
7. What two systems of temperature calibration are in use today?
8. These scales are based on the freezing and boiling points of what compound?
9. Define Btu.
10. How does heat flow?
11. What increases the rate (speed) of heat flow?
12. Name three ways heat can travel.
13. Frying eggs is an example of which method of heat travel?
14. The thermostat in an auditorium is set at 60°F (15°C). When the room is full of people, no one complains of the cold, but some complain of the heat. What method of heat travel is exemplified in this instance?
15. If heat is removed from molten metal, what happens to the metal?
16. If heat is applied to butter, what happens to the butter?
17. When enough heat is applied to a compound (or element), what occurs?
18. Can the temperature of water ever reach 250°F? Justify your answer.
19. What is the "critical temperature" of a compound or element?
20. What is the affect of extra heat applied to the critical temperature of a substance?
21. How can one increase the molecular energy of a compound?
22. Temperature alone cannot cause a change of state—what other factor is involved?
23. When molecules hurry to move in a liquid, what is occurring?
24. In the coolest part of a refrigerator, which state is the refrigerant in—a gas or a liquid?

True or False

25. Refrigeration is the process of removing heat from water.
26. In the general concept of refrigeration, we could say that if chilling happens in one part of the cycle, boiling occurs in another part.
27. Molecules in motion cause heat.
28. An iceberg is colder than a 25-pound cake of ice.
29. A Btu is the amount of energy it takes to raise 1 degree of water to a pound.

30. If room A is 75°F and the next room (B) is 55°F, there will be less draft than as though room A were 60°.
31. In the preceding problem, the draft would flow from room A to room B in either case.
32. Touching a hot iron and withdrawing your hand fast is an example of heat convection.
33. Melting a solid produces a change that is called vaporizing.
34. Changing solid to liquid or liquid to solid is called the latent heat of fusion.
35. Latent heat is the type we measure with a thermometer.
36. A good example of the latent heat of sublimation is "dry ice."
37. When heat energy can be measured, it is referred to as sensible heat.
38. Two basic factors control gas behavior.
39. "The perfect gas law" is the basis for the theory and practical application of refrigeration.
40. A good refrigerant should be odorless and completely nonabsorbent.
41. Since a refrigerant is under mechanical control at all times, stability is not a requirement.
42. Current refrigerants, accepted by the American Society of Heating, Refrigeration, and Air Conditioning Engineers, are nonflammable.
43. An important factor to the refrigeration technician is the pressure–temperature relationship of the refrigerant.
44. An exchange of latent heat would indicate that a refrigerant is going through a change of state.
45. It is possible to analyze the condition of even a sealed refrigeration system by putting gages in certain places.
46. One should install the vacuum reading gage on the output side of the compressor.
47. Mechanical refrigeration can be described as constantly condensing, circulating, and evaporating a fixed supply of gas in a system that is closed.
48. The state of the gas in the system *outside* the appliance is liquid.
49. The operating range for all appliances using a refrigerant is the same.
50. The door gasket must be completely uniform.
51. Within a refrigerator, there is always a natural and constant temperature.
52. If the condenser is on the outside of an appliance, it is unlikely to be the cause of faulty operation.

53. In testing for leaks, the pressure gage should be connected to the suction line at the compressor.

54. If you find extra high-pressure and low-suction readings on test gages, you should suspect trouble in a capillary tube.

55. There are only two reasons for entering a sealed system to test and/or service it: compressor failure and incorrect refrigerant charge.

56. The first step in sealed system servicing is to install the processing valves on the unit.

57. It is advisable to cut the dryer out of the sealed system when servicing it, but without using a torch.

58. A low-side leak should be repaired after the high side has been checked.

59. Dryer reconnects should be crimped rather than soldered.

60. It helps to use a heat gun to the various parts of the system to help evacuation.

61. Manometer reading should stabilize at 29.8+ within 15 minutes and should hold for 30 seconds.

62. The final step in servicing a sealed system is reinstallation of unit within the cabinet.

KEY

Refrigeration

1. 50°F or 10.0°C
2. Preserves them
3. Ice
4. The process of removing heat from where it is not wanted
5. Molecular, substance
6. By a thermometer
7. Fahrenheit and Celsius
8. Water
9. British thermal unit = amount of heat needed to raise the temperature of one pound of water, one degree Fahrenheit
10. From warmer to cooler
11. Greater temperature differential between two objects
12. Conduction, convection, radiation
13. Conduction
14. Radiation
15. It becomes solid
16. It becomes liquid (melts)

17. A change of state
18. No—at normal pressure
19. That temperature at which a change of state takes place
20. It speeds up its change of state
21. Apply extra heat at its critical temperature
22. Pressure
23. Increased temperature result
24. A gas
25. False
26. True
27. True
28. False
29. False
30. False
31. True
32. False
33. False
34. True
35. False
36. True
37. True
38. False
39. True
40. True
41. False
42. True
43. True
44. True
45. True
46. False
47. True
48. True
49. False
50. False
51. False
52. False
53. False
54. True
55. False
56. False
57. True
58. True
59. False
60. True
61. True
62. True

9

Air Control

Air, as man's natural environment, occurs in many moods—hot, cold, wet, dry, clear, polluted, and almost any combination of the above.

Man can adjust to and survive in a wide range of air changes, but is actually healthy and comfortable in only a narrow range of temperature, humidity, and impurities. Figure 9-1 is a relative humidity chart on which a comfort zone has been indicated.

The comfort zone is generally considered to be between 74 and 80°F (23° and 27°C) and between 40 and 60% relative humidity.

POLLUTANTS

Impurities impose a different problem, for there are many different kinds.

1. Bacteria and other airborne microorganisms are a health menace.
2. Allergens such as pollen, dust, and molds are discomforting.
3. Odors, whether resulting from food preparation, evaporative chemicals, or unclean bodies, are unpleasant.
4. Flammable and toxic gases in sufficient quantities can be dangerous.
5. Radioactive air contaminants are active in very low concentrations.
6. Solid particles, such as soot and fly ash, in addition to irritating the eyes and nose, are also hard on clothes.

Total air conditioning to control all these pollutants would require fans to move the air and refrigeration to cool and dry it, as well as filters to remove such particles as pollen, dust, and soot. Bacteria are controlled by filters and ozone or germicidal lamps. Charcoal filters (activated carbon) are a big help in absorbing odors and gases. If possible, the odors should be diluted by ventilation with air that is cleaner.

Air conditioning would be much simpler if such pollutants were controlled at the source and not allowed to escape. Moreover, living in an area that was not air conditioned would be much healthier.

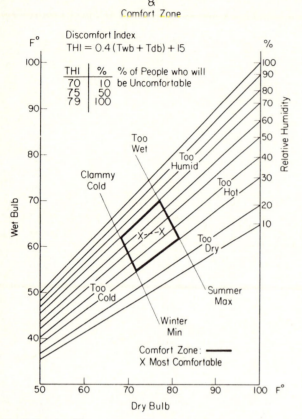

FIGURE 9-1.   Temperature humidity index = THI.

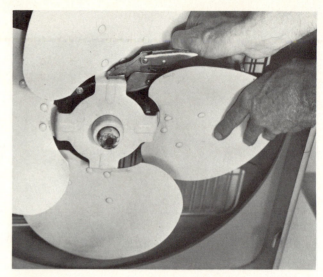

FIGURE 9-2.   Axial-flow fan. A method is shown for straightening a misaligned fan blade.

FIGURE 9-3.   Blower-type air mover. This model is designed for mounting, at its outlet, to a wall or duct.

MOVING AIR

The natural motion of air is from a high-pressure area to a lower-pressure area; and the greater the pressure difference, the faster the air will move. This is nature's way of moving air. Man uses this method, too. He creates a pressure difference either by heating the air or by blowing it.

As air is heated, the molecules expand, thereby making the heated air lighter than the surrounding air so that it rises. The cooler (heavier) surrounding air moves in to replace the heated air. The best example of this situation is the draft up a chimney that carries with it the products of combustion from a fire.

Moving air by blowing it is done by fans. There are two basic types of fans. The axial flow is the first. Here the fan blade is like a propeller, and in slicing through the air, the blade moves it openly in the direction the fan is pointing. Most cooling fans are of this kind (Fig. 9-2). The other type, a blower, is built like a squirrel cage and has the air entering at the

center and exhausting from the outside of the rotor (Fig. 9-3). It turns the air 90 degrees and is chiefly used in ducted systems because it can work effectively against the back pressure created by the duct work or chimney. Examples of blower usage would be the air supply for an oil burner or the fan that moves the air in a forced air central heating and air conditioning system.

The size of fan and motor will depend on how much air is to be moved. The air-recirculating fan in a refrigerator uses a small 4-inch-diameter fan blade driven by a $\frac{1}{150}$-hp motor. The air-circulating blower for a central heating system could have a 12-inch-diamter wheel driven by a $\frac{3}{4}$-hp motor to deliver 2000 cubic feet of air a minute.

The speed with which air moves around a person and the number of air changes per hour have a strong influence on a person's feeling of comfort. Velocities less than 15 feet per minute generally cause a feeling of air stagnation, whereas velocities greater than 65 feet per minute may result in a sensation of draft. Air velocities of 25 to 35 feet per minute were found to be most satisfactory.

The number of air changes per hour may be anywhere from $1\frac{1}{2}$ upward, depending on the number of people in the room, their body condition, and the amount of smoking.

Air within a room can be effectively moved with a ceiling fan (Fig. 9-4). This long-blade fan is an adaptation of the axial-flow fan. It usually has four blades 42 to 52 inches (105 to 130 centimeters) in diameter. The speed ranges from 50 to 200 rpm. The amount of air that can be moved depends on the blade speed, angle, and motor power.

FILTERING AIR

The purpose of an air filter is to remove the particles of dust, pollen, and smoke from the air. Two basic types of cleaners do so.

1. *Filter type:* air passes through an inter-

fering media, which traps the particles. This media may be wet or dry, stationary or moving, and may be made from any of several materials, such as fiberglass, sponge, mineral wool, or vegetable fiber. Some are discarded after use, and others can be vacuum cleaned or washed and reused.

When the pressure drop across a filter is between 0.5 and 1.0 inch of water, it is loaded and should be changed.

The charged media electronic filter type consists of a dielectric filter in contact with a wire gridwork consisting of alternately charged and grounded wires at about a 12,000-volt dc potential difference. Airborne particles are polarized and drawn to the media that collects them. This filter also has to be cleaned regularly according to manufacturer's instructions.

2. *Electronic ionizing type* (Fig. 9-5): a cleaner rather than a filter because there are no media to interrupt the airflow and slowly clog up with entrapped dust. In operation, the cleaner consists of a series of plates, charged to about 12,000 volts dc, which ionize the particles in the air. Then the air flows through a series of parallel plates, charged to about 6000 volts, which will attract the ionized particles. The plates are cleaned by washing at regular intervals. Naturally, with this high voltage, there are a number of safety features to protect the technician while he works on such a unit.

It is important to remember that the *location* of the *filter* in any air system must be at the entrance. This location makes it possible for the rest of the equipment to operate in clean air, thus prolonging the motor life of the fan, for instance.

FIGURE 9-4. Typical popular ceiling fan.

GERMICIDAL AND OZONE LAMPS

Germicidal lamps are used to kill airborne bacteria and mold spores. Ozone lamps are used

STANDARD MODEL

(1) Heavy-gauge cabinet   (2) Power Pack
(3) Ionizing-collecting cell
(4) Pre-filter   (5) Access door

OPTIONAL WATER WASH ACCESSORIES

(6) Water wash manifold with operating lever   (7) Drain fitting
(8) Water mist suppressor   (9) Water manifold swivel
(10) Water solenoid valve   (11) Mounting base

FIGURE 9-5. Typical electronic ionizing air cleaner, showing both the basic unit and the accessories for washing. (Courtesy of Whirlpool)

to remove undesirable odors. Some lamps do both. Figure 9-6 is a chart showing the relative amounts of energy released in both the germicidal and the ozone-producing regions by the G4S11 lamp shown in Fig. 9-7.

This lamp is popular in dryers, range hoods, window air filters, and washrooms. It must be located so that its direct output is shielded from the skin and eyes. Also, it should

FIGURE 9-6. Relative spectral energy distribution of the G4S11 lamp.

MICROWATTS PER SQ. CM. AT 1 METER

WAVELENGTH - NANOMETERS

184.9 OZONE - PRODUCING
253.7 GERMICIDAL - FLUOR.
296.7
312.6 - 313.2 ┐ SUN TAN
365.0 - 366.3 BLACK LIGHT - FLUOR.
435.8
VISIBLE LIGHT
546.1

MILLIWATTS TOTAL OUTPUT

FIGURE 9-7. This small lamp, shown with its ballast, uses the same-size socket as that used for outdoor Christmas tree lights.

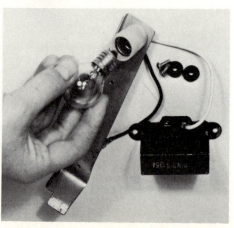

Lamp Ordering
Code: G4S11
Watts: 4
Bulb: S-11, Clear
Overall Length: 2¼''
Base: Intermediate
Approx. Life: 6000 hr

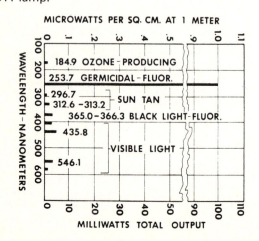

Single or multiple lamp operation with ballasts, 110-125-volt ac circuits

1, 2, 3 or 4 G4S11 lamps

89G504

6 or 7 G4S11 lamps

89G381

Single or multiple lamp operation with an incandescent lamp, 110-115-volt ac or dc circuits

1 G4S11 lamp

40-w
120-v
lamp

2, 3 or 4 G4S11 lamps

50-w
120-v
lamp

Multiple lamp operation with resistor, 110-125-volt ac or dc circuits. Not more than 8 lamps.

G4S11 lamps

Resistor

$$R = \frac{SUPPLY \ VOLTS - (10.5 \times NO. \ OF \ LAMPS)}{0.35}$$

Watts loss in R = 0.1225 x R

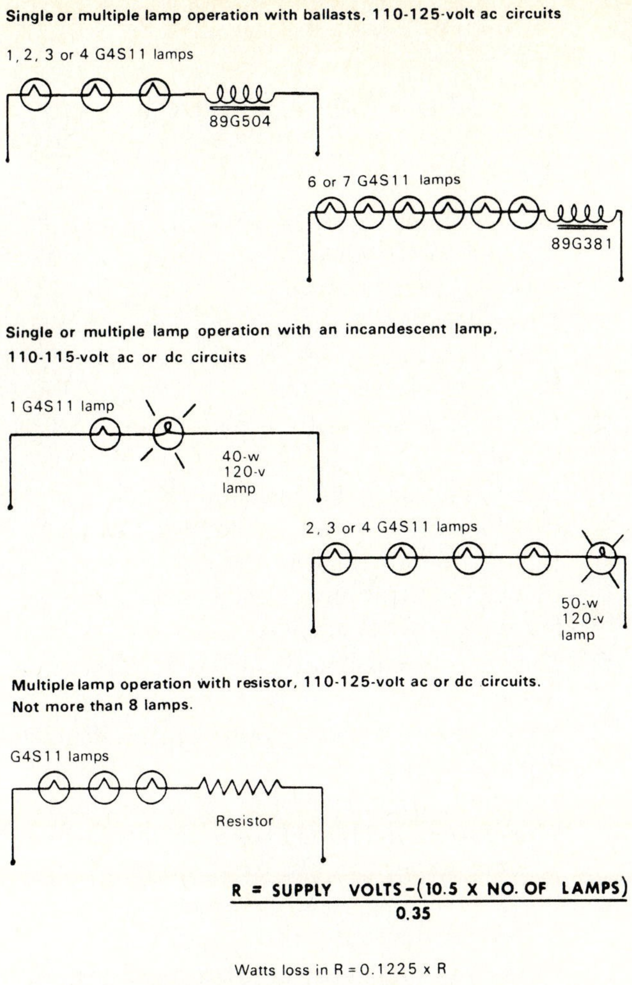

FIGURE 9-8. Wiring diagrams.

be well ventilated to allow the ozone to mix into the air effectively.

Electrically, the lamp operates at 10.5 volts, drawing 350 milliamperes ac or dc. Because this lamp operates as an arc source, it must have a higher voltage to strike the arc. The minimum supply voltage is 28 volts, and a ballast, either inductive on ac or resistive for ac or dc, must be used to limit the current flow. Where the additional light is not a factor, a 40-watt, 120-volt incandescent lamp is commonly used. Figure 9-8 illustrates a number of alternate wiring diagrams for single- or multiple-lamp usage.

Germicidal lamps used to irradiate the air are effective in microorganism kill. This effectiveness is in direct relation to intensity of exposure times the length of time. It is similar to getting a tan on the beach—you burn faster at noon than late in the afternoon. The ultraviolet energy at a 2537-Å wavelength will react to skin and eyes in the same manner as direct exposure to strong sunlight, so precaution must be taken to protect eyes and skin from excessive radiation. Usually, the light fixtures are designed to shine the light to the ceiling or high on a wall, thus allowing reflected rays to bounce around the room (see Fig. 9-9).

FIGURE 9-9. An important thing to remember in using germicidal or any other ultraviolet light source is to protect the eyes from the direct radiation. Normally, ordinary glasses will do the job, but don't push your luck. (a) In this application of germicidal lamps for food preservation, please note that the lamps may be mounted horizontally or vertically from the ceiling. Also, there is a reflector to help concentrate the rays where they will do the most good. (b) Typical uses of germicidal tubes in pharmaceutical laboratories.

Sanitary storage of meat.

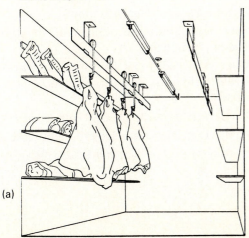

(a)

(b)

Typical uses of germicidal tubes in pharmaceutical laboratories.

Germicidal lamps are effectively used in heating and air-conditioning ducts to free the air of bacteria and other undesirable pathogens. Normally, these lamps are mounted in the duct *after* the filter and *before* any humidification device. There should be an access door or panel to allow periodic cleaning of the lamps.

TEMPERATURE AND HUMIDITY

In addition to the pollutant control discussed previously, temperature and humidity control are important to man's well-being. Details on the typical appliances that are used to make temperature and humidity changes are covered in chapters relating to them specifically.

SYSTEMS MAINTENANCE

Since air contains so many impurities that must be filtered out, filters and all other parts of an air-moving system will accumulate a coating of dust that *must* be cleaned periodically.

1. Motor and blower bearings should be lubricated.
2. Lamps should be changed. An old lamp that shows a heavy deposit inside the glass will not generate ozone or germicidal frequencies effectively.

A good preventative maintenance program calls for changing lamps annually, and, where fluorescent lamps are used, change the starter at the same time.

Figure 9-10 indicates that the general length of life for germicidal lamps is from 6000 to 7500 hours. A year is 8760 hours. Hence the annual change seems practical. If a lamp is used only a fraction of the time, it need not be changed so frequently. Shelf life is not a factor in its efficiency.

STUDY QUESTIONS

Air

1. Man's comfort zone is between 70 and 80°F (21 and 27°C) but humidity makes no difference.   True or False
2. If temperature is correct and impurities are well filtered out, man should be satisfied with the air of his environment.   True or False
3. Odors are not a health menace in air.   True or False
4. Toxic gases are a factor mainly because of the discomfort they cause.   True or False
5. Dust in the air could be dangerous as well as irritating, especially if flammable gases are present.   True or False
6. There are six pollutant factors to remember. Name them.
7. Name three sources from which offensive scents can come.
8. Two main methods used to condition air, once temperature and humidity are established, can be listed under two F's. What are they?
9. Of the preceding answers, which must always be nearest the source end of ductwork?
10. What two methods are used to control bacterial factors in air?
11. What is a good filter for odor control?
12. Where is the best place for complete air conditioning?
13. How is air changed?
14. Name two methods by which a difference in air pressure can be created.
15. What happens when air is heated?
16. Would a home refrigerator need an 8-inch fan?
17. Could a central-heating system use a 4-inch fan effectively?
18. What causes one to feel a draft?
19. Explain 25 to 35 feet per minute.
20. What determines the number of air changes needed per hour?
21. When you read about an interfering medium of vegetable or glass fiber, what is being mentioned?
22. Can all air filters be cleaned and reused?
23. What two methods are used to clean air filters?
24. What symptom indicates a loaded filter?

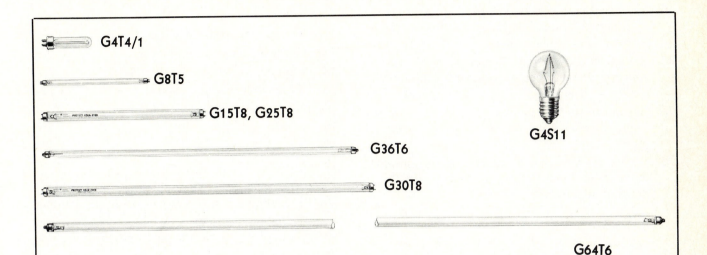

GENERAL ELECTRIC GERMICIDAL LAMPS

Lamp ordering abbreviation	G4S11	G4T4/1	G8T5	G15T8	G25T8	G30T8	G36T6	G64T6
Nominal watts	4	4	8	15	25	30	36	65
Nominal length ①	2¼''	5¾''	12''	18''	18''	36''	36''	64''
Tube diameter	1⅜''	½''②	⅝''	1''	1''	1''	¾''	¾''
Approx. lamp amps.	.350	.080	.160	.310	.600	.360	.425	.425
Approx. hours life	6000⑩	5000③	7500④	7500④	7500④	7500④	7500⑤	7500⑤
Bulb type	S-11	T-4	T-5	T-8	T-8	T-8	T-6	T-6
Effective lgth. of U-V source	. . . .	6''	8½''	14''	14''	32''	29¼''	58''
U-V output (watts 2537A) at 100 hours	.1	.5	1.3	3.6	5.0	8.3	13.1⑥	18.0⑥
Avg. U-V output (watts) through life	.08	.38	.98	2.8	4.0	6.6	10.4⑥	14.4⑥
Max. intensity perpendicular to bare tube⑦								
Watts/sq. ft. at 10 ft. ⑧	.0001	.0007	.0014	.0038	.0054	.0085	.0135⑥	.0185⑥
Watts/sq. ft. at 3 ft.	.001	.008	.015	.042	.060	.0945	.15⑥	.19⑥
Watts/sq. ft. at 1 ft.	.011	.07	.14	.38	.54	.5	.79⑥	.57⑥
Watts/sq. ft. at 8 in.	.024	.16	.315	.7	1.00	.77	1.22⑥	.86⑥
Watts/sq. ft. at 4 in.	.094	.63	.86	1.45	2.07	1.5	2.42⑥	1.72⑥
Watts/sq. ft. at 2 in.	.370	1.7	1.75	2.95	4.20	3	4.85⑥	3.40⑥
Base	Intermediate	Radio Contact	Min. Bipin	Med. Bipin	Med. Bipin	Med. Bipin	Single Pin	Single Pin
Ballast	89G504	⑬	⑭	⑮	6G1042	⑯	⑰	⑱
Starter	None	FS-5	FS-5	FS-2	FS-2	FS-4	None	None

① Nominal length includes the tube and standard lampholders.
② Bent tube construction makes tube approximately one inch in width.
③ Life under specified test conditions with tubes turned off and restarted no oftener than once every 3 burning hours.
④ Life under specified test conditions with tubes turned off and restarted no oftener than once every 8 burning hours.
⑤ This tube may be operated at 425 ma. in air duct or other applications where ambient conditions prevent excessive bulb temperature. Life will be slightly reduced.
⑥ In still air at 80°F. With optimum cooling the tube at this current will give about 6% more ultraviolet output.
⑦ Average at 100 hours life; initial ratings about 20% higher; for average intensity throughout life multiply G30T8, G15T8 and G36T6 values by 0.79 and G4T4/1, and G8T5 values by 0.75.
⑧ Multiply by 10,000 for microwatts per cm² at one meter.

⑨ Max. over-all length. Does not include Lampholders No. 95X954 (nickel-plated parts).
⑩ Life under specified test conditions, when operated continuously.
⑪ Life under specified test conditions, with 5 or more burning hours per start.
⑫ Bulb wall temperature may not exceed 790°C; seal temperature may not exceed 300°C.
⑬ 58G827–60 cy., or 89G435 which operates lamp near 140 amperes, increasing U-V output by 70%.
⑭ Same as for 8-W fluorescent lamp.
⑮ Same as for 15-W fluorescent lamp.
⑯ Same as for 30-W fluorescent lamp.
⑰ Same as for 40-W Instant-Start fluorescent lamp.
⑱ Same as for 72T12 Slimline fluorescent lamp.

FIGURE 9-10. Germicidal lamps. (Courtesy of General Electrical)

25. One type of filter requires more safety precautions. Which type is it?
26. One particular caution is given in the use of germicidal lamps. What type of energy does it deliver that needs such care?
27. If a germicidal lamp is used in the duct of a central system in conjunction with a humidifier and/or a filter, where should it be placed?
28. Is it true that as long as air-purifying lamps light, they do the job they were designed to do?
29. If a housewife uses her dryer, which contains a germicidal lamp, about 6 hours a week, how long should the lamp last?

KEY

Air

1. False
2. False
3. True
4. False
5. True
6. Bacteria, odors, allergens, gases, radioactivity, solids
7. Cooking, evaporating chem. unclean bodies
8. Fans and filters
9. Filter
10. Filters and germicidal lamps
11. Charcoal or activated carbon
12. The source of fresh air intake
13. By moving it (ventilation)
14. Heating and blowing
15. It rises
16. No
17. No
18. Air velocity greater than 65 feet per minute
19. Satisfactory air velocity of 25 to 35 feet per minute.
20. Number of people in room and their activity
21. Air filter
22. No
23. Washing and vacuuming
24. When it offers too much resistance
25. Electronic ionizing type
26. Ultraviolet
27. After a filter, before a humidifier
28. No
29. About 20 years

10

Water

Water has been called the universal solvent. In its travel it picks up many impurities that affect its usefulness to man. Some of these impurities are natural minerals gathered by the water as it flows through the ground. Others are man-made contaminants absorbed by the water in its travels.

MAN-MADE CONTAMINANTS

As water falls from the clouds through a polluted atmosphere, it picks up many airborne chemicals. These, in turn, react with the water to form acids. The acids make water more chemically active in absorbing minerals from the earth.

Man doubles his error when he dumps his wastes into pits in the ground, swamps, and streams. In all such areas water reacts to the materials, thereby adding to its impurities. The result causes serious consequences to the normal balance of nature.

For example, modern man's very cleanli-

ness has backfired in his use of detergents. Effective detergents contain two principal ingredients, a *surfactant* and *a phosphate*.

A surfactant (surface active—or wetting—agent) lowers the surface tension of water to improve its penetration into the fabric being washed. The surfactant in general use *was* ABS (alkyl benzene sulfonate), which is slow to break down, resulting in excessive foaming in sewage treatment plants and in the lakes and streams into which the effluent was discharged. The surfactant being used currently is LAS (linear alkylate sulfonate), which is readily biodegradable. Biodegradation is accomplished by the useful bacteria (commonly found in sewage, surface water and soil), which break down and utilize the surfactant as food, ultimately converting it to water and carbon dioxide. This change to LAS is bringing one form of pollution under control.

The phosphate component in a laundry detergent does several important things:

It increases the efficiency of biodegradable LAS.

It keeps the soil particles in suspension in the wash water.

It furnishes the necessary alkalinity level needed for efficient cleaning.

It emulsifies oily and greasy soils.

It aids in softening water by uniting with the hardness minerals.

It contributes to the reduction of germ levels in clothes.

Phosphates are effective in detergents, but do not decompose in normal septic tanks or sewage treatment activity and, therefore, appear in sewage effluent to lakes and streams. Phosphorus, when combined with other nutrients that are found in lakes and streams, provides excellent nutrition for plant life. In many instances too good since an abnormal growth of algae can clog streams causing the fish life to die off.

We cannot condemn the detergent manufacturers because they are supplying a needed commodity and are spending millions of dollars in research to improve their detergents. Their goal is to make cleaning agents readily decomposable without losing their cleansing powers.

Only man can control his waste materials and treat them before disposal to ensure reasonably uncontaminated air and water for his own use. Water's natural affinity for certain minerals has made it necessary to use conditioning equipment designed to remove them. The two major problems—hardness and iron—are discussed in detail.

HARD WATER

When home economists refer to water as being "hard," they mean that it contains certain minerals that make it less efficient for almost every household use.

Hardness is that characteristic or quality in water that results in formation of an insoluble sticky curd when used with soap. It is caused by the presence of calcium and magnesium, carbonates, and sulfates.

How Water Gets Hard

The constant movement of water, from the clouds to earth and back again, is called the hydrologic or water cycle (Figs. 10-1 and 10-2). Water is in constant motion, being drawn from the earth by evaporation and stored in the clouds until released as snow, hail, or rain.

Except for distilled water, this is about the only time water is in its pure state, for during precipitation it absorbs some of the various gases, dust, smoke, bacteria, and other foreign particles from the air.

Thus when water reaches the earth, it is contaminated, to some extent, and, in its contact with the earth, becomes more so, to an extent dependent on its path of travel.

Ground water, such as that obtained from a well, derives its character from the soils through which it percolates. Water, as it passes through the air, picks up some carbon dioxide and other gases, which increase its solvent action on the soluble materials in soil, rock, and mineral formations.

The quality and character of water are determined by the amount of dissolved minerals and organic matter contained in it. Among the common ingredients found in water are such dissolved or suspended matter as carbonates, chlorides, and sulfates of calcium, magnesium and sodium, and copper, iron, and manganese. Also, organic matter of vegetable or animal origin and gases may be dissolved in the water. Suspended solids, such as iron, silt, refuse, organic and animal matter, may be present.

Limited amounts of dissolved minerals are not harmful to the human system. However, these impurities affect the taste, color, odor, and general usefulness of water, often to such an extent that it must be treated to remove the harmful factors.

The more common complaints about water are hardness, turbidity, red water, odor, and taste—individually and in combinations.

Just how hard water is depends on the amount of calcium and magnesium found in the earth through which the water flows. The degree of water hardness varies in different parts of the country (Fig. 10-3). It may even vary considerably within a rather small area.

Effect of Hard Water

These minerals seriously hamper the efficiency of soaps when combined with them in water.

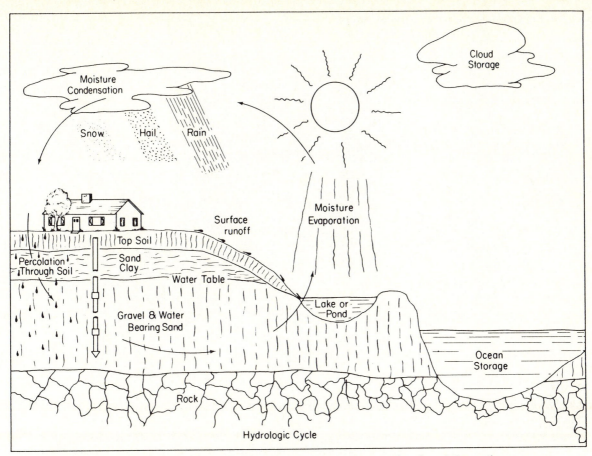

FIGURE 10-1. About the only time that water is naturally pure is when it is on the way up to the clouds. The sun is the big pump.

FIGURE 10-2. Grandma's hard water came from the same source as the modern house-wife's—minerals picked up by water as it runs underground.

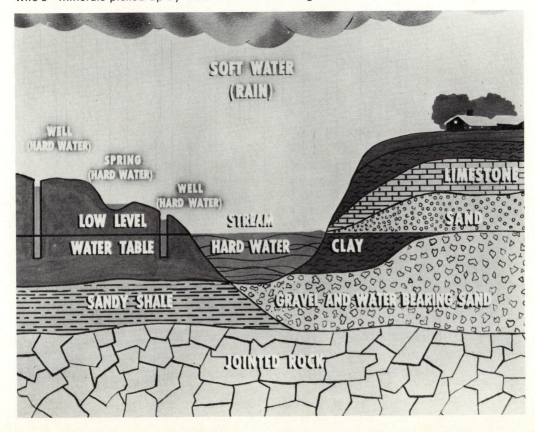

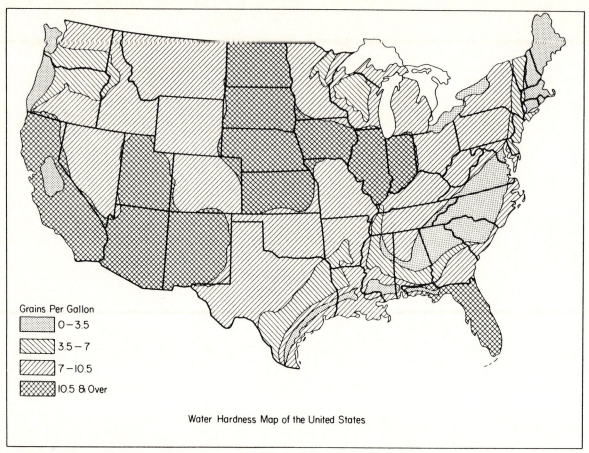

Grains Per Gallon
0-3.5
3.5-7
7-10.5
10.5 & Over

Water Hardness Map of the United States

FIGURE 10-3. Map showing the relative hardness distribution of water in the United States. The hardness divisions are classified as follows:

Soft water: 0–0.5 grain per gallon
Slightly hard water: 0.5–3.5 grains per gallon
Moderately hard water: 3.5–7.0 grains per gallon
Hard water: 7.0–10.5 grains per gallon
Very hard water: 10.5 and up grains per gallon

Lake Michigan water, for example, averages about 10 grains (hard).

The result is a grimy scum that keeps clothes and other washables gray and dingy, prevents them from ever being as soft and white as they should be. Plumbing systems become lined with scale, impairing the passage of water and causing costly repairs (Fig. 10-4).

Hard water makes living difficult in so many ways that it is almost impossible to enumerate them all. Shampoos do not leave hair clean, soft, and lustrous as they should. Skin becomes coarse and rough as a result of soap curd lodging in pores instead of washing clean.

In the kitchen, too, hard water means trouble. Dishes and glassware are streaked and dull after washing. This condition is caused by soap, which does not dissolve completely and clings to china and glass. Often, even hard rubbing will not remove it.

Foods cooked in hard water are apt to be shriveled and tasteless. In countless ways hard water makes running a household much more difficult than it should be.

Water hardness has always been a problem to man, and since the advent of modern plumbing and the increase in water usage for bathing and domestic purposes, it has become more so. It is evidenced by the "liming up" of pipes, increased soap consumption, and formation of the insoluble curd that causes bathtub rings, tattletale gray clothes, or "dishpan hands."

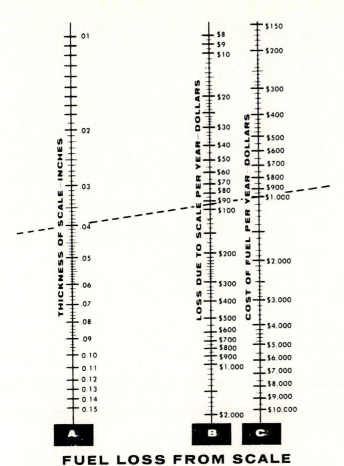

FUEL LOSS FROM SCALE

FIGURE 10-4. Fuel loss from scale. To ascertain the loss caused by scale in the boiler, find the thickness of the scale and put a dot on column A of the chart representing this thickness. Put a dot on column C to indicate your fuel cost per year. Join the two dots with a straight line. The point at which the straight line touches column B indicates your fuel loss per year.

Soaps

We have mentioned that calcium and magnesium salts have harmful effects in washing, whereas sodium salts do not.

All three can combine with vegetable oil acids to make soap. Sodium soaps, which constitute many of the popular brands on our grocery store shelves, are natural agents for washing. However, the calcium and magnesium salts found in hard water, together with vegetable oils in our household soaps, produce sticky, non-cleaning curds.

When we wash with ordinary soap in hard water, an exchange takes place, and the result is a mixture of "hard soaps" and soft water. Thus we are forced to waste the initial portion of the soap used to form a sticky, insoluble precipitate, with no lather being built up until all the hardness minerals are consumed. This forms the curds, or "scum," one often sees. Washing in this mixture is frustrating because the curds will not rinse off!

To make matters worse, rinsing must be done with more, fresh, *hard* water. The hard-water minerals combine with any free lather left in the clothes to form additional curd. This curd is one of the major causes of tattletale gray. With soft water there is no adverse mineral reaction; the rinsing merely floats the lather and dirt down the drain.

The inefficient use of soaps in hard water is also costly. A tabulation of the water hardness and the per capita consumption of soap showed that the soap cost doubled with a change in water hardness from 3 grains to 32 grains of hardness. Added to these dollar costs are the other unpleasant costs and side effects, such as skin irritation, stiff clothing, poor-tasting coffee, and plumbing damage. The uses of extra lotions, creams, and oils needed for complexion care to offset the drying effect of the film left on the skin with hard-water usage are also an added cost.

Hardness Testing

Water hardness is measured in grains per gallon. A grain is a measure of weight of scale forming solids in the water. Many people who have water softeners reach the point where they can tell by the feel of the water whether it is soft or hard. The rest of us have to test it. A number of different kits are available that will test water for hardness, iron, pH (acidity), and/or other impurities (Fig. 10-5). Our major concern, however, is with hardness, and it is comparatively easy to run your own check as to degree of hardness with a "green soap test."

One way of doing so is to measure an ounce of water into a glass jar and add tincture of green soap, shaking the water after each drop is added. Do this until suds are formed, keeping track of the number of drops you add.

Zero soft water requires only 1 drop of green soap per ounce of water to form suds. Hard water requires 1 drop per ounce of water

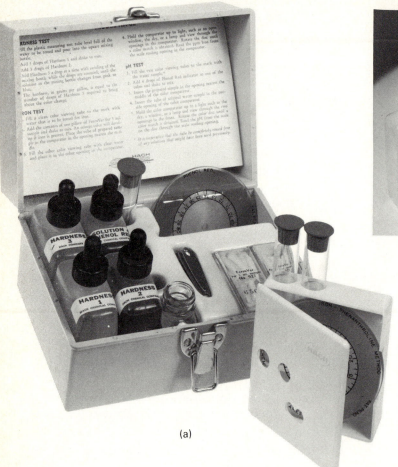

(a)

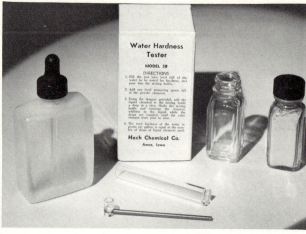

(b)

FIGURE 10-5. (a) Small, inexpensive kit that tests water for hardness only. It is good for checking water softener operations. (b) A more versatile kit that tests for hardness, iron, and pH is used by those who do a great deal of water softener service. (Courtesy of Hach Chemical Co.)

for each grain of hardness before it produces suds. If you have water that requires 15 drops per ounce, you would have 15-grain hard water.

In actual use, you will note that if water gets hard due to water-softener failure, normal lathering will not occur when washing hands and face. This is the first clue to a softener malfunction.

Water Needs

If a household runs out of soft water regularly, it would be best to check the water needs against the softener's capacity. It is possible that the family has grown and is now using more water than when the softener was installed.

The average family uses 50 gallons of water per day, per person. To find the number of grain hardness to be removed from water per day, you should multiply the number of people in the family times 50 gallons times the water hardness.

For example, a family of five with 15-grain hard water would require: $5 \times 50 \times 15 = 3750 -$ grain hardness removed per day. Dividing this figure into the softener's capacity will tell you how many days it can go between regenerations. Continuing the example: if a softener has a 15,000-grain capacity, then $15,000 \div 3750 = 4$ days between regenerations.

With a fully automatic softener, using a 7-day clock, one regeneration could be set for Sunday night and the other for Wednesday night.

Ion Exchange

The method of controlling this water hardness for domestic use is with an ion exchange water softener.

Basically, this unit consists of a tank containing resin beads liberally loaded with sodium ions. It so happens that these resins prefer calcium and magnesium ions. Therefore, as water flows through this tank of resin beads, the beads give up their sodium ions and take on the calcium and magnesium ions from the mineral in the water.

The resulting minerals do not have the harmful reaction with soap that the calcium and magnesium salts have, nor do they leave a deposit in the pipes.

Eventually the resin beads will have given up all their sodium ions and will require regeneration. This step is accomplished by flowing a brine solution through the resin. By mass action, due to the concentration of sodium ions in the brine, the sodium ions force the magnesium and calcium ions off the beads and take their place. The resultant calcium and magnesium chlorides go down the drain. The resin is then rinsed with fresh water, and the unit is ready to supply more soft water.

During both the softening and the regenerating cycles, there is no wear or chemical loss of the resin. The only requirement is that the user add salt periodically.

Only the purest type of rock or pellet salt should be used with any softener. The reason is that all impurities settle in the salt tank bottom and must be cleaned out once in a while, possibly annually, depending on the amount of salt used and impurities in it.

SOFT WATER

There are a number of questions about water softeners and soft-water usage. To help clarify some of them, we include the following miscellaneous information, which should help your understanding of water problems.

1. It will require several washings with soft water to clean out completely the embedded tattletale gray curd that has accumulated in clothing from past washings in hard water.
2. The first several hot-water draw-offs after a softener is installed will be hard. Then the inflowing soft water will dilute the hard water in the tank until all water is soft.
3. Rinse water from a softener will not harm a septic tank.

4. It is not advisable to water lawns and other outdoor plants with soft water because (a) of cost and (b) over a long period the excess sodium ions will unbalance the soil composition.
5. House plants can be watered with soft water if care is taken to completely saturate and flush the water through soil each time and not water too frequently. Some types of house plants are not affected; others are to some extent.
6. In a well-run, balanced aquarium in which water is added only to replenish loss by evaporation, soft water can be added slowly as required and the fish will not be bothered, for the dilution ratio is so small.
7. In the first washing with softened water, decrease the amount of soap or detergent which is usually used to one-third or one-quarter of that previously used. On subsequent washings, add or decrease as required to obtain sufficient suds.
8. The rust and scale accumulated in your plumbing will be loosened by the soft water. Be prepared for occasional "slugs" of rusty water until the entire system has been cleared.
9. Car batteries should not be given soft water, only distilled water.
10. Other filters and water-conditioning equipment, as needed, must be installed in the line between the pressure tank and the softener.
11. Some people do not like the taste of softened water and others, under doctor's orders, on a low-sodium diet, must not drink softened water. For these reasons, many installations include one tap from the unsoftened source.

IRON IN WATER

"Red water" is that condition in water that produces red stains on plumbing fixtures and on clothes when they are washed. It can also alter the taste and appearance of tea, coffee, and other drinks.

The color is due to the presence of iron hydroxide in the water. In some instances, the water will appear clear when drawn but will redden on exposure to air as the soluble iron changes to an insoluble iron. This is the form that does the damage.

Two methods are available to control iron in water for domestic use. The choice of method used depends on the amount of iron to be controlled. An ion-exchange water softener can handle small amounts of iron effectively, but large amounts—over 3 to 5 ppm (parts per million)—must be removed by special iron filters.

Because of the complexities in an iron-water relationship, no general recommendations can be given. Each installation must be designed for the local iron conditions. However, any iron filter must be installed between the water source and the water softener (Fig. 10-6). This point is mandatory. Locating the filter in such a position protects the softener's resin from becoming iron-fouled. Resin that is iron-fouled will no longer soften water!

Several actions can be taken to help a softener handle small quantities of iron and to protect the resin from becoming iron-fouled. One important step is to prevent the water from coming in contact with air (oxygen) before it enters the softener. This is done by using an air cell, such as an inflated bag or balloon in the pressure tank.

Another helpful method is to introduce some *sodium hydrosulfite* into the brine periodically. Its reducing action will clear the ferric hydroxide from the beads (resin), thereby allowing them to function in a normal manner. A badly fouled resin bed may require several successive treatments to clear it. Once cleared, it should perform well with only periodic, preventative treatments. Generally the amount of sodium hydrosulfite needed is 2 to 4 ounces per cubic foot of resin. Thus a 15,000-grain softener (having ½ cubic foot of resin) would require 1 to 2 ounces. This compound should be poured into the brine tank every time a 100-pound bag of salt is added.

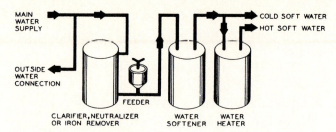

FIGURE 10-6. Proper relationship of a sequence of water treatment devices.

Iron-reducing chemicals are available under a number of different brand names: Lykopon, Fer-rid, Iron out, and Rover are some of the more common ones. These products are usually found at the stores that handle water-softener salt.

OTHER UNDESIRABLE QUALITIES IN WATER

There are other impurities in natural water. Some can be noticed by color, some by taste, and a few by odor. These impurities are evident to a lesser degree, and where they exist, there is usually some qualified personnel to help with questions and corrective measures. The superintendent of your local water works would be the most logical authority to contact.

STUDY QUESTIONS

Water

1. Water is pure until it touches the earth. True or False
2. Colored or smelly water is not necessarily harmful. True or False
3. Water with an unpleasant taste is most likely to be harmful. True or False
4. The greatest water problem the technician meets is that of _____.
5. Hard water may clog pipes and make soap scum, but it improves cooking. True or False
6. The average family uses 50 gallons of water per day. True or False

7. If you have a water softener, the first clue that it may be inoperative is (a) your skin feels dry, (b) your coffee tastes different, (c) there are curds and a scum ring around the sink, (d) you cannot get good suds when washing.

8. Regenerationing cycles of a water softener do not wear out the appliance.   True or False

9. The softening process does cause some chemical loss within the resin bed of the softener. True or False

10. Soft water is excellent for car batteries. True or False

11. If an iron filter is also used by the household, it should be connected between the water source and the softener.   True or False

KEY

Water

1. False
2. True
3. False
4. Hard water
5. False
6. False
7. d
8. True
9. False
10. False
11. True

11
Servicing Techniques

SOLDERING

The use of solder is based on a few simple principles, which, once understood, will ensure consistently successful results.

1. The ordinary soft solder used in joining electrical wiring and parts, sweating pipes, or other metal to metal is an alloy of tin and lead having a melting point ranging from 361 to about 600°F (210 to 315°C) depending on the tin–lead ratio (Fig. 11-1).

2. The success of soft solders in joining metals is due to a solvent action in which the molten solder dissolves a little of the copper or steel that is being joined. This low-temperature solvent action creates a new alloy at the boundary of the solder and the other metals with no damage to the other metals.

3. It is essential that the parts being soldered be absolutely clean or the solder will not adhere. Any nonmetallic impurity in the metal or on its surface—paint, grease, corrosion, oxides, to name some—will prevent the solder from interacting with the metal.

4. In order for the solder to interact properly with the metal, the part being soldered must be hot enough to allow the solder to remain liquid while the action goes on. But it must not be too hot, nor heated for too long a period of time, or else the oxides formed during the heating process will insulate the solder from the metal.

5. A soldering flux must be used during soldering to remove oxides and prevent more from forming while the metal is being heated. Three types of fluxes are in general use.
   (a) "Acid" or chloride type is very active and corrosive. It should be cleaned by washing with hot water or by steaming. It *cannot* be neutralized. This type of flux must never be used for radio or electrical work.

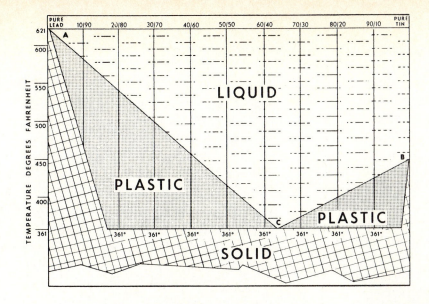

FIGURE 11-1. The tin–lead percentage in a solder will determine its melting point. The eutectic composition, 63 percent tin and 37 percent lead, has a definite melting point at 361°F with no plastic range. The $\frac{60}{40}$ alloy that is commonly used for electronic work has a very narrow "soft" or plastic range and requires only a few degrees to melt completely.

(b) Organic fluxes are not as active or as corrosive as the chloride types, and generally these fluxes char, burn, or decompose during the soldering operation. This fact makes cleanup easier.

(c) Rosin flux is the least active of the fluxes, but it is completely noncorrosive, nonhygroscopic, and electrically nonconducting. These properties make it most useful as an electrical and electronic-soldering flux.

Thus we have the three basic needs for effective soldering. The parts to be soldered must be

1. Clean
2. Hot enough to melt the solder
3. Properly fluxed

Finally, *do not disturb while it is hardening!*

The tin–lead solders most commonly used in electrical work are of the rosin-cored wire type. Tin–lead ratios of 50:50 or 60:40 in $\frac{1}{16}$ in. (0.062) diameter are in general use. The 60:40 $\frac{1}{32}$ in. (0.031) diameter is used for delicate work on printed circuit boards. Electronic work requires use of a suitable electric soldering iron since a flame could seriously damage the surrounding parts (Fig. 11-2).

A torch is generally used for soldering copper tubing or large-diameter copper wire where the surrounding area is clear of combustible material (Fig. 11-3). Another method of apply-

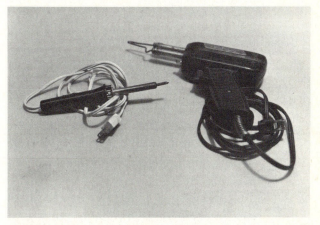

FIGURE 11-2. The small soldering iron (left) is generally used for small work, such as soldering to printed circuit boards. The solder gun (right) is used for general electrical work.

FIGURE 11-3. The propane torch kit is a handy service technician's tool that can be used for both "soft" and silver soldering.

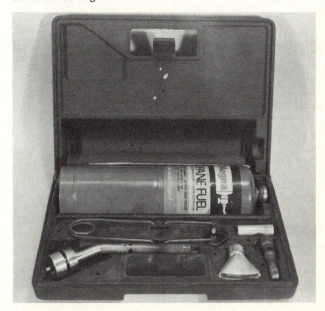

ing heat to tubing is by the electrical resistance method. This consists of clamping a pair of carbon electrodes, attached to a low-voltage transformer, around the tubing (Fig. 11-4). When turned on, the heat generated in the tubing by the resistance of the carbon electrodes will melt the solder. Solder will flow toward the heat. Thus when applying solder to a part, start at some point away from the heat source. If the work is hot enough, the solder will melt and flow the rest of the way. These methods of soldering often use a paste flux applied to the surface being soldered and a solid wire solder, commonly $\frac{1}{8}$ inch in diameter.

Soldering Procedure

1. When using a soldering iron the hot tip of the iron must be clean and well tinned with solder. Cleaning is done by wiping the tip on a damp cloth or sponge immediately before soldering.
2. Tinning is done by applying solder to a clean tip.
3. The parts being soldered must be clean and in secure mechanical contact with each other.
4. Place the tip of the soldering iron against *both* parts to be soldered (Fig. 11-5a).

5. Heat the two parts for a few seconds.
6. Apply the rosin-core solder to the side of the parts opposite the iron. The solder should be melted by the heat of the parts and will flow over them toward the iron (Fig. 11-5b).
7. Remove the solder, then the iron.
8. **Caution:** *Do not overheat or oversolder the parts being soldered!* Overheating can cause the foil on printed circuit boards to peel loose from the board. It can also damage transistors.
9. Examine the connection to be certain the solder has flowed smoothly over both parts and *has not* spread to adjacent parts to create a shorted connection. Clean up any excess flux or solder.

SILVER SOLDERING AND BRAZING

As with soft solder, the parts to be silver soldered or brazed must be clean, fitted to each other, and properly fluxed. Then they must be

FIGURE 11-5. (a) Note the mechanical contact of parts and contact of soldering iron to *both* parts. (b) Note the position of the solder and its flow toward the hot soldering iron.

FIGURE 11-4. The resistance soldering method is most useful in a place where a flame could be dangerous. Enough heat can be generated to melt silver solder.

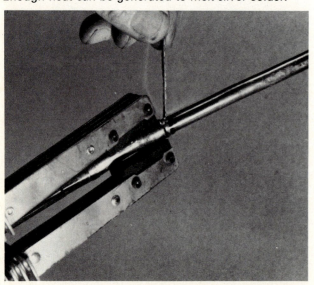

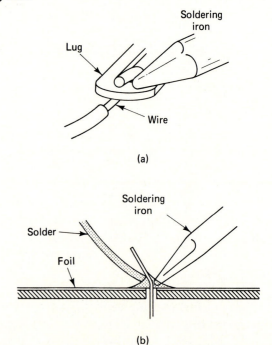

(a)

(b)

heated to a temperature that will melt the soldering or brazing alloy.

Silver solder melts at about 1150°F (620°C) and the silver–phosphorus solders will flow at about 1450°F (790°C).

Heat is usually applied with a torch set for a large soft flame and kept moving so as to avoid excessive heating in one spot and insufficient heating on the other side (Fig. 11-6).

When the part is hot enough, apply the solder to the joint and it should flow between the pieces to fill all voids and make a good strong seal. Avoid both overheating and applying an excessive amount of solder or flux.

FLARING

Many connections of fittings to copper tubing are made by flaring the tubing and clamping it between a special nut and seat on the fitting. To flare a tube,

1. Make certain that the end of the tube is square and free of burrs.
2. Install the flare nut on the tube with the threaded side facing the end.
3. Set the tube in the flaring block so that about one-third of its free height is above the block. Make certain that this is on the chamfered side of the block (Fig. 11-7).
4. Mount the flaring tool on the block, put

a drop of refrigeration oil on the spinner, and turn the handle until the flare is formed. Do not overtighten.

5. Remove tube from tools and install on fitting.

SWAGING

Swaging copper tubing permits the joining of two pieces of copper tubing of the same diameter without extra fittings. This process is done by expanding the end of one piece of tube with a special punch. The tube is expanded until its inside diameter equals the outside diameter of the other piece of tube (Fig. 11-8). The length of swage is normally equal to the outside diameter of the tube. After fitting, the tubes are soldered.

CONSTRICTING

Sometimes it becomes necessary to fit a small tube inside a much larger one. Since good soldering results require the tube diameter to match within 0.005 in., the larger tube must be constricted around the smaller one.

This step is usually done with a tool like a tubing cutter but having a dull wheel. As the

FIGURE 11-7. Imperial Hi-Duty Flaring Tool.

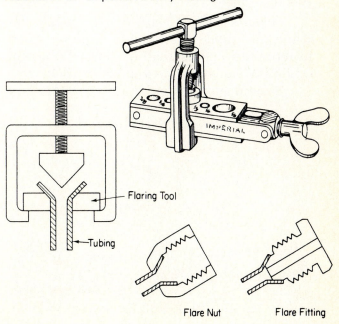

FIGURE 11-6. Silver soldering flux usually has a borax base and will turn into a glassy liquid just before brazing temperature is reached. Apply the solder to the side away from the heat and it will flow toward the heat.

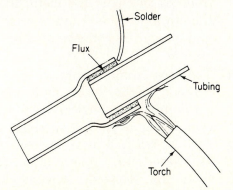

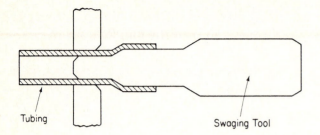

The Tubing is Held in the Same Block That is Used for Flaring

FIGURE 11-8. Tubing is held in the same block that is used for flaring.

tool is turned around the tube, the wheel indents the tube until it fits to the smaller tubing (Fig. 11-9). It can then be soldered.

Sometimes wire-terminal crimping pliers can be used to constrict the larger tube, but the best results are obtained with a roller.

WIRE SPLICING

The joining of wires to complete an electrical circuit must be accomplished in an approved manner.

The National Electrical Code® states that conductors shall be spliced or joined with approved splicing devices or by brazing, welding, or soldering with fusible metal alloy. Such solders and fluxes shall be suitable for use and not adversely affect the conductors or associated equipment.

Soldered splices shall first be made mechanically and electrically secure, then soldered. All splices shall be covered with an insulation equivalent to that of the conductor.

Several approved pressure-type connectors

FIGURE 11-9. Tube constrictor and tube cutter.

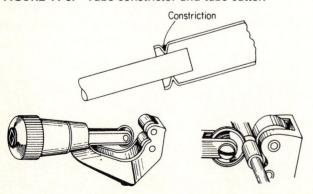

Constriction

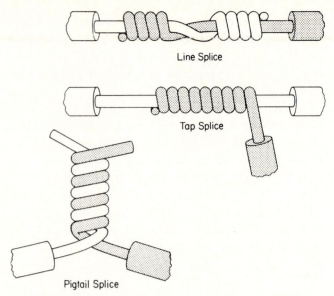

Line Splice

Tap Splice

Pigtail Splice

FIGURE 11-10. All splices must be soldered and tapped.

are illustrated in Fig. 5-7. A soldered connection is shown in Fig.11-10. The approved insulating tape in current use is a vinyl plastic type about 0.007 inch thick.

ELECTRICAL TESTING

Electric Circuit Problems

There are three different types of circuit failure, as shown in Fig. 11-11.

1. The *short circuit* occurs when the electricity finds an unauthorized path from one part of the circuit to another part of the circuit. Usually, some part of the circuit is bypassed and a higher-than-normal amount of current will flow.

2. The *grounded circuit* occurs when the electricity finds an unauthorized path from its circuit to the metal frame of the machine. This frame is conducting and should be connected to an earth ground.

3. The *open circuit* occurs when there is a break in the wiring or a part that prevents the electrical current from completing its path. A variation on the open circuit is a loose connection or a high-resistance connection which will limit the electrical current flow to

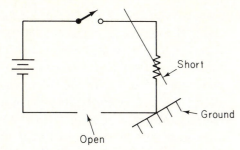

FIGURE 11-11. Basic circuit failures.

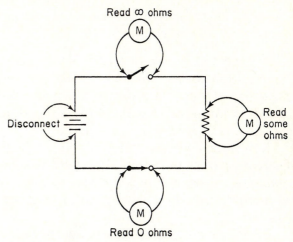

FIGURE 11-12. Ohmmeter use.

something less than the amount needed by the user.

Any one of these failures can occur in either the wiring or a part used in the circuit. Any of these electrical failures can stop a machine from doing its job. The failure must be located and the defect repaired or replaced.

Electrical Test Meters

Meters are available to test a circuit for its resistance (ohmmeter), available voltage (voltmeter), current flow (ammeter), and power used (wattmeter). Sometimes several test meters are combined in one unit called a *multimeter* (Fig. 2-17).

Testing with an Ohmmeter. Warning: Before using an ohmmeter to test a circuit, be certain that the machine is disconnected from the power source. The ohmmeter has its own built-in power supply for making resistance tests. Any additional voltage from an outside source will damage the ohmmeter!

An ohmmeter will measure the resistance from one part of a circuit to another part of that same circuit. When this measuring is done in a logical manner, it is called point-to-point testing. For this procedure to be of value, you must know what the correct resistance between the points being tested should be.

Listed below are some ohmmeter readings that can be expected under normal conditions (see Fig. 11-12).

1. When testing a connecting wire from one end to another the reading should be zero ohms. A very long wire (many feet) will show low resistance read-

ing—possible $\frac{1}{2}$ or 1 ohm. A wire this long in an appliance would be unusual. A high-resistance reading or an infinity-resistance reading would indicate a defective wire that should be replaced. If the wire is moved or shaken while connected to an ohmmeter, and the ohmmeter reading changes from 0 to infinity, the indications are a break in the conductor that makes contact only in some positions. This condition is called an intermittent break. The wire should be replaced.

2. When testing a switch from terminal to terminal, the ohmmeter should read zero (0) ohms while the switch is in the ON position. In the OFF position the ohmmeter should read an infinite amount of ohms. Any resistance reading, even 1 or 2 ohms, indicates poor switch contacts, and the switch should be replaced.

3. When using an ohmmeter to test an electricity using part such as a light bulb, heating element, motor, solenoid, etc., there must be a resistance reading in ohms. The amount depends on the device. Sometimes it can be calculated or it may appear on the wiring diagram. A zero-ohm reading taken at the terminals of the part will usually indicate an internal short circuit. An infinite-ohm reading will usually indicate an internal open circuit, sometimes

called "burned out." In either event the part should be replaced.

Testing with a Voltmeter. Before using a voltmeter make certain that the meter is set for the type of voltage (ac or dc) to be measured and that its full-scale capability is greater than the maximum voltage to be measured.

A voltmeter will measure the voltage available in any part of the circuit. Voltage measurements can be made only when the machine is connected to its power supply and the switch is turned ON.

Listed below are some voltage readings that can be expected under normal conditions (see Fig. 11-13).

1. The voltage reading at the power source to which the machine is connected should be within ±10% of the voltage rating of the machine. The machine's voltage rating will be found on its nameplate.

2. The voltage reading at the machine end of the power cord should be the same as its source voltage.

3. There should be a zero-volt reading across the terminals of any switch while it is in the ON position. Any voltage reading indicates a voltage drop due to unwanted resistance within the switch. The switch should be replaced. *Note:* An exception to this may occur with certain types of speed, dimmer, or temperature-controlling switches which would reduce the voltage available to the electricity user.

4. Except as noted above, the electricity-using part should have the line (or

source) voltage available at its terminals. If the voltage at the using-part terminals is 0 volts, there is an "open" in the circuit between that part and the source voltage. If the voltage at the using part is notably lower than the source voltage, there is a poor connection somewhere between that part and the source. If there is full voltage at the using part and it does not do its job, there is something wrong with the part and it must be more closely checked to determine whether it can be repaired or must be replaced.

Testing with an Ammeter. To measure the current flowing in a circuit, the ammeter must be connected in series with that circuit. An exception is that with an ac voltage a clamp-on type of meter can be used. This meter has a spring-loaded clamp that can encircle *one* conductor, whose magnetic field will induce a voltage in the pickup coil. This voltage is then measured on a meter calibrated in amperes (Fig. 2-28).

Discussed below are some ac ampere readings that can be expected under normal conditions (see Fig. 11-14).

With the meter connected to *one* wire from the source to the machine, with the switch ON, the meter should read within 10% of the rated amperes shown on the nameplate. If the nameplate reading is in watts, the information will have to be converted to amperes by the formula: amperes equal watts divided by the volts ($W \div V = A$). An appreciably lower reading would indicate that the user is not doing as much work as it should. (Something isn't happening.) A much higher reading than normal would indicate that the user is doing more work

FIGURE 11-13. Voltmeter use.

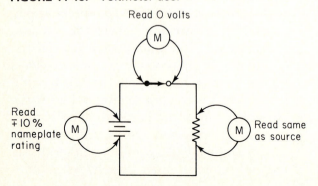

FIGURE 11-14. Ammeter use.

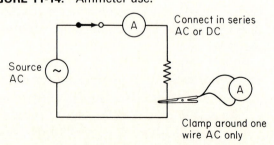

than it should (either overloaded or partially shorted). This type of test is most often used on motor-driven machines.

Testing with a Wattmeter. A wattmeter is used to determine the true power being used by a machine. It is connected between the power source and the machine as shown in Fig. 11-15. The wattage reading should be within ±10% of the nameplate rating of the machine being tested. A wattmeter reading will often differ from the wattage calculated from voltmeter and ammeter readings ($W = V \times A$), because of the power factor in the line. Since a wattmeter reading is taken while the machine is operating, it provides an indication of abnormal operating conditions. Higher-than-normal wattage use indicates an overloaded condition, and lower-than-normal wattage indicates that some work is not being done.

SUMMARY

Proper use of your electrical test equipment will give you a great deal of information on how well any electromechanical machine is working. Figure 11-16 shows how each type of meter should be connected to give you the readings it is supposed to.

1. The voltmeter is always connected across the line or part being tested for voltage.
2. The ammeter is always connected in series with, or clamped around, the

wire carrying the current to be measured.

3. The wattmeter is always connected to both the wires supplying power to the machine.
   All three are used while the machine is in operation.
4. The ohmmeter is used only when the machine power has been disconnected. It is connected across the part being tested to measure its resistance.

STUDY QUESTIONS

Soldering

1. The most satisfactory type of flux for electrical work is (a) organic, (b) rosin, (c) chloride type, (d) either (a) or (c).
2. List four factors to keep in mind when soldering.
3. The purpose of flux is chemical rather than mechanical.   True or False
4. When soldering, the torch or flame must guide the solder to where it must go.   True or False
5. It is impossible to join two pieces of similar copper tubing without extra fittings.   True or False
6. Length of swage should be equal to the inside diameter of opposing tube.   True or False
7. Good soldering requires a tolerance of only 0.005 inch between mating tubes.   True or False
8. If wires to be soldered together do not meet, it does not matter because the solder will serve as a conductor.   True or False
9. There are no approved insulating tapes. True or False
10. When soldering, one need not be concerned about too much heat.   True or False
11. It is important to control the use of solder— an excessive amount is not good.   True or False

Electrical Testing

1. Two important pieces of equipment for servicing electrical appliances are _____ and _____.

FIGURE 11-15. Wattmeter connection to machine.

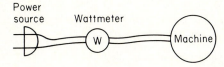

FIGURE 11-16. Test meter connections.

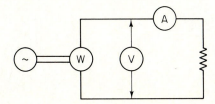

2. An appliance should remain plugged in if testing is to be done.   True or False
3. When test leads are plugged into a continuity testor, the lamp will not light on circuit continuity.   True or False
4. Meters having a scale reading up to 300 volts are adequate for most appliance testing. True or False

ELIMINATING ODORS FROM REFRIGERATORS

An understanding of the problem of odors within the storage compartment of refrigerators is based on one fundamental fact. No substance used in the construction of modern refrigerators and freezers can cause an unclean odor—this includes the refrigerant as well as the construction materials. Thus unpleasant odors contained in the storage compartment must come from some external source. Usually this external source is decayed food particles.

The solution to this problem is a complete and thorough cleansing of the refrigerator with a solution of warm water and baking soda (2 tablespoons to 1 quart) or warm water and vinegar.

Remove all foods from the refrigerator and completely defrost. Remove the shelves and clean each shelf individually in the solution. Each cross member of the shelves should receive special attention to ensure that no food particles, however minute, adhere to any part of the shelf. The interior of the food-storage compartment should be thoroughly cleansed, rinsed, and dried. The face of the cabinet, including the breaker strips, the inner door panel, the rubber gasket, and the portion of the inner door panel underneath the rubber gasket must be thoroughly cleansed, rinsed, and dried. The ice cube trays may be washed in a dishpan. Exceptionally difficult cases may require a second or even a third cleansing with the solution. It is impossible, however, to remove completely all semblances of odors from any refrigerator.

Even on new refrigerators there will be an odor which comes from the plastic parts as the residual plasticizer migrates from the part. It is like the "new car smell."

A small, porous bag of activated charcoal or an activated carbon filter can be placed in the refrigerator to help absorb some of the odors; so can an open box of baking soda.

EPOXY REPAIR SEQUENCE

The epoxy repair procedure for patching a leak in an aluminum evaporator is as follows:

1. Carefully sand the area to be patched.
2. Clean it with acetone.

FIGURE 11-17. (a) The proportions of resin to hardener is critical. Too little hardener results in a long cure time; too much results in a fast cure time. Sometimes the hardener is a different color than the resin, and mixing is complete when the colors blend. (b) Resin and hardener should be mixed quickly on a firm, clean surface that can be discarded—a paper plate, jar cover, tin can, etc. A tongue depressor or stirring stick will do for mixing.

(a)

(b)

3. Squeeze equal lengths of resin and hardener on a clean surface (Fig. 11-17a).

4. Mix thoroughly for about $1\frac{1}{2}$ to 2 minutes (Fig. 11-17b).

5. Apply to the precleaned area and smooth out. *Note:* Do not play around too long with mixed epoxy; it will set while you are working it and may not hold as well as it should.

6. Using a 250-watt heat lamp to maintain a 140°F (60°C) temperature at the patch will cure it in about an hour. Place a thermometer near the patch to check temperature. Shield the breaker strips from the heat lamp or they will melt and deform.

7. A touch-up with silver or aluminum spray paint (after the epoxy is cured) will cover the patch, and it will have a "professional" look.

ORDERING REPLACEMENT PARTS

There may come a time, in the repair of an appliance, when a new part will be needed.

No one book or source lists all the parts needed for the repair of all appliances; so you will have to get your parts from the local, authorized dealer for the particular appliance. If you do not know who the dealer is, "try the Yellow Pages" in your telephone book.

Dealers, however, cannot possibly carry in stock all parts that might be needed. The cost of inventory and the warehouse space needed would be prohibitive. They do carry a small stock of the parts most likely to need replacing.

The dealer will have a parts list and catalog on hand from which the correct part number for the needed part can be obtained. If he has the part in stock, fine; if not, it will have to be ordered. How soon you will get it depends on the dealer, the mails, the distributor, and/or the factory. Two weeks should be long enough to wait. If it takes longer, someone has "goofed"!

If there is no dealer in your area, you may order directly from the distributor or from the factory. The chances are that you will not have a parts list for the machine and will not know what to order. Try sending the old part back and saying that you want one like it. This is very often done, so do not hesitate to do it if you feel that there is no other way.

We suggest you try this step first, however. List the name of your appliance, its model number, and the serial number (Fig. 11-18). Then describe the part by its location in the machine, by the job it does, and by its condition. If the service parts department is still puzzled as to what you mean, they at least will know which parts catalog to send you so that you can order by the correct part number.

If you know the cost of the part, include a check or postal money order with your order. If you do not know the cost, you must be prepared to accept the part from the factory COD.

COD parts shipment is a practical method of handling individual orders because the extra paper work and bookkeeping involved in separate billing and collection are costly.

CALIBRATING OVEN CONTROLS

Instrumentation

A thermocouple-type test instrument is preferred to measuring oven temperatures accurately. Mercury thermometers are acceptable providing they can be proven accurate. For example, a recent check of mercury thermometers

FIGURE 11-18. When ordering a part, use its part number if available. If not, include all the information you can about the appliance.

commonly used by service organizations showed errors of 5 to 15°F.

Regardless of the type test instrument used, it is most desirable to double-check it just before making an oven temperature check. This can be done simply and quickly by placing the thermocouple tip (or immersing the entire mercury thermometer) in boiling water. *Note:* Mercury oven thermometers should be the "total immersion type." The resulting, reading should be within several degrees of 212°F (100°C), depending on altitude.

Generally, a mercury thermometer can be best used if the oven door has a glass window because the door need not be opened to check temperatures. This fact is important because many mercury thermometers will drop in excess of 25°F (14°C) within seconds after the door is opened. This condition makes it extremely difficult to get a true temperature reading.

Almost all ranges today have oven burners that cycle "on and off." The high and low points of the resulting temperature "swing" (differential) in the oven must be measured—and averaged—to determine the true operating temperature. A thermocouple-type test instrument is best suited to measure these temperature changes quickly and accurately, and without opening the oven door.

With an accurate thermocouple test instrument or mercury thermometer, it is still difficult to measure these changing oven temperatures and then average them correctly. Consequently, we recommend that the thermocouple tip or the bulb of the thermometer be "weighted." Weighting (adding mass) the test instrument compensates for oven temperature changes by making the test instrument less sensitive to these constant changes in temperature. *Note:* How this weighting can be done is discussed later.

Measuring these changing high and low temperature points in an oven is possible with either type of test device—without weighting—but is subject to possible inaccuracies. This is most true at low temperature settings because, in this area, the function of "time" becomes a factor. The simple averaging of temperatures, then, may not produce the true operating temperature. "Weighting" provides the compensation for both time and temperature that is necessary.

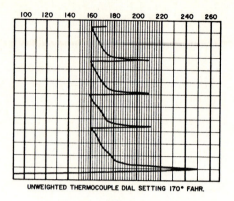

UNWEIGHTED THERMOCOUPLE DIAL SETTING 170° FAHR.

Chart 1

WEIGHTED THERMOCOUPLE DIAL SETTING 170° FAHR.

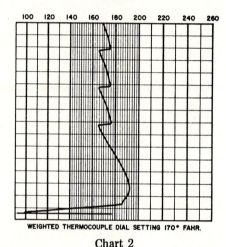

Chart 2

FIGURE 11-19

We produce above test curves showing actual results in the low-temperature area showing the difference in results when using an unweighted and a weighted thermocouple or thermometer. Tests were made with the same control, same oven, and dial at 170°F mark (not moved).

From the above, it can be seen that an error of 15°F is possible. Chart 1, unweighted thermocouple, indicates an erroneous average oven temperature of 185°F. Chart 2, weighted thermocouple, indicates the "average" or true oven temperature to be 170°F (Fig. 11-19).

Weighting

The thermocouple can be weighted by using a letter-size sheet of aluminum foil—Reynolds Wrap or equal.

Fold the foil five times, doubling the thickness with each fold. After the fifth fold, place the thermocouple tip in the center of the aluminum piece and fold once more. Finally, fold in the sides so that the foil clings to the thermocouple tip (Fig. 11-20).

A mercury thermometer can be weighted in much the same way by wrapping several layers of aluminum foil around the bulb end, thus creating the necessary mass. This procedure is a *must* if you open a windowless oven door to check temperatures.

Procedure for Checking Calibration[*]

1. Place thermocouple or mercury thermometer in center of oven.
2. With thermostat dial in the OFF position, make certain OFF mark on the dial agrees with reference point of the bezel or panel; misalignment will affect calibration.
3. Turn dial to 350°F mark. Allow oven to heat until control cycles ON and OFF thermostatically at least three times. This will allow oven temperature to stabilize and eliminate possible error resulting from initial oven temperature overshoot and/or undershoot. *Note:* The oven burner will cycle on and off at full burner rate at all dial settings below the 575°F marking. There is no bypass flame.
4. After the control has cycled thermostatically three or more times, note the oven temperature when the burner cycles off, and the oven temperature when the burner cycles on. Recalibrate only if the average of these two tem-

[*] The section on calibration appears courtesy of Robertshaw, New Stanton Division, Robertshaw Controls Company, Youngwood, PA 15697.

perature readings varies more than 25°F from the dial setting.

Example: With the dial set at 350°F, and after the system has cycled thermostatically three or more times, the highest temperature recorded (when burner cycles off) is 364°F. The lowest temperature recorded (when burner cycles on) is 356°F. The average of these two temperature readings, 360°F, is the calibration temperature. This is well within acceptable calibration tolerances. With the dial at 350°F, recalibrate only if the average of the two readings is either greater than 375°F or less than 325°F.

If the thermocouple tip or the thermometer bulb is "weighted" (it should be), there will be little or no temperature change noticed in the oven as the oven burner cycles. This makes "averaging" very simple or unnecessary.

Procedure for Calibrating

1. Turn control to OFF position and remove dial.
2. Insert screwdriver through the center of the D stem to engage the slot of the calibration screw. Using the screwdriver blade as a reference point, turn calibration screw clockwise to lower oven temperature or counterclockwise to increase oven temperature. Each calibration mark on front of calibration plate represents 25°F. Make certain that D stem does not move during this adjustment.
3. Replace dial.
4. Recheck calibration by setting dial at 400°F mark and repeating steps 3 and 4 of "To Check Calibration."

REMOVING STAINS FROM WASHABLE FABRICS

Most stains can be removed easily when they are fresh, but become difficult or impossible to

FIGURE 11-20

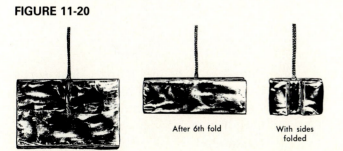

After 5 folds

After 6th fold

With sides folded

remove later. Learn the simple methods for removing stains at home and act promptly. Start to remove stains before washing, for hot, soapy water can set many stains permanently. Cool-water presoaking is a safe and effective procedure.

Identify the stain if possible. The treatment for one kind of stain may set another. If you cannot determine what caused the stain, it will help if you can tell whether it is a greasy stain, a non-greasy stain, or a combination of the two. Before using any stain remover, be sure it will not harm the fabric. If in doubt about the fabric, test a hidden part before applying to the entire garment.

Enzyme presoakers are excellent stain removers. Some do well on protein soils; others do well on carbohydrate soils; a few do well on both. Check your local supermarket for one that will do both and that does not contain a chlorine bleach. It should be compatible with all fabrics except wool and silk, which are protein material. The soak period may range from ½ hour to overnight depending on the stubbornness of the stain. After stain removal the fabric may be washed in the usual manner. Some enzyme removable stains may be: blood, milk, eggs, gravy, cocoa, catsup, spaghetti sauce, perspiration, grass, baby formula, ice cream, soft drinks, coffee, tea, salad dressing, and urine.

Other stains may be removed by the methods listed in Table 11-1.

TOUCH-UP PROCEDURE FOR REPAIRING DAMAGED FINISH WITH AEROSOL SPRAY TOUCH-UP PAINT

See Figure 11-21. Preparing the Surface

Prepare the surface to be refinished by sanding lightly with a No. 400 wet or dry sandpaper and water- or oil-free naphtha. Feather the edges outward for about ½ inch, using a light rotary motion tapering to the full paint thickness. In so far as possible, avoid sanding down to bare metal.

When area seems smooth and is blended in, wipe it clean with a lint-free cloth moistened with an oil-free thinner. Mask the surrounding area with tape and paper.

Applying the Primer

A primer should be used when the bare metal has been exposed or there was a primer under the original finish. Apply primer by spraying several light coats, allowing at least 5 minutes between coats.

When primer is dry, wet sand lightly with No. 400 wet or dry sandpaper and water- or oil-free naphtha. Remove any primer "overspray."

Applying a Filler

When primer fails to fill the damaged area, the depression should be filled with a pyroxylin putty. Allow the putty to dry at least an hour or two—air drying at room temperature. Sand smooth as before and apply a coat of primer.

Finishing

To obtain a satisfactory finish:

1. The temperature of the touch-up paint container must be about 70°F (21°C)! A cold container can be warmed by setting it in hot tap water for 5 minutes before painting. Do not exceed 120°F (50°C). The surface to be painted must be clean, dry, and not too much colder than the paint.

2. Shake well before using! Most spray cans include a steel agitator ball to assist in mixing the pigment. When the ball "rattles" freely, the paint can be considered well mixed. Insufficient

TABLE 11-1
Stain removal chart

Stain	Treatment
Adhesive tape Chewing gum Candle wax Paraffin	Using ice or cold water to stiffen the soil scrape away as much as possible, then soak in a grease solvent. Dry by blotting with cloth or paper towel. Launder.
Cosmetics Greases and oils Crayon Deodorants	Pretreat with a generous amount of detergent and hot water. Soak remaining stain in a grease solvent or industrial-type cleaner. Launder.
Pencil marks Carbon paper Duplicating carbon Ballpoint printing Writing inks Shoe polish	Treat with a liquid organic surfactant, or trichloroethane, soak in detergent solution. Remaining stain, if any, may be treated with turpentine, ammonia, or rubbing alcohol. Launder.
Fruits and berries	Sponge peach, pear, cherry, and plum stains at once with cool water and rub with glycerine. After 2 hours, apply a few drops of vinegar for a minute or two; then rinse and launder in warm suds. For other fruits, stretch the stained portion of fabric over a bowl and fasten with an elastic band or string. Pour boiling water through it from a height; launder in suds.
Mildew	Wash with strong detergent; dry in sun. If stain remains, treat with sodium perborate bleach or hydrogen peroxide. Mildew spots must be treated while fresh, before the mold growth can weaken the fabric.
Mustard	Soak in an industrial cleanser or a liquid organic surfactant. Launder. If stain persists, sponge with rubbing alcohol.
Nail polish Hair sprays and lacquers	*Do not use polish remover.* Sponge with chemically pure amyl acetate or acetone. Launder. If stain persists, sponge with rubbing alcohol to which a few drops of ammonia have been added, but first pretest to make sure that alcohol does not harm the dye or fabric, especially if it is on acetate fabric.
Perspiration	Wash or sponge thoroughly with detergent and warm water. If fabric is discolored, sponge fresh stains with ammonia and old stains with vinegar. For oily stains, use naphtha or another grease solvent.
Rust	1. Moisten stain with oxalic acid solution. (Do not use oxalic acid on nylon.) *or* 2. Use a solution of sodium hydrosulfite or a similar nonacid rust solvent. For serious rust discoloration use a reducing bleach in the wash water. Do not use a chlorine bleach at the same time.
Scorch	Rub with detergent and bleach in the sun, or dampen with hydrogen peroxide. Launder.
Suntan lotion	Sponge with carbon tetrachloride, followed by a suds solution with vinegar added. Tannic acid stains may be bleached out of white cotton or linens. Launder with clean suds.
Tar and asphalt	Must be done before stain is dry. Sponge with turpentine, trichlorethelene, or a petroleum distillate solvent. (Stain may be impossible to remove.)

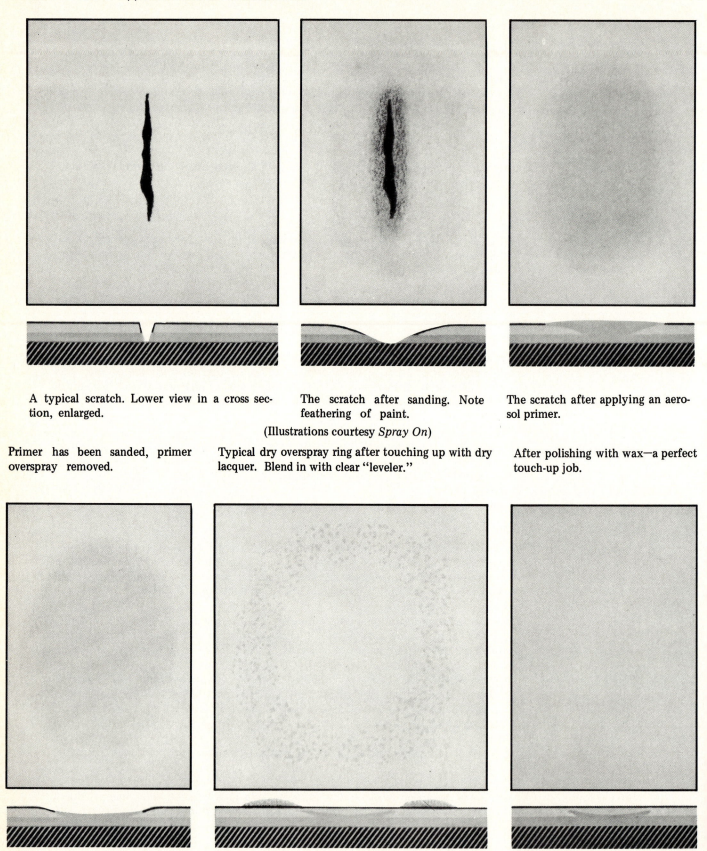

A typical scratch. Lower view in a cross section, enlarged.

The scratch after sanding. Note feathering of paint.

The scratch after applying an aerosol primer.

(Illustrations courtesy *Spray On*)

Primer has been sanded, primer overspray removed.

Typical dry overspray ring after touching up with dry lacquer. Blend in with clear "leveler."

After polishing with wax—a perfect touch-up job.

FIGURE 11-21 (Courtesy of Spray On)

mixing may cause an off-color finish.

3. Paint with short burst of spray. Holding the valve open continuously will produce an uneven spray job. Open the valve completely each time in order to obtain proper pressure.

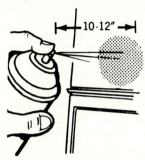

4. Hold the spray valve about 10 to 12 inches from the work surface for the best result! Too far will allow thin, spotty coating. Too close will cause paint pileup and bubble. The can may be held in a vertical to horizontal position, but not upside down.

5. Avoid drafts! Moving air will cause the spray to blow away from the work surface and onto some other surface. Having the work horizontal will help prevent sags and runs.

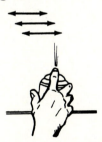

6. Apply the spray in short, steady strokes! Shut the valve off at the end of each stroke. Apply repeated light mist coats until the color and finish blend into the surrounding area. Allow enough time between coats for the paint to become tacky.

Problems

Poor color match may be due to insufficient agitation of the paint *or* not enough light coats of paint to build up to the correct color value. Runs, sags, or curtains may be caused by any one of the following:

1. Holding the valve open too long
2. Holding the can too close to the work surface
3. Not moving the can while spraying
4. Moving the can too slowly
5. Applying successive coats without allowing the previous coat to dry or at least become tacky

Precautions

1. When using lacquer, allow about 5 minutes for the finish coat to set up; then spray with leveler (a clear solvent in a pressurized container) to blend the overspray ring. Apply lightly; allow an hour for drying. Any lacquer dust still remaining around the touched-up area can be removed with a rubbing compound and polishing wax.

2. White enamels tend to acquire many color variants depending on baking cycle, exposure to light, and aging. Touch-up white may not match on an older machine. To retain the "orange peel" effect, *do not* rub down the touched-up area. Overspray can be blended by a light application of leveler.

3. After completion of a spray job, invert the can and open the valve to clear it of paint. When plain gas flows and the passages are clear, close the valve,

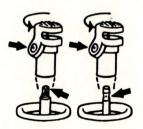

GENERAL TROUBLESHOOTING GUIDE

Trouble diagnosis is a matter of applied logic. You must integrate your observations with your knowledge of what the machine is supposed to do. Complete product knowledge but haphazard observation will not provide the most efficient service combination; nor will careful observation technique and scanty product knowledge. Knowledgeable product background and a comprehensive observation procedure are needed for good diagnosis.

The manufacturers try to help by publishing trouble diagnosis charts for their products. These serve as reminders of what to look for under various operating conditions. If we throw them all in a pot and stir them, we come up with a universal trouble diagnosis sequence which will guide the service technician to the trouble area.

Obviously, we must make certain that there is a problem—at least the customer thinks there is one or she wouldn't have called—but is it a real problem or a misunderstanding about what the machine can do?

For convenience we are grouping problems in two major areas: external and internal areas of the appliance.

Meyerink's law: If it works, don't fix it!

A. EXTERNAL FACTORS

Problem Area	Observed Condition
Location	Too wet
	Too cold
	Too hot
	Too drafty
	Too weak—springy floor
	Too inconvenient—crowded
	Too crooked—not level
Utilities	Low voltage or defective receptacle
	Incorrect gas pressure or supply
	Poor water supply
	Inadequate drain
	Insufficient clean air
Usage	Machine overloaded
	Incorrect operation
	Careless handling
	Dirty environment
Corrective action:	1. Instruct customer
	2. Move appliance.
	3. Correct condition.

GENERAL TROUBLESHOOTING GUIDE

B. INTERNAL FACTORS

Problem Area	Condition to Observe
Utility supply and connection	Electrical: plug, cord Gas: Leak, restriction Water: Leak, restriction Drain: Leak, restriction Air: Draft, restriction
Connectives	Electrical: wiring and terminals: loose, open, shorted, grounded Hoses: loose, leaky, restricted
Power transmission	Belts: loose, broken, slippery Gears, Bearings: noisy, worn, broken Seals, etc.: frozen, jammed, leaky
Housing—frame	Broken, cracked, bent, warped, rusted, corroded, damaged or worn door hinges and latches, and gaskets
Electrical controls (operating and safety)	Switch: open, short, damaged Timer: motor, cams, switch contacts Solenoid, valve: open coil, sticky or clogged valve Thermostat: sensor not responding, switch open or shorted
Workers (electrical)	Motor: open, shorted, frozen Capacitor: open, shorted Start relay: burned contacts, sticky Heater: open, grounded Lamp: open Solenoid: open, sticky
Gas burner	Flame color and character Pilot outage and ignition method Primary air adjustment Secondary air availability Pressure regulator Solenoid valves
Water System	Fill valve, water level switch, pump, drain
Air system	Fan/blower, filter, diverter, duct
Refrigerant system	Leak, restriction, inefficient compressor

Corrective action: Repair or replace part.

Note: We must assume that the technician knows what the correct condition should be. That is what this program is all about.

place the cover over it, and store at room temperature. Never expose an aerosol spray can to over 120°F (50°C)!

4. Make certain the pilot light on gas-fired appliances is turned off before painting.

5. *Warning!* Never puncture or incinerate an aerosol can. The solvents in paints and lacquers will burn easily.

STUDY QUESTIONS

Refinishing

1. The most important approach to good refinishing is (a) knowledge of available materials, (b) environmental conditions when applying, (c) a good match with existing finish, (d) proper preparation of area to be done.

2. Levelers have a tendency to (a) soften the primer, (b) provide too much of a buildup, (c) dry too slowly, (d) leave a rough finish.

3. When using putty (a) it may be dried with a heat lamp, (b) it should be sanded lightly, (c) it must be covered with primer, (d) all of these.

4. Besides the various liquid materials and cleaning rags, what other tool is needed for most finishing jobs?

5. Temperature is not a factor in the use of modern finishes.   True or False

6. Breezes help a finish dry but are a hindrance during application.   True or False

7. The best position for the surface to be refinished is vertical.   True or False

8. It is best to apply each coat immediately after the one before.   True or False

9. The closer you hold the can to the area to be refinished, the thicker and better the job will be.   True or False

10. The most important practice a technician can bring to a refinishing job is practice.   True or False

KEY

Soldering

1. b
2. Clean parts, enough heat, proper flux, hardening cooling time
3. True
4. False
5. False
6. False
7. True
8. False
9. False
10. False
11. True

Electrical Testing

1. Voltmeter and ohmmeter
2. False
3. False
4. True

Refinishing

1. d
2. a
3. d
4. Sandpaper
5. False
6. True
7. False
8. False
9. False
10. True

12

The Gas Flame

HISTORY

History tells us that the Chinese used gas for both cooking and heating more than 2000 years ago. Gas was tested experimentally for street lighting as early as 1812 in Newport, Rhode Island. In 1816, Baltimore became the first American city to use manufactured gas in lighting its streets. Gas was originally a light-producing industry. It became a heat-producing industry when electric light bulbs were introduced at the turn of the century. Electricity, and the resulting competition, has helped the consumer to obtain the efficient, low-cost heating that gas offers today.

STANDARDS

The American Gas Association (AGA) was organized in 1918 to promote and develop the gas industry in order that it might serve the best interests of the public. Recognizing that unless the industry policed itself it would invite exter-

nal controls, members of the gas industry have successfully relied on voluntary cooperation to accomplish its aims. In 1925 the AGA Testing Laboratories were formed to establish standards for gas equipment. Today, before any manufacturer can hope to offer his appliances for sale, he must submit his equipment to the AGA Testing Laboratory. If it meets the laboratory's rigorous safety and dependability requirements, it is then authorized to carry the Blue Star, or seal, of the AGA, which reads, "certified—complies with national safety requirements."

The standards demanded by AGA are strict and have been determined by continuing tests and thorough inspection in the laboratories. Production models of certified equipment are not inspected once and then forgotten. They are inspected several times a year not only at manufacturers' plants but also on sales floors, at the warehouses of the distributors, even in the customers' homes. This watchful method of "spot checking" ensures that gas appliances are constructed in exact conformity with the original approved models.

199

Committees of the AGA were formed to draw up requirements that would ensure three specific ends:

1. *Safety*.
2. A reasonable degree of *durability*.
3. Gas appliances must operate with *efficiency when properly installed* and in actual use. Because proper operation is impossible without proper installation, AGA further requires that equipment manufacturers enclose complete installation instructions with each piece of equipment.

THE DIFFERENT GASES

The gases generally used in the home are natural gas and bottled gas. Natural gas is normally piped into the home from a gas main owned and operated by a public utility—"the gas company." Bottled gas is usually piped into the home from a tank in the backyard. This tank may be refilled as needed either by exchanging tanks or by pumping more into the tank from a truck. Bottled gas is also called LP (liquefied petroleum) gas.

Natural Gas

Natural gas is the name given to all gas obtained from wells. It is a mixture of various gases. The following tabulation is a recent analysis of natural gas as supplied in Michigan:

Methane CH_4	Ethane C_2H_6	Propane C_3H_8	Butane C_4H_{10}	Nitrogen N_2	Carbon Dioxide CO_2
84.1%	5.8%	1.8%	0.4%	7.3%	0.6%

The four "CH" gases give off heat when burned. The heavier hydrocarbons—ethane, propane, etc.—have a faint sweetish odor; the other gases are odorless. All the above-listed gases are colorless.

Carbon dioxide and nitrogen do not produce heat and are classed as inerts. Each is also odorless and colorless, so an obnoxious, artificial odor is added in order to warn people of a leak.

None of these gases is poisonous; and, for this reason, natural gas as a whole is not poisonous. If it is breathed steadily for several hours—the time depending on the percentage of gas in the air—a person would smother and die from lack of oxygen, for the gas does not contain oxygen to supply the needs of the body.

Natural gas has a specific gravity of about 0.65. Therefore, being lighter than air, it will rise into the atmosphere to be dissipated by air currents.

LP Gas

The need for gas in bottles for use in areas remote from a central supply line was recognized as early as 1810. However, it was not until just prior to 1920, when the LP gases were developed, that a successful bottled gas program could be started. Records of LP gas began to be kept by the Bureau of Mines in 1922.

The liquefied petroleum gases are extracted from crude petroleum and must have the following features:

1. They can be burned.
2. They can be liquefied under moderate pressure and stored as a liquid at most outdoor temperatures.
3. They can be changed from liquid to gas at most outdoor temperatures by reducing the pressure on them.

The two most commonly used are propane (C_3H_8) and butane (C_4H_{10}). Both are colorless, odorless, and tasteless and need an odorizing agent to warn of leaks. They are not poisonous but, as with natural gas, could suffocate a person by depriving him of oxygen.

The specific gravity of LP gases is greater than air; thus they tend to collect in low spots. However, they do disperse in air as long as there is air motion. See Table 12-1 for other properties.

When handling LP gas, you should know the following facts:

TABLE 12-1
Properties of gases

Property	Propane	Butane	Natural
Formula	C_3H_8	C_4H_{10}	CH_4
Molecular weight	44.06	58.08	Mostly
Specific gravity at 60°F (17°C)			
Liquid	0.509	0.584	—
(water = 1.0)			
Gas	1.522	2.006	0.65
(air = 1.0)			
Boiling point at atmospheric pressure [water = +212°F (100°C)]	−44°F (−42.2°C)	+32°F (0°C)	−258°F (−161°C)
Range of flammability (percent gas in air)	2.4–9.5%	1.8–8.4%	4–14%
Vapor pressure			
At 60°F (17°C)	92 psi	12 psi	—
At 100°F (38°C)	172 psi	37 psi	—
Minimum ignition temperature			1100°F (593°C)
Heat output [at 60°F (17°C)]			
Btu per			
Gallon	91,300	103,000	—
Pound	21,600	21,300	—
Cubic foot	2,520	3,270	1,000

1. It will dissolve rubber, so use copper tubing.

2. It will dissolve some pipe joint compounds. Make certain that the one you use is gas resistant.

3. When LP gases change from their liquid state to vapor, they absorb heat. Exposing skin to escaping liquid LP can cause severe frostbite.

4. The vapor pressure increases as the surrounding temperature is raised. **Keep the tanks away from fire and excessive heat.**

5. **Never fill a tank full up with liquid LP.** It must have expansion space. Use only tested and approved LP storage bottles.

GAS AS A FUEL

The purpose of burning gas is to produce heat that can be used to raise the temperature of a home, water, food, or anything else man wants at a higher temperature. In this country heat is measured in British thermal units (Btus). One Btu is the amount of heat energy required to raise 1 pound of water 1 degree Fahrenheit. Table 12-2 lists some of the more common fuels and their heat output.

COMBUSTION

Combustion may be defined as the rapid combination of a fuel with oxygen, producing heat and light. In connection with any burning process, it is necessary to provide heat (ignition), adequate movement of the air, and ventilation so that the products of combustion may be carried away.

For every cubic foot of natural gas burned, 10 cubic feet of air are required. (For LP gases see Table 12-2.) This combustion is accomplished by a "burner," which combines the proper ratio of gas and air. Air is one part oxygen to four parts of nitrogen; and since nitrogen is an inert gas, playing no part in the combustion, only 2 cubic feet of oxygen is used with a cubic foot of natural gas. Perfect combustion

TABLE 12-2
Heating-value comparison of some more common fuels[a]

Fuel	Unit	Average Btu	Percent Efficiency	Cubic Feet of Air Required	Net Btu
Oil, No. 2	1 gallon	139,000	70	1,224	97,300
Coal	1 pound	13,000	60	108	7,800
Electricity	1 kilowatt	3,415	98	0	3,347
Gas					
Butane	1 cubic foot	3,200	80	31	2,560
Propane	1 cubic foot	2,550	80	24	1,995
Natural	1 cubic foot	1,000	80	10	800

[a] A fire cannot occur unless enough fuel, oxygen, and heat are present at the same time.

produces two products—carbon dioxide and water vapor as well as heat.

When excess oxygen is present in combustion, the temperature of the flame is somewhat reduced. In heating units, the result is a slight waste of fuel. However, it is difficult to obtain perfect combustion. To be on the safe side, burners are usually adjusted with about 20 percent excess of air.

The Yellow Flame

The early gas burners were small holes or slits in a pipe that would allow the gas to escape and mix with the surrounding air. When ignited by a match or spark, the gas would burn with a yellow flame, which produced light as well as heat. Figure 12-1a illustrates the combustion process. This flame was the basis of heat and light in the early days of gas usage.

The Blue-Flame Burner

In practically all gas burners now in use, the gas and part (about 50 percent) of the air (primary air) are mixed before burning. This mixture passes through the burner head, where the burning takes place. The rest of the air (secondary air) flows in around the flame. The primary gas-air mixture burns first to form aldehydes and alcohol in the inner cone; then the aldehydes and alcohols burn with secondary air to form carbon dioxide and water vapor. This is a clean, blue flame that will never create soot when properly adjusted (Fig. 12-1b; also see Fig. 12-2).

Incomplete Combustion

Some of the common causes of incomplete combustion and resultant formation of carbon monoxide with domestic gas appliances include the following:

1. Lack of primary or secondary air. The first condition is denoted by a long, streaming flame with an indistinct inner cone and possibly a yellow-tipped flame (Fig. 12-1c). The second condition will usually be recognized. A part or all of the flame tends to float away from the burner (Fig. 12-1d). Location of appliances in confined spaces like closets can produce these conditions (Fig. 12-1e).
2. Overrating of burners—too much gas—causing the flames to impinge on solid objects or resulting in too much gas for the limited amount of air available.
3. Flames striking each other or impinging on cold surfaces (Fig. 12-1f). The first may be due to insufficient clearance between burner parts, which prevents circulation of air between these parts. The second may be caused by improper distance between burner parts and whatever is being heated. Figure 12-1g shows correct utensil location.

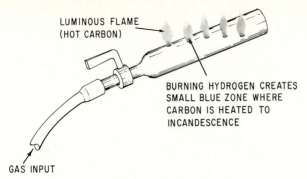

LUMINOUS FLAME
(HOT CARBON)

BURNING HYDROGEN CREATES
SMALL BLUE ZONE WHERE
CARBON IS HEATED TO
INCANDESCENCE

GAS INPUT

(a) The yellow flame results from the glowing carbon particles that complete their combustion on contact with the surrounding air. There is no primary air used with this type of burner.

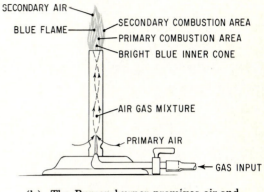

SECONDARY AIR

BLUE FLAME

SECONDARY COMBUSTION AREA
PRIMARY COMBUSTION AREA
BRIGHT BLUE INNER CONE

AIR GAS MIXTURE

PRIMARY AIR

GAS INPUT

(b) The Bunsen burner premixes air and gas *before* burning. Burning gas in this manner provides a cleaner, faster, and higher temperature heat.

(c) A lack of primary air will produce a flame with a yellow tip due to unburned carbon particles. This flame will produce soot! As the primary air is *decreased* the amount of yellow in the flame will *increase*.

(d) A lack of secondary air causes a lazy floating flame that does not have complete combustion and produces CO and obnoxious aldehydes.

(e) An excess of primary air will produce a noisy blowing flame that lifts off the burner and could blow itself out. Incomplete combustion can form CO and obnoxious aldehydes.

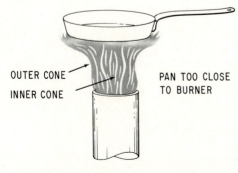

OUTER CONE
INNER CONE

PAN TOO CLOSE
TO BURNER

(f) When the inner cone comes in contact with a cold surface, the cone's temperature is lowered, combustion is incomplete, releasing CO, aldehydes, and unburned carbon that will produce soot.

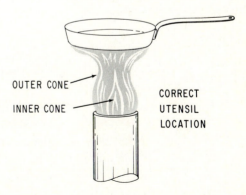

OUTER CONE
INNER CONE

CORRECT
UTENSIL
LOCATION

(g) The outer cone may safely come in contact with a colder surface without affecting combustion.

FIGURE 12-1.

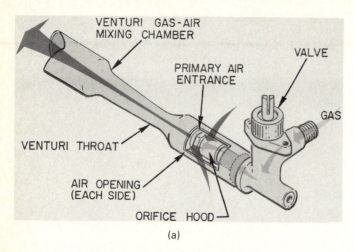

(a)

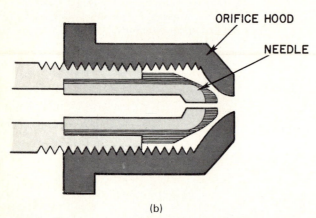

(b)

FIGURE 12-2. (a) Valve and venturi assembly illustrating the relationship between the valve, venturi, and primary air entrance. The primary air entrance has an adjustable shutter to provide the most satisfactory flame. Also make certain that the venturi is aligned with the valve. (b) Adjustable orifice used on many "universal" ranges and can be used with any type of gas. When tightened down, it will adjust to a smaller orifice suitable for LP gas; when the hood is turned outward, the orifice size increases, becoming suitable for natural gas.

4. External drafts striking the appliance combustion area.

5. Lack of draft through the appliance to carry off products of combustion and provide an adequate amount of secondary air. This condition may be caused by restricted vent openings, flue passages, or secondary air openings. These openings may be too small or may be partially closed because of warpage or collection of lint, dirt, or

other foreign matter. The natural draft through the appliance may be adversely affected if the appliance is connected to a chimney without the use of a properly designed draft hood. In case of down draft or blocked chimney, the burner flames may be entirely extinguished.

The burning of gas in properly adjusted gas appliances under ordinary circumstances does not produce any dangerous or disagreeable substances, nor does it cause substances to be given off, that will discolor the walls in time. Neither is the atmosphere appreciably changed by the amount of air used in the combustion process if the room is reasonably well ventilated.

Formation of Water Vapor

Burning 1000 cubic feet of natural gas produces approximately 2000 cubic feet of water vapor, which, when condensed, makes about 11 gallons of water. This water vapor, like steam, will pass off and merge with the surrounding atmosphere and be absorbed by it. However, if the walls of the room are cold and the room is not properly ventilated, condensation of water vapor takes place, causing excessive dampness. This condition can be remedied by proper venting of the heater or better ventilation of the house itself.

Some water vapor is a decided advantage in maintaining comfortable room temperatures, especially in cold, dry weather. This same amount of water vapor added to a room on a humid summer day can add to the discomfort.

GAS EXPLOSIVE LIMITS

People in the gas business are very concerned about explosions. If a customer suffers property damage and/or possible injury, the company has claims to settle. Also, the public becomes wary of gas as a fuel. Therefore it is to everyone's advantage to take care in gas installations and usage. Leaks must be detected early.

Most gas companies have a combustible gas indicator (leak detector) to determine the presence or absence of combustible gas. The instrument will directly indicate percentages of gas up to the lower explosive limit and will approximately indicate higher percentages.

The percentage of gas in air that will burn will vary as shown in Table 12-1. Below the lower limit, the gas will not burn. The mixture is too lean. Above the upper limit, the mixture is too rich to burn. We must remember that as long as a combustible fuel and oxygen are present in a room, a little *heat* [above 900°F (482°C)] is all that is needed to ignite the fuel. Whether it explodes or burns will depend on the mixture and its surrounding conditions. So—do not test for gas leaks with a match!

THE DANGER IN CARBON MONOXIDE

Incompletely burned gas will produce carbon monoxide (CO), which is extremely poisonous to the human body. The red corpuscles of the blood *prefer* CO 300 times more than oxygen, so even low concentrations of CO can quickly saturate the blood, preventing it from carrying oxygen to the various parts of the body. An atmosphere containing one part of CO to 2000 parts air (0.05 percent) could kill a person in 3 hours of exposure.

Since carbon monoxide is a colorless and odorless gas and gives no warning of its presence, it can be present in dangerous quantities without being noticed. Often the production of carbon monoxide is accompanied by other gases known as aldehydes, which have a very disagreeable and penetrating odor, irritable to both nose and eyes. If the combustion products possess a sharp and pungent odor, carbon monoxide is invariably present. Properly designed and adjusted gas appliances will not produce monoxide (except possibly for the first few moments after lighting the burner) and therefore are perfectly safe in this respect if reasonable care is exercised in their use.

INPUT RATING

In discussing the operation of an appliance we must consider its input rating. This rating is dependent upon the following factors:

1. Heat value of the gas
2. Specific gravity of the gas
3. Size of the orifice
4. Pressure at the orifice

The service technician has no control over either the heat value or the specific gravity of the gas being supplied to an appliance. He must know the heat output of the gas being used (Table 12-1) and the correct size of the orifice to match the gas being used to the burner rating as shown in Table 12-3.

Incorrect orifice size will result in improper combustion and all of its attendant problems.

The pressure of the gas at the orifice is subject to many variables; some can be corrected by the service technician.

EFFECT OF PRESSURE ON GAS

Gas flows through piping, meters, appliances, etc., because of its pressure. An increase in the gas pressure increases the flow of gas and a decrease in the pressure decreases the flow (Fig. 12-3).

Variation in gas pressure at the appliance may be due to several causes.

1. Partial stoppage of the gas line will reduce the pressure.
2. The service regulator may be out of adjustment.
3. Water trapped in the gas line can cause a fluctuating gas pressure.
4. Gas pressure may be low if house piping is undersized.

The maintenance of a constant pressure at the appliance is one of the most important

TABLE 12-3
Orifice capacity

a. LP gases (Btu per hour at sea level)

	Propane	Butane
Btu per cubic foot	2,500	3,175
Specific gravity	1.53	2.00
Pressure at orifice (in.)		
Water column	11	11
Orifice coefficient	0.8	0.8

Orifice/ Drill Size	Propane	Butane or Butane–Propane Mixtures	Drill Size	Propane	Butane or Butane–Propane Mixtures
0.008	445	492	51	31,400	35,000
0.009	570	630	50	34,200	38,000
0.010	703	778	49	37,200	40,300
0.011	845	936	48	40,400	44,700
0.012	1,005	1,110	47	43,000	47,600
80	1,270	1,410	46	45,800	50,700
79	1,470	1,625	45	47,000	52,000
78	1,790	1,980	44	51,600	57,200
77	2,260	2,500	43	55,300	61,300
76	2,790	3,090	42	61,100	67,700
75	3,080	3,410	41	64,400	71,300
74	3,540	3,920	40	67,000	74,200
73	4,020	4,450	39	69,200	76,600
72	4,370	4,840	38	72,000	79,600
71	4,730	5,240	37	75,500	83,600
70	5,490	6,070	36	79,300	87,800
69	5,960	6,600	35	84,500	93,600
68	6,720	7,440	34	86,200	95,300
67	7,150	7,920	33	89,800	99,500
66	7,600	8,420	32	94,000	104,000
65	8,560	9,480	31	100,600	111,500
64	9,050	10,030	30	115,300	127,600
63	9,570	10,600	29	129,500	145,200
62	10,100	11,140	28	137,500	152,500
61	10,600	11,800	27	145,000	160,000
60	11,170	12,300	26	151,000	167,000
59	11,750	13,000	25	156,000	173,000
58	12,300	13,600	24	161,500	179,200
57	12,930	14,300	23	166,000	183,500
56	15,100	16,700	22	172,000	190,700
55	18,850	20,900	21	176,500	195,700
54	21,200	23,400	20	181,100	200,000
53	24,700	27,400	19	193,000	215,000
52	28,200	31,200	18	200,500	222,000

TABLE 12-3 (Continued)

b. Natural Gas—1100 Btu/ft^3)
(Btu per hour at sea level)

Drill Size	Orifice/Drill Size					Drill Size	Orifice/Drill Size				
	3 In.	4 In.	5 In.	6 In.	7 In.[a]		3 In.	4 In.	5 In.	6 In.	7 In.[a]
80	445	516	584	638	683	40	23,300	27,200	30,800	33,700	36,100
79	513	595	674	738	788	39	24,300	28,100	31,700	34,800	37,200
78	653	725	820	897	960	38	25,200	29,200	33,000	36,200	38,700
77	820	916	1,040	1,140	1,210	37	26,100	30,700	34,500	38,000	40,600
76	975	1,130	1,280	1,410	1,500	36	27,700	32,200	36,400	39,800	42,600
75	1,075	1,250	1,410	1,530	1,666	35	29,600	34,300	38,800	42,500	45,400
74	1,240	1,440	1,620	1,780	1,900	34	30,200	34,900	39,500	43,300	46,300
73	1,405	1,630	1,850	2,020	2,100	33	31,100	36,100	40,800	43,700	47,900
72	1,525	1,770	2,000	2,230	2,340	32	33,000	38,100	43,300	47,300	50,500
71	1,695	1,915	2,170	2,370	2,540	31	35,200	40,800	46,200	50,600	54,000
70	1,915	2,220	2,510	2,760	2,940	30	40,400	46,900	53,100	58,100	62,100
69	2,080	2,420	2,740	3,000	3,200	29	45,100	52,300	59,200	64,800	69,300
68	2,350	2,720	3,080	3,380	3,610	28	48,200	55,900	63,300	69,300	74,000
67	2,500	2,900	3,280	3,600	3,840	27	50,700	58,800	66,690	72,900	77,800
66	2,660	3,090	3,490	3,820	4,080	26	52,800	61,300	69,400	76,000	81,200
65	2,920	3,470	3,930	4,300	4,500	25	54,800	63,500	71,800	78,800	84,400
64	3,170	3,680	4,170	4,560	4,870	24	56,600	65,700	74,300	81,500	87,000
63	3,380	3,900	4,420	4,830	5,160	23	57,800	67,200	76,000	83,200	88,000
62	3,520	4,080	4,620	5,050	5,400	22	60,400	70,000	79,200	86,800	92,700
61	3,740	4,330	4,900	5,370	5,730	21	61,900	71,800	81,300	89,000	95,200
60	3,920	4,555	5,140	5,540	6,020	20	63,500	73,600	83,300	91,200	97,500
59	4,100	4,760	5,380	5,900	6,310	19	67,200	77,900	88,200	96,700	103,000
58	4,330	5,020	5,670	6,222	6,640	18	70,300	81,500	93,300	101,000	108,000
57	4,500	5,230	5,920	6,480	6,920	17	73,100	84,800	96,000	105,000	112,000
56	5,500	6,130	6,940	7,600	8,120	16	76,500	88,800	100,500	110,000	118,000
55	6,600	7,650	8,660	9,480	10,100	15	79,300	92,000	104,000	114,000	122,000
54	7,402	8,590	9,730	10,600	11,400	14	80,000	93,800	106,300	116,000	124,000
53	8,650	10,050	11,350	12,400	13,300	13	83,700	97,100	110,000	120,000	129,000
52	9,870	11,450	12,900	14,200	15,200	12	87,400	101,000	115,000	126,000	134,000
51	11,000	12,740	14,400	15,800	16,900	11	89,300	103,500	117,300	128,000	137,000
50	12,000	13,900	15,700	17,200	18,400	10	91,500	108,200	120,300	132,000	141,000
49	13,100	15,150	17,100	18,800	20,100	9	94,000	109,000	123,400	135,000	144,000
48	14,100	16,400	18,500	20,300	21,700	8	96,700	112,300	127,000	139,000	149,000
47	15,100	17,500	19,800	21,300	23,200	7	98,600	114,500	129,600	142,000	152,000
46	16,000	18,600	21,000	23,000	24,600	6	101,600	118,000	133,600	146,000	156,000
45	16,500	19,100	21,500	23,600	25,300	5	103,400	120,000	135,700	149,000	159,000
44	18,100	21,000	23,800	26,000	27,800	4	106,800	124,000	140,000	153,000	164,000
43	19,300	22,500	25,400	27,800	29,800	3	111,000	128,600	145,500	159,000	170,000
42	21,500	24,800	28,000	30,700	32,800	2	119,600	138,700	157,000	172,000	184,000
41	22,400	26,200	29,500	32,400	34,600	1	127,000	147,300	167,000	183,000	195,000

[a] Pressure at orifice, inches water column.

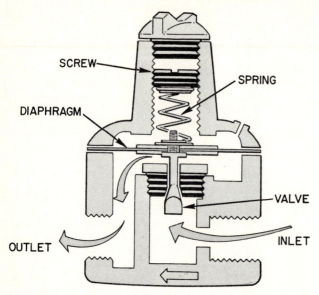

FIGURE 12-3. Gas pressure regulator. The pressure of gas flowing in the direction of the arrow lifts the diaphragm against the spring to achieve a balance at a preset pressure. A surge in pressure will lift the diaphragm, pulling the valve closer to its seat, thus decreasing the opening through which gas can flow.

factors determining the quality of service rendered by the appliance. All modern gas appliances are designed by their manufacturers to have a definite "input rating" expressed in Btu per hour. This rating is usually indicated on the appliance except in case of ranges, which have a more or less standard rating. The input rating indicates almost the highest rate at which heat can be liberated within the appliance and still give service that is satisfactory and safe in all respects. If the gas rate to the burners is adjusted much less than their normal rate or if the pressure is below the required setting for some reason, operation of an appliance becomes very inefficient. Time and/or heat can be lost. If the burners are overrated or overgased, trouble is encountered from incomplete combustion or from the overheating of the appliance. In the majority of cases, the factor that limits the rating is the danger of incomplete combustion and the consequent release of carbon monoxide and sometimes soot. Therefore it is essential that the appliance be properly adjusted and that all reasonable precautions be exercised in the installation of piping, meters, regulators, etc., to ensure constant gas pressure to the appliance.

PIPING SERVICE

Services are sized to carry the ultimate load to the customer. Usually the meter is sized for the load with a 2-inch allowable pressure drop through the meter connections. (Some are sized for a 0.5-inch drop.)

Normally, the gas meter pressure available to a home is about 7-inch water column, and many appliances have built-in pressure regulator set for $3\frac{1}{2}$- to 4-inch pressure. This arrangement allows some pressure drop in the pipelines to the appliance.

LP gas is installed with an 11-inch water column pressure from the tank and no regulator at the appliance (Fig. 12-4).

House piping is sized to take the full load with consideration for the various takeoffs for appliances as shown in Table 12-4.

Often a separate line is run from the meter to each appliance even though the appliances are placed next to each other. This is done to prevent the sudden pressure drop that occurs when appliance No. 2 turns on while No. 1 is

TABLE 12-4

Approximate gas input for some common appliances[a]

Appliance	Input (Approx.) (Btu/Hr)
Range, free standing, domestic	65,000
Built-in oven or broiler unit, domestic	25,000
Built-in top unit, domestic	40,000
Water heater, automatic storage	
30- to 40-gal tank	45,000
50-gal tank	55,000
Water heater, automatic instantaneous	
Capacity	
2 gal/min	142,800
4 gal/min	285,000
6 gal/min	428,400
Water heater, domestic, circulating or side-arm	35,000
Refrigerator	3,000
Clothes dryer, type 1 (domestic)	35,000
Gas light	2,500
Incinerator, domestic	35,000

[a] For specific appliances or appliances not shown above, the input should be determined from the manufacturer's rating.

TABLE 12-5
Maximum capacity of pipe in cubic feet per hour for gas pressures of 0.5 Psi or less and a pressure drop of 0.3 inch water column based on a 0.60 specific gravity (natural gas)

Nominal Iron Pipe Size (in.)	Internal Diameter (in.)	Length of Pipe (Ft)													
		10	20	30	40	50	60	70	80	90	100	125	150	175	200
$\frac{1}{4}$	0.364	32	22	18	15	14	12	11	11	10	9	8	8	7	6
$\frac{3}{8}$	0.493	72	49	40	34	30	27	25	23	22	21	18	17	15	14
$\frac{1}{2}$	0.622	132	92	73	63	56	50	46	43	40	38	34	31	28	26
$\frac{3}{4}$	0.824	278	190	152	130	115	105	96	90	84	79	72	64	59	55
1	1.049	520	350	285	245	215	195	180	170	160	150	130	120	110	100
$1\frac{1}{4}$	1.380	1,050	730	590	500	440	400	370	350	320	305	275	250	225	210
$1\frac{1}{2}$	1.610	1,600	1,100	890	760	670	610	560	530	490	460	410	380	350	320
2	2.067	3,050	2,100	1,650	1,450	1,270	1,150	1,050	990	930	870	780	710	650	610
$2\frac{1}{2}$	2.469	4,800	3,300	2,700	2,300	2,000	1,850	1,700	1,600	1,500	1,400	1,250	1,130	1,050	980
3	3.068	8,500	5,900	4,700	4,100	3,600	3,250	3,000	2,800	2,600	2,500	2,200	2,000	1,850	1,700
4	4.026	17,500	12,000	9,700	8,300	7,400	6,800	6,200	5,800	5,400	5,100	4,500	4,100	3,800	3,500

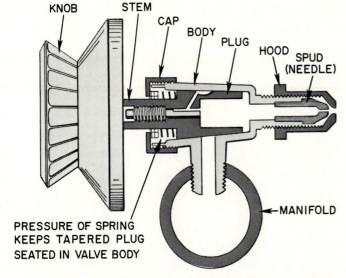

FIGURE 12-4. Gas valve used to control the flow of gas to the burner. Rotating the knob will change the area of an opening in the side of the plug in relation to the passageway from the manifold. This valve is shown in the OFF position.

operating, causing No. 1 to go out. Table 12-5 lists the maximum gas capacity for various pipe sizes.

At the time of installation, when the gas is first turned on, *all* pipe joints must be tested for leaks. The easiest way to do this is to swab a soap suds solution around the joint and watch for soap bubble movement while the system is under pressure.

One final word of caution—**No smoking on the job!**

STUDY QUESTIONS

Gas

1. All gases derived from wells are called natural gas. True or False
2. Propane and butane are colorless and odorless. True or False
3. Propane gas is lighter than air, hence will dissipate quickly. True or False
4. Perfect combustion produces water vapor and carbon dioxide. True or False
5. List five causes of incomplete combustion.
6. What four factors must be considered in regard to the input rating of a gas appliance?

KEY

Gas

1. True
2. True
3. False
4. True
5. Lack of air, overrated burners, flames hitting cold surface, too much external draft, too little draft, pressure at orifice
6. Heat value, specific gravity, size of orifice, pressure at orifice

13

Customer Relations

FROM A CUSTOMER

Mr. Author, Sir, you claim that you are writing about what service technicians have to know. Wouldn't you agree that 99 percent of the time a technician is working on someone else's machines? The covering name for that "someone else" is **customer.** Well, I am a customer. May I speak for myself? Unlike the appliances and machines, I have feelings and reactions. Mistakes in handling me can't be repaired with some solder and paint. If I am damaged, the technician can't replace me; I'll replace the technician—you can bet on it! So, in all fairness to your readers, you had better let 'em have it "straight from the horse's mouth."

Let's see, where shall I begin? I need to have my washer serviced. Who shall I call? I don't want to call the dealer. The machine is out of warranty and that store has changed hands. Everyone says their service prices are— sky high! At the club the other day someone mentioned Bill's Service. I could try them but

someone else said that clean uniforms and a flashy truck was all they had. Their men couldn't even follow directions as to how to find the house.

Neatness and an imaginative presentation, such as uniforms, a sharp truck, or a catchy slogan are all good attractors; but word of mouth goes faster than the trucks, and it can go more places. It's rather like trying to tell a book by its cover or a present from its wrappings.

The yellow pages are a help. Sometimes I can tell from the way my call is received. If there is one factor that turns me off, it's poor telephone courtesy. I expect the 'phone to be answered with identification clearly stated. I like to have an attentive listener on the other end of the line that sounds as though he recognized me as a human being in need of help. I like honest answers and honest estimates. I appreciate being asked pertinent questions—all that are needed—so as to save everyone's time. Of course, I don't mind spelling my name or

FIGURE 13-1. A customer's first impression of your organization is made by your telephone manner.

helping the listener learn to pronounce it correctly. However, if I do that, then I expect to hear my name said correctly (Fig. 13-1).

If an appointment is made, I expect to keep it and I think the technician should do likewise. When he comes to my home, I presume he will have brushed up on the kind of washer that I said I had so that he need not waste time figuring out how to get to the innards.

One point that really turns me off is the attitude some technicians have that shows they think women are incapable of intelligent communication regarding the nuts and bolts of their appliances. That goes for the office machines as well as those in the household. I admit that some women are not mechanically inclined. Not only can I describe the difference between "squeak" and "thump," but I can usually give a reliable clue as to which part of my machine is inoperable. I can usually understand when diagnosis suggests new parts needed, too. Try me.

But don't try to sell me a new machine! And keep your opinions as to "my" brand to yourself, *unless* you agree with mine. Techni-

cians often work in front of my young children and we are trying to teach good manners. Therefore, I take a dim view of derogatory remarks and careless, uncouth language. My little ones may get in your hair, but you should have learned how to handle them before you started making house calls. A gentle reminder to me will help you out.

When a technician arrives at my door on time, neat in dress, with a pleasant attitude pronouncing my name correctly, and obviously knowledgeable about *my* machine, he gains my confidence immediately. I'll probably usher him to the offending appliance, point out my observations, shoo the children away, and leave him to his "fixing."

I *will* expect an explanation of what has been done and a look at any old parts replaced. If an expensive part is needed, I would expect to be consulted *before* it is put into service. I will also expect any major mess the technician had to make to be cleaned up by him before he leaves. And another thing. . . .

Well, I *could* go on, but it's *your* paper. Far be it from me to do unto others what I *don't* like done to me—insult intelligence. I hope I have given you a clear picture.

Sincerely,
A. Customer

CUSTOMER RELATIONS

This letter from A. Customer must be seriously considered as being the feelings of a special customer—yours, and not that of an average customer. In fact, there is no average customer. Averages are for statistical purposes and, believe me, a customer does not want to be considered a statistic.

When something goes wrong with her appliance, your customer is upset and unhappy. If she cannot get prompt, reliable service, she becomes more so and is very willing to tell the neighborhood about her appliance problems and the poor service that she got from the company. Make certain that it is not your company she is complaining about.

A technician faces two problems when making a service call:

1. To ease the customer's hurt, anxiety, or fears about her appliance
2. To diagnose the cause of appliance failure

Fixing the appliance is often easier than placating the customer.

MANUFACTURERS ARE HELPING

The major appliance manufacturers have not only spent much time and money to provide technical service literature and training aids relating to their products, but they are also providing training material to improve the service technician's attitude to the job and to the customer. They realize that their product's performance will be only as good as the person who installed it and future sales depend on the ability of the service technician to keep the customer happy with their brand of appliances.

You must believe as they do—that without customers you are out of a job. Every time you mistreat and lose a customer you are decreasing your job security.

WHAT MAKES A GOOD TECHNICIAN

Very simply stated, a good technician has two things going for him: product knowledge and attitude. Basic product knowledge is covered in the rest of this book. It must be kept up to date by reading trade journals and studying the service literature relating to new models and products. Include in product knowledge any special tools and test equipment needed to service efficiently those products for which you have a responsiblity (Fig. 13-2).

Attitude is: your approach to each day—all people—their machinery—shown by your appearance and actions. A positive attitude can be developed with practice and by holding the right thoughts. Attitude may make the difference between success and failure in a service business (Fig. 13-3).

FIGURE 13-2. The successful technician never allows himself to become obsolete.

THREE OBSERVABLE AREAS

Your family, fellow workers, employer, and customers judge your nontechnical abilities/traits in three major areas:

FIGURE 13-3. Set a good positive attitude—then work for it! Think of what you want your future to be—the best in the area!

Appearance
Work habits
Ability to communicate

These are with you all the time for everyone to see and judge. To be successful, you must cultivate those abilities that satisfy the people you are with.

Let's examine each in more detail.

Appearance

This is the way you look, smell, dress, and move. Also, it is the way you keep your tools and truck and the way you approach other people. *It is the way other people see you.*

Would you like to pick you up as a hitch-hiker? Or closer to home, would you like you to make a service call to your home?

People like people who

1. Are clean-shaven, washed, and wearing clean clothes, or at least as clean as can be expected for the hour of the day and the type of work.

2. Are neat—clothes, shoes, truck, tools should all start the day clean and in order. Here, too, the time of day will affect this factor, but there is never any excuse for sloppiness. Some good service technicians carry a change of clothes or a shop coat in their truck to maintain the neat, clean image. All should carry a waterless hand soap and some paper towels or shop cloths in the truck to allow a quick cleanup after a call.

3. Are pleasant—a smile, confidence of action, posture, consideration for the other person all make an impression on the customer and must not be overlooked.

4. Some no-no's:
   (a) No smoking on the job.
   (b) No drinking on the job.
   (c) No beer, onion, garlic, or other smelly food for lunch.

Work Habits

After you have entered the house and start to examine the appliance, the way you set your toolbox down and proceed to examine the machine tells the customer much about your ability. At this point, your care for the customer's home and its furnishings, plus your confident, knowledgeable approach, will do much to make the customer respect your ability. Placing a clean drop cloth or newspapers on the floor to protect it, an orderly arrangement of tools to be used, a neat and careful placing aside of parts removed all contribute to customer confidence (Fig. 13-4).

When the job has been completed and the appliance tested, make certain that you clean up *all* your tools, drop cloth, and anything else that was used to make the repair. Finally, *clean the machine;* remove all dirt and fingerprints. A little polish at this time will do much to improve customer relations.

FIGURE 13-4. Your neatness and care for the machine and customer's home will do much to inspire customer confidence.

Ability to Communicate

The need to communicate starts when the customer makes the telephone call requesting your services. A cheerful voice clearly identifying the company, asking a few pertinent questions, and setting a realistic time that the technician can be at the home are all important. The technician must make every effort to arrive at the appointed time. If for some reason he cannot, he must call the customer with an explanation and then set up a new time. The technician must remember that the customer may have a schedule, too, and can only be at home during specific hours.

Upon arriving at the house, identify yourself, calling the customer by name and mentioning the appliance you are to service.

Once in the house, *listen* to the customer's complaint and ask a few related questions. It is possible that the customer knows exactly what is wrong but does not have the tools or parts to fix it. If she mentioned her suspicions over the phone, the wise service technician will have brought a new part with him to save another trip. In this way the customer gets faster service and feels that her intelligence was not insulted by ignoring her diagnosis.

If the machine needs a part that must be ordered, or if it has to be returned to the shop for a more complete evaluation, you must tell the customer what you are doing and why. If at all possible, try to give the customer an idea of how much the job will cost, how long it will take, and when her appliance will be returned.

At the completion of the job, the customer deserves an explanation of what was wrong, and if it was a matter of careless usage, how further trouble can be avoided.

All this communication with the customer must be carried on in polite, nontechnical language.

GENERAL CONSIDERATIONS

The customer expects the service technician to be an expert in his area, but she does not want the technician to make a guess as to the trouble, even though you know from past experience on that model what the trouble may be. You must make a thorough check of the appliance before you commit yourself. The customer is paying for your time and should get her money's worth.

When the technician finds a major, expensive problem on an older appliance—for example, a defective compressor on a 10-year-old refrigerator—he must tell the customer how much it will cost. He must not say "you need a new refrigerator." He can suggest that the service cost is a lot of money for a refrigerator that has given so many years of service. He can suggest that the customer think it over before making the repair. Only when asked for an opinion can he honestly say that he would recommend replacement rather than repair.

If the service technician has had good relations with the customer, he will not have to do any selling or even mention that his organization sells refrigerators. A customer knows that the satisfactory performance of her appliance depends largely on the quality of service it receives, and she will be loyal to the company that serves her well.

When the new appliance is delivered, the installer must take time to demonstrate the machine and explain to the customer its features and any differences from the old machine. He must make certain that the machine is correctly installed and running properly and that the customer has all the literature that belongs with it. He must also acquaint the customer with the warranty policy on that particular machine so as to avoid misunderstandings at a future date.

Naturally, the service organization expects to be paid for the work it does. Warranty work is usually paid for by the manufacturer, but out-of-warranty work is paid for by the customer.

At the time the customer makes the initial call for service, the person taking the call should acquaint the customer with the company's policy on service rates and collections. Most service organizations operate on a COD basis, especially on the first service call from that customer. Later, after the customer's integrity has been established, work may be charged. The COD procedure has much merit in that it saves billing and bookkeeping costs and decreases the

the customer politely to remove the pets and children because they are getting in his way and may get hurt.

SUMMARY

Good customer relations are the lifeblood of a service organization, and your ability to cope with customer objections will do much to keep it flowing.

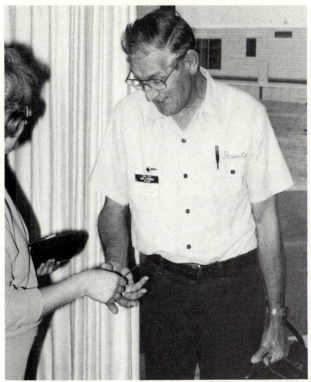

FIGURE 13.5. When a customer calls for service, she knows what it will cost and is prepared to pay. Collect for the job when it is completed.

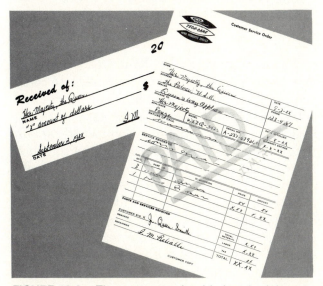

FIGURE 13-6. The customer should sign the job order to indicate satisfaction with the work done. You should mark the copy "PAID." Be businesslike.

S ERVICE
A SSOCIATES

route no. 1 box 325
st. joseph, mich.
49085
ph. 616-429-9449

Dear Mr. Author:

You asked for an opinion, but you'll have to take a few of my comments along with it. I'll buy the customer's statements, can't blame her for her attitude; and what you have said about the service technician's attitude and actions at a house call are good. But—you're a good Boy Scout and an idealist. People aren't all that good. This is true for both the service personnel and customers.

There is no question that some servicemen are dishonest and will charge for parts that aren't used or will magnify the problem to charge a higher fee. Some just don't know enough about the appliance and take more time to do the job than they should. Others are just downright discourteous and careless about the customer's property. Fortunately, the customer soon learns and the quacks and clods don't last long in this business.

On the other side of the coin, some customers are not too honest with the servicemen. They forget to pay their bills or invent excuses to find something wrong with the work in order to avoid payment or to get extra work done. They will demand arrival for a service call at a specific time, then not be home when the man arrives. Or how about the lady who says, "I hope you don't mind Johnny watching while

risks and costs of nonpaying customers (Figs. 13-5 and 13-6).

There are times when the customer, her children, and/or pets will annoy a man trying to do a good job. The best he can do is to ask you work, he's home from school with the mumps." Now she tells me! (See Fig. 13-7.)

Sometimes even the best customers get trapped in an embarrassing situation that they feel reflects on their housekeeping. Very few customers know enough to clean the condenser of their refrigerator semiannually, less know how to clean excess dust and lint from their dryer or washer. In fact, they might not be able to move the machine to get at it for a thorough cleaning. These conditions make good nesting places for mice and cockroaches that the customer may not know she has. But when you, as a service technician, move the machine away from the wall and see these animals scurrying around—what do you do?

Let your students kick this one around for a while.

Yours for better-qualified service technicians,
Joe

THE ETHIC AND THE MORAL!

Joe has some good points in his comments on dishonesty and his question on what is the right thing to do when you find an abnormal condition that reflects on the customer's standards.

There is no pat answer on how to handle a roach infestation. You'll have to play it by ear. Keep in mind the overall housekeeping and degree of infestation. Will a quick spray of roach killer clean it up, or is it likely to spread them? Many service technicians will not work on a machine with roaches until the customer has called in the exterminator. Suppose they get into the toolbox and get carried to the next customer's home or to the technician's own home?

How about the sick child? Will you pick up his "bug" and get it or carry it on to the next customer?

How about housekeeping? You are not to judge the customer's housekeeping *except* as it affects the operation, performance, and safety

FIGURE 13-7. Customers' children like to help the service technician, no matter what he is working on.

of the appliance. Even though the customer may not be able to move the appliance to clean behind it, she should be told of the effects and potential danger of dust, lint, or paper accumulation behind the appliance. You should explain that blocking air from a refrigerator's condensing coil will lower the machine's cooling efficiency and build up external heat; or that blocking the airflow into a dryer will prevent it from drying the clothes and allow it to overheat. Any accumulation of combustible material around an appliance that prevents a free flow of cooling air is a fire hazard.

Think these situations over and apply them to your home. How would you or your family like to be treated? Remember the *Golden Rule*.

14

Safety

Anyone working with electromechanical equipment is exposed to potentially dangerous conditions which in a moment of carelessness could cause an accident. On-the-job safety must be practiced constantly by all workmen. The following information should serve to help you recognize a potential danger.

ELECTRICAL SAFETY

When working with any electrical equipment the only time power should be on is while you are testing it for operation or making voltage tests. The power should be turned off when making continuity tests or replacing parts.

The power can be turned off three ways:

1. By turning off the appliance at its switch
2. By disconnecting from the power source
3. By removing the fuse or breaking the circuit for the circuit supplying power

to the machine. On heavy-duty and high-voltage machinery, the switch should be *locked* in the OFF position while it is being serviced.

When working with someone else, always warn him when you are going to turn the power on—*and make certain that he answers before you do turn it on!*

Working Conditions

Make certain that there is ample light in all areas in which you will be working. You must be able to see where your hands will be in respect to electrical wiring, terminals, and possible moving parts.

The work area should be clean, dry, and clear of unnecessary material. The floor should be free of oil, grease, soap, or anything that would make it slippery or that could make your footing insecure.

All tools and auxiliary equipment used should be clean, secure, and in good condition.

Avoid the use of unsafe or questionable equipment.

Wear properly fitting clothes. Avoid loose and flapping clothing that could get caught in a machine. When working with live electrical circuits, remove wristwatches, ID bracelets, rings, and any other metal objects from your hands and wrists. They could touch a live circuit and do damage. Wear safety glasses whenever working around machines that can produce chips, sparks, or splashes; also when using refrigerants.

If necessary, improve the working conditions before you start a job.

Equipment Ground

All electrical machinery, tools, and appliances should have their nonconducting metal parts connected to a safety ground. The term *ground* refers to a pipe or metal rod buried in the ground. The metal water supply pipe to the building may be used, *but* a gas supply pipe *must not* be used!

Connection to this ground is made to the wiring system with a special wire connecting from the fuse panel to the buried pipe. One side of the normal 120-volt system and a grounding wire is connected to this grounding system. The grounding wire connects to the metal frame of the appliance through the green wire in its power cord. The resistance from any exposed metal part on the machine to the grounding (round) terminal on the cord plug (three-prong type) should be less than 1 ohm.

Electrical Shock

All the discussion on care in working with electricity is to avoid shocking the user. Electric shock can kill and it does not require very much current.

It is the electrical current flowing through the body that does the harm. The voltage pushes the current against the body's resistance. Thus either a higher voltage or a lower body resistance will allow more current to flow—increasing the danger. Current as low as 1/1000 (0.001) of an ampere can be felt. Currents of 0.075 to 0.100 ampere or more are

nearly always fatal. A current as low as 0.012 to 0.015 ampere or more will usually cause the muscles to freeze, so that the person cannot let go of the connection (see Figure 14-1).

Total body resistance will vary from a few thousand ohms to 500,000 ohms depending on the skin dryness and its condition. Resistance of the flesh, as at a cut, can be less than 1000 ohms. Therefore, carefully protect your skin so that it can protect you.

An ac voltage as low as 50 volts can be fatal when the skin or body resistance is low. With normally dry skin and dry conditions, a

FIGURE 14-1. Physiological effects of electric currents.

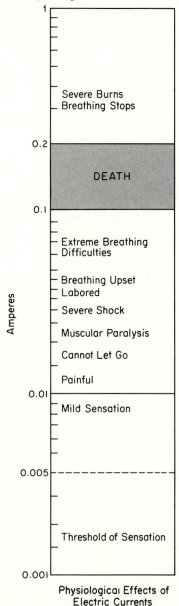

Physiological Effects of Electric Currents

120-volt shock will cause a skin burn but need not be fatal. A 240-volt or higher voltage condition can be very dangerous, so **be extra careful.**

Any appliance that has a current leakage over 0.005 ampere (5 milliamperes) will give the user a shock and should be checked for leaky insulation, and proper repairs made.

Precautions

1. Do not service high-voltage or dangerous equipment alone.
2. Make certain that the system is turned off or disconnected from its power source before changing any part.
3. Keep dry.
4. Do not ground yourself.
5. Make certain that your test equipment can withstand the voltage.
6. Keep your hands away from bare wires and live terminals.
7. Warn everybody nearby when power is about to be returned.
8. If possible, use tools with insulated handles.

Storage Batteries

Storage batteries require a liquid electrolyte in each cell. This is a sulfuric acid and water mixture which can burn skin and make holes in clothes when spilled. *Good* practice would be to wear eyeshields, rubber gloves, rubber apron, and rubber boots when doing any amount of work on storage batteries. At least, when testing a battery or adding water to it, wear eye protection and be careful; *do not splash!*

Any spillage or cell overflow should be flushed off with water or neutralized by spreading baking soda over the area. This can be washed off later.

During the charging cycle, hydrogen gas is generated within the cell. The cell fill caps must be removed to prevent an explosive accumulation of this gas. The escaping hydrogen gas will burn and/or explode when it is contacted by a flame or spark. So *do not* smoke around a storage battery. Also—turn the power

to the charger *off* before disconnecting its leads from the battery terminals. The spark that occurs when pulling a live clip from a battery terminal may be all that is needed to cause an explosion.

A battery is heavy and has small (if any) handles. If possible, use a battery-strap type of carrier when removing or installing a storage battery. If not, grasp it firmly in both hands, lift, and carry it so as not to spill any of the acid. Observe all the precautions that should be used when lifting a heavy object.

Power Tools

All portable electrically powered tools should have a three-wire cord in which the third wire is connected to the nonelectrical metal part of the tool (the other two are the current-carrying wires). The other end of the third wire (the grounding wire) must be connected to a good electrical ground. If this is not available at the outlet into which the tool is connected, an extra wire should be provided to a grounded pipe (*not a gas pipe*). This is particularly important when working out-of-doors or in a damp location!

Before using a tool, its power cord should be inspected for damaged insulation and condition of the attachment cap. These should be clean, dry, and free of defects. The tool should be clean and dry. A slippery, oily tool is dangerous. Clogged ventilation holes will allow the motor to overheat and shorten its life. The switch should be tested for operation. It should feel secure with a positive connection.

Permanently installed electrical tools usually have no trouble with their connecting wiring, but it should be checked in case someone got careless. Make certain that there is adequate work area around the machine and that the floor is dry and secure. Never operate a machine without its safety guard in position, and do not start a machine until you have checked to see if anyone is in the way. Always turn the machine off when the work is finished or when you have to inspect or repair the machine.

When installing or connecting a machine to the power line, make certain that the supply voltage matches the nameplate requirements.

GASES/VAPORS/FUMES

Volatile liquids tend to evaporate at low temperatures to produce vapors that can be as dangerous as gases escaping from their containers. They may be explosive, toxic, or asphyxiating. They are rarely harmless.

As long as the containers are correctly sealed and stored, there will be no danger, but when the fumes escape, a danger exists.

Explosive Vapors

The vapors produced by liquids, such as gasoline, alcohol, paint thinners, certain cleaning fluids, and other hydrocarbons having a low flash point are combustible. When mixed with the right percentage of air, any excess heat— a match, the spark of a switch, or hot metal— will cause the mixture to ignite. It will either burn or explode. Adequate ventilation to prevent a dangerous vapor accumulation is the best safeguard. In addition, *do not smoke,* and use vaporproof electrical switches and lighting in the area where these materials are stored.

Toxic Vapors

Fumes from other easily evaporated liquids, such as chlorinated solvents or ketones, and some gases will cause people to have headaches, nausea, dizziness, and generally cause normal body functions to lose control. Damage may be temporary, limited, permanent, or cumulative. Symptoms may be fast or slow in showing themselves. Avoid exposure to all questionable vapors. Again, as with the explosive vapors, good ventilation is the best safeguard.

Asphyxiating Vapors

Many fumes, gases, or vapors are not explosive or toxic, but if they are present in sufficient volume to displace the normal air supply, they can suffocate a person. A person needs oxygen for survival and if the gas concentration is so great that the oxygen supply is decreased, a person could "drown" in a "safe" gas atmosphere.

This is particularly important to remember when working in a confined space such as a tank, manhole, or small room with inadequate ventilation. A "fresh air" supply is essential for life.

Refrigerants

Of the more commonly used refrigerants, ammonia has a distinctive odor and is irritating to the eyes, lungs, and nasal passages. One percent of ammonia fumes in air could be fatal if a person were to breathe it for about a $\frac{1}{2}$ hour. In the event of a leak, get out of the area at once and let someone with a gas mask do the repair job. Ammonia is used primarily in commercial installations.

The halide refrigerants, such as R12 and R22 (commonly called Freon), are normally not toxic or explosive. They are colorless and almost odorless, thus requiring special test equipment for locating a leak. Since the halides are heavier than air, they tend to settle in a room. When a halide gas is exposed to a flame or glowing electric heater element, the gas tends to break down. When the contact is hot copper, it will produce a green color in the flame. The products formed from the decomposition of these gases have a sharp, pungent odor and will be harmful to the respiratory system.

Should a large leak of refrigerant develop:

1. Make certain that all gas pilot lights and electric heater elements are turned off.
2. Open the windows for ventilation.
3. Remove canaries and other pets from the area.

If you have to remove the refrigerant from a system, connect a hose to the system and run it outside the building.

Since refrigerants absorb heat when they change from a liquid to a gas, they get very cold. Refrigerant liquid spilled or sprayed on the skin will freeze it. A drop on an eyeball could be very dangerous. Safety glasses should be worn whenever handling refrigerants.

Fuel Gases

The two most common types of fuel gas in use are natural gas and bottled (LP) gas. Both are colorless and odorless, but natural gas is treated with an obnoxious artificial odor to warn of leakage. These gases are not toxic, but if they are in high concentration in a room and breathed steadily for awhile, they could cause death by suffocation.

When these gases are present in a room, there is danger that any flame or spark could trigger an explosion. To be safe, anytime you smell gas or suspect a leak, *do not* light a match or turn on a light switch to look for the leak. First ventilate the area, then locate the leak.

Combustion Products

Whenever a fuel is burned, smoke and other products of combustion go up the chimney. Carbon monoxide is one of the products of incomplete combustion. An improperly adjusted and vented heater could produce enough carbon monoxide in a room to be fatal.

The other problem with open heaters in a closed room is that they use the oxygen, and a person could suffocate due to lack of oxygen. Always make certain that an ample air supply is available for any fuel-burning heater and that the products of combustion can escape from the room.

FIRE HAZARDS

A fire needs:

1. Fuel
2. Heat
3. Oxygen

The fuel could be paint thinner, solvents, gas, paper, wood, greasy rags, or anything else that will burn.

The heat could come from a match, an electrical spark, a hot wire, sunlight, hot water, a steam pipe, or heat generated by decomposing vegetation. Some fuels do not require a very high temperature to start a fire. The oxygen would come from the surrounding air.

The best safety precaution is to keep the fuels and the heat away from each other. Any storage area should be well ventilated to *prevent heat from accumulating*. Do not allow piles of rags, papers, and such trash to collect in an area where someone might drop a match or cigarette, or near a hot or warm machine.

Fire Fighting

Most fires, if caught soon enough, can be controlled with a hand extinguisher. These should be conveniently located throughout the home and in factories, and there should be one in the car or truck.

CO_2 or dry chemical extinguishers are the most satisfactory for general use. Water will spread an oil, gasoline, or paint fire, and could be dangerous in an electrical fire unless all the power has been turned off.

FIRST AID

Any wound that breaks the skin should be treated. Minor cuts should be washed with soap and water, dried, treated with a mild antiseptic, and covered. A puncture wound should be squeezed to make it bleed to flush out any dirt that may have been on the nail or whatever caused the wound. Wash the area around the wound with soap and water, apply a mild antiseptic, and cover it. A deep puncture wound must be treated by a doctor, including steps taken to protect against tetanus. Splinters should be removed with tweezers or a sterile needle, then treated as a puncture wound.

Minor burns should be treated with cold water, either by submerging the burn or by applying a cold wet compress. Keep the burn area cool, wet, and covered until the pain has stopped, then cover the area with a sterile dressing to protect it. Major burns should be treated by a doctor, but keep them covered with cold, wet cloths until a doctor is available.

Any skin break should be watched carefully for signs of infection, which may appear several days after the wound was made. Signs of infection are:

1. The area around the wound will redden and feel warm. There may be some swelling and pain.
2. Red streaks may radiate from the wound.
3. Chills or fever.

Any infection should be treated by a doctor at once.

Severe Wounds

An animal bite is a puncture wound and should be treated by a doctor immediately. Wash the wound and the surrounding area very carefully with soap and water to remove the animal's saliva, then treat as a puncture wound. If the animal is available, it should be turned over to the police and checked for rabies. A doctor will determine the best method to protect the patient against rabies and tetanus.

Severe bleeding can usually be controlled by pressing a gauze pad or other *clean* cloth directly over the wound and securing it in position. Try to keep the patient in a horizontal position and watch carefully for signs of shock until he is under a doctor's care.

Shock

Any serious injury will cause some degree of shock to the system, so you must do what you can to lessen it. Symptoms of shock are:

1. The skin is pale, cold, and clammy.
2. The pulse is rapid.
3. Breathing is shallow, rapid, or irregular.
4. The injured person may be frightened, worried, and either restless or numb.

Treatment for shock is as follows:

1. Keep the person lying down with the head lower than the feet *except* for head or chest injuries; then keep the head and shoulders above the feet.
2. Loosen all tight clothing.
3. Keep the victim warm and comfortable.
4. Act calm and reassure the victim.

Electric Shock

Before touching the victim, make certain that he is no longer in contact with the electric circuit! If necessary, break the source of power from the victim by any *safe* means, such as:

1. Turn off at switch.
2. Unplug the machine.
3. Remove a fuse.
4. Use a *dry,* nonmetallic pole, rope, or cloth to separate a wire from the victim.

Protect yourself by standing on dry or insulated ground, touching and using only *dry insulated* materials.

After the victim is clear of the electrical source, check for pulse and breathing. If necessary, apply mouth-to-mouth breathing.

Moving a Victim

Do not move an injured person until you are certain that there are no neck or back injuries!

A broken arm or leg must be supported before moving the person; use a splint securely tied to the broken limb on either side of the break. If splints are not available, a rolled-up newspaper or anything else that is reasonably stiff will do.

The easiest method of moving a person is by stretcher. A wide board or door can be used if it seems necessary to move the victim. Dragging him on a blanket will be better than pulling him by an arm or leg. Try to keep the victim in a reclining position.

In all cases keep the victim quiet, warm, and reassured. Treat him gently and keep yourself calm.

STUDY QUESTIONS

Safety

1. Who should practice safety?
2. When should safety be practiced?
3. What is done with the power supply when replacing parts or making a continuity test on electrical equipment?
4. Name three ways that power can be turned off.
5. When warning people nearby, what two things should you be sure of before you turn on a circuit?
6. It is not necessary to see what you are repairing as long as you can feel it.   True or False
7. It is important to be sure work area is clean *before* you start working on equipment. True or False
8. Keep all tools in good condition.   True or False
9. Repair jobs are often dirty so wear old, loose clothes to fix equipment.   True or False
10. Jewelry is no problem—it is better to keep it on than lose it by forgetting where you put it.   True or False
11. If you wear glasses anyway, safety glasses are not needed.   True or False
12. Some metal part of the machine frame should be connected to the nearest pipe as "ground." True or False
13. The round prong in a three-pronged plug is the "ground" connection.   True or False
14. When measuring the resistance of metal on machinery to ground, the reading should not be more than 1 ohm.   True or False
15. Only high voltage will harm or kill a person. True or False
16. List at least five precautions you should take when working on electrical equipment.
17. Batteries are filled with distilled water so there is no need to be careful of the liquid in them. True or False
18. Smoking is not advised when working on batteries.   True or False
19. If you are familiar with your machine or equipment it is okay to operate it without the safety guard.   True or False
20. It is important to see that the power supply of any equipment matches the nameplate requirements.   True or False

21. If you are handling volatile liquids, you need plenty of room.   True or False
22. The only types of vapors you need to worry about are the explosive ones.   True or False
23. Refrigerants are usually nonexplosive and nonpoisonous, so they are quite safe.   True or False
24. A helpful piece of equipment when working with gas, fumes, or vapors is a gas mask. True or False
25. When suspecting a gas leak, use a flashlight rather than a wall switch to see where you are going.   True or False
26. Carbon monoxide is the product of incomplete combustion of a heater.   True or False
27. What could happen when you use an open heater in a closed room?
28. What three things are needed for a fire?
29. CO_2 or dry chemical should not be used on electrical fires.   True or False
30. What is the first thing that should be done to a wound?
31. Minor burns should be treated with warm water.   True or False
32. Name three signs of infection.
33. All wounds, no matter how small, should be treated by a doctor.   True or False
34. You should wait 24 hours to see a doctor if bitten by an animal.   True or False
35. Can severe bleeding usually be controlled? True or False
36. Name any two symptoms of shock.
37. Name any two treatments for shock.
38. Before helping someone who has had an electric shock, you should unplug the machine or remove the fuse.   True or False
39. After the victim is clear of the electrical source, what two things should you check for?
40. A broken leg should be supported before moving the victim.   True or False
41. Before moving a victim, what two things should you check first?
42. What is the easiest way to transport a person who is badly hurt?

KEY

Safety

1. Everyone
2. All the time

3. Turn it off.
4. Turn off at switch.
   Disconnect from source.
   Remove fuse or break circuit.
5. Be sure that they answer you.
   Know that they are clear from harm.
6. False
7. True
8. True
9. False
10. False
11. Only if your glasses are safety type
12. False
13. True
14. True
15. False
16. Keep power off, keep dry, keep hands away from bare wire and electrical terminals, use insulated tools, be sure test equipment is suitable.
17. False
18. True
19. False
20. True
21. False
22. False
23. False
24. True
25. False
26. True
27. A person could suffocate
28. Fuel, oxygen, heat
29. False
30. Wash with soap and water.
31. False
32. Chills and fever
    Redness and swelling
    Red streaks radiating from the wound
33. True
34. False
35. True
36. Cold and clammy
    Rapid pulse
    Rapid or irregular breathing
    Restless or numb
37. Keep head lower than body.
    Loosen all tight clothing.
    Keep warm.
    Keep victim calm.
38. True
39. Pulse and breathing
40. True
41. No neck or back injuries
42. By stretcher

15

Room Air Conditioners

Information on selecting the correct size air conditioner for a particular room is included in Chapter 4. So are general installation data and electrical requirements.

The room air conditioner (or cooler) is pretty much a combination of fans and filters, plus a refrigeration unit. Therefore what has been said about them also applies to this type of appliance.

As with all refrigeration systems, this, too, is a heat-moving machine whose purpose is to remove excess heat from a room.

The room or window air conditioner is arranged so that the evaporator is located on the room side of the window, and the condenser is on the outside of the window, each having its own fan. A partition inside the unit separates the room air from the outside air. On the outside of the unit, the rubber seals and panels at the window prevent the exchange of inside and outside air (Fig. 15-1).

Within the air conditioner the two separate air systems (the outside and the inside Fig.

15-2) can be joined to allow air intermixing in controlled amounts, thus bringing some fresh outside air into the room.

This extra outside air, being warmer than the inside air, will tend to raise the room temperature somewhat; unless the cooling air thermostat is set for a lower temperature to compensate. The resultant air blend will then be at the desired room temperature.

The capacity or amount of cooling that an air conditioner can do limits the size of room and the number of people that it can keep comfortably cool. So before blaming the air conditioner for not doing its job, make certain that it is working within its limits. Its rating plate will tell you what you should expect, in the way of performance, from the unit. Comparison of one machine to another will be misleading because usage, conditions, and machine capacities will often vary considerably.

Make certain that the wiring is adequate and that some other, heavy, current-using device has not been plugged in on the air condi-

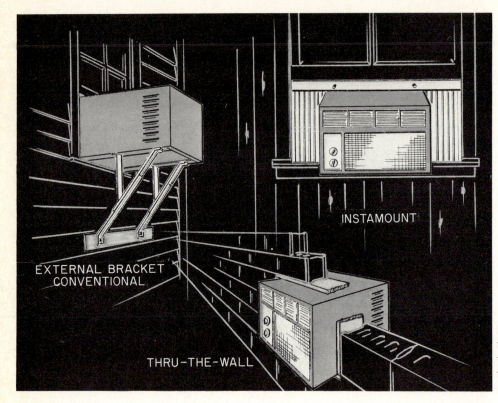

FIGURE 15-1. Window air conditioners may be mounted in many different ways. You must make certain that they are secure and that all surrounding air leaks are closed off.

tioner circuit. Be sure, too, that someone has not opened a window or a door, expecting to cool the entire house.

Clogged filters are a prime cause for service on air conditioners. Instruct the customer in a regular cleaning and/or replacement sched-

FIGURE 15-2. Typical schematic diagram for the air and refrigerant systems in an air conditioner.

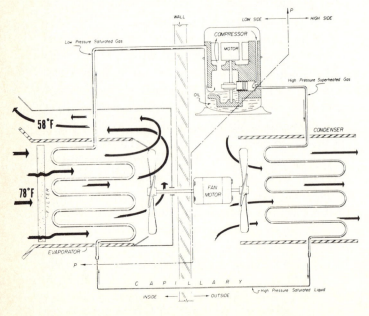

ule for the filters in his machine. How frequently this is done depends completely on local external and internal condition, such as the amount of dust in the air and hours of usage. Specific information on filter cleaning and handling is included in Chapter 9 and in the owner's manual for the model being serviced.

The air conditioner, like the dehumidifier, extracts a lot of water from the air, as part of its cooling action. This extra work requires some of the unit's capacity, especially on a humid day. The result could leave insufficient capacity for cooling down the room. The reduction of humidity by an air conditioner is known as the "latent load," while the reduction of temperature is known as the "sensible load."

This "thermometer-measurable" reduction of air temperature is the cooling effect the customer expects from his air conditioner. The removal of latent heat, by reducing the amount of water vapor in the air, causes a secondary cooling effect because it permits more evaporation of moisture from the skin.

Removing moisture from the air requires approximately 1000 Btu per pound (pint) of water to make the change from vapor to liquid state. Thus every pint of water removed per hour reduces the cooling rating of an air conditioner by 1000 Btu. This fact could make as much as 10°F (6°C) difference in room temperature reduction between a humid day and a nonhumid day for an air conditioner that is working up to its capacity on an average-humidity day.

The moisture removed from the air drains to the condenser sump where it is picked up by the condenser fan and blown through the condenser coil to the outside air. Any of this moisture that lands on the condenser coil helps to cool the coil. It is important to keep the path

FIGURE 15-3. Typical wiring diagrams for a window air conditioner: (a) pictorial diagram; (b) schematic for the same machine having the fan switching options listed.

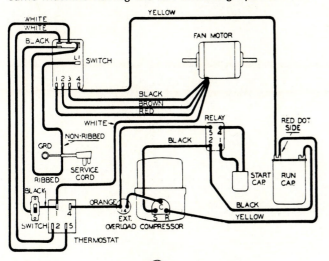

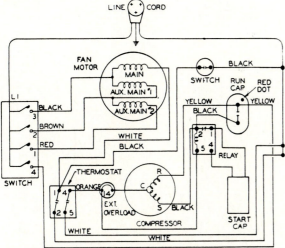

SWITCH POSITION	CONTACTS CLOSED
OFF	NONE
FAN	L1 to 2
HI-COOL	L1 to 3 & 4
MED-COOL	L1 to 2 & 4
LO-COOL	L1 to 1 & 4

this condensed moisture must take open or else water will drip into the room.

If you have to check a part for electrical continuity, it is best to disconnect it from its wiring (Fig. 15-3). When you do, make a sketch showing which wires connect to which terminals. Doing so will ensure correct reconnection.

Check the switch with your continuity tester (Fig. 15-4). There should be no continuity between the line and any other contact when the switch is in the OFF position. Operate the switch to each position and check for continuity. Each position, in turn, must show continuity. A defective switch must be replaced.

The motor-starting relay must show continuity when it is disconnected from the circuit. The contacts will open when the motor reaches normal speed. Your ohmmeter will indicate this point. Make certain that the motor relay is mounted in its correct position or else it will not operate properly.

The motor-starting capacitor can be tested with an ohmmeter. Disconnect the capacitor leads and short its terminals in order to discharge it. You can get a good shock if you fail to do so. Set your ohmmeter at its highest scale and connect to the capacitor terminals. A low-resistance reading will indicate a leaky or

FIGURE 15-4. A test cord of this nature is used to test the compressor independently of the rest of the air conditioner. It differs from the standard refrigeration test cord in that it has an extra set of clips for the run capacitor, when it is used.

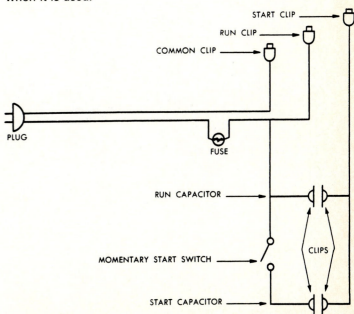

shorted capacitor. On a good capacitor, the needle will swing to the zero ohms position but then slowly change (within a few seconds) to a resistance reading of over 100,000 ohms.

The running capacitor is checked in this same manner. Of course, a capacitor tester is the best instrument for testing a capacitor. The overload protector should show continuity under normal conditions. Check the fan in the air conditioner as outlined in Chapter 9.

The defrost thermostat should show continuity at temperatures above 60°F (16°C). It normally opens at 30°F (−1°C) to stop the compressor. So, for a test, it must show no continuity below 30°F (−1°C). Between 30 and 60°F (−1

and 16°C), it will show no continuity while it is cooling.

The control thermostat will show continuity as its knob is turned, depending on room temperature. Failure to get a continuity change at somewhere near the room temperature, as the knob is turned, is reason to suspect the thermostat. *Caution:* Make certain that the room temperature, or thermostat-feeler-bulb temperature, is within the thermostat's operating range. Any temperature between 65 and 75°F (18 and 24°C) will surely be within this range.

When checking out an air conditioner, be guided by the owner's manual and the wiring diagram included in the unit (Fig. 15-5). Use

FIGURE 15-5. Exploded-view drawing and parts list for an air-conditioner chassis. This does not include the cabinet, bulkhead, and trim, which are shown in separate listings. This information should help acquaint you with part relationships and names.

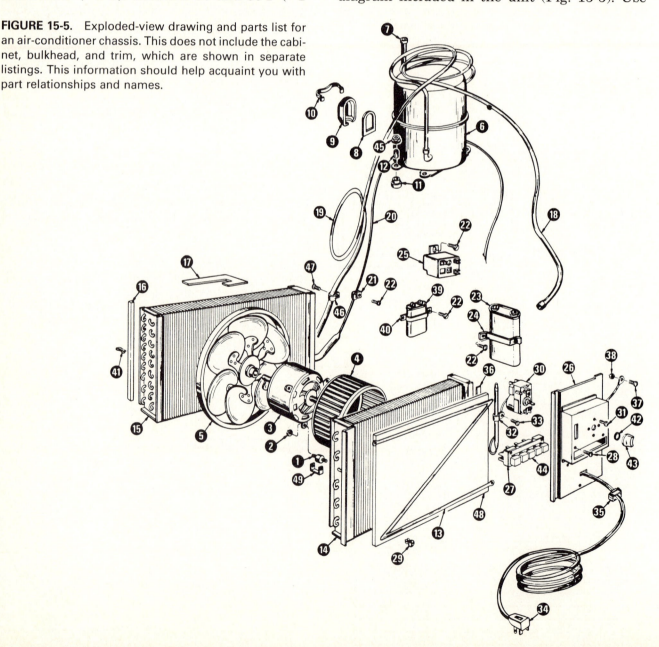

FIGURE 15-5 (Continued)

CHASSIS PARTS FOR "C" SERIES FOR MODELS ACJ-1C18-18-2S, ACJ-2C18-18-2S,
ACJ-1C21-18-2S AND ACJ-2C21-18-2S

REF. NO.	NAME OF PART	QTY.	PART NUMBER
1	Fan Motor Isolator	3	20-3038
2	1/4 – 20 Keps Hex Nut	6	**25-0225
3	Fan Motor (ACJ-1C18-18-2S and ACJ-2C18-18-2S)	1	21-1506
	Fan Motor (ACJ-1C21-18-2S and ACJ-2C21-18-2S)	1	21-1507
	Fan Motor (Opt.) (ACJ-1C21-18-2S and ACJ-2C21-18-2S)	1	21-1523
4	Evaporator Fan (ACJ-1C18-18-2S and ACJ-2C18-18-2S)	1	20-8173
	Evaporator Fan (ACJ-1C21-18-2S and ACJ-2C21-18-2S)	1	20-8174
5	Condenser Fan	1	21-1510
6	Compressor (ACJ-1C18-18-2S and ACJ-2C18-18-2S)	1	20-6714
	Compressor (Copeland) (ACJ-1C-18-18-2S and ACJ-2C18-18-2S)	1	21-2240-1
	Compressor (ACJ-1C21-18-2S and ACJ-2C21-18-2S)	1	20-8093
7	Compressor Charging Tube	1	20-3246
8	Cover Seal	1	20-5535
9	Terminal Cover	1	20-5534
	Terminal Cover (Copeland)	1	20-8557
10	Cover Strap	1	20-5536
	Cover Strap (Copeland)	1	20-8558
11	Compressor Isolator	4	20-5530
	Compressor Isolator (Copeland)	4	*21-2474
12	Compressor Spacer	4	**25-3541
	Compressor Spacer (Copeland)	4	*21-2492
13	Filter Retainer	1	20-8202
14	Evaporator Coil	1	21-1443
	Evaporator Coil (Opt.)	1	21-1442
15	Condenser Coil (ACJ-1C18-18-2S and ACJ-2C18-18-2S)	1	21-1445
	Condenser Coil (Opt.) (ACJ-1C18-18-2S and ACJ-2C18-18-2S)	1	21-1444
	Condenser Coil (ACJ-1C21-18-2S and ACJ-2C21-18-2S)	1	21-1447
	Condenser Coil (Opt.) (ACJ-1C21-18-2S and ACJ-2C21-18-2S)	1	21-1446
16	Condenser Coil Seal	2	20-3052
17	Condenser Air Baffle	1	20-3108
18	Suction Tube	1	21-1617
	Suction Tube (Copeland)	1	*21-2481
19	Discharge Tube Assembly	1	21-1623
	Discharge Tube Assembly (Copeland)	1	*21-2482
20	Restrictor Tube Assembly (ACJ-1C18-18-2S and ACJ-2C18-18-2S)	1	*21-2158
	Restrictor Tube Assembly (ACJ-1C21-18-2S and ACJ-2C21-18-2S)	1	*21-2159
21	Restrictor Tube Clamp	2	20-3079
22	No. 8 - 15 x 3/8" Hex Washer Head Tapping Screw	A.R.	**25-2134
23	Capacitor (ACJ-1C18-18-2S and ACJ-2C18-18-2S)	1	21-0818
	Capacitor (ACJ-1C21-18-2S and ACJ-2C21-18-2S)	1	21-1517
24	Capacitor Strap	1	20-4310
25	Relay	1	20-5547
26	Panel Assembly	1	21-0483
27	Pushbutton Switch	1	21-1532
28	No. 8 - 32 x 3/8" Phillips Round Head Machine Screw	2	**25-8205
29	Ratchet Rivet	2	**25-3563
30	Thermostat	1	20-6083
	Thermostat (Opt.)	1	20-8089
31	No. 6 - 32 x 3/16" Phillips Round Head Machine Screw	2	**25-2203
32	Plastic Clip	2	**25-3605
33	No. 8 - 15 x 1/2" Hex Washer Head Tapping Screw	2	**25-3107
34	Power Cord	1	20-9172
35	Strain Relief	1	20-7529
36	Evaporator Seal	1	20-3049
37	No. 8 - 32 x 1/2" Phillips Round Head Machine Screw	1	**25-8207
38	No. 8 - 32 Keps Hex Nut	1	**25-0223
39	Start Capacitor	1	20-4114
40	Start Capacitor Strap	1	20-3284
41	"U" Type Clip	2	**25-3500
42	Thermostat Knob Seal	1	21-1307
43	Thermostat and Switch Knob (ACJ-2C18-18-2S and ACJ-2C21-18-2S)	1	21-0426
	Thermostat Knob (ACJ-1C18-18-2S and ACJ-1C21-18-2S)	1	*21-2145
44	Pushbutton	5	21-1534
45	5/16 - 18 Lock Nut	4	**25-3532
46	Tube Clamp	1	**25-3535
47	No. 8 x 3/4" Hex Sheet Metal Screw	1	**25-9144
48	Unit Filter	1	20-7693
49	Motor Ground Strap	1	20-3039

*NOTE: New parts appearing in parts list.
**NOTE: Purchase locally if available.

the following troubleshooting chart as a guide to your orderly investigation of the possible cause of failure. Do not get involved with the compressor, the evaporator, or the condensor until all external factors have been checked.

If you do have to enter the cooling system,

do it in accordance with the information contained in the refrigeration section of this book. The specific manufacturer's service instructions for the particular model is your best source of information and should be followed whenever available.

Troubleshooting Chart AIR CONDITIONERS

Problem	Possible Cause	Corrective Action
Unit does not run.	No power at unit.	Check power outlet for voltage. If none, replace fuse. If okay, check power cord.
	Low line voltage.	Have customer contact electrician if voltage is under 10 percent of the nameplate rating.
	Defective part or wiring.	Repair wiring or replace part.
Unit blows fuses.	Incorrect fuses.	Replace with correct size and of time delay type.
	Improper voltage/wiring.	Having customer contact an electrician to make proper corrections.
	Defective part or wiring.	Repair wiring or replace part.
	Trying to start unit too soon after shut off.	Instruct customer to wait 2 minutes.
Fan runs, compressor does not.	Incorrect power supply.	Have customer contact electrician.
	Defective part or wiring in the compressor circuit.	Repair wiring or replace defective part.
Compressor runs, fan does not.	Blower wheel or fan blade binding.	Relocate fan assembly or shroud to clear.
	Defective part or wiring in the fan circuit.	Repair wiring or replace defective part.
No cooling-compressor and fan operating.	Airflow restriction.	Replace filter. Clean air passages.
	Loss of refrigerant.	Repair leak and recharge system.
	Inefficient compressor.	Replace compressor.
Insufficient cooling—both fan and compressor operating.	Improper use, excessive load, undersize unit, incorrect thermostat setting.	Instruct customer in unit's limitations and correct thermostat setting.
	Excessive air leakage in room.	Correct sealing in window around unit. Instruct customer to keep doors and windows closed.
	Improper airflow.	Install new filters if needed.
	Slow fan motor.	Clean condensor, evaporator, and other air passages. Check exhaust door position.

***Troubleshooting Chart* AIR CONDITIONERS (Continued)**

Problem	Possible Cause	Correction Action
	Partial loss of refrigerant.	Repair leak and recharge.
	Low line voltage.	Notify customer.
	Inefficient compressor.	Replace compressor.
Compressor stops and starts. Short run time. Cycles on overload.	Attempting to start unit too soon after shut off.	Instruct customer to wait 2 minutes before turning unit on again.
	Overheated compressor.	Allow unit to cool down.
	Low line voltage.	Notify customer to contact an electrician.
	Thermostat set too warm.	Instruct customer.
	Defective thermostat.	Replace thermostat.
	Defective capacitor.	Replace capacitor.
	Defective start relay.	Replace start relay.
	Defective compressor.	Replace compressor.
	Defective overload.	Replace overload.
	Slow fan speed.	Lubricate bearings.
	Improperly located temperature anticipator.	Relocate to a warmer portion of evaporator.
Compressor runs continuously.	Excessive heat load.	If heat load is temporary it will correct itself. If permanent advise customer a larger unit is needed.
	Partial restriction in liquid line.	Remove restriction and recharge unit.
	Loss of refrigerant.	Repair leak and recharge.
	Defective thermostat or incorrect setting.	Check setting. If unit will not shut off and temperature is satisfied replace thermostat.
Noisy operation	Loose parts.	Remove loose parts and tape tubing to solid support.
	Insecure mounting area.	Add additional supports or secure those in use.
	Loose fan blade, dry bearing or motor mount.	Tighten fan blade, lubricate motor, secure its mounting.
	Noisy compressor.	Replace compressor.
Evaporator frosts up.	Low outside air temperature.	Instruct customer not to use unit when outside air is below 70°F (21°C).
	Thermostat set too cold or stuck on.	Adjust or replace thermostat as needed.
	Improper airflow.	Replace filter. Clear condensor and evaporator free of dust and dirt.
	Slow or inefficient fan.	Lubricate fan motor or clear bind.
Moisture dripping into room.	Improper leveling.	Cabinet should slope $\frac{1}{4}$ inch down on the outside.
	Plugged drain holes.	Clear drain holes.
	Extreme humidity.	Advise customer.

16

Blenders

This handy gadget will chop, blend, grind, shred, and liquidize any foodstuffs fed to it. Also, it does a nice job on tall, cool, adult drinks as well as on malts for the children. There are two basic sections to a blender: the base or power unit and the blending container (Fig. 16-1).

Aside from dropping the glass container and breaking it, about the only thing that can go wrong in this area is damage to the cutting blades. These blades will chip or curl back if large chunks of ice, bones, or silverware have been used in the container.

Leakage may be due to a poor seal between the jar and the bushing assembly. Sometimes the bushing can be tightened, but usually it is best to replace the seal. If leakage is due to a cracked jar, the jar should be discarded—it is too dangerous to use.

Typically, to change blades, hold the knife shaft with a wrench and turn the nut clockwise. Once the shaft and blades are out, it is simple to remove the nut and bushing in order to replace a gasket or seal.

Use a tester to determine the condition of the cord. A break in the wire, held in place by insulation, will cause intermittent operation. Shaking the cord while the tester is connected will quickly locate the break.

If the cord has to be changed, it must be freed from the base first by squeezing the strain relief, then by pulling outward. Make certain that pressure is applied on the insert section of the strain relief. The cord ends can be disconnected and the cord replaced.

Reference to the exploded view (Fig. 16-2) will serve to locate parts.

In order to remove and replace the switch, it is usually necessary to first remove the switch escutcheon plate. Once the base and base cover have been separated, it is relatively easy to examine the motor and switch assemblies. Do not be too quick to condemn the motor. Check the switch and control circuitry first.

The motor brushes can be examined by removing two brush caps and pulling out the brushes. The curved surface at the end must

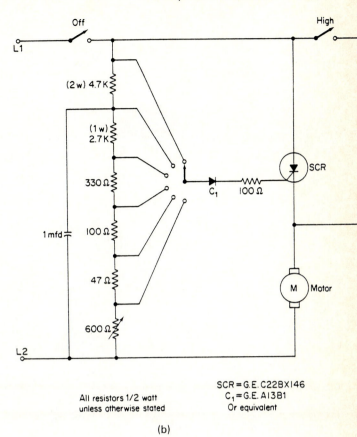

All resistors 1/2 watt
unless otherwise stated

SCR = G.E. C22BXI46
C_1 = G.E. A13B1
Or equivalent

(b)

FIGURE 16-1. (a) The Lady Vanity solid-state blender uses an eight-position General Electric pushbutton switch for speed control. The top no-load speed is 21,000 rpm and reduces approximately 2000 rpm per switch step. (b) Schematic wiring diagram for the Lady Vanity blender. In the high position, all speed control is shorted from the circuit.

(a)

be clean, dry, and undamaged. The carbon must extend at least $\frac{1}{4}$ inch beyond the spring. Short or damaged brushes must be replaced. If they are burned, pitted, or greasy, the commutator on the motor armature must be checked.

The motor housing and field assembly are separated from the base by taking out the screws holding them to the base plate. The entire wired assembly or switch, resistor, and motor field coil can be lifted completely out of the base.

From this point on you will have to follow the diagram for the particular make and model you are working on. If it is not available, Fig. 16-2 should provide some information on basic construction.

Do not fool around with the bearings un-less you have to change them. The retainers are often staked in place and restaking the retainer must be done carefully.

When reassembling the motor, use a few drops of a good, light spindle oil on the felt pads around the bearings. You had better put a drop or two on each end of the shaft also. Make certain that each bearing is free on the shaft but not sloppily loose. While tightening the screws on the motor housing, spin the armature with your fingers to ensure that it is not being bound.

Many blenders have a paper sleeve around the motor. This must be replaced to prevent wiring from getting caught in the fan and armature.

The troubleshooting chart should cover most of the more common problems.

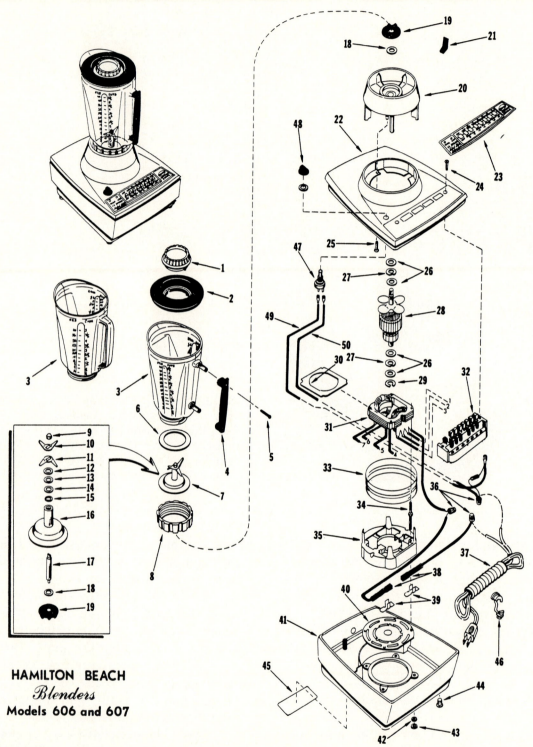

HAMILTON BEACH
Blenders
Models 606 and 607

FIGURE 16-2. Speed control for this blender is by tapped field (31) winding leads connecting directly to matching numbers on the pushbutton switch (32). The phase switch (47) inserts the diode (49) into the circuit at low speed, cutting the normal speed in half. (Courtesy of Hamilton Beach)

FIGURE 16-2 (Continued)

HAMILTON BEACH MODELS 606 and 607 *Blenders*

DESCRIPTION	REF. NO.	PART NUMBER	DESCRIPTION	REF. NO.	PART NUMBER
FILLER CAP	1	3108-215-0000	WASHER (8 Used)	26	4402-322-2000
CONTAINER COVER	2	3107-218-0100	SPRING WASHER (2 Used)	27	4403-327-2600
CONTAINER	3	3110-208-0002	ARMATURE	28	6001-260-0010
GLASS CONTAINER	3A	3310-208-0000	RETAINING RING	29	3075-260-0000
HANDLE	4	3081-218-0100	FIELD CORE BAFFLE	30	3043-612-0000
SCREW (2 Used)	5	4106-008-1802	FIELD COMPLETE	31	6048-606-0010
GASKET	6	3109-206-0000	SWITCH	32	3086-626-0000
CUTTING UNIT COMPLETE	7	6119-265-0000	BAFFLE SLEEVE COMPLETE	33	6033-612-0000
CONTAINER BASE	8	3121-625-0100	FIELD STUD (4 Used)	34	3066-612-0500
NUT	9	4203-250-3201	BRUSH HOLDER HOUSING	35	3040-612-0500
UPPER BLADE	10	3129-265-0000	WIRE NUT (3 Used)	36	3125-451-0000
LOWER BLADE	11	3128-265-0000	CORD	37	3084-211-0100
WASHER	12	3132-209-0000	BRUSH COMPLETE (2 Used)	38	6060-260-0000
WASHER	13	8133-201-0501	BRUSH CAP (2 Used)	39	3039-260-0000
WASHER	14	4402-314-2100	MOTOR CAP	40	6023-612-0000
WASHER	15	4402-320-2100	BLENDER BASE-WHITE	41	3115-612-0102
SUPPORT ASSEMBLY	16	6120-265-0000	BLENDER BASE-HARVEST GOLD	41	3115-612-0402
CUTTER SHAFT	17	3123-209-0000	BLENDER BASE-AVOCADO	41	3115-612-1002
THRUST WASHER (2 Used)	18	4402-252-2100	LOCKWASHER (4 Used)	42	4401-168-2000
CLUTCH (2 Used)	19	3124-201-0000	NUT (4 Used)	43	4201-008-3201
MOTOR COVER	20	6044-612-0100	BUMPER (4 Used)	44	3204-210-0100
CONTAINER REST (4 Used)	21	3111-210-0501	NOMENCLATURE INSERT (606)	45	3098-606-0010
BASE COVER - White	22	8100-606-0100	NOMENCLATURE INSERT (607)	45	3098-607-0010
BASE COVER - Harvest Gold	22	8100-606-0400	STRAIN RELIEF	46	3083-275-0100
BASE COVER - Avacado	22	8100-606-1000	PHASE SWITCH	47	3086-606-0000
SWITCH PLATE	23	3090-606-0000	SWITCH KNOB	48	3087-214-0500
SWITCH SCREW (2 Used)	24	4102-006-2004	LEAD COMPLETE WITH DIODE	49	6155-606-0510
SCREW (4 Used)	25	4101-006-3221	LEAD COMPLETE	50	6157-606-0500

```
WARRANTY
1 YEAR PARTS AND LABOR
CODE DATE ON MOTOR CAP
```

Troubleshooting Chart BLENDERS

Problem	Possible Cause	Corrective Action
Motor won't run.	Blown fuse.	Replace fuse and check wiring.
	Defective line cord.	Repair or replace line cord.
	Defective switch.	Repair or replace switch.
	Burned-out motor.	Replace armature or field coil.
	Frozen bearings.	Free and lubricate.
	Armature hitting due to worn bearing.	Replace bearing.
	Open diode or scr.	Replace.
Motor runs but blades do not turn.	Broken belt (GE).	Replace belt.
	Incorrect placement of container.	Relocate container on base.
	Defective motor coupling.	Replace motor coupling.
Blade edges damaged.	Using large solid ice cubes, bones, or hitting a spoon.	Replace blades.
Runs at high speed only.	Defective switch.	Replace switch.
	Open resistor.	Replace resistor.
	Defective section of field coil.	Replace field coil.
	Shorted scr.	Replace scr.
Container leaks.	Cracked glass jar.	Replace glass jar.
	Poor seal between jar and bushing.	Tighten bushing or replace seal.
Abnormal noise.	Bent cutter blade hitting jar or seal.	Replace cutter blade.
	Motor fan blade hitting.	Straighten fan blade.
	Loose coupling between motor and container.	Reseat the container.
	Loose cutter blade assembly.	Tighten cap nut.

17

Ceiling Fans

Much of man's comfort depends on the air surrounding him: its cleanliness, humidity, temperature, and motion. In the preelectricity days, man used hand-operated fans to move the air around him. In the early days of available electricity, public places such as restaurants, hotels, etc., had large-bladed fans suspended from the ceiling. These were highly satisfactory until air conditioning became popular. Then ceiling-fan popularity faded.

Now fans are proving to be an energy-efficient means of improving the room air circulation in homes and businesses. Although they do not actually raise or lower the room temperature, the air within the room is moved to provide gently changing air around the body. In the summer this will improve the moisture evaporation from the body, making it feel about 6 to 8°F (3.4 to 4.5°C) cooler. Also, the extra air circulation will assist in cooling the room a few degrees. This will allow the air conditioner to be set a few degrees higher so that it will not need to run as much.

In the winter the ceiling fan will move the

hot air that has layered next to the ceiling down into the room living area, raising the temperature a few degrees. This will allow the thermostat to be set a few degrees lower to save fuel.

SIZE

The size of the fan will depend on the size of the room, its usage, and the fan's appearance. A fan that is too large for a room or too small for a room will not look right. A 52-inch-diameter fan will look right in a room 12 feet by 12 feet or larger and will easily move enough air. Long, narrow rooms, or extra-large rooms may require more than one fan to provide proper air movement. A large-diameter fan on slow speed will move as much air as will a smaller fan at a higher speed. People are most comfortable with air movement of between 25 and 35 feet per minute. Too much air movement will feel drafty and possibly generate too much air noise. For these reasons ceiling fans have large blades and are slow moving. They generally move between 50 and 250 rpm. Table 17-1 is a summary of specification listings for several different fans. The four-blade fan is the most commonly used type, and the 52-inch and the 42-inch are the most popular diameters.

Other factors that affect fan performance are the number of blades and their pitch angle. Fans are designed with three, four, five, or six blades, and the pitch-angle design can be from about 10 to about 15 degrees.

Most fans are designed with a reversible permanent split-capacitor (PSC) motor that can be switched for three different speeds. They will also operate from a wall-mounted variable-speed control switch like a light dimmer switch.

LOCATION AND INSTALLATION

The location of the fan has much to do with its efficiency in a given room:

1. The nearer the room center in the ceiling, the more uniform the airflow.
2. The blade tips of the fan must be at least 10 inches from a wall or any obstruction.

TABLE 17-1

Summary of operating data for several different-diameter four-blade fans[a]

Size	Speed	Cfm	Rpm	Watts
52	H	6000	190	115
	M	3150	130	70
	L	1600	55	50
48	H	7200	220	80
	M	3500	100	25
	L	1500	50	10
42	H	5500	240	70
	M	4000	180	60
	L	1300	80	40
38	H	4600	240	82
	M	2860	159	56
	L	1020	61	27

Source: Compiled from Hunter and Fasco literature.

[a] Size, blade diameter in inches; speed: H, high, M, medium, L, low; Cfm, cubic feet per minute air movement; rpm, revolutions per minute blade speed; watts, power consumption at 120 volts, 60 hertz.

3. The fan should not be mounted over a path to a doorway or window area. Air turbulence from these areas may disrupt the airflow from the ceiling fan.
4. The minimum clearance from the floor to blade tip is 7 feet, but 8 feet is better.
5. If lights are mounted under the fan blades, their lowest point should be at least 7 feet from the floor. More clearance is better (Fig. 17-1).

FIGURE 17-1. Light kits, in many different styles, are available for most fans. Watch your head clearance!

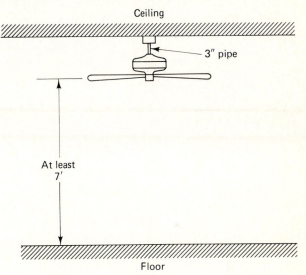

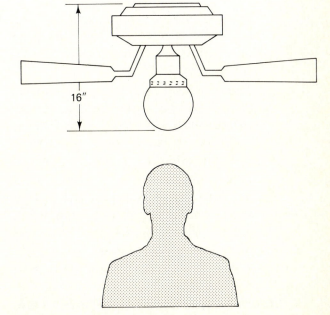

(a)

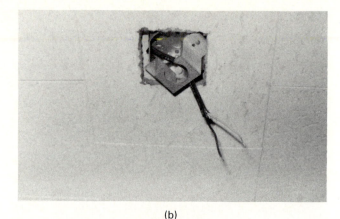

(b)

(c)

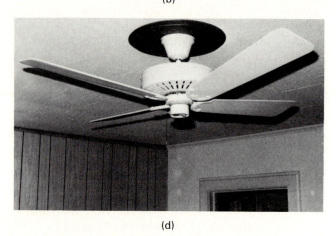

(d)

FIGURE 17.2 (a) Fans may often replace a ceiling light. If a light is needed, it can be mounted under the fan. Light kit assemblies are available for fan use. (b) A special hanger bracket is attached to the ceiling outlet box. This assembly must be secure enough to support a 50-pound live load. (c) Motor assembly and canopy secured to the ceiling. For a soft ceiling such as Celotex, the medallion shown in part (d) is recommended. (d) Final assembly with blades attached and medallion installed. Sometimes the medallion makes the installation more attractive.

6. Fans should be mounted so that the blade tops are at least 10 inches from the ceiling for best air-moving efficiency (Fig. 17-2).

7. The electrical outlet box from which the fan will be suspended must be secure enough to support 50 pounds. It must be supplied with 15-ampere, 120-volt, 60-hertz ac circuit.

8. Each manufacturer of ceiling fans provides an assembly and installation owner's manual with the fan when purchased. Follow those instructions when assembling and installing a fan. Try to double check them when servicing a fan.

9. All models of ceiling fans have a slide-type reversing switch located in the switch housing. The fan switch must be in the OFF position and the blades not turning when this reversing switch is used. The fan switch, also enclosed in the housing, is a pull-chain type having four speed positions: OFF, HIGH, MEDIUM, and LOW. These control the amount of air being moved.

10. Additional wiring can be added to control the fan's operation from a wall outlet. This wiring must be done in accordance with the national and local electrical codes (Fig. 17-3). Some companies are developing a remote-control unit similar to that used with television sets.

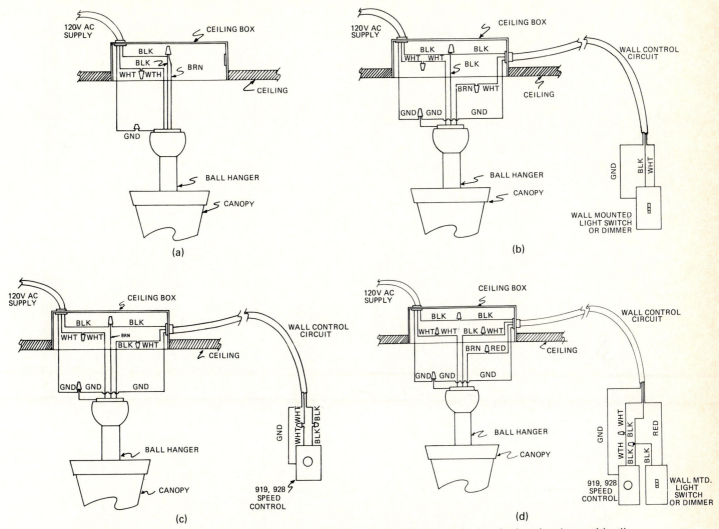

FIGURE 17-3.   (a) *Option 1:* fan and light controlled at unit. This installation is the simplest, with all controls mounted on fan as shipped. Light fixture, if used, is controlled with the pull-chain switch on the light. (b) *Option 2:* fan controlled at unit, light controlled at wall with switch or dimmer. This option is the same as option 1, except that the light is controlled from an existing wall-mounted switch or light dimmer. This option is designed for installations in which the fan is installed in place of a ceiling light or chandelier controlled by a wall switch. No additional wiring is necessary. (c) *Option 3:* light controlled at unit, fan controlled at wall by Model 919 or 928 speed control. This option utilizes the speed-control device to operate the fan from a remote wall location. The light is controlled at the fan. A wall control replaces the existing wall switch. No additional wiring is necessary. Fan switch on unit must be left in high-speed position for control to function properly. To eliminate confusion, the pull-chain of the switch should be cut off after high-speed position is established. (d) *Option 4:* light controlled at wall, fan controlled at wall by Model 919 or 928 speed control. This option can be used when the fan is installed in place of an existing ceiling light fixture. The speed-control device will replace the light switch (or dimmer) in the wall that controlled the ceiling light. An additional box and switch must be installed for this option. (Courtesy of Fasco Industries)

FAN MOTOR SPEED

The basic speed of an alternating-current induction motor is governed by the power supply frequency and the number of poles in the motor. For revolutions per minute from a 60-hertz power supply, the formula is

$$\text{rpm} = \frac{\text{hertz} \times \text{time}}{\text{poles}/2} = \frac{60 \times 60 \times 2}{\text{poles}}$$

Ceiling-fan motors are designed to operate

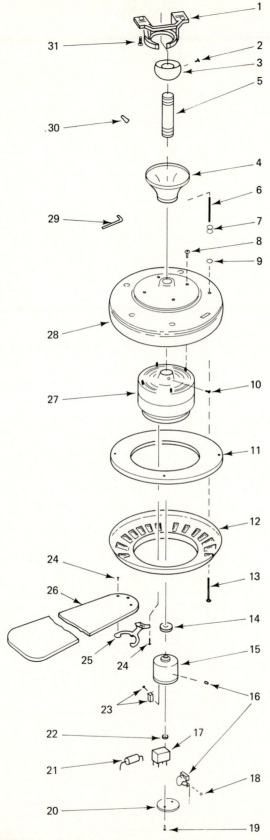

Key No.	Description
1	Bracket, hanger
2	Screw, pan head, 8-32 × 3/8"
3	Hanger ball
4	Hanger pipe
5	Cover, ceiling
6	Stud, threaded, 8-32 × 1-1/4"
7	Knob, knurled
8	Screw, pan head, 8-32 × 3/8"
9	Nut, Hex, 8-32
10	Setscrew, 5/16-18 × 1/4"
11	Insert, foam
12	Housing, bottom
13	Screw, oval head, 8-32 × 3-1/8"
14	Washer 5/8 I.D.
15	Housing, switch
16	Switch, fan speed
17	Capacitor 7 MFD
18	Washer, internal tooth lock
19	Screw, round head, 8-32 × 3/8"
20	Cover, switch housing
21	Capacitor 5 MFD
22	Nut, lock 5/8"
23	Switch, reversing screw, 6-32 × 3/8"
24	Screw, pan head, 10-32 × 5/8"
*25	Flange (set of 4)
*26	Blade (set of 4)
27	Motor
28	Housing, top
29	Wrench, setscrew 5/32"
30	Nut, wire
31	Screw, round head, 8-32 × 3/4"
—	Owners manual

FIGURE 17-4.   Exploded view and parts list for a 52-inch ceiling fan. (Courtesy of Emerson Environmental Products)

at much slower speeds than are normal for most appliance motors. Thus they will have more poles. They are designed for the highest speed. The medium and low speeds are obtained by adding series resistors to lower the voltage.

If the fan operates at high speed but not at one of the lower speeds, the problem is either a defective switch or a defective resistor.

CARE AND USE

Manufacturers are very careful to balance the fan blades and other moving parts of the fan. The technician should also be careful in assembling and working around the fan so as not to damage the blades or disturb their balance.

Carelessness of handling will cause the blades to wobble or vibrate, which will be noisy and shorten the life of the fan. Check the blade attachment screws periodically since a loose screw can cause noise and/or wobble.

The fan housing and blades should be cleaned several times a year. Use a soft brush or lint-free cloth. An occasional application of furniture polish to the wooden blades will help keep them looking good and protect their finish.

The exploded drawing and parts list will show the relationship of parts to each other and their commonly used names (Fig. 17-4). The troubleshooting chart is included to provide a systematic approach to locating an operational problem and correcting it.

Troubleshooting Chart   FANS

Problem	Possible Cause	Corrective Action
Fan will not start.	Fuse or circuit breaker blown.	Check main and branch circuit fuses or circuit breakers.
	Loose power line connections to the fan, or loose switch wire connections in the switch housing.	Check line wire connections to fan and switch wire connections in switch housing. **Caution**: *Make sure main power is turned off.*
	Reversing switch in neutral position.	Make sure reversing switch position is all the way to one side.
Fan sounds noisy.	Blades not attached to fan. Loose screws in motor housing.	Attach blades to fan before operating. Check to make sure all screws in motor housing are snug (not over-tight).
	Screws securing fan blade flanges to motor hub are loose.	Check to make sure the screws which attach the fan blade flanges to the motor hub are tight.
	Switch housing binding against blade hub.	Check to make sure that the switch housing does not bind against the blade hub.
	Wire nuts inside switch housing rattling.	Check to make sure wire nut connectors in switch housing are not rattling against each other or against the interior wall of the switch housing. **Caution**: *Make sure main power is turned off.*
	Motor noise caused by solid-state variable-speed control.	Some fan motors are sensitive to signals from Solid-State variable speed controls. If Solid-State Control is used and motor noise results choose an alternate control method.

Troubleshooting Chart   FANS (Continued)

Problem	Possible Cause	Corrective Action
Fan wobbles excessively.	Screws securing fan blade flanges to blade hub are loose.	Check to be sure that the screws which attach the fan blade flanges to the motor hub are tight.
	Fan blade flanges not seated properly.	Check to be sure that the fan blade flanges seat firmly and uniformly to the surface of the motor hub. If flanges are seated incorrectly, loosen the flange screws and retighten.
	Hanger bracket and/or ceiling junction box is not securely fastened.	Tighten the hanger bracket screws to the junction box, and/or secure the junction box.
	Fan blades out of balance.	Interchanging an adjacent (side-by-side) blade pair can redistribute the weight and result in smoother operation.

Source: Emerson Environmental Products.

18

Coffee Makers

For most of us that first cup of coffee in the morning is rather important. It can influence our mood for the day. So let us see what makes a good cup of coffee and how to keep the coffee maker perking.

We cannot presume to tell you how to make a good cup of coffee, but we can say that the coffeepot must be clean and that the water used must be between 175 and 195°F (80° to 91°C) at the time it passes over the coffee. Any lower temperature will produce coffee that is too weak; a higher temperature will bring out a bitterness in your coffee. Water that is too hot will extract more of the bitter oils from the coffee bean. If it is too cold, it will not extract all the flavor from the bean.

Much bitter coffee results from using a pot that has collected a layer of cooked-on deposits. This condition may result from the water used (lime in hard-water areas) or from coffee solids (containing some of the bitter oils) that accumulated in this lime deposit. A regular schedule of boiling out the pot with baking soda will help to keep it "sweet." (A strong vinegar solution

can be used to dissolve the lime.) *Note:* Dip-It, a commercial cleaner, does an excellent job.

Many people use hot water in making coffee in order to save time. This step results in weak coffee when a percolator is used because the "perking" time is shortened. Also, with a percolator, any coffee grounds that spill down the stem and get into the pump area may hold the pump valve open, thus producing poor pumping action.

Several kinds of automatic coffee makers are in use, so we will look at representatives of these types.

PERCOLATOR OPERATION

A percolator type of coffee maker uses steam to push the water up the tube. The water then splashes down over the coffee grounds, extracting the flavor from them.

Steam is generated under the perk tube in the pump-heater well. Its flow is controlled by the pump check (Fig. 18-1), which is the loose

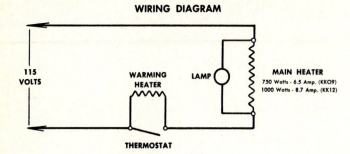

WIRING DIAGRAM

115 VOLTS

WARMING HEATER

LAMP

MAIN HEATER
750 Watts - 6.5 Amp. (KKO9)
1000 Watts - 8.7 Amp. (KK12)

THERMOSTAT

FIGURE 18-1. Service warning holds for any major repair involving the main heater which is brazed to the perk well and body. (Courtesy of Presto)

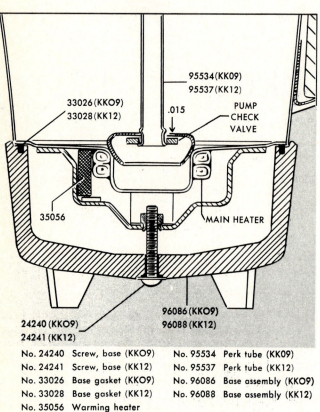

33026 (KKO9)
33028 (KK12)

95534 (KK09)
95537 (KK12)

.015

PUMP CHECK VALVE

35056

MAIN HEATER

24240 (KKO9)
24241 (KK12)

96086 (KKO9)
96088 (KK12)

No. 24240	Screw, base (KKO9)	No. 95534	Perk tube (KK09)
No. 24241	Screw, base (KK12)	No. 95537	Perk tube (KK12)
No. 33026	Base gasket (KKO9)	No. 96086	Base assembly (KKO9)
No. 33028	Base gasket (KK12)	No. 96088	Base assembly (KK12)
No. 35056	Warming heater		

**CROSS SECTIONAL VIEW OF
LOWER PART OF COFFEEMAKER**

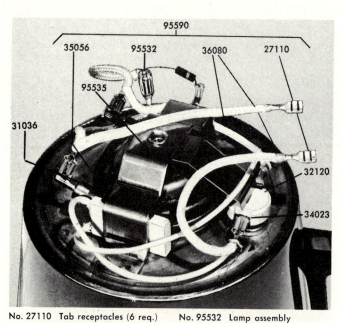

95590

35056   95532   36080   27110

95535

31036

32120

34023

No. 27110	Tab receptacles (6 req.)	No. 95532	Lamp assembly
No. 31036	Spring clip (warming heater hold-down)	No. 95535	Bracket assembly (not a separable part — listed for identification only)
No. 32120	Cap, thermostat	No. 95590	Wiring assembly, includes following parts: 27110, 35056, 36080 and 95332
No. 34023	Thermostat		
No. 35056	Warming heater		
No. 36080	Wire lead (3 req.)		

**VIEW OF BOTTOM AND WIRING OF
KKO9-A AND KK12-A COFFEEMAKERS**

disk at the bottom of the stem. The clearance between the disk and its seat should be about 0.010 to 0.015 inches. A 0.020 feeler gage would not fit in this space.

In operation, the check valve allows water to flow through the holes in the stem base, past the disk into the pump-heater well, then up inside the stem. When the water in the well gets hot enough to steam, this steam will expand, forcing the water out of the well. It can go two ways, up the tube or out the base through

which it entered. As the water tries to escape through the entrance holes, it lifts the check valve, thus closing these holes. The only escape for the steam is through the stem, so up it goes, lifting the standing water in the stem with it. As the steam pushes the slug of water out of the tube, we hear the welcome sound of percolating coffee. Of course, when the steam pressure in the well has been relieved by the preceding action, more water runs in and the cycle starts all over again.

Operating Temperatures

If you have to adjust the thermostat, set up the percolator with a good thermometer or thermocouple immersed in the water, *but not touching the metal*. Use an ac ammeter or watt meter in the line circuit. During the percolating cycle, with the pump heater on, the wattage reading should be about ±5 percent of the rated wattage. The wattage rating for any coffee maker is shown on the specification plate on the base of the coffee maker. During the "keep warm" cycle the wattage should be about 60 ± 10 watts. When the control is set at the "Strong" position the temperature, at which the "keep warm" element cuts into the circuit, should be between 180 and 195°F (82° and 90°C).

Adjustment of the thermostat is made by unlocking the lock nut and (Fig. 18-2) turning the adjusting screw on the thermostat. After adjustment, be sure to tighten the lock nut. Turning the screw clockwise lowers the temperature, counterclockwise raises it. A one-quarter turn will change the temperature about 20°F (7°C).

If the thermostat has to be replaced, position the new one exactly as the old one was; otherwise there may be interference with the base.

Replacement of the "keep warm" element consists of loosening a screw at each end of the element, removing the old element, and installing a new one.

The pump-heater element requires a large socket wrench to remove the ring nut inside the bottom of the body assembly (Fig. 18-3). When reassembling, a Silicone sealer must be placed on all the mating surfaces to prevent leakage. A Silicone grease must be applied to the pump heater and ring nut threads to prevent them from corroding together.

This GE, along with many other electric coffee makers, is not submersible! That means, don't get the heater element, thermostat, and wiring wet! If you do, you will get a nasty shock and possibly burn out the unit.

The Presto KK09A is a submersible coffee maker. All the electrical parts and wiring are contained in a sealed base (see Fig. 18-1).

The screw holding the base to the body

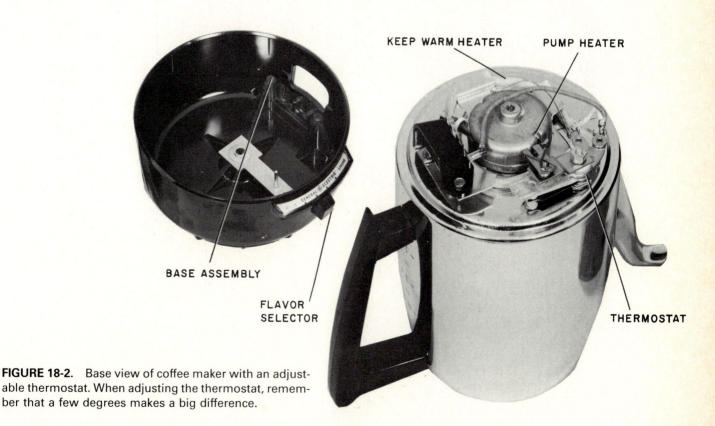

FIGURE 18-2. Base view of coffee maker with an adjustable thermostat. When adjusting the thermostat, remember that a few degrees makes a big difference.

has a gasket under its head to prevent leakage around it. A gasket seals the base to the body. When removing the base, be careful not to damage these gaskets. If there is any doubt as to their condition, replace them with new ones.

Note that the thermostat for this coffee maker is not adjustable. It is designed to operate at a fixed temperature of about $195 \pm 6°F$ ($90°C$). If the coffee gets hotter than this, check

to make certain that the thermostat is making good contact to the bottom of the coffeepot. If that is good and the coffee temperature is still not right, the thermostat will have to be changed.

The neon pilot light is connected in parallel with the main heater. This light is ON while the unit is perking and will go OFF when the coffee is ready to serve. The thermostat is con-

FIGURE 18-3. This General Electric coffee urn also uses a fixed thermostat, but the main heater and perk well (No. XP10 × 22) are removable for service. (Courtesy of General Electric)

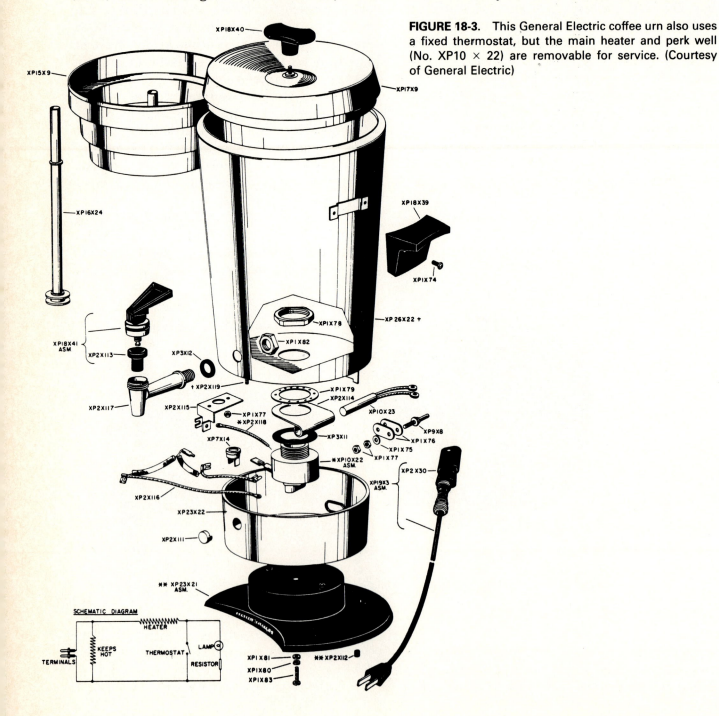

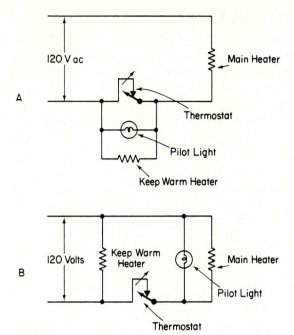

Please note:

1. The thermostat is always in series with the main heater and turns it *Off* whenever the pre-set temperature has been reached.

2. The keep warm heater may be connected across the thermostat (Fig. A), to come *On* when the thermostat opens; or across the line (Fig. B), to be *On* at all times.

3. The pilot light may be connected in parallel to the main heater (Fig. B) to turn *Off* when the coffee is ready or it may be connected across the thermostat (Fig. A) to turn *On* when the coffee is ready.

FIGURE 18-4. Schematic wiring diagram for a typical coffee maker.

nected in parallel with the warming heater element, as shown in the schematic wiring diagram (Fig. 18-4).

The warmer the water gets, the faster steam is generated and the faster the coffee perks.

For all coffee percolators having a pump, it is important to keep the disk and valve seat clean, smooth, and unscratched. Anything that will prevent the disk from seating properly will affect the coffee making.

The glass lid knob sometimes breaks or becomes loose because the lid has been bent. Usually a new knob can be purchased at the store where the percolator was sold. Most hardware stores carry an assortment of knobs. A word to the wise: Bring the lid along to ensure a correct fit.

A broken handle on most General Electrics can be replaced by removing a $\frac{5}{16}$-inch hexagonal-head screw from the top inside with a right-

angle socket wrench; then, from the bottom of the handle outside, remove the round, slotted nut with a special screwdriver.

All cords with plugs at each end can be tested for continuity with your tester. Connect one test lead to one attachment cap prong and the other lead to first one then the other plug terminal. There must be continuity at one connection but not at the other. When the unit is cold, the ohmmeter should read from 12 to 24 ohms from one terminal to the other. This is the main heater resistance. When the unit is hot (thermostat open), the resistance reading should be 270 ohms ± 10 percent. This is the "keep warm" heater resistance. These are approximate resistances, which will vary from model to model.

Heat, to produce the steam, comes from a thermostatically controlled heater element. All of this is available for tests and service by removing the base.

You will note that there are two heater elements and a thermostat. In order to test the main or pump heater element, it is necessary to open the thermostat switch by lifting the spring element.

In operation, the thermostat remains closed to allow full line voltage to reach the pump heater so it can boil water quickly in the pump area. As percolating continues, the water in the rest of the pot will get hot. When the temperature reaches the preset range between 180 and 195°F (82 to 90°C), the thermostat opens to connect the "keep hot" heater in series with the pump heater. This series connection decreases the voltage available to the pump heater holding the contents at the present temperatures. (This is below boiling temperatures.) The total resistance is enough to allow the coffee to be kept hot enough for drinking.

Do not fool around with the thermostat adjustment if enough control of coffee quality and temperature can be had by changing the "Mild" to "Strong" setting.

When reassembling the base, make certain that the base gasket is positioned correctly in its groove and that the gasket around the screw is in good condition. Any water leakage around the gasket or through a crack in the plastic will cause electrical leakage that can shock the user.

Figure 18-5 shows the exploded view and parts list for the Sunbeam percolator-type coffee maker.

VACUUM OPERATION

The Sunbeam is a vacuum type (Fig. 18-6). Coffee grounds are placed in the upper bowl, which seals to the lower bowl containing water. Heat

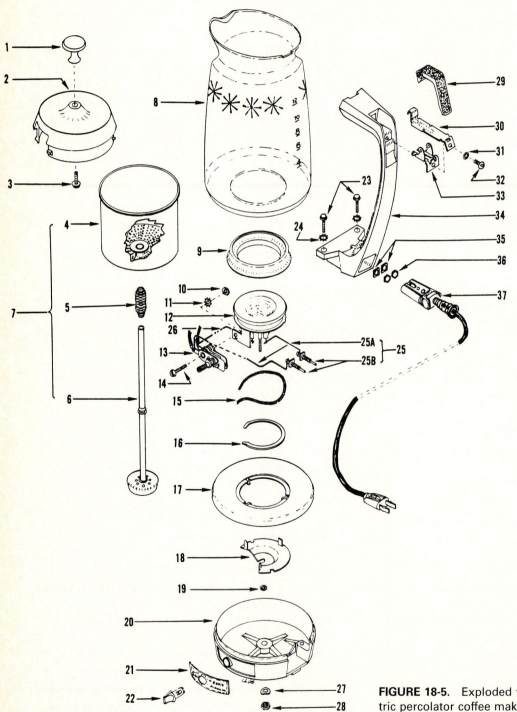

FIGURE 18-5.   Exploded view and parts list for an electric percolator coffee maker. (Courtesy of Sunbeam)

FIGURE 18-5 (Continued)

Key No.	Description	Stock No.
1	Knob, cover	37-85838
2	Cover	37-89483
3	Screw, cover	37-89569
4	Basket assembly	37-89526
5	Spring, basket	37-89522
6	Pump and stem assembly	37-89632
7	Basket and pump assembly (incl. keys 4, 5, and 6)	37-90157
8	Vessel, glass	ZAP80
9	Seal	37-89572
10	Nut, thermostat	37-4341
11	Lock washer, thermostat	37-22249
12	Element assembly (600 watts)	37-89574
13	Thermostat	37-89523
14	Screw, thermostat	37-1377-1
15	Auxiliary heater	37-89527
16	Clamp, auxiliary heater	37-89634
17	Cover, outer base	37-89575
18	Clamp base	37-89633
19	Speed nut	37-22891
20	Base assembly (incl. key 21)	37-89474
21	Nameplate	37-89520
22	Control knob	37-89532
23	Screw, handle	37-89567
24	Washer, handle	37-86870
25	Lead wire and terminal assembly (incl. key 25A & 25B)	37-89693
25A	Lead wire	37-89692
25B	Terminal stud	37-24127
26	Lead wire	37-89691
27	Spring washer	37-86885
28	Nut	37-24287
29	Latch lever	37-89534
30	Latch spring	37-89533
31	Lock washer, latch spring	37-1691
32	Screw, latch spring	37-80513
33	Bracket	37-89524
34	Handle	37-89482
35	Terminal lock	37-24096
36	Terminal nut	37-5593
37	Cord assembly	37-1570

boils the water and steam pressure in the lower bowl forces the water up. When the lower bowl has cooled, the steam pressure is reduced and the coffee flows down into it. The "low heat" element keeps the coffee warm and the thermostat keeps the high heat element from reboiling the coffee.

DRIP OPERATION

The modern drip coffee maker heats the water, then allows it to flow through the grounds and filter into a serving vessel. The serving vessel, or carafe, sits on a keep-warm heating element

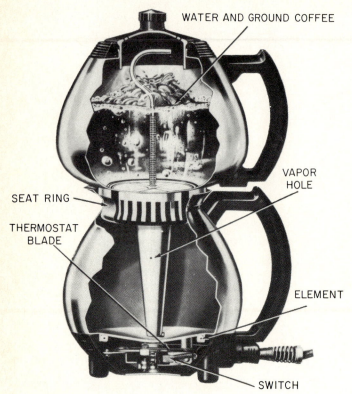

FIGURE 18-6. Sectional view of a typical vacuum coffee maker. (Courtesy of Sunbeam)

which maintains the brewed coffee at the proper serving temperature.

Most units are portable. Some, such as the Sunbeam shown in Fig. 18-7a, can be mounted under a kitchen cabinet with only the carafe and filter holder as the removable parts.

The drip-type coffee makers require that the user pour water into the machine. This ensures that the correct amount of water will be used for the grounds in the basket.

The typical operating procedure is to fill the carafe with cold tap water to the desired cup level. (The recommended minimum is two cups.) Then pour this water into the filling compartment (Fig. 18-7b). This may be an opening in the top of the machine or a drawer at the

FIGURE 18-7.   (a) When mounted under a cabinet this unit frees counter space for other usage. It can be set up the night before to provide freshly brewed coffee the next morning; (b) The pull-out water drawer makes for easy filling.

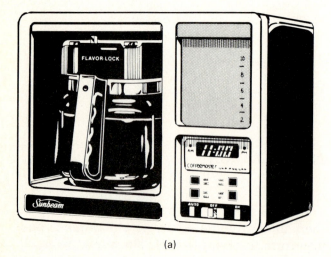

(a)

(b)

top that pulls forward for filling and is closed for the brewing cycle. The empty carafe is then placed on the keep-warm element to be ready to receive the brewed coffee.

A filter paper and the desired amount of coffee is placed in the filter holder, which is positioned over the carafe. When the switch is turned ON, coffee will begin to brew as soon as the water is hot enough. When brewing is completed, the coffee maker will automatically switch to the keep-warm cycle and will remain ON until the switch is turned OFF. Some models have a built-in timer that can be preset to start the brewing cycle at a desired time, such as "wake-up" time (Fig. 18-7a).

The exploded view of a Sunbeam unit (Fig. 18-8) and its parts is included to illustrate general construction and assembly. Whenever possible, use the information that is specific for the machine you need to service since each model will have features that make it different from others.

The West Bend Quik Drip II coffee maker has two warming plates, one to keep the coffee warm while a fresh pot is brewing on the other plate. Figure 18-9 illustrates this unit. The exploded drawing and parts list are included to show the relative location of parts and their names. Basically, this is the same as the unit shown in Fig. 18-10 with an extra separately switched, 75-watt keep-warm element.

As mentioned earlier, a clean coffee maker is important to brewing good coffee. The cleaning instructions from West Bend will serve as a procedure for other drip coffee makers (Fig. 18-11).

ELECTRICAL TESTING

In all electric coffee makers the thermostat is connected in series with the main heater element and is normally closed (making contact) at room temperature. Thus there must always be a complete circuit from one contact of the cord plug to the other, typically about 14 to 15 ohms for a 1000-watt coffee maker.

If there is no continuity, disconnect the cord from the coffee maker and test it with an ohmmeter from terminal to terminal (Fig. 18-12). Shake the cord while testing. Continuity here would indicate the cord to be good. The next sequence of testing would be to test the coffee maker by leaving one test lead connected to a terminal and follow the wiring from the other terminal back to the original, testing at every connection. As long as there is no continuity, the defect lies between the probing test lead and terminal 1. When you do get a continuity reading you have just passed the defective part. That part will have to be replaced.

One final word of caution: After you have repaired a coffee maker, test the circuit for continuity and electrical leakage from each prong of the cord terminal block to the metal shell. If you have an ohmmeter, check the resistance from terminal to shell, which should be over 600,000 ohms. If you have a current leakage meter, use it to check that the reading from a live metal part to any nonconducting metal parts does not exceed 0.2 milliampere when the coffee maker is operating.

The manufacturers of coffee makers have done a great deal of research in the making of good coffee. The proper performance of the coffee maker is just one part. The choice of coffee blend and quantity used, as well as *the condition of the water used,* all contribute to that good cup of coffee.

If the coffee maker is clean and working right and the customer is still not satisfied with the results, it is time to reread the manufacturer's instructions on the care and use of this particular unit (or else it's the customer's sense of taste).

Good luck and a better cup of coffee!

The troubleshooting chart should be used to determine the area in which to look for a problem. Often, careful looking will locate the problem and the method of repair will be self-evident. Repair or replace! As with all appliances, after servicing, test the unit for operation, insulation resistance, and current leakage before returning it to the customer.

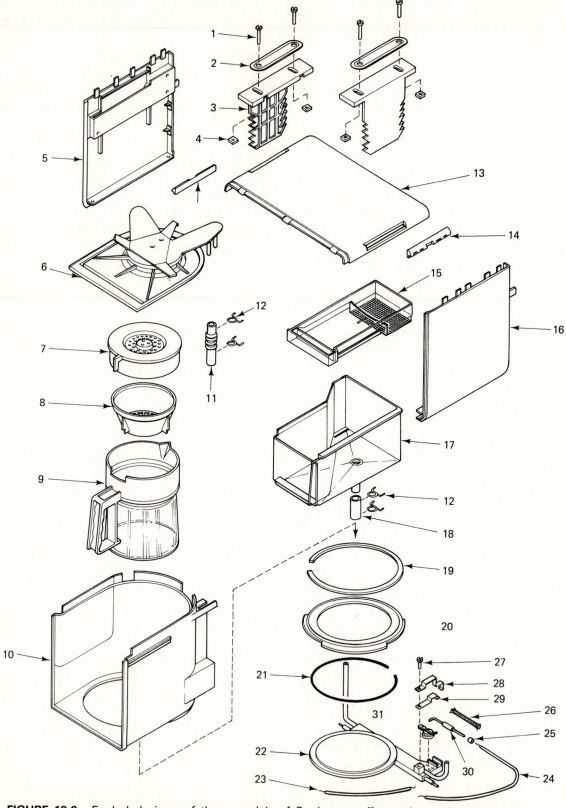

FIGURE 18-8. Exploded views of three models of Sunbeam coffee makers with wiring diagrams and parts list. The Sunbeam under-cabinet drip coffee maker has a digital clock and timer that can be preset up to 24 hours for automatic brewing. The locking cover will keep all the condensation, flavor, and aroma inside the carafe. (Courtesy of Sunbeam)

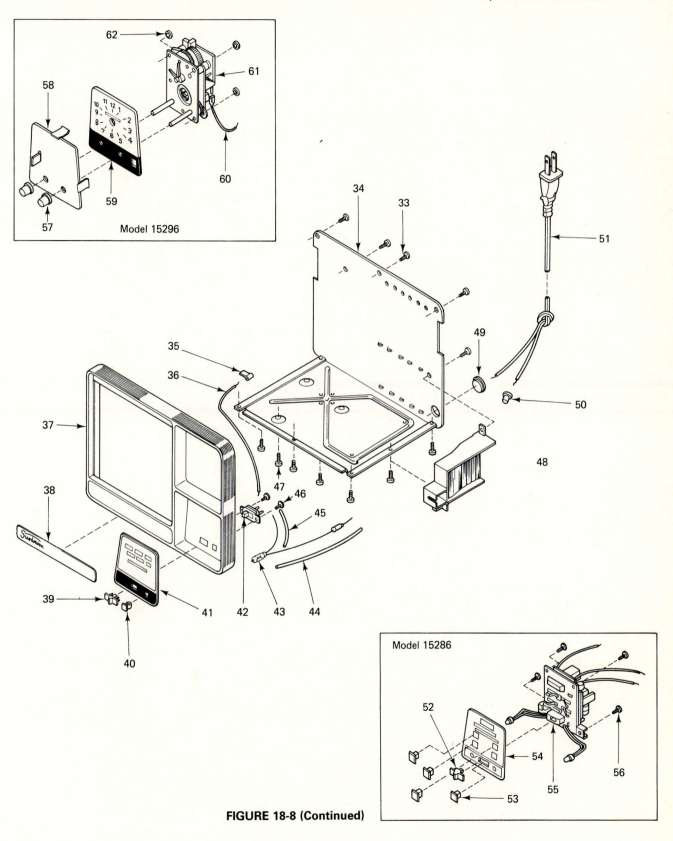

FIGURE 18-8 (Continued)

Wiring diagrams

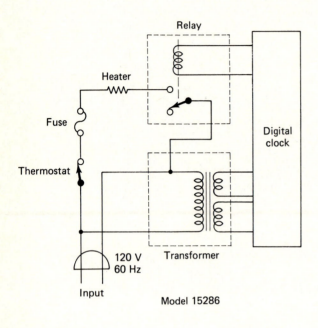

Model 15286

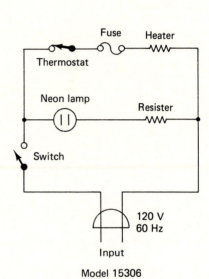

Model 15306

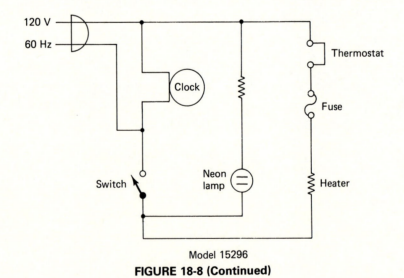

Model 15296

FIGURE 18-8 (Continued)

Important notice: Order parts by STOCK NUMBER in Fourth Column below. That will help us supply the correct part and give you quick, accurate service. Service Information contained in Service Bulletin #473.

Number Item	Description	Model Used On	Sunbeam Part Number
1	Cabinet screw (4 req'd)	15-30AC, 15-28AD, 15-29AE	102900
		15-30AG, 15-28AF, 15-29AH	102900
2	Cabinet screw plates (2 req'd)	15-30AC, 15-28AD, 15-29AE	102901
		15-30AG, 15-28AF, 15-29AH	102901
3	Cabinet bracket (2 req'd)	15-30AC, 15-28AD, 15-29AE	102902
		15-30AG, 15-28AF, 15-29AH	102902
4	Cabinet nuts (4 req'd)	15-30AC, 15-28AD, 15-29AE	102903
		15-30AG, 15-28AF, 15-29AH	102903
5	Left side panel	15-30AC, 15-28AD, 15-29AE	102904
		15-30AG, 15-28AF, 15-29AH	102904
6	Shower head ass'y		
	Old	15-30AC, 15-28AD, 15-29AE	102906
	New	15-29AH, 15-30AG, 15-29AF	105026
7	Cap cover	15-30AC, 15-28AD, 15-29AE	102907
		15-30AG, 15-28AF, 15-29AH	102907
8	Basket	15-30AC, 15-28AD, 15-29AE	102908
		15-30AG, 15-28AF, 15-29AH	102908
9	Vessel assembly	15-30AC, 15-28AD, 15-29AE	See note
		15-30AG, 15-28AF, 15-29AH	See note
10	Base		
	Old style	15-30AC, 15-28AD, 15-29AE	102909
	New style	15-30AG, 15-28AF, 15-29AH	105027
11	Bellows	All models	105033
12	Clamp tube (4 req'd)	All models	102911
13	Top panel	15-30AC, 15-28AD, 15-29AE	102912
		15-30AG, 15-28AF, 15-29AH	102912
14	Bracket cover (2 req'd)	15-30AC, 15-28AD, 15-29AE	102913
		15-29AH, 15-30AG, 15-28AF	102913
15	Water drawer	15-30AC, 15-28AD, 15-29AE	102914
		15-30AG, 15-28AF, 15-29AH	102914
16	Right side panel	15-30AC, 15-28AD, 15-29AE	102915
		15-30AG, 15-28AF, 15-29AH	102915
17	Reservoir ass'y	15-30AC, 15-28AD, 15-29AE	102916
		15-30AG, 15-28AF, 15-29AH	102916
18	Water tube B	15-30AC, 15-28AD, 15-29AE	102917
		15-30AG, 15-28AF, 15-29AH	102917
19	Rubber seal A	All models	102918
20	Heat cup	15-30AC, 15-28AD, 15-29AE	102919
		15-30AG, 15-28AF, 15-29AH	102919
21	Rubber seal B	All models	102920
22	Heater unit ass'y	15-30AC, 15-28AD, 15-29AE	102921
		15-30AG, 15-28AF, 15-29AH	102921
23	Lead wire A	All models	102922
24	Lead wire B	All models	102923
25	Terminal	Not for service	
26	Sleeving	All models	102924
27	Screw (thermostat)	All models	102925
28	Bracket holder B	15-30AC, 15-28AD, 15-29AE	102926
		15-30AG, 15-28AF, 15-29AH	102926
29	Bracket holder A	15-30AC, 15-28AD, 15-29AE	102927
		15-30AG, 15-28AF, 15-29AH	102927

FIGURE 18-8 (Continued)

Important notice: Order parts by STOCK NUMBER in Fourth Column below. That will help us supply the correct part and give you quick, accurate service. Service Information contained in Service Bulletin #473.

Number Item	Description	Model Used On	Sunbeam Part Number
30	Fuse	15-30AC, 15-28AD, 15-29AE	102928
		15-30AG, 15-28AF, 15-29AH	102928
31	Snap-disc-thermostat	15-30AC, 15-28AD, 15-29AE	102929
		15-30AG, 15-28AF, 15-29AH	102929
32	Not used	Not for service	
33	Screw (11 req'd)	All models	102930
34	Back panel	15-30AC, 15-28AD, 15-29AE	102931
		15-30AG, 15-28AF, 15-29AH	102931
35	Terminal	Not for service	
36	Lead wire C	15-30AC, 15-28AD, 15-29AE	102932
		15-30AG, 15-28AF, 15-29AH	102932
37	Front (frame) panel		
	On-off	15-30AC, 15-30AG	102933
	Automatic	15-28AD, 15-28AF	105034
	Analog	15-29AE, 15-29AH	105035
38	Decal (sunbeam) B	All models	102934
39	On-off switch/knob	15-30AC, 15-30AG	102935
40	Lens	15-30AC, 15-30AG	102936
		15-29AE, 15-29AH	102936
41	Switch decal A	15-30AC, 15-30AG	102937
42	On-off switch	15-30AC, 15-30AG	102938
43	Lamp and resistor ass'y	15-30AC, 15-29AE	102939
		15-30AG, 15-29AH	102939
44	PVC tube A	15-30AC, 15-30AG	102940
		15-29AE, 15-29AH	102940
45	PVC tube B	15-30AC, 15-30AG	102941
		15-29AE, 15-29AH	102941
46	Screw (2 req'd)	15-30AC, 15-30AG	102942
47	Screw (1 req'd)	All models	102943
48	Cord storage plate	All models	102944
49	Cord guide	All models	102945
50	Connectors	All models	SASRB4
51	Line cord	All models	102946
52	On-off switch/knob	15-30AC, 15-30AG	102935
		15-28AD, 15-28AF	102935
53	Clock buttons (4 req'd)	15-28AD, 15-28AF	102947
54	Decal A	15-28AD, 15-28AF	102948
55	P.C. board	15-28AD, 15-28AF	102949
56	Screw (4 req'd)	15-28AD, 15-28AF	102950
57	Analog clock knob (2 req'd)	Not for service	
58	Analog clock lens	15-29AE, 15-29AH	105030
59	Decal analog clock	15-29AE, 15-29AH	105032
60	Lead wire (black)	15-29AE, 15-29AH	105031
61	Analog clock with hands and knobs	15-29AE, 15-29AH	105028
62	Retainer ring (3 req'd)	15-29AE, 15-29AH	105029

Note: Vessel assembly (R94670) must be ordered as an accessory.

FIGURE 18-8 (Continued)

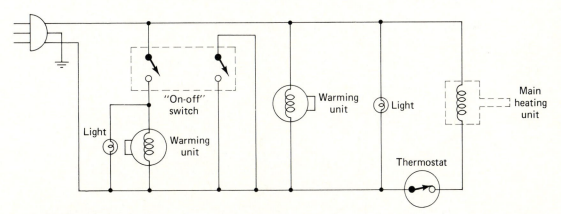

FIGURE 18-9. West Bend coffee maker and its wiring diagram, exploded view, and parts list.

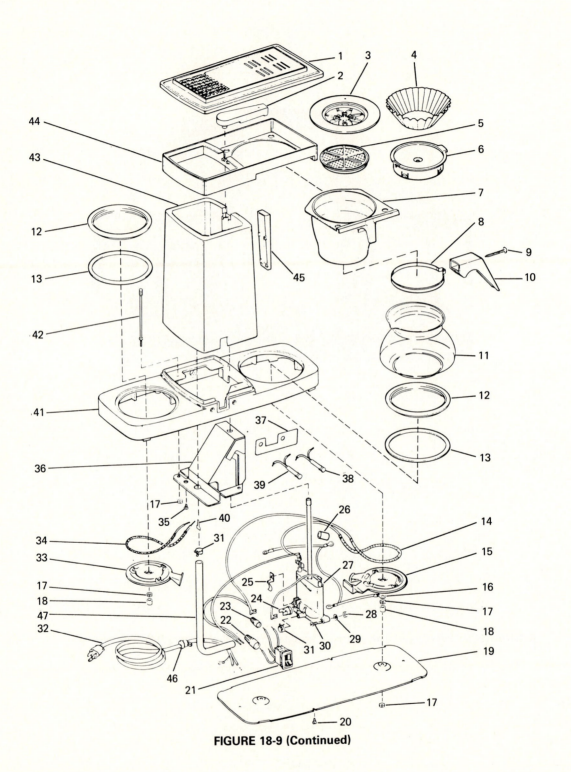

FIGURE 18-9 (Continued)

Item No.	Part No.	Description
1	P226-2	Cover
2	P110-7	Spout
3	P77-43	Spreader
4	**P19-69	Filter (*pack of 50*) sub w/P23-69
5	P2-187	Screen
6	P227-2	Cover
7	P92-45	Basket
8	P4-136	Band
9	**P508-127	Screw Sub w/P535-127
10	P863-20	Handle
11	5963	Carafe
12	P331-10D	Plate assembly
13	P81-72	Gasket
14	P211-109	Unit (*rope*)
15	P201-42	Bracket
16	P119-90	Lead
17	P57-81	Nut (*10-24*)
18	P56-146	Spacer
19	P265-9	Bottom
20	P482-127	Screw
21	**P22-86	Switch sub w/P31-86
22	P175-81	Nut
23	**P192-81	Nut sub w/P228-81
24	P77-97	Thermostat
25	P205-42	Bracket
26	P8-93	Hose
27	P225-109E	Unit assembly (*w/thermostat and P205-42 bracket*)
28	P54-81	Nut (*8-32*)
29	P220-47	Washer
30	P530-127	Screw (*8-32*)
31	P47-5	Clamp
32	P84-74	Cord
33	P179-42	Bracket
34	P177-109	Unit (*rope*)
35	P339-127	Screw
36	P18-199D	Housing
37	P328-10	Plate
38	P70-73	Lamp
39	P69-73	Lamp
40	P18-38	Blade
41	**P355-3	Base sub w/P397-3
42	P30-57	Bolt
43	P105-1	Outside
44	P30-200	Support
45	P364-10	Plate
46	P46-5	Clamp
47	P10-93	Hose

* Discontinued—No Longer Available
** Discontinued—Substitute

FIGURE 18-9 (Continued)

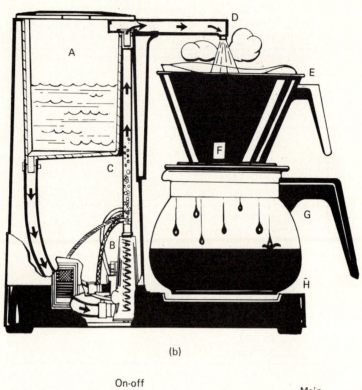

(b)

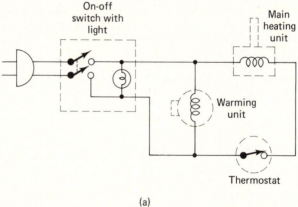

(a)

FIGURE 18-10. Gravity feeds water from the reservoir (A) to the internal tank heater (B). With the touch of a switch water begins heating, and within seconds it is forced up the tube (C) and out the dripper spout (D). The scientifically designed conical basket distributes water around the ground coffee, through the disposable filter (F), and into the glass carafe (G). The warming plate (H) keeps coffee serving hot to the last drop. *Caution:* Do not plug in coffee maker in line voltage without water in reservoir. (Courtesy of West Bend)

Water used for making coffee contains dissolved minerals. When heated, minerals deposit on the main unit, in the tubes and spout. The rate at which the deposit forms will vary according to the hardness of the water used and the frequency with which the coffeemaker is used. If not removed, efficiency of the coffeemaker will be impaired; excessive steaming will occur and unpumped water will remain in the reservoir. Special cleaning with brush (which is provided) and with vinegar is necessary every 30 cycles, or twice a month if the coffeemaker is used twice daily.

1. Unplug cord from wall outlet. Remove carafe-basket-spreader assembly from coffeemaker.

2. Detach spout from coffeemaker. Brush out spout under running water; shake dry. Brush out transfer tube to which spout connects, inverting coffeemaker to remove mineral deposit.

3. Pour household vinegar (5% acetic acid) into reservoir to the 4-cup level. Add cold water to vinegar in reservoir, filling to the 6-cup level.

4. Replace spout. Place carafe-basket-spreader assembly under spout on warming unit. Plug cord into outlet.

5. Turn switch on. "Pump" one cup vinegar-water solution into carafe. Turn switch off. Let stand for 15 minutes.

6. Turn switch on. "Pump" the rest of the solution into the carafe. Discard solution. Remove any remaining mineral residue from spreader with a dish cloth.

7. Unplug cord from wall outlet. Rinse reservoir with tap water. Invert coffeemaker removing tap water and remaining vinegar solution from the "pump." (*Vinegar should not remain in unit beyond the cleaning time*)

8. For final rinsing, fill reservoir with cold water to the 4-cup level. Turn switch ON to pump water throught basket-spreader assembly into carafe. Turn switch off.

NOTE: *Repeat Step 5; four or five times if:*
1. Excessive steaming has occurred and unpumped water remained in reservoir.
2. Your water is particularly hard (greater than 22 grain hardness).
3. You have not cleaned the coffeemaker as often as indicated in the instructions.

FIGURE 18-11. Instructions for special cleaning. (Courtesy of West Bend)

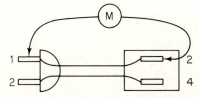

FIGURE 18-12. Terminals 2 through 4 may be attached to the appliance. To test a coffee maker cord there must be continuity from points 1 to 2 and from 3 to 4 even while the cord is being shaken. There must be *no* continuity between 1–2 and 3–4.

Troubleshooting Chart COFFEE MAKERS

Problem	Possible Cause	Corrective Action
Unit is slow in brewing coffee.	Low line voltage.	Check line voltage; if it is low, notify the power company.
Does not operate at all.	Switch is OFF. No voltage at outlet. Defective power cord. Defective pump heater element. Defective thermostat.	Check fuse and replace if necessary. Repair or replace. Replace heater element. Replace thermostat.
Gets warm but does not percolate.	Defective pump. Defective thermostat. Incorrect setting of thermostat.	Replace pump. Repair or replace thermostat. Reset thermostat.
Coffee tastes bitter.	Accumulated residue in coffee maker. Lime deposit	Clean with baking soda or other mild kitchen cleanser. Operate with 50 percent vinegar solution.
Weak coffee.	Not enough grounds. Control not set right. Starting with hot water. Pump valve stuck.	Use more. Reset control. Use cold water. Clean valve for free action or replace.
Coffee boils.	Incorrect thermostat setting. Defective thermostat.	Adjust thermostat setting. Replace thermostat.
Leaks.	Loose nut holding element. Defective gasket. Loose handle stud. Loose hose fitting.	Tighten nut. Replace. Tighten. Tighten clamp.
Does not stay warm.	Open fuse. Defective thermostat. Open element.	Replace part.
Unit will not start on timer.	Timer not set. Timer or printed circuit board defective. Thermostat or fuse defective.	Reset timer. Replace part. Replace part.
Coffee not hot enough.	Cold carafe. Second warming place cool.	Warm carafe before brewing. Check element, switch, fuse.
Unit pumps lime.	Hard water.	Clean according to special instructions.
Does not pump all water out of reservoir.	Pinched hose. Lime deposit in system. Stuck check valve.	Clear it. Clean according to instructions.

19

Compactors

The compactor (Fig. 19-1) is an appliance designed to reduce the volume of normal trash and prepare it for refuse pickup. It will accept cans, bottles, cartons, paper, and garbage, although garbage is handled better by a disposer.

INSTALLATION

The compactor can be located where convenient, provided that the floor is secure and the leveling legs have been adjusted to ensure that the compactor is level and set firmly on the floor. It should *not* be on the same circuit as any appliance with electronic controls! It requires a 120-volt, 60-hertz ac electrical outlet with provision for a grounding terminal.

OPERATION

Trash is put into the machine through an access door. Compacting action begins when the start button or lever is activated. A motor-driven mechanism forces the ram about two-thirds of the way into the container (drawer or basket) to compress the waste (Figs. 19-2 and 19-3). It then moves upward to its original position. As the container fills, the ram will travel until it exerts about 2000 pounds of pressure (Fig. 19-4). Then the motor automatically reverses to retract the ram.

Once the compacting cycle has started, neither the loading door nor the access door can be opened. The compacting cycle cannot be started if either door is not latched securely. There are a number of limit switches in the electrical circuit which prevent the motor from starting unless all the doors, the collector (container), and the ram are in the correct position. These safety features are designed to protect the user from harm.

With normal usage this operation is trouble free. However, with careless loading a hard object might lodge against the side of the drawer and could wedge between the container and the ram, causing a jam in a down position.

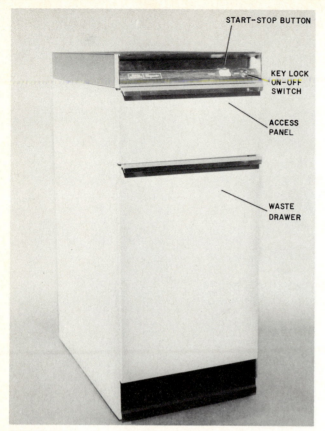

FIGURE 19-1. To open the waste drawer, lift up slightly on the handle and pull out to its STOP position. The waste drawer must be half open before the access door is opened.

JAMMED RAM

On some Whirlpool models any jamming action that could force the drawer out of position will open the drawer tilt/safety switch. This will cut the power to the motor. Sometimes, pushing the drawer in while pressing the start button will reactivate the circuit and the ram will retract. If this does not work, unplug the unit from its outlet and remove the escutcheon to expose the start switch. Make certain that the start switch is in the OFF position. Then connect the black test cord lead to the yellow wire at the start switch and the white test cord lead to the white wire terminal of the compactor power cord. Plug the test cord into the outlet and the ram should retract. At the end of the return stroke, disconnect the test cord at the outlet.

SANITIZING

Some compactors have provisions for mounting a can of deodorant that automatically sprays into the refuse when the door is opened. Others have directions that recommend use of a spray

FIGURE 19-2. On Whirlpool models built before 1972, the motor drives the power screws through a cog belt. On later models the cog belt has been eliminated and replaced by a direct gear to gear drive. Both systems provide a 6-to-1 torque advantage. An additional 55-to-1 force advantage is provided by the power screws. The force supplied by the ram is the product of these two forces, which is approximately 330 times greater than the torque supplied by the motor. (Courtesy of Whirlpool)

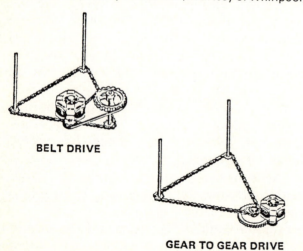

BELT DRIVE

GEAR TO GEAR DRIVE

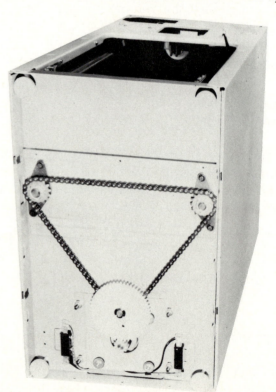

FIGURE 19-3. In-Sink-Erator drive. The motor drives a screw through a speed-reduction gear, which, in turn, opens and closes the scissors linkage to the RAM, thus lowering it and raising it. (Courtesy of In-Sink-Erator)

disinfectant that eliminates unpleasant odors generated by garbage or unclean tins and bottles. Still others recommend using a "stick-up" deodorizer.

Even though it is not recommended, many people will compact garbage with their other trash. With this in mind, the compactor manufacturers recommend the use of their special sanitized and moisture-proof bags to contain the compacted waste. These bags are strong enough to withstand the compacting action.

CLEANING

Cleanliness is important. The outside of the compactor can be cleaned as you would any other appliance. The inside can be cleaned after

FIGURE 19-4. Effect of voltage on ram power. This voltage–force curve indicates the pressure at which the motor will stall. When the motor stalls, the centrifugal switch closes, reconnecting the motor start winding. Because the directional switch has changed position, the motor direction will reverse to move the ram upward. (Courtesy of Whirlpool)

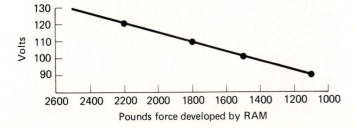

removing the trash container. For the In-Sink-Erator unit, simply removing the refuse container will provide room enough to clean. For the Whirlpool and the Sears units, the drawer can be removed by pulling it all the way out to its stop position, then pushing down on the drawer to release the tabs, which are located one on each side of the drawer, near the cabinet bottom.

The ram and other inside areas that have been exposed to trash can now be disinfected and/or cleaned with detergent and water. The cleaning process will have to be done more frequently if any amounts of unwrapped garbage are added to the trash.

For your safety it is advisable to disconnect the compactor from its electrical outlet while you are working on the unit.

SERVICE

Most service work can be performed by either removing the drawer or container, or by removing the rear service panel. On Whirlpool compactors built after 1972, switches are accessible by removing the escutcheon, which is held in place by two screws. Figures 19-5 and 19-6 should illustrate the parts location and the method of operation for each unit shown.

ELECTRICAL

The wiring diagrams (Fig. 19-7) show the interlocking safety switches, reversing switches, and the motor connections. These diagrams may be used for reference, but as with other appliances, there should be a wiring diagram for your model attached to the back panel of your machine. Use that diagram.

You will note from the wiring diagrams that a quick test can be made to determine whether a "no operation" problem is due to a motor or a switch failure. Connect a jumper wire from the L1 (black power cord) terminal to the motor main (run) winding terminal. If the motor runs, the trouble is in the switching circuit.

Most of these tests can be made from the front when the escutcheon is removed. Figure

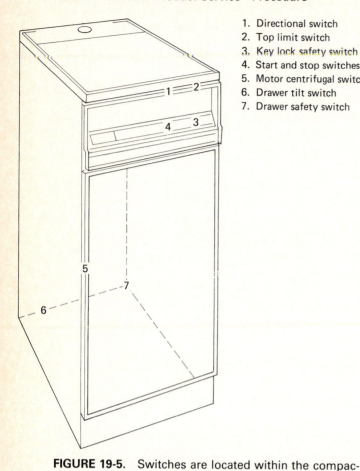

1. Directional switch
2. Top limit switch
3. Key lock safety switch
4. Start and stop switches
5. Motor centrifugal switch
6. Drawer tilt switch
7. Drawer safety switch

FIGURE 19-5. Switches are located within the compactor, where they can do their job most effectively. (Courtesy of Whirlpool)

19-7(a) indicates the physical location of the switches, but the wiring is generally available at the upper set of switches.

As with all electrical appliances, make certain that voltage is available to the machine. This test should be made with a voltmeter at the machine end of the power cord. If voltage is available there, the safest thing to do is to disconnect the power cord from the wall outlet and test each part with an ohmmeter for proper operation. Follow each circuit and test it from point to point. The motor windings should show continuity, and a switch will change from a 0 (zero) to an ∞ (infinity) reading as it is turned on and off.

Pay particular attention to the limit switches. These switches operate *only* when all moving parts (drawer, basket, ram, etc.) are in their correct position, that is, when all switches are adjusted to make firm contact to the switch actuating lever. Either the switch or the part may be out of position. Check the switch mounting first. The motor and drawer safety switches are accessible from the rear.

Service parts for compactors are available through the local dealer for the particular make and model you have. The following troubleshooting chart should serve as a guide in locating operational problems.

FIGURE 19-6. Limit and directional switches are adjustable so that they can be accurately set at the end of the ram's travel. (Courtesy of Whirlpool)

FIGURE 19-7. (a) With the outer casing removed, all the working parts are accessible in this Whirlpool compactor. The motor and its wiring are accessible from the back, the ram screw drive train from the bottom. (Courtesy of Whirlpool)

(a)

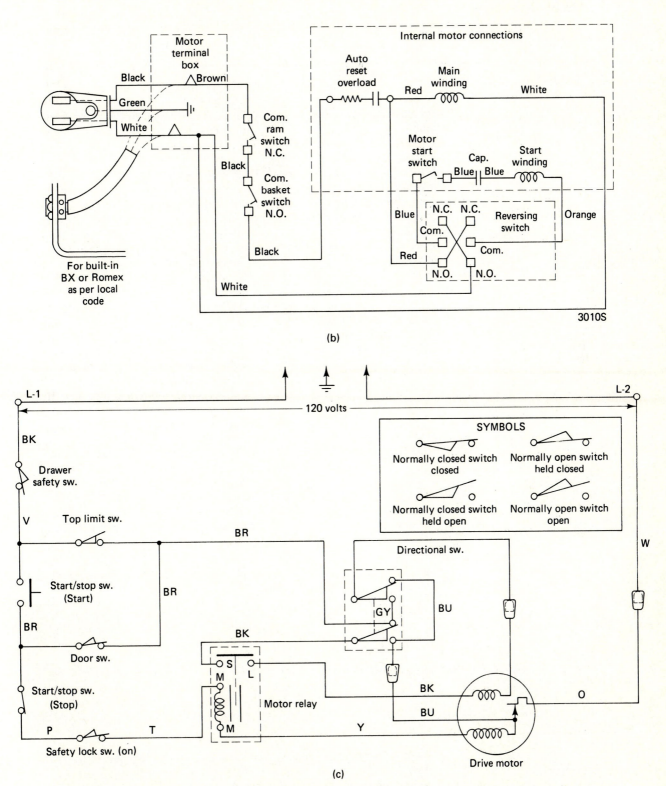

FIGURE 19-7. (Continued) Compactor wiring diagrams. [(b) Courtesy of In-Sink-Erator; (c) Courtesy of Whirlpool.]

Troubleshooting Chart **COMPACTORS**

Problem	Possible Cause	Corrective Action
A. Unit will not operate.	1. Unit not plugged into electrical outlet.	Check to make sure power cord is plugged into outlet.
	2. Branch circuit breaker or fuse "open."	Check fuse or circuit breaker. Replace fuse or reset circuit breaker.
	3. Safety key lock is in the OFF position.	Turn key to ON position.
	4. Drawer is more than $\frac{1}{4}$ in. open (door switch open) due to improper closing or object wedged in front of ram and container.	Push start switch and hold switch until drawer closes. Release start switch immediately. Unit will continue upward travel and shut off. Open drawer.
	5. Motor out on overload (ram down).	Wait 10 minutes for motor to cool. Check for low voltage, tight belt, tight chain, defective directional switch, motor relay, broken power screw, or power nut.
	6. One or more of the following switches defective:	
	Start/stop switch.	With the start side of the start/stop switch depressed, there should be continuity between terminals BR and V with wires removed. Also check terminals BR and P on the "stop" side of the switch; there should be continuity.
	Safety lock switch.	With the key in the ON position (switch plunger depressed), there should be continuity between the two terminals on the safety lock switch. If no continuity, change switch.
	Door switch.	With the door switch plunger depressed, there should be continuity between the two terminals on the switch. If no continuity, change switch.
	Drawer safety switch.	With the drawer closed, there should be continuity between terminals BK and V of the drawer safety switch. Adjust switch or replace switch if it is defective.
	7. Loose connections or broken wires.	Check all connections to make sure they are tight. Also check for broken wires. Replace connectors or broken wires.
B. Unit starts cycle and then stops.	1. Branch circuit breaker or fuse "open."	Check circuit breaker or fuse. Reset or replace fuse if necessary.
	2. Drawer open more than $\frac{1}{4}$ in.	Refer to Item A4.
	3. Drive motor out on overload.	Refer to Item A5.
	4. Defective top limit switch.	Check continuity between V and BR.

Troubleshooting Chart **COMPACTORS** *(Continued)*

Problem	Possible Cause	Corrective Action
C. Drive motor runs but unit will not compact trash.	1. Drive belt broken or teeth missing.	Check drive belt. Replace if necessary.
	2. Drive chain broken.	Check drive chain and replace broken link with a master link Part No. 775306 if necessary.
	3. Motor pulley loose on motor shaft.	Check motor pulley for tightness on drive motor shaft. If loose, tighten.
	4. Drive pulley loose on drive shaft.	Check drive pulley for tightness on drive shaft. If loose, tighten.
	5. Drive sprocket roll pin sheared or missing.	Check drive sprocket and replace roll pin if necessary.
	6. Drive shaft broken.	Check drive shaft and replace if broken.
	7. Ram power nuts stripped.	Check power nuts on ram and replace if necessary.
	8. Directional switch contacts not in proper position (i.e., contacts are positioned or fused so that ram will always go up).	Check adjustment or change switch.
D. Unit operates but is noisy.	1. Drive chain too tight or loose.	Check drive chain tightness. Chain should have about $\frac{1}{4}$ inch deflection.
	2. Belt too tight or loose.	Adjust belt.
	3. Power screws dry.	Lubricate.
	4. Loose screws, panels, etc.	Tighten as required.
E. Aerosol spray can will not work.	1. Aerosol spray can is empty.	Replace with full spray can.
	2. Spray can nozzle plugged.	Unplug nozzle or replace spray can.
	3. Aerosol spray can actuator defective.	Check the aerosol spray can actuator for proper movement. If defective, replace or repair the actuator. Also, check to make sure the actuator arm is hitting on the cams of the bag retainer.
F. Drawer will not open.	1. Ram is partway down.	a. Check electrical outlets.
		b. Check to see that drawer is not open more than $\frac{1}{4}$ inch.
		c. Check to verify that key is turned on.
		d. Door or drawer safety switch not closed.
		e. Motor out on overload.
		f. Check for broken belt, chain, pulleys, etc.
G. Unit will not shut off.	1. Defective top limit switch.	a. Push start/stop switch to STOP or turn key to OFF position.
		b. Then check continuity of the top limit switch (wires removed) with the ram at the top of the stroke.

Troubleshooting Chart   COMPACTORS (Continued)

Problem	Possible Cause	Corrective Action
		Switch should be open; if not, replace.
	2. Start side of start/stop switch remains closed (contacts fused).	Turn key to OFF position. Check continuity of the V and BR terminals on the start/stop switch. Contacts should be open, replace switch if defective.
	3. Defective directional switch.	If unit continues to run up and down without stopping at top of cycle, replace the directional switch (contacts of switch are defective internally).

Adjustments

Component	Adjustment	Adjustment Specification
Belt	Loosen the four nuts that secure the motor mount assembly.	Tighten until belt will deflect $\frac{1}{8}$ inch.
Chain	Loosen the four bolts that secure the motor mount assembly.	Tighten until chain will deflect $\frac{1}{4}$ inch across the drive sprocket.
Directional switch and top limit switch.	Loosen the two screws that hold the switch to the mounting bracket.	Adjust top limit switch so it opens when the ram is $\frac{1}{2}$ inch below the top of ram guide posts (frame). The directional switch should actuate $\frac{1}{4}$ inch lower than the top limit switch ($\frac{3}{4}$ inch from the top).

Lubrication

Component	Lubricant	Instructions
Chain	Chain lubricant.	Usually not needed.
Power screws	Multipurpose, Extreme Pressure Moly-Lithium Grease Part No. 674792.	Apply in case of noise or when replacing component.
Power nuts	SAE 30 oil.	Apply in case of noise or when replacing component.
Triangular bearing	SAE 30 oil.	Apply in case of noise.
Ram pressure pads	Multipurpose, Extreme Pressure Moly-Lithium Grease Part No. 674792.	Apply in case of noisy pressure pads or when replacing pads or ram.
Drawer slide rollers	SAE 30 oil.	Apply if rollers become noisy or hard to roll.
Motor	None.	Prelubed additional lubrication not needed.

20

Dehumidifiers

At certain times during the year, some geographic areas are too humid for comfort. If you live in one of these regions, you will probably be called on to service a dehumidifier. There may be times when the dehumidifier cannot seem to do the job, when it just is not removing the moisture fast enough to suit your customer.

Be slow to blame the appliance. Check its working conditions first. These include size of room, relative humidity, ventilation, and such. It is possible that weather or household conditions are producing more moisture than is normal. Good spot ventilation, such as exhaust fans, that will remove excessive moisture-ladened air, as it is produced in the laundry, kitchen, and bathroom, will do a great deal to decrease the working load on a dehumidifier.

TYPES

Basically there are two types of dehumidifiers. One operates on the absorption principle, drawing air through a chemical drying agent that soaks up the moisture. When it has picked up all the moisture it can hold, this chemical is dried out (regenerated) by an electric heater. This type of machine is not portable in that it requires an outdoor vent. For an equal capacity it will cost more than the refrigerant type. Its prime advantage is that it will effectively dry air to as low as 20 to 30 percent of relative humidity—even at low temperatures!

The other type of dehumidifier is the refrigerant type (Fig. 20-1). This unit operates on the principle that moisture will condense on a cool surface. Here the fan draws the air over a cool coil of tubing (Fig. 20-2), and the condensed moisture either collects in a pan, which must be emptied regularly, or it runs through a drain. This machine is very efficient on hot humid days, but its efficiency drops noticeably as the surrounding temperature decreases. For example, at 90°F (32°C) a machine may remove 3 gallons of water per day, at 80°F (26°C) it will remove about 2 gallons, and at 70°F (21°C) it will remove only about 2 quarts of water. Also, this machine is not able to reduce the relative humidity much below 40 to 50 percent because the coils would have to be so cold that the moisture would freeze on them.

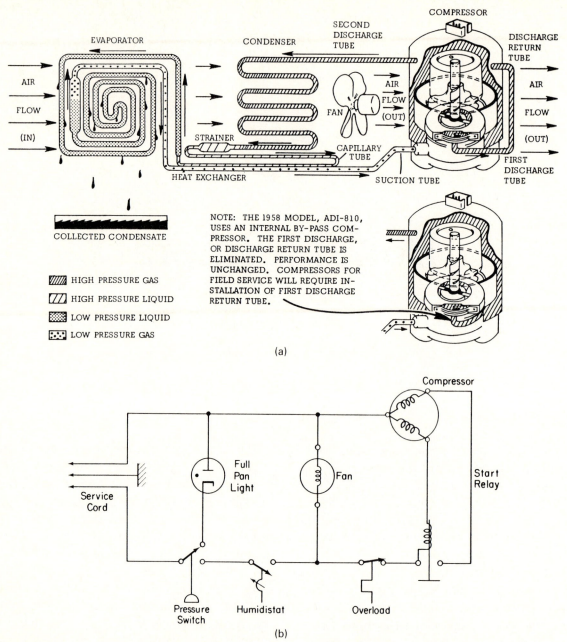

COLLECTED CONDENSATE

NOTE: THE 1958 MODEL, ADI-810, USES AN INTERNAL BY-PASS COMPRESSOR. THE FIRST DISCHARGE, OR DISCHARGE RETURN TUBE IS ELIMINATED. PERFORMANCE IS UNCHANGED. COMPRESSORS FOR FIELD SERVICE WILL REQUIRE INSTALLATION OF FIRST DISCHARGE RETURN TUBE.

▨ HIGH PRESSURE GAS
▨ HIGH PRESSURE LIQUID
▨ LOW PRESSURE LIQUID
▨ LOW PRESSURE GAS

(a)

(b)

FIGURE 20-1. (a) The refrigeration system used in a dehumidifier is essentially the same as the sealed system described in Chapter 32. The information included there applies here as well. Practically, the capillary tube used in a dehumidifier is larger in diameter, allowing a greater flow of refrigerant and less pressure differential than those of the capillary tubes used in refrigerators. (b) The wiring diagram includes about all the electrical features that can be on a dehumidifier. The pressure switch shuts the unit off and turns on a signal light when the water tank is full. The humidistat stops and starts both the air circulating fan and the compressor as the moisture content of the air varies. Some dehumidifiers do not have all these controls. Others have none at all and run continuously.

CAPACITY

The amount of air a dehumidifier can handle depends on its size and local conditions. Most units are rated between 10,000 to 15,000 cubic feet of air. Thus a 12,000-cubic foot unit could handle an 8-foot-high area with 1500 square feet of floor space, or an average ranch-type house. If you find that a dehumidifier is large enough for a home but does not seem to be doing

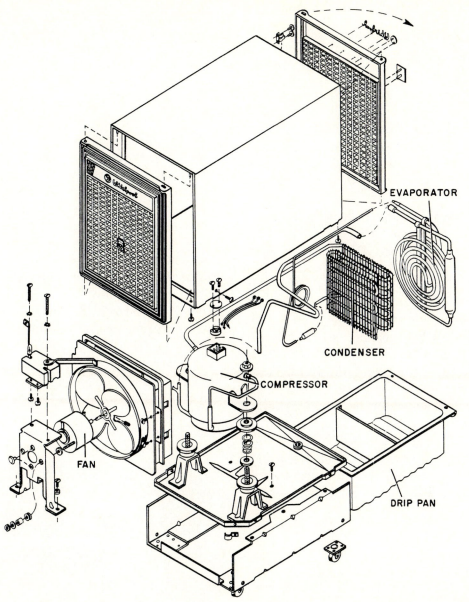

FIGURE 20-2. The airflow is straight through the cabinet in a typical dehumidifier. It passes over the cooling coil (evaporator) to give up its moisture. Then it passes over the condenser and compressor to keep them cool.

the job, you must consider the sources of humidity.

Trying to dry the air often becomes a major project, like bailing out a leaky boat! Walls, windows, doors, or any source of fresh air coming into the home will allow moisture to enter with it. In addition, there is the steam from cooking, shower, clothes drying, and so on. Consider all these factors before checking the dehumidifier itself for trouble.

To check for humidity, you need a wet and a dry bulb thermometer and a relative humidity chart. Both come in a special tools category. Whether humidity is too high can also be determined in other ways.

Doors and drawers swell and stick.

Shoes and other leather items mildew in the closet.

Cold-water pipes sweat (drip water).

Tools get rusty.

Exterior paint blisters on siding.

The piano goes sharp in tone.

Foods mold quickly.

Last, but hardly least—YOU feel hot and sticky!

Check the owner's manual for specific information on the temperature–humidity range and capacity of a dehumidifier. Also, be sure to follow the manufacturer's advice on emptying the water pan.

Incidentally, this water is completely mineral free and can be used in a steam iron or in the car's storage battery. Make certain that it has not collected dust from the air. If it has, filter it out.

A dirty or clogged condenser will cause poor performance. It should be inspected and vacuum cleaned or brushed periodically to remove any accumulation of dust. Note (Fig. 20-3) the accumulation of dust and lint on the evaporating coil (one-half is cleaned). This coil is wet with condensed water, and dust passing through it sticks.

A dehumidifier should be so located in the room that its end grilles are free of obstructions (at least 6 inches) to allow for unhampered air circulation. The grilles should be cleaned when the coil is cleaned.

Only air that passes through the dehumidifier will be dried. That is why its location in the room is important.

TESTING

A dehumidifier is quiet under normal operation, making no more noise than a fan. Any rattles or unusual noises due to vibration can be stopped. Loose wires can be taped to the nearest firm part away from the fan. Tubing can be gently bent away from the parts it might hit. The fan blade must be balanced.

Automatic control is attained with a humidistat, which consists of a snap-action switch actuated by human hair or by a plastic ribbon (Fig. 20-4). Moisture affects the element by causing it to shorten with decreasing humidity and to stretch with increased humidity.

The relative humidity that is to be maintained is set by rotating a knob, thus controlling the linkage between the element and switch.

FIGURE 20-4. This humidistat is actuated by the change in hair length with change in humidity.

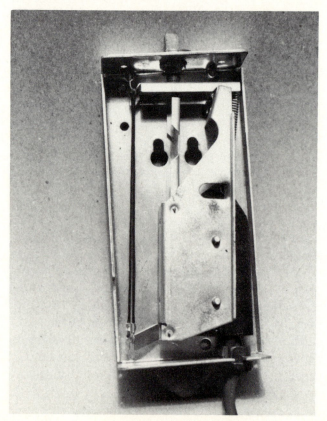

FIGURE 20-3. Dust in the air passing over a wet evaporator will stick to it. The half-clean section shows how the coils should look.

The humidistat is reasonably troublefree, so don't be too quick to blame it. Check other operating conditions first.

To check the humidistat, disconnect its leads from the humidifier and connect them to a test light or ohmmeter. Rotate the knob from the WET, or 80 percent marking, toward the DRY, 20 percent marking, until a click is heard in the switch. From this point to the DRY end there should be continuity. Rotate the knob back to the WET end until a click is heard in the switch. From this point to the WET end there should be no continuity. Repeat this pro-

cess several times to be sure that the switch is operating correctly. Note the relative humidity reading at the time the click is heard. This reading should be within + or −10% of the actual room humidity.

If the preceding test shows a defect, replace the humidistat. The dehumidifier can be operated manually until a new humidistat is installed.

Any fan problems should be handled in the manner described in the section on fans. The cooling comes from a small refrigeration unit that can be tested as such.

Troubleshooting Chart   DEHUMIDIFIERS

Problem	Possible Cause	Corrective Action
Unit does not run.	Defective motor-starting relay.	Replace relay.
	Humidistat set wrong.	Adjust for desired humidity.
	Faulty humidistat.	Replace humidistat.
	Defective motor or compressor.	Replace refrigerant unit.
	No power to unit.	Check fuse, outlet, cord, and cord plug.
	Low voltage.	Overloaded branch circuit.
		Incorrect wiring.
		Incorrect voltage for machine's rating.
	Loose wiring.	Check all wiring connections.
	Defective switch.	Replace switch.
	Defective motor overload protector.	Replace protector.
	Defective fan motor.	Replace fan motor.
Unit turns off and on too frequently.	Defective humidistat.	Replace humidistat.
	Failure in refrigerant system.	Check compressor.
Unit runs but does not dehumidify.	Abnormal conditions.	Check operating conditions.
	Poor location and air circulation.	Relocate with more clearance.
	Defective fan motor.	Replace fan motor.
	Refrigerant low in system.	Locate leak, repair and add more refrigerant.
Unit runs but evaporator frosts.	Abnormal conditions.	Check operating conditions.
	Poor location and air circulation.	Move unit.
Fan not running.	Defective fan motor.	Replace fan motor.
	Jammed fan blade.	Straighten fan blade.
	Motor relay defective.	Replace motor relay.
Noisy operation.	Fan blade hitting.	Straighten fan blade.
	Tubing hitting.	Rebend tubing to clear object being hit.

Troubleshooting Chart   DEHUMIDIFIERS (Continued)

Problem	Possible Cause	Corrective Action
	Loose cabinet, etc.	Tighten loose parts that might vibrate.
Fans run but compressor does not.	Defective compressor. Defective overload protector. Motor capacitor defective. Thermostat set too warm. Thermostat defective. Defrost thermostat defective. Defective switch.	Replace compressor. Replace overload protector. Replace capacitor. Reset thermostat. Replace thermostat. Replace defrost thermostat. Replace switch.
Compressor runs but fan does not.	Defective switch. Fan-speed control reactor defective. Defective fan motor. Defective fan motor capacitor. Fan blades or shaft binding.	Replace switch. Replace reactor. Replace motor. Replace capacitor. Straighten blades or free shaft.
No cooling; both fan and compressor running.	Clogged air passages. Compressor not pumping (coils will not be cool).	Clean filters. If coils are iced, allow to defrost. Clear any foreign obstructions from air passages. Check sealed system.
Insufficient cooling, both fan and compressor running.	Excessive load due to doors and windows being open; above normal temperature, etc. Partially clogged filters or air passages. Check duct air damper position. Fan motor speed set too low. Fan dirty, blades loaded with dust. Compressor not pumping at full capacity. Insulating seals out of place.	Check room conditions. Clean or replace filters. Clean air passages. Check control's position and owner's manual for correct setting. Check fan speed and adjust if a variable-speed fan is used. Clean fan and blades; make certain fan blades are tight on shaft. Replace compressor. Replace insulation.

21

Dishwashers

When people tell me that they are not happy with their dishwasher, my first thought is: "What are you doing wrong?" Very often the person who does the dishes was not instructed in how to use the machine correctly.

The dishwasher, more than most appliances, gets blamed for poor performance that is not necessarily its fault. Perhaps the water is not hot enough or the dishes were not loaded correctly. Hard water will leave spots on glasses and silverware. (They are on dishes, too, but it is harder to see them there.)

So before we get into machine failure, let us quickly review what the machine needs in order to do a good job of dishwashing.

1. There must be enough electrical power at the machine. Low-line voltage, or running the machine from a long (or light-duty extension) cord, will prevent motors from operating at their normal rated capacity. If this is the case, washing action and water circulation will be weak.

2. The machine must be properly installed. If there is any question in your mind about the installation, check the instructions that came with the machine. Specific requirements for each brand of built-in unit will vary to some degree. But they all need to be level, secure, ventilated, and have very definite draining specifications. Portable machines must be on a level floor.

3. The water supply must be good. There must be an available flow of 3 gallons per minute of clean water that is at least 150°F (66°C). If the water pressure is under 15 psi, with the water running, the chances are that you will not get enough water flow to provide the correct fill level. If the pressure is over 120 psi, the fill valve and connecting hoses are apt to become damaged. A pressure regulator in the water line will correct the high-pressure problem, but the low-pressure and low-flow problems call for a careful study of the water supply system.

The water temperature range of 140 to 160°F (60 to 70°C) in the machine during both the wash and final rinse cycles is important. Dishwasher detergents contain active ingredients that will not dissolve until they reach this temperature. This temperature is also needed to assist the drying.

Poor water quality, resulting from dissolved minerals, can spoil all other factors that make a good dishwashing job. The spots on silverware and cloudy appearance on glasses are the mineral deposits left after the water has evaporated in drying. Since 80 percent of the water in this country is classified as "hard," most people will have this problem. The best solution is to install a water softener in the water line.

Instruct the customer: To clean those spots off glasses and flatware, run them through a vinegar wash in this manner:

(a) Run the dish load through a normal wash cycle up to the drying period.

(b) Advance the timer to the wash cycle and allow the machine to fill.

(c) Add two cups of vinegar—*no detergent.*

(d) Allow the machine to complete its cycle.

It may be necessary to repeat this cycle in stubborn cases. If this program is followed at regular intervals, the lime deposit will not have a chance to build up to an unsightly level.

4. Load dishes in the machine according to the manufacturer's instructions. Each dishwasher has its own definite pattern of water action and thus requires a specific loading pattern to maintain the best results. It may help you to remember that

(a) Dishes are washed by the scrubbing action of the water being sprayed (Fig. 21-1a) or splashed (Fig. 21-1b) on them.

(b) This water must drain off the

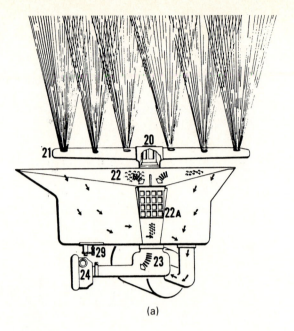

(a)

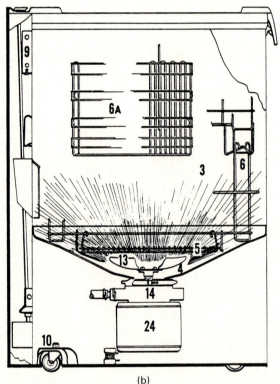

(b)

FIGURE 21-1. (a) Water action for the spray-type dishwasher is from the pump up through the hub (20) out the swirl arm (21), which rotates on the hub. The return path is through the filter screen (22) and scrap basket (22A) to the pump. During the drain cycle the water travels through the drain boot (23) to the drain pump (24) and out. (b) Water action for the splash-type dishwasher is from the impeller (13) through the impeller guard (5) and against the dishes in racks (6 and 6A), then by gravity back to the pool from which the impeller picks up the water. The water level should not be more than $1\frac{1}{2}$ inches above the bottom of the impeller when the machine is in use.

dishes by gravity. Therefore, load the dishes in such a way that water can reach all dish surfaces. Do not block off an area in the upper rack by placing a big bowl in the lower rack. The water simply will not go through the bowl. Place all cups and glasses upside down so that they do not collect water.

5. Use the detergent recommended by the manufacturer in the quantities suggested. Too much detergent, or a sudsing detergent, will generate so many suds that the pump will clog and you will lose the washing action. Another thing suds do is to act like a blanket over the water, softening the washing action, and, in extreme cases, all the water energy is spent in trying to get through the suds barrier.

Properly used, a dishwasher will give a better, more sanitary job than hand washing. Try to see that the machine is used as the manufacturer intended it to be used.

The major recent changes in dishwasher design have been in its water distribution and cycle control. To get water up to the upper dish rack, extra spray arms have been mounted under the rack as shown in Fig. 21-2. Dishes must still be loaded so as not to interfere with the spray arm action.

Cycle control has progressed from a clock timer operated from a control knob (Fig. 21-3), through a series of pushbuttons to control the clock, to a touch-pad-controlled electronic timer (Figs. 21-4 and 21-5).

The electronic controls are subject to a different set of problems than those of clock-motor-operated timers. These are mentioned in Chapter 6.

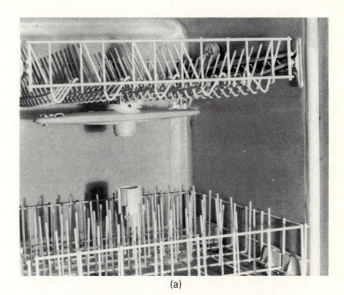
(a)

(b)

FIGURE 21-2. (a) Care must be taken that no item projects through the bottom of the top rack. This would prevent the spray arm from functioning. (b) Some models shoot a concentrated stream of water directly at the spray arm under the top rack.

SERVICE PROCEDURE

Now that we have discussed the external factors, let us look at some of the things that can happen to the machine itself.

The problem locater chart should be used in conjunction with the schematic wiring diagram. This diagram (Fig. 21-6) can normally be found pasted on the inside of a removable panel of the machine. It will be your guide to finding the defective part. Match your problem to the chart; check each probable cause and make the correction outlined.

The operation of the dishwasher is controlled by a timer, whose operating sequence is shown in Fig. 21-7.

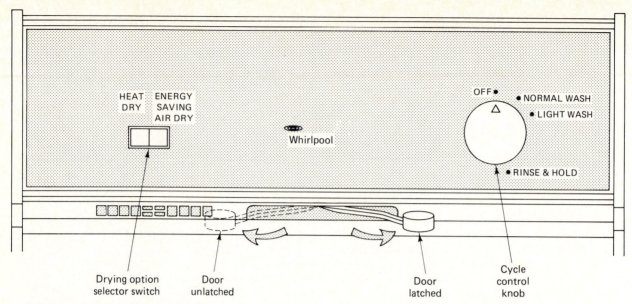

FIGURE 21-3. Single control knob for clock timer control of a dishwasher cycle sequence. (Courtesy of Whirlpool)

FIGURE 21-4. (a) Touch panel operator for an all-electronically controlled dishwasher, Model SHU9900; (b) cycle sequence chart showing what action to expect during operation of each selected option of the Model SHU9900; (c) schematic wiring diagram for the Model SHU9900. Note here that electronic timer and control system interfaces to the working parts through relays.

[(a) (b) and (c) Courtesy of Whirlpool]

(a)

Notes.

Ⓐ Only 1 min. fill in "china crystal".

Ⓑ Wet. agt. dispenses when both heater and det. disp. relays are energized.

Ⓒ Pots and pans only.

Ⓓ "High temp. washing only".

Ⓔ Heater off when "air dry" is selected. Heater cycles on and off in "china crystals".

Ⓕ Drain and fill to purge water lines in "quick wash" only.

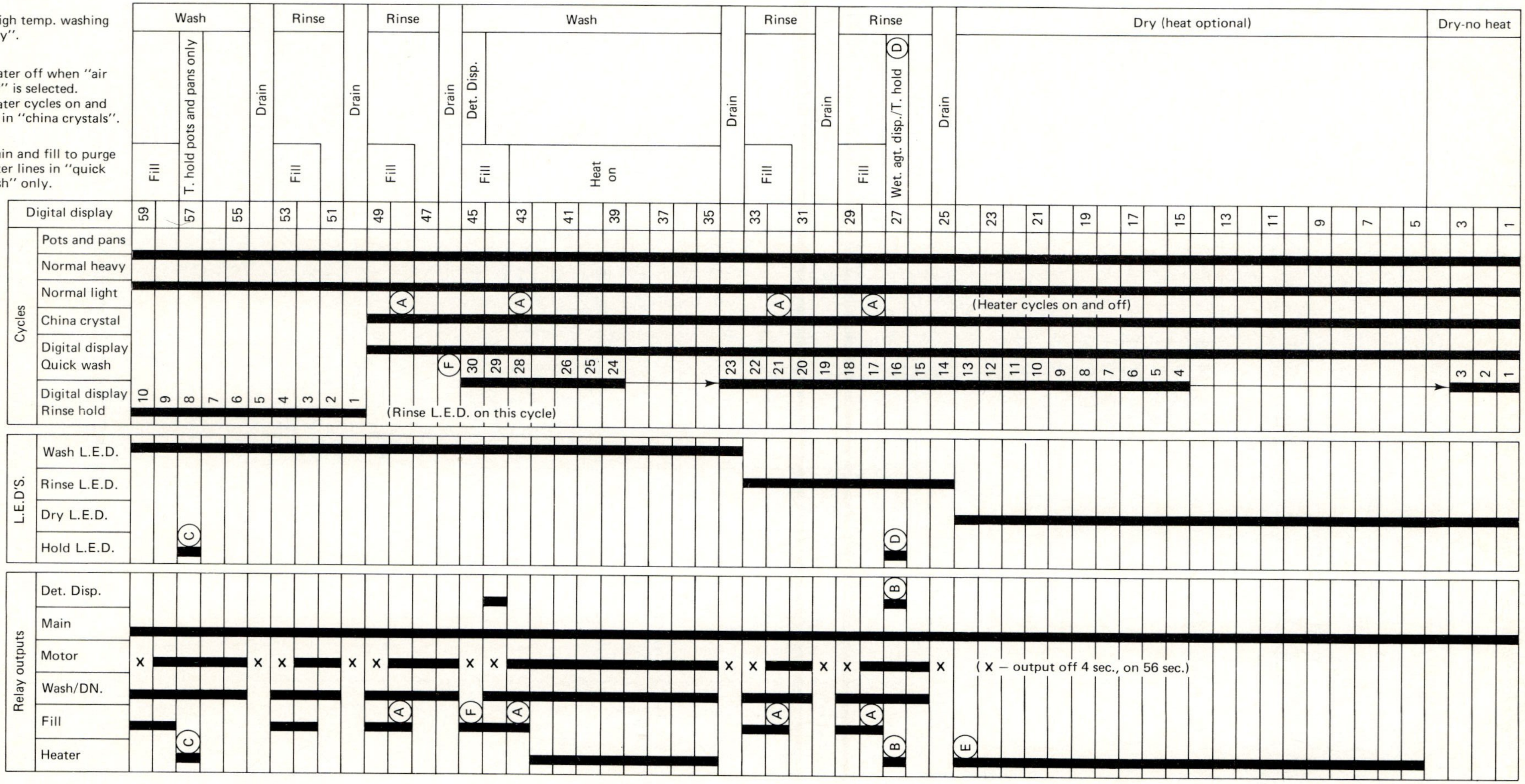

FIGURE 21-4. (Continued)

(b)

281

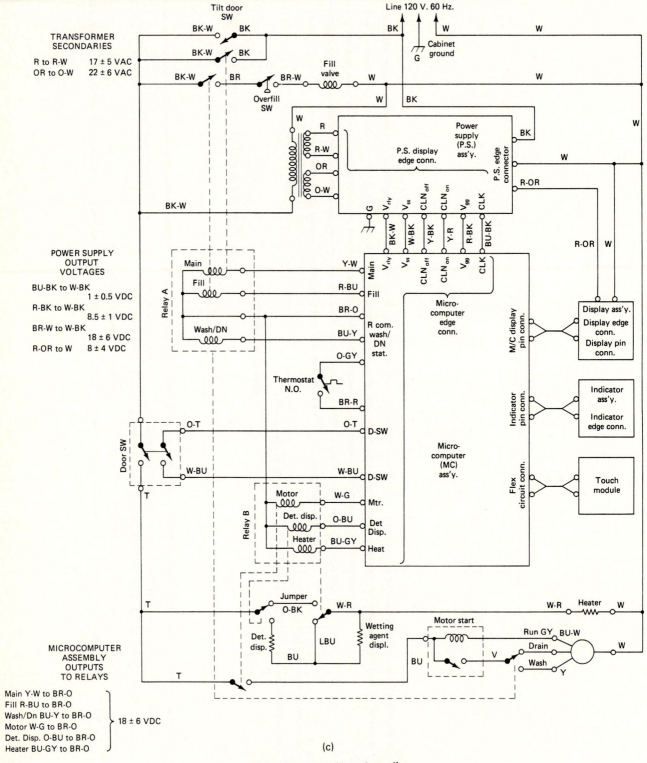

FIGURE 21-4. **(Continued)**

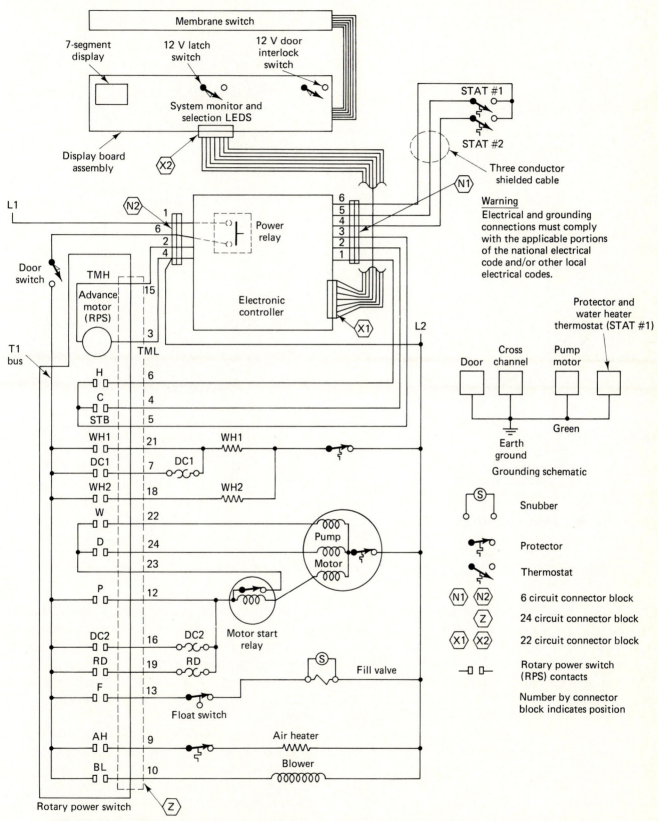

FIGURE 21-5. Schematic wiring diagram shows an electronic touch pad and control module which operates a rotory switch to control the machine action. (Courtesy of Maytag)

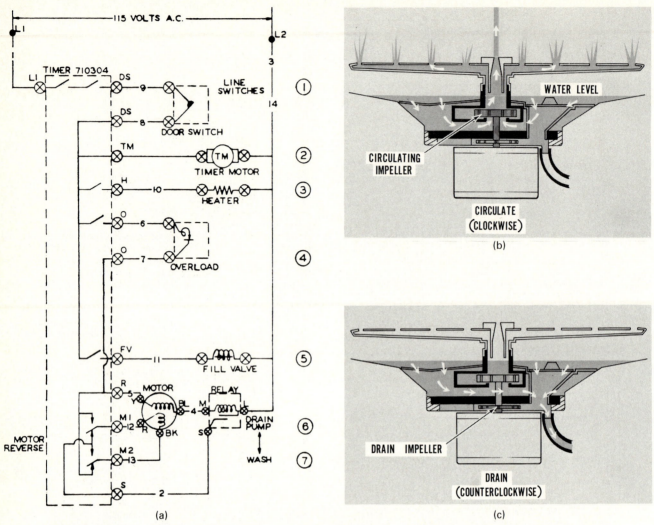

FIGURE 21-6. (a) Typical dishwasher schematic diagram. A reversible motor is used for a drain pump (circuits 6 and 7). (b) When the motor runs clockwise, it drives the impeller to circulate the water in the dishwasher. The central vertical stream impinges on the upper spray arm to be redistributed among the dishes in the upper rack. (c) When the motor runs counterclockwise, the drain impeller pumps water out of the machine.

Use a test light or voltmeter and follow the wiring diagram to check the circuit, and the cycle chart will tell when each part is in operation. You can determine whether the trouble is in the part or the circuit to the part by the test light action.

If the bulb lights when connected across the terminals of a valve, solenoid, or motor and that part does not work, you know that there is trouble in the part.

If the bulb does not light when there is supposed to be electricity available to that part, you must look for trouble in the circuit to that part.

Before condemning the timer, make cer-

tain that the lid or door switch (Fig. 21-8) is working and that there is power to the machine.

Water leakage and mechanical failure of parts usually can be seen. More rust than usual, a broken part, or a part that binds are signs pointing to a specific area that should be checked.

Often the problem is how to approach a machine. Most portables have a back that can be removed to expose the working parts. Built-in machines have a removable panel under the door. Figures 21-9 to 21-11 show the component locations for various machines.

The timer knob must be removed before the back can be taken off the Whirlpool ma-

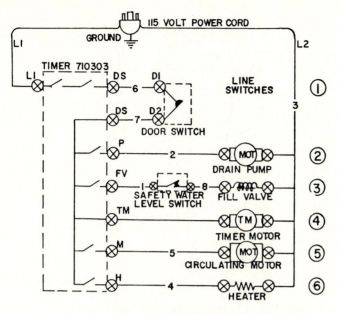

ABOVE IS FOR MODELS 50 SERIES. MODEL FP 50-0 IS SAME EXCEPT SAFETY WATER LEVEL SWITCH IS OMITTED AND WIRE #1 CONNECTS DIRECTLY FROM FV ON TIMER TO THE FILL VALVE.

TIMER CIRCUIT FUNCTION   FP-50 SERIES

MACHINE FUNCTION		TIMER INCREMENT	ACTIVE CIRCUITS
OFF		0-1	
1st PRE RINSE	DRAIN	2	1 2 4
	FILL	3	1 3 4 5 6
	RINSE	4-5	1 4 5 6
	DRAIN	6	1 2 4 5
PRE WASH	FILL	7	1 3 4 5 6
	WASH	8-10	1 4 5 6
	DRAIN	11	1 2 4 5
2nd PRE RINSE	FILL	12	1 3 4 5 6
	RINSE	13-15	1 4 5 6
	DRAIN	16	1 2 4
WASH	FILL	17	1 3 4 6
	WASH	18-30	1 4 5 6
	DRAIN	31	1 2 4 5
1st RINSE	FILL	32	1 3 4 5 6
	RINSE	33-34	1 4 5 6
	DRAIN	35	1 2 4 5
2nd RINSE	FILL	36	1 3 4 5 6
	RINSE	37-41	1 4 5 6
	DRAIN	42	1 2 4 5
DRY	DRY	43-59	1 4 6
	DRAIN	60	1 2 4 6
OFF			

NOTE 1: NUMBERS SET IN WIRES REFER TO TABS ON INDIVIDUAL LEADS IN WIRING HARNESS.

NOTE 2: ABOVE TIMING IS USED FOR MODELS FP 50 SERIES

NOTE 3: EACH TIMER INCREMENT - 45 SECONDS.

NOTE 4: ACTIVE CIRCUITS COLUMN REFERS TO ENCIRCLED NUMBERS ON DIAGRAM.

FIGURE 21-7. Typical dishwasher diagram and cycle chart for machine using separate drain pump.

(a)

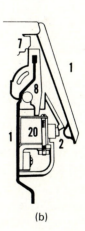

(b)

FIGURE 21-8. Lid switch. (a) All dishwashers have a lid switch to turn off the motor when lid is opened. The usual location is under the handle. (b) Lid switch identification: 1, lid latch and handle; 2, lid striker; 7, lid gasket; 8, tub rin gasket; 20, lid switch.

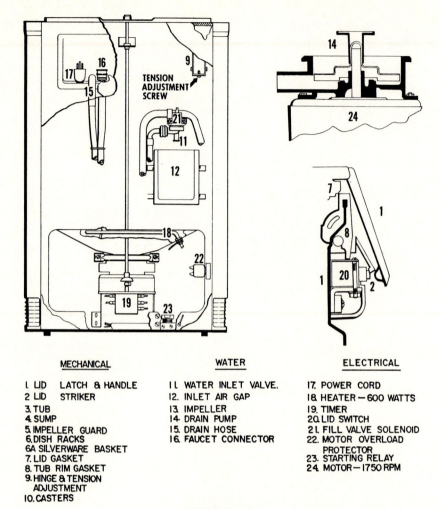

MECHANICAL

I. LID LATCH & HANDLE
2 LID STRIKER
3. TUB
4. SUMP
5. IMPELLER GUARD
6. DISH RACKS
6A SILVERWARE BASKET
7. LID GASKET
8. TUB RIM GASKET
9. HINGE & TENSION
    ADJUSTMENT
10. CASTERS

WATER

I1. WATER INLET VALVE.
12. INLET AIR GAP
13. IMPELLER
14. DRAIN PUMP
15. DRAIN HOSE
16. FAUCET CONNECTOR

ELECTRICAL

17. POWER CORD
18. HEATER — 600 WATTS
19. TIMER
20. LID SWITCH
21. FILL VALVE SOLENOID
22. MOTOR OVERLOAD
    PROTECTOR
23. STARTING RELAY
24. MOTOR — 1750 RPM

FIGURE 21-9.

chine. With the back set to one side, it is now possible to (1) adjust the lid spring tension, (2) check the fill-valve filter screen for stoppage, and (3) test each part for its proper function. The only part not available from here is the lid latch switch, which is on the front of the machine.

Sometimes a little carelessness in use can snowball into a major project. Consider a dish, with a small bone still on it, placed in the dishwasher. Here is a case of incomplete, or careless, scraping of the dinnerware before loading. Sometime during the cycle this bone will probably get stuck in the drain pump. Then (1) the pump jams, (2) the soiled water does not get pumped out so that the dishes are washed in dirty water, (3) (depending on which cycle the pump jammed) there will be too much water

in the tub, thus poor washing action, and (4) with the pump jammed, the motor coil will overheat, possibly burning out. If the motor has a built-in thermostat, this will function to protect the motor.

The jamming bone must be removed from the pump before the machine can be used. If the pump is as illustrated in Fig. 21-12, remove the four Phillips head screws holding the pump housing to the frame, to allow disassembly. Better have a bucket or large shallow pan under the machine to catch the water; using a bucket is easier than mopping up afterward.

One last word:

Be sure that the trouble is in the machine before taking it apart. Only 25 percent of the dishwasher service calls are due to machine part failure!

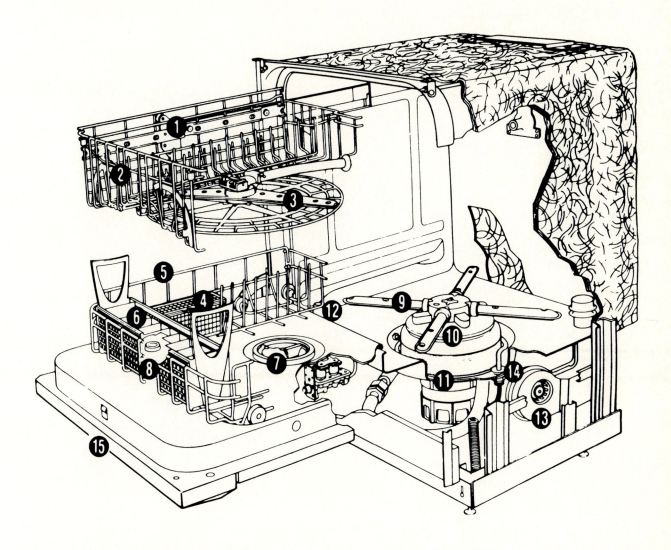

The Inside Story: KDSS-21

1. 16-Position Adjustable Upper Rack
2. Stay-Put Flex-O-Dividers
3. Upper Level Power Wash With China Guard
4. Small Items Basket
5. Lower Rack
6. Cutlery/Silverware Basket
7. Detergent Dispenser
8. Automatic Rinse Agent Dispenser
9. 4-Way Hydro Sweep Wash
10. Self-Cleaning Filter With Jet Spray
11. Removable Coarse Strainer
12. Overflow Protection Float
13. Flo-thru Drying Fan/Heater
14. Water-Heating Element
15. Solid State Electronic Controls

FIGURE 21-10. (Courtesy of Maytag)

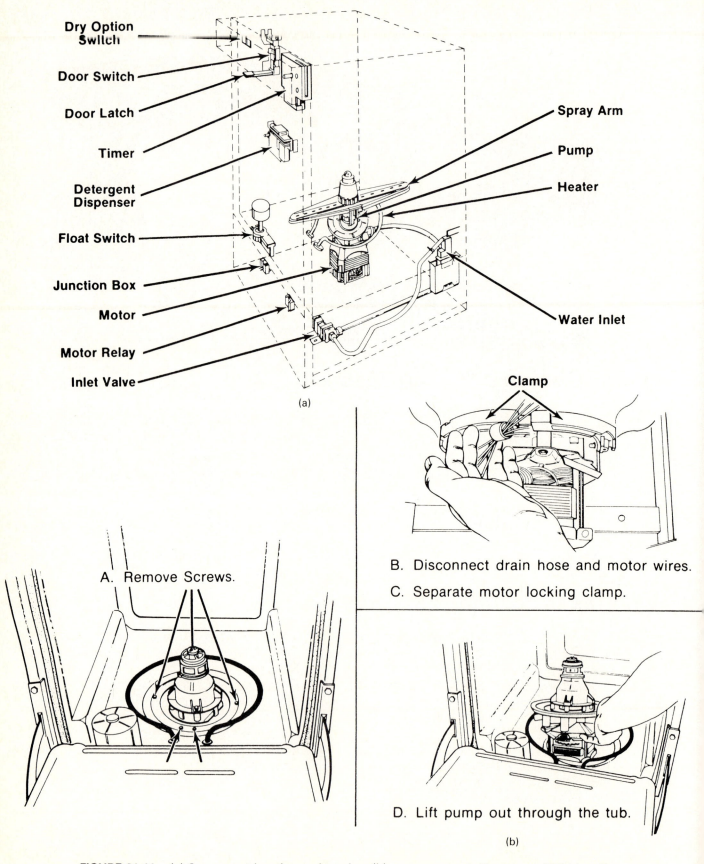

FIGURE 21-11.   (a) Component location and service; (b) pump removal. (Courtesy of Whirlpool)

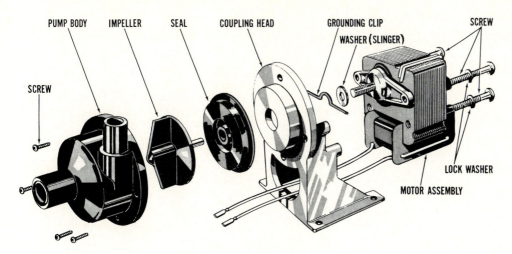

PUMP BODY   IMPELLER   SEAL   COUPLING HEAD   GROUNDING CLIP   SCREW
WASHER (SLINGER)

SCREW

LOCK WASHER

MOTOR ASSEMBLY

FIGURE 21-12. This is a popular type of pump for use as a drain pump in dishwashers and other appliances requiring a small separate drain pump.

Troubleshooting Chart **DISHWASHERS**

Problem	*Possible Cause*	*Corrective Action*
Machine will not run at all.	No power to machine.	Check for power at outlet.
	Loose leads.	Check and secure all leads.
	Door switch not operating.	Check door adjustment and continuity through switch.
	Manual reset open.	Push reset button.
	Timer not operating.	Check timer.
Does not make a complete cycle.	No power to component.	Check for power at component.
	Timer working erratically.	Check operation against sequence chart on wiring diagram.
	Loose lead.	Check and secure all leads.
	Defective component.	Check for proper operation.
Water does not enter machine.	Supply valves closed.	Open valves.
	Open circuit in wiring.	Check continuity.
	Timer not operating.	Check timer.
	Solenoid not operating.	Check leads and operation.
	Supply line restricted.	Check for kinks and foreign matter in lines.
Water does not drain from machine.	Restricted lines.	Check for kinks and foreign matter.
	Pump jammed.	Remove foreign material.
	Motor not reversing (some impeller machines).	Check timer switching (pump should run counterclockwise for pump-out).
Water leakage.	Poor door seal.	Adjust gasket, latch. Check level of machine.
	Splash at fill valve.	Check alignment and tube end for burrs or deposits left by hard water.

Troubleshooting Chart   DISHWASHERS   *(Continued)*

Problem	Possible Cause	Corrective Action
	Split hose or loose clamps.	Check condition of hose and clamps.
	Overfill (undercounter machines).	Check operation of pressure switch timer and inlet valve.
	Tub leaks.	Repair with patch kit.
Water leakage in gravity-drain.	Timer switch sticks in closed position.	Replace timer.
	Leak in tub.	Repair or replace tub.
	Drain plug leaks.	Check O-ring for damage or foreign particles. Clean or replace O-ring.
	Drain plug sticks open.	Clean lime or foreign matter from plug. Also, clean inside of drain opening.
	Solenoid plunger sticking because of dirt or bent bracket.	Remove plunger and coil and clean, straighten, or replace bracket.
Unsatisfactory drying.	Low water temperature.	Should be 140°F (60°C) in machine at last rinse.
	Heater element inoperative.	Check electrical circuit; replace if defective.
	Impatient user.	Wait for end of cycle.
Poor washability.	Improper water level.	Check flow, washer water pressure, and installation of drain.
	Improper water temperature.	Check temperature in machine during last rinse (140 to 160°F) (60 to 70°C).
	Undesirable water conditions (hardness or excess iron).	See about a water softener.
	Improper loading.	Use proper procedure.
	Pump motor not operating (defective starting relay).	Check for restrictions; replace electrical circuit if necessary.
	Spray arm not turning.	Clean under bearing, check clearance between arm and basket rail, etc.
	Impeller loose or damaged.	Tighten and/or replace.
	Detergent dispenser not dumping.	Check solenoid; adjust linkage; avoid blocking cups.
	Loose or dirty filter screen.	Refit screen to eliminate gaps; wash out well.
	Drain pump inefficient.	See if water is being properly evacuated. Look for kinks, obstruction, foreign matter.
	Back siphoning from sink.	Check installation drain loop and/or air gap.
	Incorrect timer function.	Check timer against sequence chart.

Troubleshooting Chart   DISHWASHERS   (Continued)

Problem	Possible Cause	Corrective Action
Abnormal noise (other than water hitting dishes or small item knocked loose).	Spray arm or impeller hitting.	Check clearances.
	Foreign matter in tub.	Check for broken dish.
	Loose parts.	Check impeller and pump operation.
	Low water level.	Check for proper fill and water pressure.
	Improper loading.	Read owner's manual.
	Machine not level and solid.	Level and fasten.
Touch pads in control panel fail to work.	No power to machine.	Reset fuse or circuit breaker.
	Loose connection at a connection block.	Secure connector blocks.
	Faulty controller or display board.	Replace electronic controller or display board.

22

Disposers

The food waste disposer is designed to break up and grind food into particles small enough to be acceptable to the household drain system. And it wants only food waste! Silverware will not survive a session in the disposer; nor should one expect it to grind glass, metal, rubber, foil, paper, cloth, string, etc. However, it should handle almost all food wastes.

LOCATION

It is installed in the drain outlet of the kitchen sink and connected directly to the household drain. The electrical supply must be a 120-volt, 60-hertz branch circuit protected with a 15-ampere Fusetron (time delay) type of fuse (Fig. 22-1).

During operation it is important that a coldwater flow of at least 2 gallons per minute be maintained. The flow rate must be strong enough to flush the ground wastes all the way down the drain. You can check the flow rate by filling a 2-quart (½-gallon) milk container in

15 seconds. Remember how far the faucet must be turned on to get this much water flow; then, during disposing, always turn the faucet at least that far. The water must be cold enough to congeal greases to make them firm for grinding. Hot water will soften any grease in the disposer and allow it to cool and congeal in the drain lines. Eventually such grease accumulation could block the pipes.

For customers who have a septic tank, it is advisable to keep insoluble wastes, such as clam, oyster, and lobster shells, and cigar and cigarette butts, out of their disposers. The reason is mainly because of the limitations of septic tank action.

Note: Disposers should *not* be located on the same circuit as an appliance with electronic controls!

OPERATION

Grinding is accomplished by the action of a rotating disk on which a projection forcefully im-

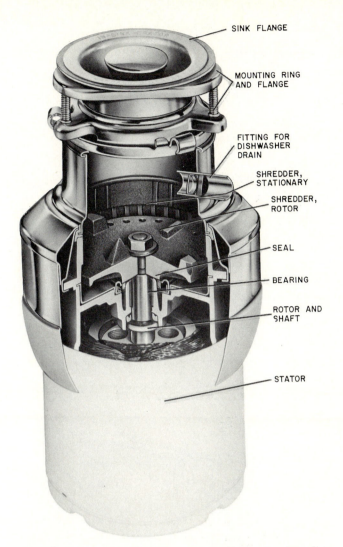

SINK FLANGE

MOUNTING RING
AND FLANGE

FITTING FOR
DISHWASHER
DRAIN

SHREDDER,
STATIONARY

SHREDDER,
ROTOR

SEAL

BEARING

ROTOR AND
SHAFT

STATOR

FIGURE 22-1. Cutaway view of a disposer showing the relationship of the essential parts. Other parts can be identified from the exploded view and parts listing. (Courtesy of In-Sink-Erator)

pels waste matter against fixed "spurs." Constant repetition of this action breaks the waste into small particles, which are then flushed through a screen by the water flow. The water flow must continue for 15 to 20 seconds after all wastes have been disposed of in order to flush the drain lines completely.

The projecting vanes or hammers on the rotating disk (or impeller) may be either fixed or pivoted, depending on the particular disposer. Both fixed and pivoted designs are in common usage. In each case, there must be "vanes" (or "spurs" or "hammers") on the disk, near the outer edge, to move the waste against or into fixed projections. If any of these items

are broken, there will be little or no disposal action.

Two different disposer operating methods exist. Each has its own merits.

1. *Continuous feed.* First turn on the cold water; then turn on the electricity. The user can then start scraping food wastes into the disposer opening. Grinding occurs as the food wastes are being fed into the disposer. A rubber splash guard prevents the ground particles from being tossed back into the sink.

2. *Batch feed.* Food wastes are scraped into this type of disposer to collect in the chamber. When ready to dispose of these wastes, cold water should be turned on and left running until after all waste has been removed. Then the control cover should be set in the opening to operate the switch that starts the machine. Grinding occurs with the cover in position (Fig. 22-2).

FIGURE 22-2. Typical ON–OFF switch location and plunger protection method used in batch feed disposers. The cover activates the switch when it is turned in a clockwise direction.

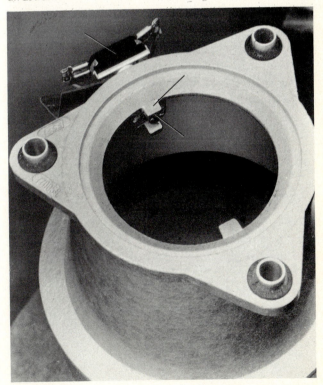

Both methods do an equally good job of disposing. The continuous-feed type is faster because it grinds as it is fed. The batch-feed model is safer for the operator because the closed cover prevents "stuff" (and fingers) from getting in during the grinding process.

Do not operate any disposer unless water is flowing! The shaft seal depends on water for lubrication and cooling. Also, if you grind without water, your machine will clog and jam because water is necessary to carry off the ground wastes. It is a messy job to disassemble and clean a disposer that is clogged or jammed.

Overloading a batch-feed-type disposer or loading a continuous-feed type too rapidly is another way jamming may occur.

All disposer motors have overload protection. Some have an automatic reset device and others have a manual reset device (Fig. 22-3). In either case, turn off the disposer and allow the motor to cool. This should take only a few minutes. By that time the automatic reset will have returned to normal, or the manual reset button can be pressed to return it to normal.

CLEARING A "JAM"

Some disposers have a reversing switch that will back the flywheel and impeller away from jamming objects, thus clearing the way. On some models this is manual; on others it is automatic (Fig. 22-4). The owner's manual will tell you which type it is.

Some disposers are provided with an "unjamming" tool to "back up" the impeller. The owner's manual should tell you how to use it. If the machine has no provision or tool for unjamming, here is what to do:

First, make sure the power is off! Second, insert a broom handle or a long, heavy screwdriver into the chamber. Engage the end against the flywheel projection and pry it counterclockwise until it moves freely. Remove the tool, turn on the water, turn on the power, and you should be all set. If it rejams, repeat the procedure but this time remove the object that caused the jam.

A word to the wise: Overloading a disposer

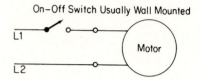

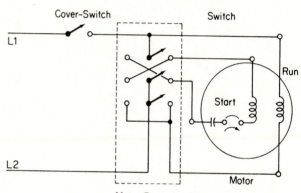

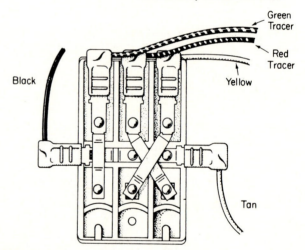

FIGURE 22-3. Typical wiring diagrams.

with wastes or being "stingy" with the water supply (flow) causes more trouble than any other factors!

NOISE

All disposers generate some noise. The amount of and the kind of noise chiefly depend on the type of material being ground. While grinding,

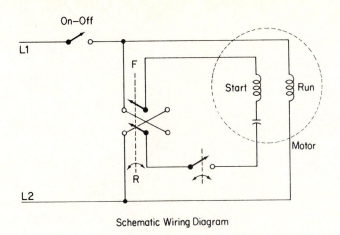

Schematic Wiring Diagram

Wiring Diagrams for Household Model

In—Sink—Erator®

Electrical Assembly for
Model 17LC Begins with Serial No. 515621
Model 17—1
Model 107
Model 107—1
Model 207

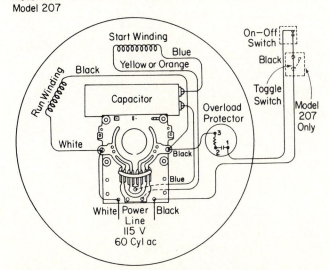

FIGURE 22-4. The In-Sink-Erator automatic reversible action is obtained by incorporating a double-pole doublethrow (DPDT) switch into the motor centrifugal switch assembly. This is in addition to the normal start winding disconnect switch. The DPDT reversing switch is transferred by the frictional action of the two spring-loaded, leather-faced detents in the shaft, which shift the switch hub by about one-sixth turn in the direction in which the shaft is moving. Thus, when the motor starts, the start winding is reversed and the armature will turn in the opposite direction the next time it starts. (Courtesy of In-Sink-Erator)

an assortment of noises is normal—starting loud, becoming more uniform, then, finally, quieting as the load is disposed. The quiet disposers are insulated to prevent radiation of noise and rubber mounted to decrease noise transmission.

A piece of undisposable matter will continue to thump or clank about within the chamber long after everything else is gone. It will have to be removed by hand *after the machine is turned off.* A pair of long tongs is best for this removal job.

Use of drain-clearing chemicals may corrode or damage the disposer!

TROUBLE DIAGNOSIS

Water-leakage troubles are usually obvious on inspection. If they are due to poor installation, tightening the mounting screws will help to correct the leak. A word of caution: Before tightening the mounting screws, be sure that the disposer is properly and squarely in position and that the gaskets are undamaged and in their place.

All disposers have a seal between the motor and the impeller chamber to prevent water from leaking into the motor and damaging it. If this seal fails, water has one more chance to escape before dripping into the motor—through a drain hole. The drain hole is a duct or hole in the motor housing that runs from the seal and/or bearing cavity to the outside of the housing.

Thus if you see water dripping from the drain hole, you should suspect a seal failure. In order to replace the seal, the flywheel must be removed. Since this procedure varies from model to model, it is best to follow the manufacturer's methods as described in the service manual.

There are several reasons why a disposer will not run. However, a logical approach and time taken for analysis of the condition before using any tools will save much effort.

First, make certain that the rotor (impeller) is free to turn. Clear any jams; rule out defective bearings, rusted motor, or shaft. This

FIGURE 22-5. Exploded view and parts list of a continuous-feed disposer. (Courtesy of In-Sink-Erator)

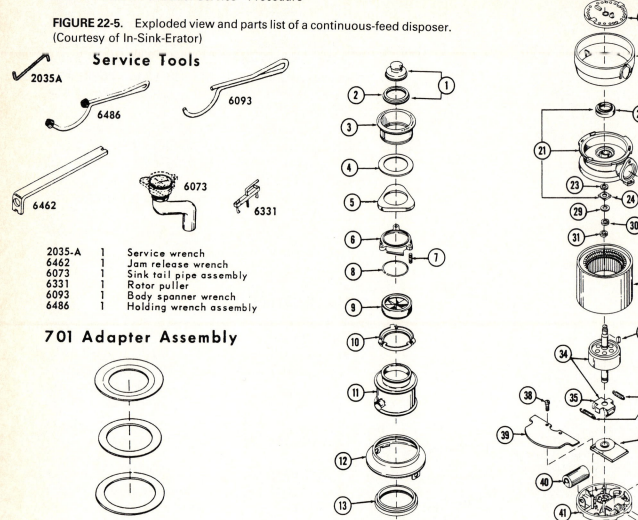

Service Tools

2035-A	1	Service wrench
6462	1	Jam release wrench
6073	1	Sink tail pipe assembly
6331	1	Rotor puller
6093	1	Body spanner wrench
6486	1	Holding wrench assembly

701 Adapter Assembly

1   Adapter assembly for 5'' opening. (Specify type of sink, whether cast iron or pressed steel).

Parts List

Code No.	Part No.	No. Req.	Description	Code No.	Part No.	No. Req.	Description
1	5032	1	Stopper assembly	26	5487	1	Tail pipe
2	5029	1	Grommet - stopper	27	5490	1	Flange - tail pipe
3	5170-A	1	Flange, strainer	28	1491	1	Bolt - tail pipe
4	1160	1	Gasket, sink hub	29	5311	2	Washer, fibre thrust
5	5150-A	1	Ring, backup	30	5332-A	1	Washer, thrust
6	5141-A	1	Flange, mounting	31	5369	1	Snap ring
7	5147	3	Screw, mounting	32	1042	1	Stator
8	5166	1	Snap ring - strainer flange	33	5330	1	Key, woodruff
9	1010-A	1	Gasket, mounting	34	1038-B	1	Rotor and shaft assembly (Incl. Code Nos. 35, 36)
10	5298-A	1	Flange, body				
11	1014	1	Body, container	35	5554	1	Switch assembly, centrifugal
12	1037-E	1	Body, lower	36	7738	2	Spring - centrifugal switch
13	1036	1	Seal - stationary shredder	37	5379-B	1	Switch, start
14	5250	1	Shredder, stationary	38	7422	4	Screw - start switch
15	5345	1	Mounting nut - shredder	39	6618	1	Insulation shield
16	11351	1	Washer - mounting nut	40	5861	1	Capacitor
17	12141	1	Gasket - mounting nut	41	5094-G	1	End bell assembly, lower
18	5314	1	Shredder, rotor	42	5765	1	Screw, ground
19	5096-C	1	Band, trim	43	5091	1	Plate, terminal
20	5097	1	Clip - trim band	44	5765	1	Screw - terminal plate
21	5469-D	1	End bell assembly, upper	45	5949	4	Bolt, thru
22	5373	1	Seal, face	46	5075	2	Screw - BX clamp
23	5052	1	Seal, oil retainer	47	5060	1	Clamp, BX
24	6557	1	Washer, thrust	48	5324-B	1	Protector, overload
25	1470	1	Washer - tailpipe				

must be done before making any electrical tests. If the rotor is not free to turn, the motor will never start. If all is clear, push the reset button on machines that have one. If there is no manual reset, then there is probably an automatic one. This reset will operate as soon as the motor cools off. Allow a few minutes for it to act.

Next, be sure that electricity is available at the disposer. A fuse may be blown, a switch may be off, or a wire connection may be loose. Low line voltage (105 volts or less) will make a motor overheat, thus causing the thermal overload protector to stop the motor frequently. A test light connected across the line terminals at the motor will prove availability of power.

CHECKING THE DISPOSER

For the continuous feed type, all wiring is in the motor (Fig. 22-5). For the batch-feed models, test the cover control switch and the reversing switch (if there is one) before entering the motor. When checking the cover control, be sure that the cover guiding tabs are not broken or worn. These tabs position the cover to actuate the starting switch. If they do not do their job, the disposer will not start. This is a visual test (Fig. 22-6).

The safest way to test switches is to disconnect the power (remove fuse) and then check with a continuity tester. If the switches prove good, check the motor.

Once trouble has been found, the remedy may suggest itself. Use the trouble shooting chart shown as a guide in your search for cause and probable solution. *Remember:* Major parts failures are rare in this type of appliance; what's more, you are working with an electrical "gadget"! So be very careful!

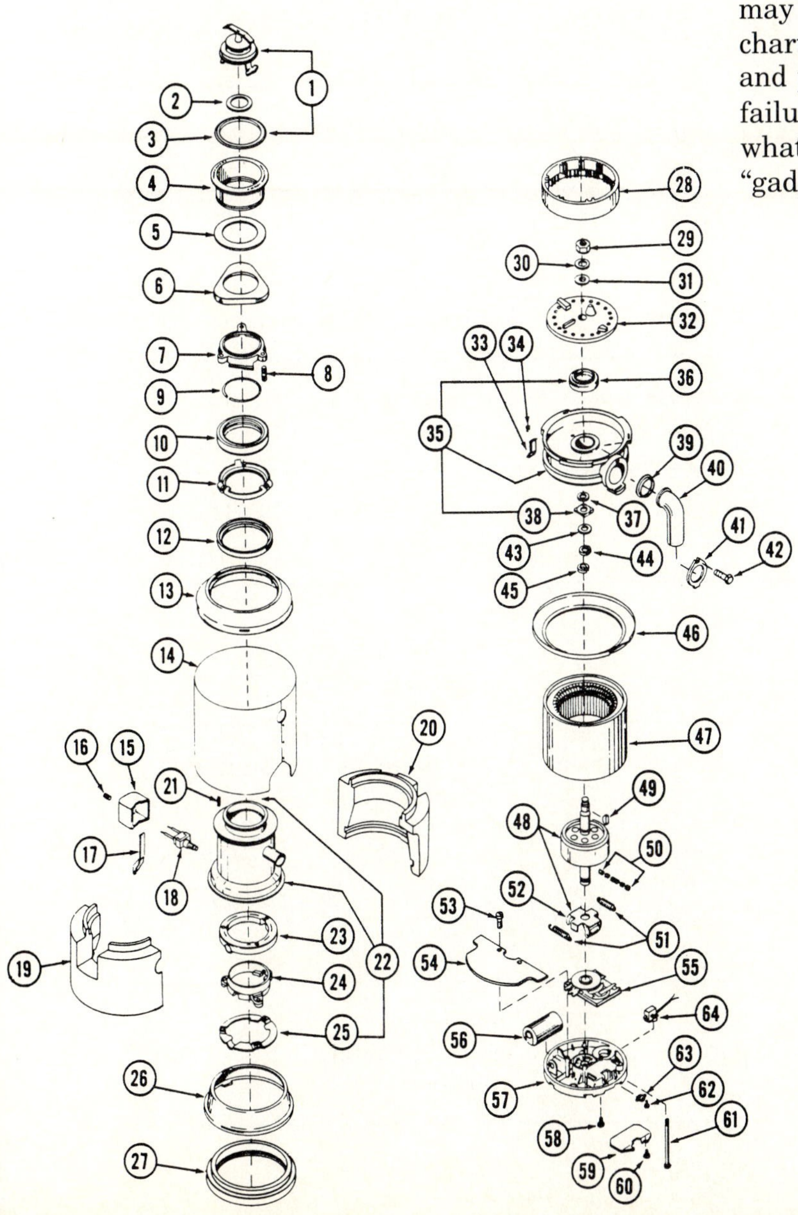

FIGURE 22-6. Exploded view list for a batch feed disposer having automatic reversal. (Courtesy of In-Sink-Erator)

Troubleshooting Chart   DISPOSERS

Problem	Possible Cause	Corrective Action
Leaks at sink flange.	Loose mounting screws.	Tighten flange screws.
Leaks at drain gasket.	Loose flange. Improper gasket.	Tighten screws. Replace.
Leaks between chamber and sink flange.	Loose mounting. Defective gasket.	Tighten nuts. Replace.
Abnormal noise and/or vibration.	Undisposable matter in chamber. Flange or tailpipe gasket improperly placed. Broken impeller vane. Motor bearings may be damaged.	Clean out chamber. Replace. Replace impeller. Replace bearings.
Erratic operation.	Loose wiring, switch, motor, or power connection.	Locate and reconnect.
Slow grinding.	Undisposable matter. Damaged impeller. Dull shredder. Insufficient water flow.	Remove. Replace. Replace. Minimum 2 gallons per minute.
Slow drain-out.	Partially clogged drain. Clogged shredder teeth.	Check plumbing. Remove and clean.
Jammed.	Something stuck between impeller and shredder.	Move impeller backward until free.
Will not stop.	Defective switch. Short in wiring. Incorrect wiring.	Replace. Clear short and insulate. Reconnect properly.
Will not run.	Overload protector has tripped. Blown fuse. Defective switch. Burned-out motor winding. Open or shorted wiring. Inoperative centrifugal switch in motor.	Wait for motor to cool; reset. Replace. Replace. Replace stator. Repair. Replace.
Cover does not control.	Defective switch. Broken guide tab. Worn cover.	Replace. Replace housing. Replace.

23

Electric Dryers

Automatic dryers need not be very complex appliances if you remember that they use heat, air motion, and that they tumble the clothes.

Heat is obtained from either an electric heating element or a gas burner. Air motion is obtained from a fan.

The tumbling of the clothes goes on in a revolving drum, driven by an electric motor.

The overall operation is controlled by a timer and protected by thermostats. Most dryers have their wiring diagrams pasted on the back. Use that diagram when looking for trouble.

You must know how to operate the machine, to know what to expect. Read the owner's manual; familiarize yourself with the controls, and always follow the fabric manufacturers' instructions for washing and drying, where given. Some of the current models have rather complex controls to provide the special drying cycles and temperatures that are recommended for the various new fabrics.

The machine is quiet in normal use. The major random noises will be due to buttons,

hooks, and other hard objects that might be tumbling in the drum. So if you hear an unusual noise, it would be wise to stop the machine and check it. Free, hard objects, such as nails, tops, knives, and lipsticks, should not be left in the machine, for they can damage the drum. Long, thin metallic items can poke through the drum and hit the shroud or heater element, causing serious trouble.

Lint should be blown through the vent to the outdoors, carried by the warm, moist air. However, some will collect in a lint filter or trap. This should be cleaned frequently, depending on usage. There is always some stray lint that collects within the dryer shell (Fig. 23-1). This can accumulate and cause trouble, possibly even a flash fire that will be confined to a small area within the dryer. You may never see it but may smell it and wonder what is burning.

A semiannual (depending on machine use) vacuuming of a dryer's interior will do much to prevent trouble. Remove the back of the machine and pick up all the dust and lint you can reach. You will find the "crevice tool" of a vac-

EXHAUST

WIRING
TERMINALS

FAN

DRUM
DRIVE
PULLEY

MOTOR

BELT TENSION
SPRING

HEATER
CONNECTIONS

FIGURE 23-1. An accumulation of lint inside a dryer is dangerous. On every service call, check for this and clean it out if needed.

uum a big help. *Play safe!* Disconnect the power to the dryer before this thorough cleaning!

We cannot emphasize strongly enough the importance of this periodic cleanup. Many dryer problems result from lint accumulating around and inside a part, thereby preventing it from doing its job.

All pulley bearings are prelubricated and should require no additional oil. In fact, excessive oiling will do more harm than good, for the extra oil can collect dust and lint.

In order to correct a squeak that is due to a dry bearing, it is best to remove the pulley or shaft from the bearing, clean off the old grease or oil, and re-oil. Use a good grade of light engine oil and allow it to soak well into the bearing. After a short wait, wipe off the excess oil and carbon that has soaked loose. Re-oil and reassemble. It is important that the bearing and shaft be absolutely clean before reassembly.

If the belt does not drive a drum or blower properly, it is because the component is not turning freely. This may be due to something that has jammed the machine or it may be due to a frozen bearing. In general, a frozen bearing can be cleaned, relubricated, and used. At least

it is worth an attempt before replacing it.

On most models the belt tension is automatically controlled by a spring-loaded pulley. An oily or greasy belt will slip even though the tension is correct. It is better to replace such a belt than to try to clean it. Before installing a new one, wipe off the pulleys carefully to remove all excess oil. On some models belts are changed by releasing the idler pulley tension and lifting off the belt.

On others, the suction fan belt tension is from the rear by removing the panel and loosening the motor-bracket lock bolt. Raise the motor slightly; then allow it to settle to a position where the weight of the motor is supported by the fan belt. Tighten the motor-bracket lock bolt securely. The drum-drive belt and the idler-drive belt are removed by running the belt over the lip of the larger pulley while slowly turning the pulley by hand.

Replacing belts on dryers is, in general, a reasonably simple procedure. Remove the back and examine the motor mounting area and the idler mounting area for adjustments (Fig. 23-2). The tension adjustment, either automatic (by spring action) or manually secured (by a bolt), can be seen easily (Fig. 23-3).

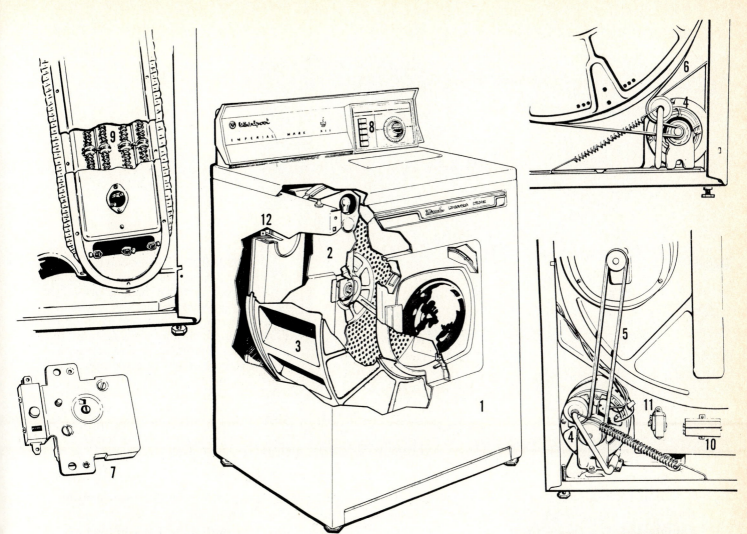

MECHANICAL SYSTEM

- **CABINET (1)**
  - U-Type
  - Removable front panel
  - Hinged Top
  - Glass port in door
  - Porcelain bearing blocks

- **BULKHEAD (2)**
  - Removable - Drum shaft welded to bulkhead

- **DRUM ASSEMBLY (3)**
  - Rolled seam
  - Stamped baffle
  - Front and rear bearing support
  - Rear self-aligning bearing
  - Delrin bearing ring - front

- **DRIVE SYSTEM**
  - Double end Shaft Motor (4)
    - Two poly "V" belts
  - Direct Blower Drive (5)
  - Direct Drum Drive (6)

ELECTRICAL COMPONENTS

- **TIMER ASSEMBLY (7)**
  - Automatic Master Touch Control Button
    - Closes start switch to machine motor
    - Depress timer reset button
    - Ten-Minute timer motor operation in a cycle

- **PUSH BUTTON SWITCH (8)**
  - Has five buttons - Controls four cycles and console light

- **HEAT ELEMENT (9)**
  - Two Sections - One 2,000-watt and one 3,600-watt - three lead terminals

- **HOT WIRE RELAY (10) and TRANSFORMER (11)**
  - Relay has an 11-volt hot wire element --
    - Operates contacts from contraction and expansion of hot wire.
  - Transformer --
    - Step down 230 to 11-volt output

- **SENSOR THERMOSTAT (12)**
  - One inlet - heater box - 200° reset, normally closed
  - One exhaust - sensor fan scroll - 145° open, normally closed

FIGURE 23-2. (Courtesy of Whirlpool)

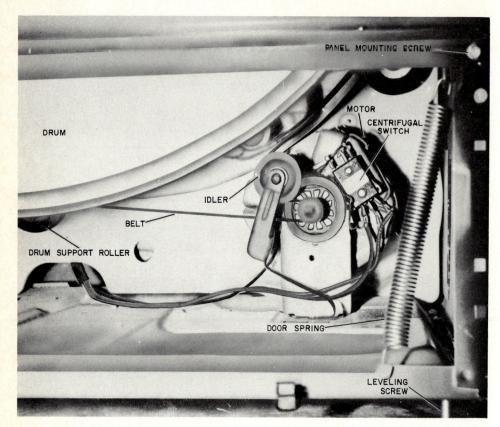

FIGURE 23-3. Removing the access panel (under the door) exposes the motor and belt on this Whirlpool dryer. Note that the belt is wrapped around the drum and is threaded through the bracket and under the idler pulley to fit on the motor pulley. To change this belt, the front panel of this machine will have to be removed. Two screws are at the bottom of the panel and two are near the top. The upper ones are accessible by pulling the top forward and lifting it up out of the way.

Belt-tension adjustment is important! If the belt is adjusted too tight, bearings will be damaged. If too loose, there will be slippage, resulting in poor drying.

Always, when placing a dryer in position, check it for being level, firm, and properly vented. These machines have leveling legs that can be screwed in or out to allow for differences in floor surface. Correct leveling is an important factor in maintaining the quiet operation of the machine.

Proper venting means exhausting the hot, humid air to the out of doors rather than into the room with the dryer. Allowing the exhaust air to pass into the room will raise the humidity of the air that is being drawn into the dryer; therefore, lengthen the drying time. Also, this exhaust air carries lint, which will be deposited within the dryer to cause future trouble.

Installation should be close to an outside wall so as to require the minimum amount of ducting. *Do not exhaust into a chimney!* Lint is bound to collect there and will eventually become a fire hazard. Long and/or winding ducts decrease the flow of air and also allow

lint to collect. This in turn will decrease drying efficiency.

Aluminum or galvanized duct materials are satisfactory, as is flexible tubing, where local codes permit.

When exhausting to the outside, make certain that there is enough air entry to the room to replace that being drawn out by the dryer. Starving a dryer for air will also result in inefficient drying. There are a number of approved ducting methods.

Electrical trouble can be traced by use of a test light and by following the wiring diagram. This procedure is explained in the section on electrical testing. As always, when checking an appliance, make sure that there is power available at the outlet. Then check the fuses within the appliance.

Most dryers have fuses to protect the motor and control circuits. Usually these are accessible from the front of the machine. The GE has its fuse in the toe plate, under a separate cover. When you remove the one screw, this cover comes off easily.

Remember: For the electric dryer, heat is

FIGURE 23-4. Typical electric heater unit that includes two separate resistance elements.

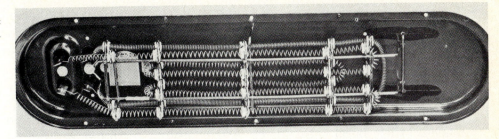

supplied by a resistance-type heating element operating at 230 volts (Fig. 23-4). Use a 230-volt bulb in your test light. A word to the wise: Do not connect your 110-volt test light across the terminals; it will blow out every time.

Dryers operate a centrifugal safety switch connected, in series, with the heater. If the motor does not run, there will be no electricity to the heater. This is also true with gas dryers. Thus, with no motor operation, there will be no heat—therefore no drying. Often this switch is incorporated within the motor as part of the motor-starting centrifugal switch. Some older dryers have this switch mounted on the fan. So if the fan does not operate, there will be no heat. (Earlier models had the switch in the motor.)

Drying depends on hot air in motion. Anything (a clogged lint screen, poor seal, air leakage, slipping belt, etc.) that decreases air motion will increase the drying time, thereby making the machine much less efficient. Let us look at some of these parts.

Both high-limit and low-limit operating thermostats are used in dryers to control the air temperature that passes through the clothes. Generally these are located in the exhaust housing and can be visually checked for correct operation by connecting the leads of a 230-volt test light across the terminals. On a double-throw thermostat, having three terminals, connect the test light across terminals 1 and 2. As the thermostat cycles, the test light will light each time the contact opens.

Safety thermostats should show continuity between their terminals at normal room temperatures. Holding a small flame (like a match or a lighter) close under the thermostat should cause it to operate. It will click and the light should go out (Fig. 23-5). Defective thermostats must be replaced.

The thermostat (temperature control) on the General Electric is accessible by lifting the top. To do this, remove the upper section of the back panel; then pry the top off at the seam between it and the outer shell. Be careful not to chip the finish of your customer's machine. (We suggest using a putty knife or a broad-bladed screwdriver and a cloth pad.) This makes the fusible link in the door switch-motor circuit accessible, also. At normal room temperature, both parts should show continuity across their terminals.

The door switch can be reached from this position. It is mounted on the front panel, near the upper door hinge.

Heater elements are accessible by removing the rear panel. A test light (230 volts) will tell you whether voltage is available to them.

FIGURE 23-5. A safety thermostat must *open* when heat is applied.

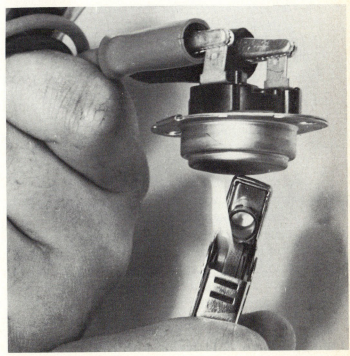

A continuity test will determine which element, if any, has failed. An element that has failed must be replaced, it can not be repaired. Be careful in checking this out! Be certain that the power is disconnected when making a continuity test! Another thing to watch for is the possibility of the heater sagging so that it touches the metal shroud or some other grounded metal part. This could give you a false reading, so look at any broken element carefully for grounding. Any grounded element must be replaced.

In the GE, the brace across the top back of the cabinet will have to be removed before the heater element assembly can be taken from the machine.

If the resistance element is sagged, stretched, or burned out, do not waste time trying to repair it. There will be trouble in the near future if you do. It must be replaced. A break near the terminal (within 4 inches of straight resistance wire) of an otherwise good element might be reconnected to the terminals. Make certain that all connections are tight!

Many dryers have an ozone bulb to help condition the air. These bulbs require either a ballast coil or a ballast bulb. This bulb must be a 40-watt bulb. Any other size will affect the ozone bulb's operation and life. The wiring diagram will indicate whether a coil or a bulb is used.

Basically dryer timers are fairly simple to service. A broken dial knob can be replaced. If it is binding on the frame, it can be shifted to clear. A defective motor—one that will not operate or drive the switching cams—can be replaced. If the switching mechanism, cams, or switches break, burn out, or otherwise fail, the most practical thing to do is replace the timer. The condition of the timer switch contacts can be checked by a continuity tester. Poor or weak switch contact will result in a burned point that will look blackened and/or melted.

Some of the newer dryers, with push button controls, become a little more complex to service. Unless the trouble is apparent and the remedy is obvious, it is wise to leave them alone, or replace the entire assembly.

Dryers using moisture-sensing devices to terminate the drying cycle operate in the same manner as the machines already described, except for the timing or dryness control (Figs. 23-6 and 23-7).

The moisture-sensing device is usually a pair of metal electrodes insulated from each other (Fig. 23-8), so located that the tumbling clothes will fall on them. The damp clothes allow a minute electric current to flow between the contacts, thus controlling an electronic circuit. Two examples, each using a different principle, are described in Fig. 23-9.

Some models of dryers, as with some models of other appliances, have microcomputers instead of clock timers to control the selection of events in the drying cycle. The working parts of the dryer will remain the same. The difference is in the control circuitry and in the selections of operations available to the user (Fig. 23-10).

Touch control panels and the electronic control modules are not readily serviceable. The recommendations are that they be replaced and returned to the manufacturer. More information on electronic controls appears in Chapter 6.

The important thing for the service technician is to determine whether the problem is in the control unit or somewhere else. Since the microcomputer output is low-voltage direct current it must operate relays to switch on the motor and the heater element (Fig. 23-11). A good volt ohmmeter having a 20,000-ohm per volt dc or better sensitivity must be used when testing the control unit.

If the control voltage is there and the relays operate, the problem is probably in one of the working circuits. If there are no control voltages, the problem could be in the electronic circuit or its power supply. Of course, there must be voltage to the machine and the power switch must be ON. You also should know how to operate a dryer.

The troubleshooting chart is basically accurate for all electric dryers. There may be minor variations in controls that are peculiar to specific models and manufacturers, but a careful analysis of the wiring diagram of a machine will guide you through the circuits.

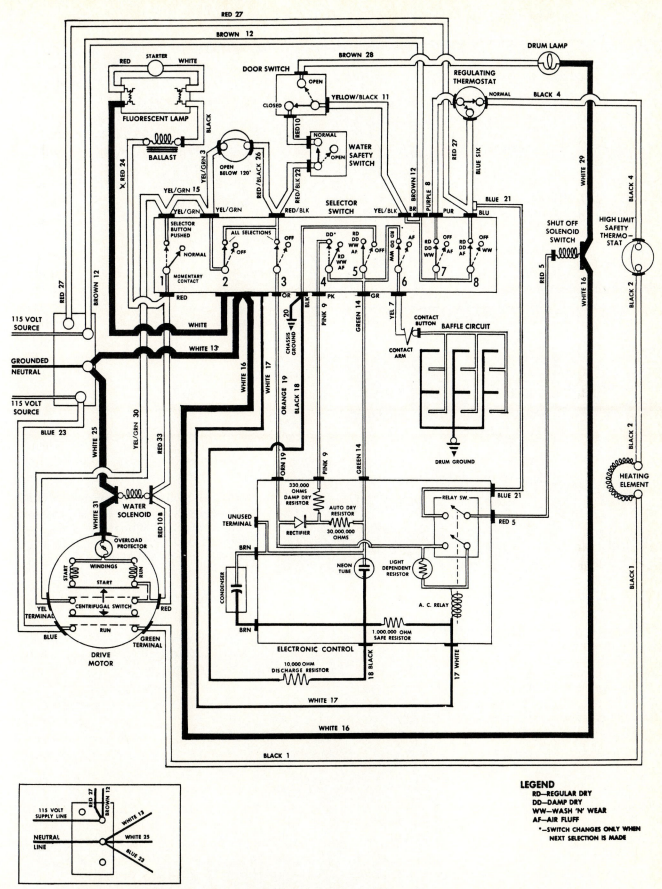

FIGURE 23-6. A Maytag electric dryer complete with wiring diagram indicates the relationship of the electronic dryness control circuit to the rest of the machine. (Courtesy of Maytag)

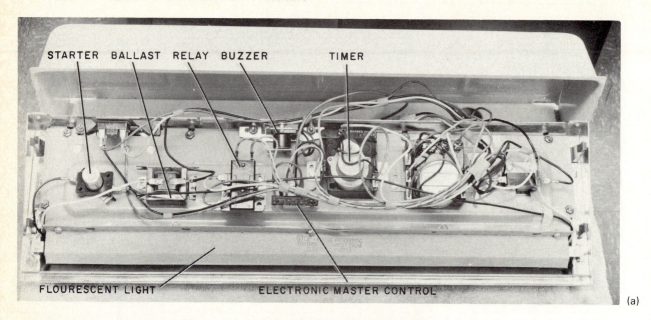

STARTER BALLAST RELAY BUZZER   TIMER

FLOURESCENT LIGHT   ELECTRONIC MASTER CONTROL

(a)

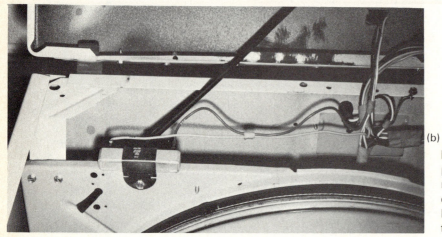

(b)

FIGURE 23-7. The circuit board for Whirlpool's dryness control, labeled "electronic master control," is mounted in back of the control panel. Its 720-ohm power resistor is located on the bulkhead of the dryer. (Courtesy of Whirlpool)

FIGURE 23-8. The sensing units for an electric dryness control are of many sizes and shapes, but all are mounted so that the clothes will contact the two metal strips. These strips are insulated from each other and from the dryer drum and backplate.

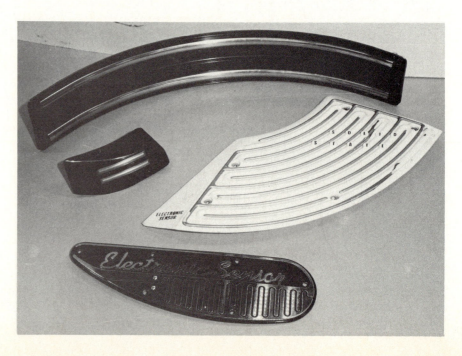

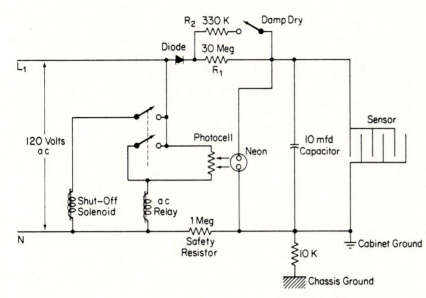

ELECTRONIC DRYNESS CONTROL – MAYTAG

As long as wet or damp clothes are touching the sensor contacts, their low resistance will serve as a "leakage" or conductance path for the current flow through the diode, R1/R2, and the sensor to the cabinet ground. This effectively prevents a voltage build up in the capacitor.

The rate of current flow through the clothes is controlled by the user's choice of R1, or R2 depending on how much dryness is desired. R2, the lower value resistor will allow more current to flow which will build up the charge in the capacitor faster than when R1, the higher value of resistance would. The longer the time the dryer the clothes.

As the clothes get dryer their resistance will increase allowing a charge to build up in the capacitor. When this charge builds up to about 70 volts the capacitor will discharge through the neon lamp, creating a flash of light.

This light, striking the photocell, will lower its resistance allowing it to conduct enough current to activate the ac relay. The ac relay will latch itself *on* and turn on the shut off solenoid.

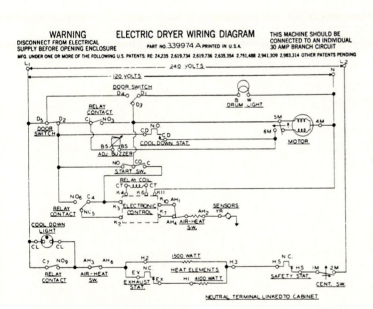

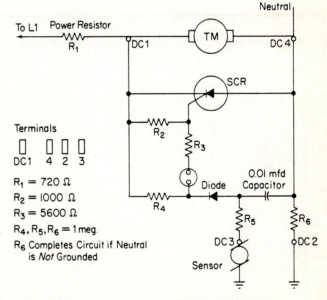

ELECTRONIC DRYNESS CONTROL – WHIRLPOOL

As long as wet or damp clothes are touching the sensor contacts, current will flow from L1 through R1,, R4, the diode, and R5. When the voltage across R4 exceeds 70 volts the neon lamp will ionize allowing current to flow through R2 and R3. This will put a + voltage on the SCR gate causing it to turn on. When the SCR conducts, it creates a direct current by-pass circuit across the timer motor, lowering its voltage and causing it to stall.

As the clothes get dryer their resistance increases, reducing the current flow through the sensing circuit until the current lowers to a level that will not support the 70 volt drop across R4. Then the neon lamp will stop conducting causing the SCR to turn off and the timer motor will start.

FIGURE 23-9. An electric wiring diagram showing the relation of the electronic dryness control to the rest of the machine. (Courtesy of Whirlpool)

TOUCH MODULE CONTINUITY DIAGRAM
EXAMPLE OF USE: WHEN "NORMAL DRY" IS TOUCHED AND HELD, CONTINUITY WILL BE
OBSERVED BETWEEN 6 AND 10 ON FLEXIBLE CIRCUIT CONNECTOR.

CONTACT \ CONTACT	8	9	10	11
7	NOT USED	KNITS GENTLE LOW	LESS DRY	20
6	PERM'T PRESS MED	REG HEAVY HIGH	NORMAL DRY	30
5	FLUFF AIR	NOT USED	MORE DRY	40
4	TUMBLE PRESS	NOT USED	VERY DRY	60
3	NOT USED	FINISH GUARD	DAMP DRY	80

FLEXIBLE CIRCUIT CONNECTOR

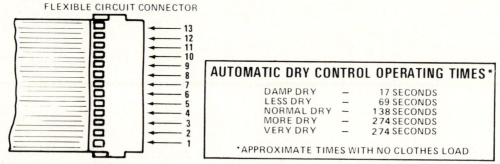

13
12
11
10
9
8
7
6
5
4
3
2
1

AUTOMATIC DRY CONTROL OPERATING TIMES*

DAMP DRY	—	17 SECONDS
LESS DRY	—	69 SECONDS
NORMAL DRY	—	138 SECONDS
MORE DRY	—	274 SECONDS
VERY DRY	—	274 SECONDS

*APPROXIMATE TIMES WITH NO CLOTHES LOAD

FIGURE 23-10. When checking the touch control pads there will be continuity between the intersecting horizontal and vertical numbers as shown on the flexible circuit connector. (Courtesy of Whirlpool)

FIGURE 23-11. The operating voltage at the microcomputer can be tested at the edge connector. The required voltage readings are shown on the chart. (Courtesy of Whirlpool)

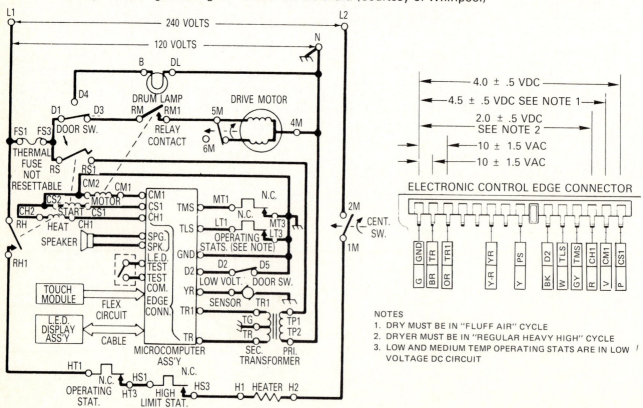

NOTES
1. DRY MUST BE IN "FLUFF AIR" CYCLE
2. DRYER MUST BE IN "REGULAR HEAVY HIGH" CYCLE
3. LOW AND MEDIUM TEMP OPERATING STATS ARE IN LOW VOLTAGE DC CIRCUIT

Troubleshooting Chart   ELECTRIC DRYERS

Problem	Possible Cause	Corrective Action
Will not run.	No power.	Check fuse and power supply.
	Loose wiring.	Check terminals and wiring.
	Door switch.	Make certain door closed properly to actuate switch.
	Defective motor.	Check motor.
	Defective timer.	Replace timer.
	Start relay.	Replace.
Runs but will not heat.	Loose wiring.	Check terminals and wiring.
	Defective thermostat.	Replace thermostat.
	Defective centrifugal switch in motor.	Replace switch (check linkage to motor).
	Defective timer.	Replace timer.
	Open heater element.	Replace heater element.
	Heat switch set to OFF.	Set switch for desired heat.
	Heat relay.	Replace.
Drum will not rotate.	Broken or slipping belt.	Replace belt.
	Jammed.	Check for foreign article between drum and shroud.
		Replace defective bearing or bearing support which allows drum to sag and hit.
	Loose pulley.	Tighten pulley set screw.
Clothes not drying, but dryer runs.	Defective operating thermostat.	Replace thermostat.
	Fan loose on shaft; no air motion.	Tighten set screw.
	Clogged lint screen.	Clean out.
	Leaky door seal; air leaks.	Replace door seal.
	Incorrect heat or timer selection.	Reset timer and/or heat control.
	Clothes too wet when placed in machine.	Wring out or extract water before placing in dryer.
Will not shut off.	Defective timer.	Replace timer.
	Fused relay contacts.	Replace.
	Shorted sensor YR to ground.	Replace.
	Microcomputer defective.	Replace.
Blows fuses.	Electrical ground.	Check heater element for foreign matter or sagging drum.
		Check wiring for bare spots touching the frame.
Motor runs when door is open.	Defective door switch.	Replace door switch.

Troubleshooting Chart   ELECTRIC DRYERS   (Continued)

Problem	Possible Cause	Corrective Action
Bulbs do not light.	Defective bulb. Loose wiring.	Replace bulb. Check and reconnect.
Timer fails to advance.	Dial binds. Timer motor defective. Door switch open (same models).	Relocate on shaft. Replace timer motor. Close door or replace door switch.

24

Gas Dryers

All that has been said about the electric dryer holds true for the gas dryer. The only difference is in the source of heat and its controls (Fig. 24-1).

It is most probable that the dryer is connected to the gas line through approved flexible tubing. This tubing should be long enough to allow you to pull the dryer away from the wall to remove the rear panel and work on the machine. However, do not over do it and "spring a leak"! If you do, turn the gas off at once and ventilate the room. (Then you'll feel it would have been smarter to have turned off the gas before starting.) If you are lucky, there will be a shut-off valve at the end of the pipe or tubing to which the dryer is connected; otherwise you will have to go back to the main shut-off valve. You should know where this is before you start working on a dryer. In fact, you should know where the gas shutoff valves are, in any case!

If you have to disconnect a dryer for any reason, check the following:

1. Use an approved type sealer on all gas pipe threads, a type not dissolved by gas.

2. Bleed air and sediment from the gas supply line until you can smell the gas before tightening the connection at your dryer.

3. Test all joints and connections for leaks with a soapsuds solution. (Paint a heavy soapsuds solution around the joints. Leaking gas will move the suds.)

4. Keep all open flames away from your work! *No smoking on the job!*

5. When restarting a gas dryer, it will be necessary to hold the pilot-light valve button down longer than normal—to bleed air out of the gas line. Release the button slowly, about 30 seconds after the pilot is lighted.

6. When you are ready to test the dryer and use it, make certain that the gas is turned back on.

7. Recheck the owner's manual for the correct pilot-lighting procedure.

Several different kinds of controls are in use among the various dryers. Each is located differently in the machine. The operating instructions should provide enough information

FIGURE 24-1. Cutaway view of gas dryer. (Courtesy of Whirlpool)

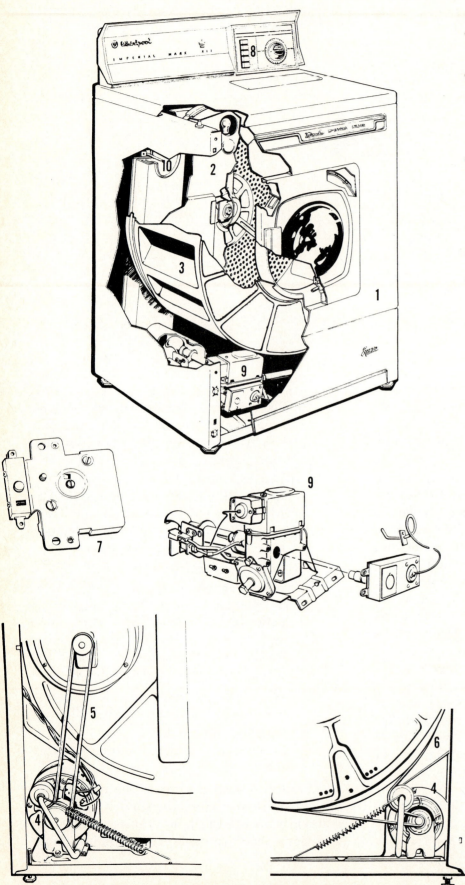

MECHANICAL SYSTEM

- **CABINET (1)**
    U–Type
    Removable front panel
    Hinged Top
    Glass port in door
    Porcelain bearing blocks

- **BULKHEAD (2)**
    Removable - Drum shaft welded to bulkhead

- **DRUM ASSEMBLY (3)**
    Rolled seam
    Stamped baffle
    Front and rear bearing support
    Rear self-aligning bearing
    Delrin bearing ring - front

- **DRIVE SYSTEM**
    Double end Shaft Motor (4)
        Two poly "V" belts
    Direct Blower Drive (5)
    Direct Drum Drive (6)

ELECTRICAL COMPONENTS

- **TIMER ASSEMBLY (7)**
    Automatic Master Touch Control Button
        Closes start switch to machine motor
        Depress timer reset button
        Ten-Minute timer motor operation in a cycle

- **PUSH BUTTON SWITCH (8)**
    Has Five Buttons
    Controls Four Cycles and Console Light

- **GAS BURNER ASSEMBLY (9)**
    New Modulator Type
        Input ranges from 37,000 to 9,000 BTU
        Modulates between 155° and 175° exhaust temperature
        Modulator sensor bulb located in fan scroll

- **INLET SENSOR (10)**
    Controls Timer Circuit in Cool Down
        Minimum heat input allows sensor thermostat contacts to close, completing timer motor circuits
        Above 220° contacts open-below 220° contacts close

- **HEATER BOX COVER (11)**
    New Heater Box Cover Design for Modulator Burner

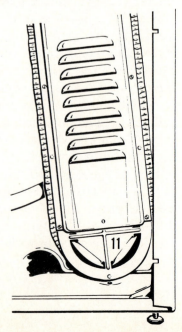

to locate the controls, and guide you, should you need to relight the pilot light on a machine.

If the dryer has recently been moved into your area, check the size of the burner hole to be sure it is appropriate for the local gas supply. This point is important because different kinds of gas require different-size burner orifices, the difference being the size of the holes through which the gas flows. Manufactured gas requires the largest hole, natural gas the "middle-sized" hole, and LP gas (liquefied petroleum) the smallest.

It is not practical to service gas burners in a home, but there are a few adjustments that can be made.

Pilot burners have an adjusting screw that will decrease the flame when turned clockwise and that will increase the flame when turned counterclockwise. You should recognize a flame condition by its appearance and size. A correct-size pilot flame is important. Once set for a particular location and use, there should be no need to change it. Its job is to ignite the main burner.

The main burner flame is controlled by the amount of air it receives. The adjustment is at the air shutter (Fig. 24-2). Moving it to the right to open and to the left to close is what controls the amount of available air. A normal gas flame is bright blue in color. A bright yellow flame indicates lack of air. Too much air will produce a roaring, distorted flame. Once adjusted, this flame, too, should be left alone.

Control valves are solenoid operated, and a timer controls their operation, as in the electric dryer. Many solenoids must have their cover on to complete the magnetic circuit, so do not take them apart to test. Usually, you can hear them click into position and thus assume that they are all right.

When troubleshooting, check the trouble diagnosis chart to locate the area of trouble; then use the wiring diagram, attached to your dryer, to check out the faulty circuit. Don't be too hasty to condemn a control or valve until you are certain that there is a problem. Given a reasonable break in installation and care, a dryer should provide years of troublefree service. Make certain that it is level and secure, and remove the excess lint frequently!

Lint, dust, cobwebs, and such airborne particles get into screens, restrict airflow, and cause a failure in operation. Check this first! The following explanation on burner operation should be carefully studied to understand the need for care and cleanliness.

We mentioned earlier that pilot burners have an adjusting screw to control the flame. When a dryer was first installed, it operated satisfactorily for some period of time. Now it is starting to give trouble. The pilot burner goes out or the flame is small and weak.

This condition is caused by incorrect combustion or a change in the air–gas ratio. The correct ratio is about 12 to 16 percent gas—the rest air. Thus as the air-filtering screen clogs with lint, dust, and so forth, it will restrict the airflow, changing this ratio to a higher percentage gas. Eventually there will be more gas available than can be completely used by the amount of air coming to it and there will be incomplete combustion. Results are a yellow flame, not hot enough to do its job, and one that will deposit unburned carbon (soot) on anything in the area. Signs of carbon or soot around a burner indicate there is incorrect air adjustment.

Therefore, readjusting the pilot flame to improve the air–gas ratio is not a "cure-all." It might be a temporary help (unless the screen is too badly clogged). You must clean the screen and orifice to allow the full flow of air! This step can best be done with carbon tetrachloride. Remove all foreign matter and carbon. If you cannot clean the screen, replace it with a new one. Do not operate your burner with the screen removed!

There is a filter in the pilot gas line, too, and it can also become clogged. The symptom of this would be insufficient gas, which prevents the pilot from igniting. A clogged filter will have to be replaced.

The main burner of your dryer must also be clean or it will generate a sooty flame due to insufficient air. Basically the flame should be a clear, bright blue in color. Small, clean yellow tails on the flame are all right. In fact, they are characteristic of some burners. As observed for the pilot burner, look for carbon or soot deposits as an indication of poor combustion. Then clean up and adjust the supply.

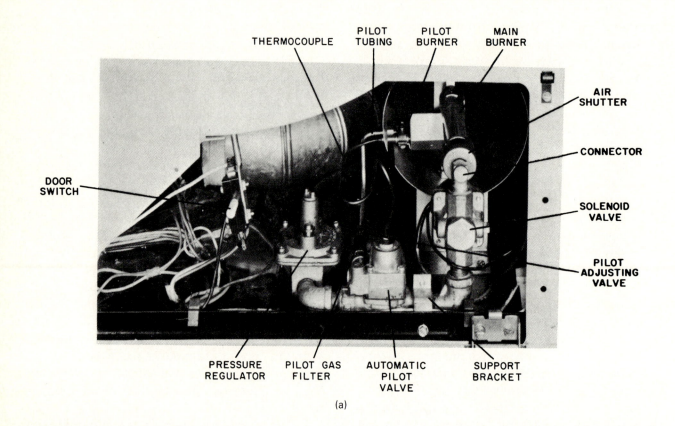

(a)

(b)

FIGURE 24-2. (a) A typical gas burner and its associated controls. (b) Gas burner controls are accessible from the front of the Whirlpool by removing the access panel under the door. (Courtesy of Whirlpool)

Troubleshooting Chart **GAS DRYERS**

Problem	Possible Cause	Corrective Action
Will not run.	No power Broken wire or loose terminal. Door switch. Inoperative motor. Defective timer or timer circuit.	Check fuse and power supply. Check terminals and wiring for continuity. Make sure door closes properly to actuate switch. Check motor thermal protector. Check motor centrifugal switch. Check motor wiring. Check timer wiring. Replace timer motor or timer switch as needed.
Motor runs, but main burner will not ignite.	Heat switch set wrong. Defective timer or timer circuit. Defective centrifugal switch in motor Gas supply valve closed. Inoperative solenoid coil.	Correct switch setting. Check timer wiring. Replace timer motor or switch as needed. Clean switch or replace it as needed. Open gas supply valve. Test coil for open circuit and the voltage to it.
Motor runs, but drum does not turn.	Broken or slipping belt. Loose pulley. Drum jammed or bearing frozen.	Replace belt. Tighten pulley on shaft. Check for foreign object and clear it. Clean or replace bearing.
Clothes not drying, but dryer runs and heats.	Defective operating thermostat. Fan or its pulley loose on shaft. Incorrect heat or timer setting. Clothes too wet when placed in machine. Clogged lint screen or duct (little air motion).	Test thermostat; replace if defective. Tighten setscrew. Reset timer. Wring out or extract water from clothes before placing in dryer. Clean out lint screen and duct.
Will not shut off.	Defective timer.	Replace timer motor or switches as needed.
Motor runs when door is open; timer fails to advance.	Defective door switch. Dial binds. Timer motor defective. Door switch open.	Replace door switch. Relocate dial on shaft. Replace timer motor. Close door more tightly or replace door switch.
Pilot burner goes out.	Insufficient gas supply.	Low gas pressure. Clogged pilot filter. Partially closed gas valve.

Troubleshooting Chart   GAS DRYERS   (Continued)

Problem	Possible Cause	Corrective Action
	Carbon deposit on thermocouple tube.	Clean tube and check pilot burner for proper combustion.
	Improperly adjusted pilot flame.	Adjust pilot flame until faint yellow tip begins to appear.
	Faulty ignition or reset valve or thermocouple.	Replace defective part.
Pilot burner does not light.	No gas to the burner.	Gas shutoff valve closed; open it.
	Defective safety thermostat.	Test thermostat, replace if defective.
	Switch set for "air."	Reset controls.
Main burner does not light, but pilot is okay.	Insufficient gas supply.	Partially closed shut-valve or low gas pressure.
	Defective safety thermostat.	Test thermostat. Replace if defective.
	Incorrect air supply.	Adjust air shutter.
	Main valve solenoid defective.	Replace solenoid.
Main burner cycles.	Loose connection.	Check wiring and terminals; tighten.
	Defective timer.	Replace timer.
	Defective operating thermostat.	Replace thermostat.
	Safety thermostat cycling due to restricted airflow.	Clean out lint and other obstructions.
Main burner goes out.	Insufficient gas supply.	Partially closed gas shutoff valve or low pressure.
	Incorrect primary air adjustment.	Adjust air shutter for clean, bright blue flame.
	Loose connection in wiring.	Tighten all terminals.

25

Food Mixers

There are few things to go wrong with a mixer. Accidental dropping, catching a tool in the beaters, and damaging the cord are the main actions to avoid.

Mixers are usually powered by a common series motor which runs at high speed. It is reduced through a worm gear assembly. Motor speed can be controlled to suit the user's needs.

The West Bend food preparation system (Fig. 25-1) consists of a pedestal power center with an adjustable arm and a group of attachments. These are shown in the parts lists and exploded drawings of Figs. 25-2 to 25-6.

The motor housing can be adjusted through several positions to accommodate the various attachments. The electronic speed control, included in the wiring diagram (Fig. 25-7), allows the four different speeds required for the attachments. It compensates for the variations in speed caused by load increases. These preselected speeds are shown in Table 25-1.

There are several safety features built into this appliance, so make certain that they are checked before trying to service the mixer. The safety cutout switch will show as a red button at the bottom of the housing. It can be reset manually after the motor control has been turned off and the machine given time to cool down. The meat grinder feed screw, the blender, and the slicer-shredder cutter assembly have breakaway protectors in case they become overloaded (see items 43, 46, and 55 in Figs. 25-4 to 25-6). If these parts are broken, they can be replaced, but the motor will not be damaged.

SERVICE

It is always better to be familiar with the appliance being serviced. Know what it is supposed to do and how it does it. This information for a specific model is available from the manufacturer. Use it if you have it. Even though there is much basic similarity in these appliances, there is a difference in their assembly and control systems. Usually, careful study of the exploded drawings and parts list will provide clues to parts location and assembly methods.

FIGURE 25-1. Representative pedestal mixer. (Courtesy of West Bend)

FIGURE 25-2. Tower and motor housing. (Courtesy of West Bend)

Item no.	W.B. part no.	Description
1	P7–174	Release button
2	P34–199	Housing
3	P64–134	Spring
4	P37–137	Locking device
5	P36–199	Motor and gear housing (2 pcs)
6	P36–137	Lever
7	P407–19	Speed control knob
8	P2–148	Steel ball
9	P65–134	Spring
10	P38–80	Resilient pad
11	P152–23	"O" ring
12	P256–47	Felt seal
13	P153–23	Special rubber seal
14	P39–80	Rubber pad (bottom)
15	P87–77	Rubber foot (bottom)
16	P40–86	Safety cut-out switch
17	P257–42	Support bracket
18	P135–74	Power cord
19	L3721	Operating manual (not shown)

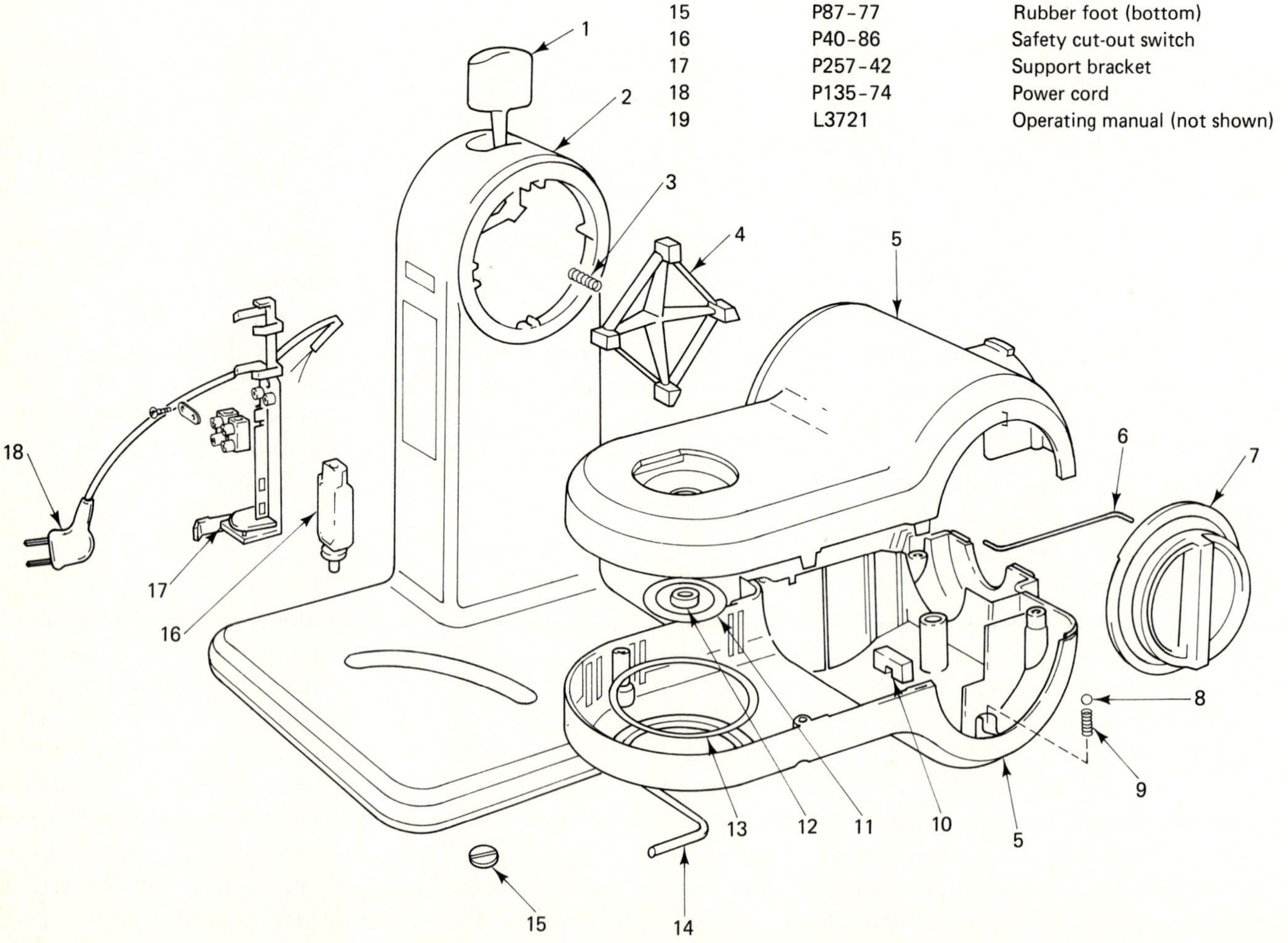

ITEM NO.	W.B. PART NO.	DESCRIPTION
20	P8-147	Ball Bearing (Small)
21	P8-155	Gear Wheel
22	P9-155	Gear Set (3 Gears, Support & Sleeve)
23	P8-59	Carbon Brush
24	P67-134	Spring
25	P35-199	Gear Cover
26	P110-72	Gasket
27	P9-147	Ball Bearing (Large)
28	P10-212	Electronic Speed Control
29	P2-141	Armature Assembly (Includes Armature, Fan Blade/Drive, Bronze Bearing Rivets & Washers)
30	P73-201	Motor and Gear Assembly Complete (Includes Electronic Speed Control)

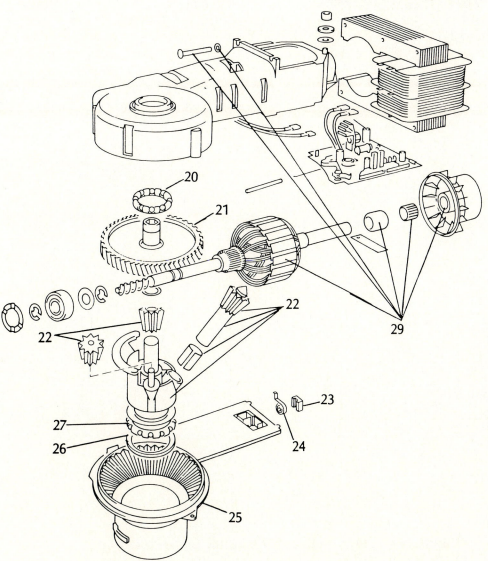

FIGURE 25-3. Motor and gear assembly. (Courtesy of West Bend)

ITEM NO.	W.B. PART NO.	DESCRIPTION
31	P132-1	Bowl
32	P11-189	Dough Deflector
33	P10-31	Dough Hook
34	P3-144	Stirrer
35	P4-144	Whisk
36	P80-21	Measuring Cup
37	P348-2	Cover
38	P111-72	Gasket (For Cover)
39	P7-138	Blender Jar
40	P112-72	Gasket (For Base)
41	P36-38	Cutter Assembly (Gasket, Blade, Casting, Driver)
42	P493-3	Blender Jar Holder With Handle
43	P81-21	Driver (Safety Coupling)

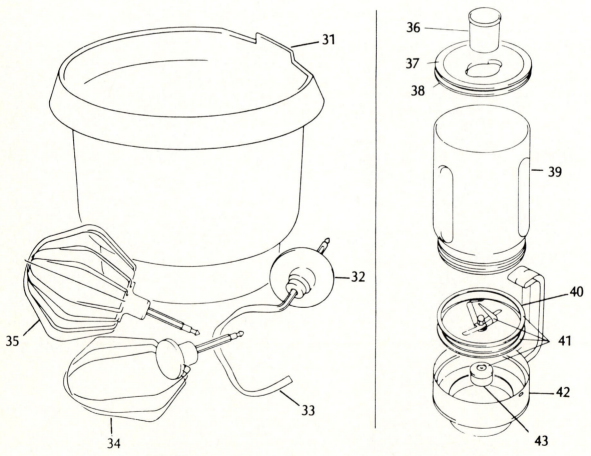

FIGURE 25-4. Mixer and blender. (Courtesy of West Bend)

ITEM NO.	W.B. PART NO.	DESCRIPTION
44	P4-102	Food Pusher
45	P148-12	Cover with Feed Tube
46	P35-38	Driver
47	P492-3	Base with Discharge Tube
48	P28-38	Thick Slice Blade
49	P30-38	Thin Slice Blade
50	P29-38	Coarse Shredding Blade
51	P31-38	Fine Shredding Blade
52	P32-38	Grating Blade

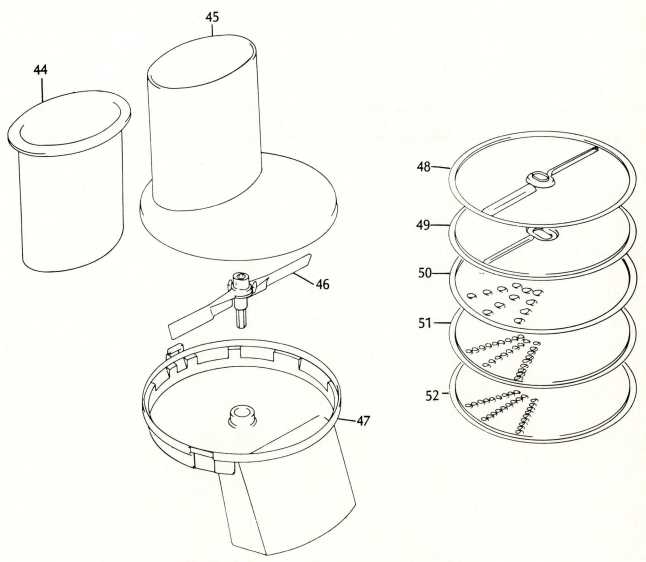

FIGURE 25-5. Slicer/shredder. (Courtesy of West Bend)

ITEM NO.	W.B. PART NO.	DESCRIPTION
53	P3-102	Food Pusher
54	P13-15	Food Tray
55	P2-115	Driver (Safety Coupling)
56	P634-127	Feed Screw
57	P66-134	Spring
58	P34-38	Cutting Blade
59	P156-79	Grinding Disc
60	P87-89	Threaded Cap
61	P8-130	Rest
62	P635-127	Screw

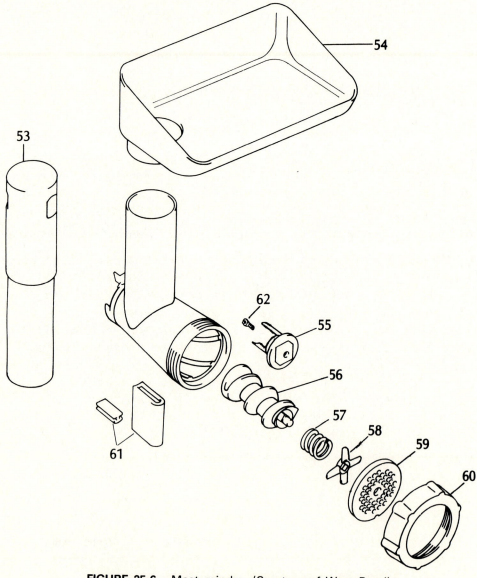

FIGURE 25-6.   Meat grinder. (Courtesy of West Bend)

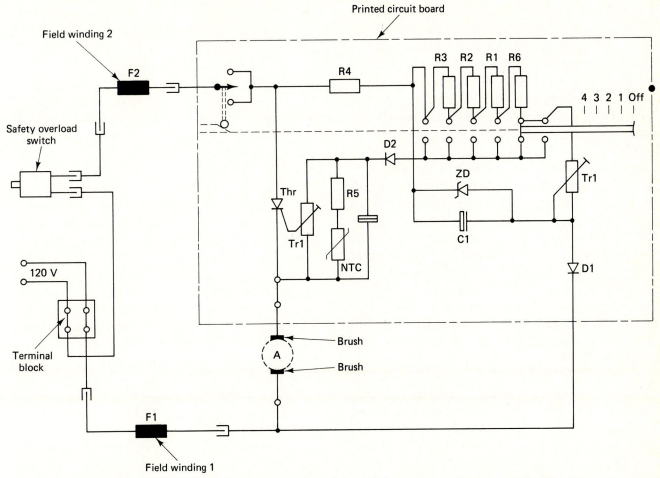

FIGURE. 25-7. Wiring diagram for the West Bend pedestal mixer. (Courtesy of West Bend)

The following is general for the small hand mixers in common use. The cord wiring is made accessible by removing the bottom cover and nameplate. To test the cord, plug it into your tester, remove the connectors, and connect a jumper wire across the exposed wire end. The test bulb should light. Work the cord in your hands, from one end to the other. As you approach the break, the bulb should flicker. If the break is near the end, and the rest of the cord is in good condition, the cord can be shortened to just beyond the break and can then be reconnected. Otherwise replace with a new cord.

A shorted line condenser will blow a fuse and an open condenser will cause very excessive radio interference. In either case, the condenser will have to be replaced.

Motor brushes are checked by removing

TABLE 25-1
Expected no-load wattage and speeds for attachments

	Switch Speed Setting			
	1	*2*	*3*	*4*
No-load wattage: acceptable range	12–40	16–70	45–80	70–110
Output (rpm)				
Blender	2500–3500	5000–6500	10,000–12,500	12,000–16,500
Slicer/shredder	105	220	480	600
Meat grinder	43	67	148	185
Mixer arm	43	67	148	185

Source: West Bend.

the brush screws and lifting the brushes out of their holders. The brush surface that comes in contact with the commutator must be smooth, clean, and curved to the commutator's surface. Any other condition is cause to suspect poor brush contact to the armature and calls for further inspection. A worn brush, or one with a damaged or weak spring, must be replaced. A brush should be considered worn when the carbon has shortened to within $\frac{5}{16}$ inch of the spring end. It is good policy to replace both brushes at the same time. (The same holds true for governor brushes.)

A defective armature or field coil should be the last thing to suspect in your electrical testing. If you should find that the armature or field assembly needs replacing and that there is something else wrong, too, you had better check the cost of a new mixer against the cost of replacement parts—and advise your customer.

There is a basic difference in the method of speed control between the large pedestal-type mixer and the portable hand mixer. Some hand mixers have three speeds: high, medium, and low. The pedestal mixers have a governor control to provide an infinite speed range between high and low.

The three speeds normal to a hand mixer are obtained by switching the amount of motor field coil used (Fig. 25-8). Part of the field winding has a higher resistance than the rest, thus reducing the available current to the armature.

From this you can see that if the mixer works in high but not in medium and low, the trouble must be either in the switch or section A, coil 2. If it works in medium but not in low, you must suspect the switch, or section B of coil 2. Of course, if the mixer does not run at all, the rest of the electrical system is open to suspicion, but don't jump to conclusions (Fig. 25-9).

Unless the trouble is obvious on inspection, use a trouble diagnosis chart to locate the possible faults. Then use one of the following procedures to replace the faulty part.

A disassembled view of the Hamilton Beach Model 75 (Fig. 25-10) mixette showing the relative location of the parts illustrates hand-mixer construction. Screws 34 and 35 are

the cover and handle-retaining screws. They serve to hold the entire assembly together and must be the first ones removed to expose the motor assembly and wiring.

When ordering beater gears, or worm wheels for any mixer, be sure to specify whether it is the left-hand (LH) or right-hand (RH) unit that you need. A word of caution: it is usually safer to replace both at the same time. This holds for motor brushes, too.

The Waring hand mixer has a removable bottom section that exposes the motor. The motor assembly has to be removed by taking out three mounting screws. This step will give access to all the wiring.

A break in the cord can be located by plugging the cord in the tester outlet, turning the switch on to high, and moving the cord while watching the test light. The most probable place for a cord to break is at the point where it enters the mixer. This point is near enough to the end to be cut off, shortened, and reconnected.

The motor can be tested for continuity by following the circuit. With the cord plugged into the tester outlet, there will be no light with the switch in OFF position. There will be light at each of the other positions, increasing in brightness as the switch is moved to high. If your tester has a 50-watt bulb or larger, the mixer will run.

The handle is held in place by two screws. They are accessible after the motor has been removed.

The switch, located in the handle, must have its leads disconnected before it can be removed. The two Phillips head screws holding the switch escutcheon in place must be removed. Then the escutcheon can be lifted up and the switch will pull out of the handle.

The service approach to some mixers is different because the shell pulls apart "fore-and-aft" rather than top and bottom.

The motor brushes are accessible from the outside of the case by removing the black plastic brush caps, one on each side of the case near the rear. If the brushes are badly worn, they must be replaced. Wear down to within $\frac{1}{4}$ inch of the spring is far enough.

Oily, pitted, or burned brushes are reason to suspect the commutator's condition. If the condition is suspect, the motor should be checked and repaired before new brushes are installed.

Disassembly proceeds as follows: Pry the escutcheon free, then take off the beater ejector, which is held in place by two screws. Remove the rear-handle screw and, with a knife, pry out the button at the top front of the handle. This exposes the front-handle screw, which must be removed. With the handle off, the speed control knob screw can be unfastened to free the knob.

The hex nut around the switch shaft must be loosened a few turns to loosen the switch from the case. The front case section is held in place with three Phillips head screws, which are removed.

The case halves pull apart. The motor and switch remain in the rear half and the beater goes in the front half. The beater gearbox cover must be removed to gain access to the gears.

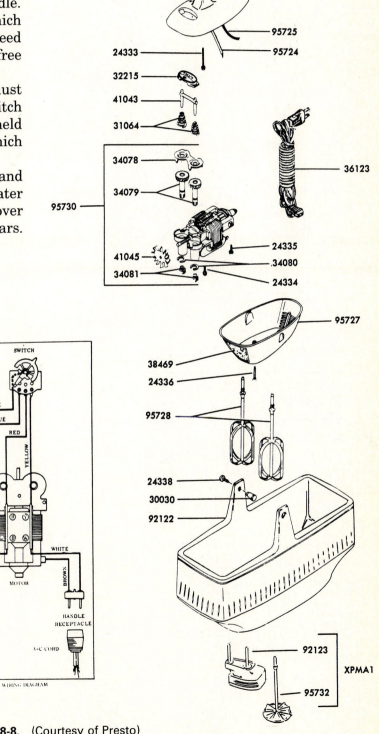

Part No.	Description
24333	Screw, handle cover
24334	Screw, motor mounting (2 req.)
24335	Screw, motor mounting
24336	Screw, base mounting
24338	Screw, pivot button (2 req.)
30030	Pivot button (bowl)
31064	Spring, ejector
32189	Cover, handle
32215	Ejector
34078	Gear cover
34079	Gear and spindle assembly
34080	Spindle "C" ring
34081	Spindle spring
36123	Cord set
38469	Grille
41043	Ejector arm
41045	Fan
92122	Bowl
92123	Knife sharpener
95724	Wire lead assembly (white)
95725	Wire lead assembly (brown)
95727	Base assembly
95728	Beater
95729	Switch assembly
95730	Motor and fan assembly
95732	Drink mixer
95803	Handle sub-assembly, includes leads
XPMA1	Mixer accessory kit

FIGURE 28-8. (Courtesy of Presto)

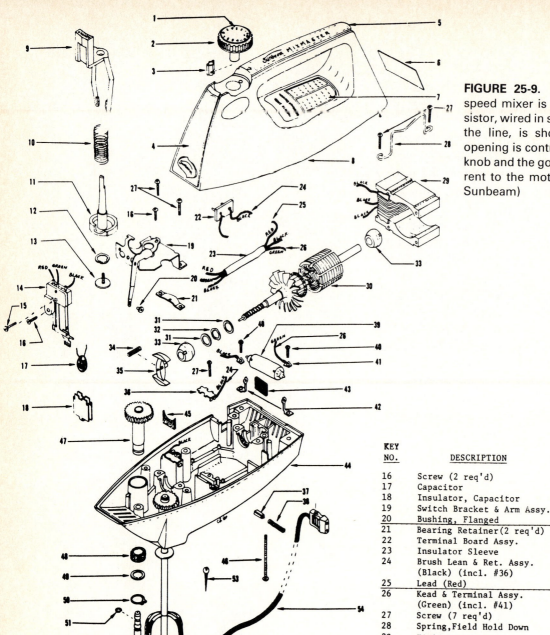

FIGURE 25-9. This 12-position, infinite-speed mixer is governor controlled. A resistor, wired in series with the field coil and the line, is shorted by a switch whose opening is controlled by the speed control knob and the governor. In this way the current to the motor is limited. (Courtesy of Sunbeam)

KEY NO.	DESCRIPTION	STOCK NO.	MODELS USED ON
16	Screw (2 req'd)	89452	HM-Q
17	Capacitor	89404	HM-Q
18	Insulator, Capacitor	89422	HM-Q
19	Switch Bracket & Arm Assy.	90142	HM-Q
20	Bushing, Flanged	89427	HM-Q
21	Bearing Retainer(2 req'd)	19794	HM-Q
22	Terminal Board Assy.	83320	HM-Q
23	Insulator Sleeve	18139	HM-Q
24	Brush Lean & Ret. Assy. (Black) (incl. #36)	89428-1	HM-Q
25	Lead (Red)	89450-4	HM-Q
26	Kead & Terminal Assy. (Green) (incl. #41)	89449-2	HM-Q
27	Screw (7 req'd)	89451	HM-Q
28	Spring,Field Hold Down	89411	HM-Q
29	Field & Coil Assy.	89445	HM-Q
30	Armature & Fan Assy.	89426	HM-Q
31	Flat Washer, Armature (2 req'd)	2340	HM-Q
32	Spring Washer	2339	HM-Q
33	Bearing (2 req'd)	2556	HM-Q
34	Spring, Governor	89403	HM-Q
35	Governor	89438	HM-Q
36	Brush Retainer(2 req'd)	89414	HM-Q
37	Brush (2 req'd)	81325-3	HM-Q
38	Brush Spring(2 req'd)	6545	HM-Q
39	Resistor	1094-1	HM-Q
40	Screw-Resistor Bracket (2 req'd)	87511	HM-Q
41	Terminal (2 req'd)	1207	HM-Q
42	Resistor Bracket(2 req'd)	89432	HM-Q
43	Thrust Pad	84472	HM-Q
44	Base, Motor Unit	89446	HM-Q
45	Bearing Wick(2 req'd)	84475	HM-Q
46	Screw, Base(4 req'd)	81354-1	HM-Q
47	Worm Gear (2 req'd)	83304	HM-Q
48	Spindle Seal(2 req'd)	83325	HM-Q
49	Flat Washer (2 req'd)	83323	HM-Q
50	Retaining Ring (2 req'd)	83326	HM-Q
51	Spring,Beater Retaining (2 req'd)	83329	HM-Q
52	Beater (2 req'd)	83322-1	HM-Q
53	Screw, Wall Mounting	18169	HM-Q
54	Cord Assy.	19828-6	HM-Q

KEY NO.	DESCRIPTION	STOCK NO.	MODELS USED ON
1	Knob Insert	89431	HM-Q
2	Knob Assy.(incl.#3)	89430	HM-Q
3	Clip, Knob	89400	HM-Q
4	Nameplate	89412	HM-Q
5	Escutcheon Handle	89413	HM-Q
6	Tape, Heat Insulating	not req'd	HM-Q
7	Speed Guide	89423	HM-Q
8	Housing (White)	89447-1	HM-Q
9	Ejector	89443	HM-Q
10	Ejector Spring	89402	HM-Q
11	Cam Shaft	89441	HM-Q
12	Spring Washer	89405	HM-Q
13	Screw & Washer	23308	HM-Q
14	Switch Assy.	89440	HM-Q
15	Screw	89455	HM-Q

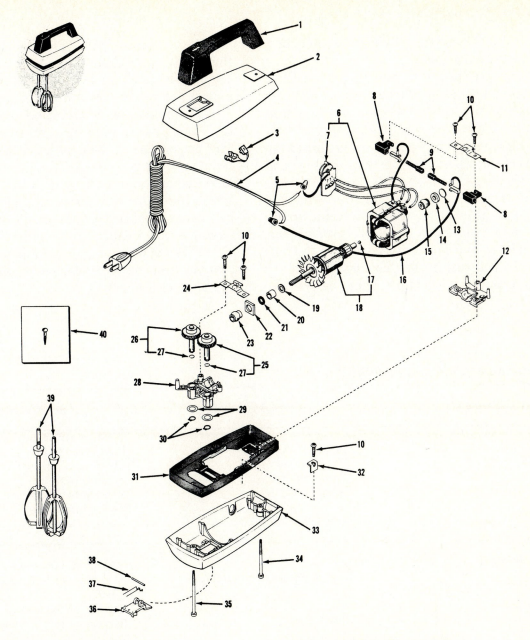

Handle	1	75FM-81
Upper Motor Cover	2	75FM-124
Cord Strain Relief	3	11FM-83
Cord	4	66FMC-82
Wire Nut (2 used) each	5	HV-125
Field Complete 115V	6	6048-075-0012
Switch & Leads Complete	7	75FMC-86
Brush Holder Housing (2 used) each	8	65FM-40B
Carbon Brush & Spring (2 used) each	9	65FMC-60
Screw (6 used) each	10	75FM-41
Rear Bearing Retaining Strap	11	65FM-27
Rear Bearing Housing	12	75FM-34
Thrust Disc	13	JM-31
Rear Bearing Felt	14	KM-88
Rear Bearing	15	65FM-26
Lead & Brush Holder Liner Complete	16	75FMC-57
Thrust Ball	17	35V-28
Armature Complete 115V	18	75FMC-1R
Armature Spacing Washer	19	65FM-73
Spacer	20	75FM-17
Armature Spacing Washer (2 used) each	21	DM-20

Front Bearing Felt	22	75FM-35
Front Bearing	23	OM-36B
Front Bearing Retaining Strap	24	75FM-37
Worm Wheel Shaft Complete--LH	25	75FMC-93
Worm Wheel Shaft Complete--RH	26	75FMC-92
Beater Retaining Ring (2 used) each	27	55FM-195
Front Bearing Housing	28	75FM-25
Worm Wheel Thrust Washer (2 used) each	29	75FM-69
Worm Wheel Retaining Ring (2 used) each	30	75FM-75
Motor Frame	31	75FM-24
Field Retaining Strap (2 used) each	32	75FM-66
Lower Motor Cover	33	75FM-145
Cover & Handle Retaining Screw	34	75FM-84
Cover & Handle Retaining Screw	35	75FM-85
Beater Ejector	36	75FM-197
Beater Ejector Retaining Spring	37	75FM-198A
Beater Ejector Pin	38	75FM-196
Beater Complete (2 used) each	39	55FMC-190
Wall Mounting Screw	40	75FM-96

FIGURE 25-10. (Courtesy of Hamilton Beach)

Be sure to pack some fresh light grease or solidified oil around the gears.

When you put the two halves together again, you must "time" the beaters before their gears mesh with the worm gear on the armature shaft. The easiest way to do so is to install the beaters in their sockets, set one beater 45 degrees from the other, and hold them in this position until the gears mesh with the shaft worm.

The switch and motor can be tested for continuity with your tester to determine which part needs replacement. A defective switch can be unsoldered and replaced. A defect in the motor or cord will require further disassembly.

First remove the motor brushes before taking the armature out. This step will prevent their getting broken. Then the front-bearing plate, held in place by two nuts, must be removed before the armature can be taken out.

The field assembly must be removed before the wiring and power cord are free to change. The power cord is held in place with a strain relief that can be removed by pinching the movable section and pulling outward. This may sound like a lot of work to change a cord, but it is just a short project and much cheaper than buying a new mixer.

When you find major trouble in the mixer, compare the cost of replacement parts against the price of a new mixer.

Troubleshooting Chart FOOD MIXERS

Problem	Possible Cause	Corrective Action
Motor does not run.	No power at outlet. Defective cord. Worn brushes. Open field coil. Open armature winding. Defective switch.	Check fuse in that circuit. Check cord for breaks. Replace brushes. Replace field coil. Replace armature. Replace switch.
Motor does not run and blows fuses.	Frozen bearing. Bent shaft, jamming armature. Defective armature or field coil. Shorted cord.	Free shaft in bearing and lubricate. Straighten or replace shaft, clear jam. Replace or repair cord. Replace or repair cord.
Motor runs; beaters do not run.	Stripped beater gears.	Replace beater gears.
Erratic operation and speed.	Defective cord. Worn motor brushes. Loose connection. Defective switch contact.	Check cord for intermittent break and replace or repair. Replace motor brushes. Repair connection. Replace switch.
Slow speed or weak power.	Incorrect speed setting for the work being done. Too heavy mix, beyond the power range of the mixer. Worn brushes. Bind in shaft.	Change speed control to correct setting. Not a fault of the mixer being used. Replace brushes. Clear bind.

Troubleshooting Chart FOOD MIXERS (Continued)

Problem	Possible Cause	Corrective Action
Motor runs hot.	Bind in shaft. Shorted winding in armature. Shorted field coil.	Clear bind. Replace armature. Replace field coil.
Noisy.	Armature hitting field. Bent cooling fan blade. Dry gears or bearing. Bent beaters hitting each other.	Replace worn bearing. Straighten fan blade to clear. Lubricate as required. Replace or straighten beaters.
Motor runs only on high speed.	Defective speed control. Linkage loose.	Replace. Reconnect.
Motor will not shut off.	Linkage loose.	Reconnect.
Bad sparking at motor brushes.	Stuck or worn brushes. Weak springs. Rough commutator.	Clear brush holder. Replace brushes. Turn commutator. Replace armature.

26

Humidifiers

The air-moisture content in a house drops rapidly as the outside temperature drops unless some way is found to put more moisture into the air. Relative humidity as low as 20 percent has been found in houses without humidification or with a humidifier that is not operating correctly.

Low humidity has several effects on a house and its contents. For instance,

Houseplants will die more readily.

Rug fibers become brittle and wear faster.

Boards shrink, causing windows and doors to rattle.

Cracks appear in floorboard seams.

A piano will go flat.

Leather shoes dry out.

Food, if uncovered, dries out.

Wooden furniture becomes unglued.

Your nose feels dry and you get colds more often.

You create static by walking across a carpet.

You feel the need to turn the thermostat as much as 5 percent higher than normal and still feel cold.

These are all signs that indicate the need of a humidifier. Proper humidity will save money on the heating bill and will provide a healthier environment.

The primary purpose of a humidifier is to put moisture into the air. This might easily mean a gallon of water per day per room. On a very cold day it could mean even more, possibly as much as a gallon per hour for the average house.

Many humidifiers cannot handle that big a job, so do not condemn the machine if it does not produce during extremes of temperature. An owner should be grateful that it is putting *some* moisture into the air. Almost the only certain way of knowing how well a humidifier is doing is by using a sling psychrometer or a wet and dry bulb thermometer with a chart to convert the readings to percent of relative humidity.

Lacking the technical aids just mentioned, you can still tell if the air is too dry if you get a static shock when you walk across a carpet and touch metal. On the other hand, if you fix an iced drink and the glass sweats, the air is humid enough.

Moisture can be put into the air in a number of different ways. These methods range from setting a pan of water in a room and allowing it to evaporate to spraying a fine mist of water into a fan's air stream.

Hard water—in fact, water containing any minerals—is an enemy of the humidifier. If you know that the water is hard (even though a water softener is in use), it is good practice to clean out the water reservoirs of all appliances. Do the same for the associated water-dispensing parts. Developing the habit of doing so will save much future servicing.

If someone fails to fill the reservoir, the unit will cycle on the limit switch, but there will be no humidification of the air.

Remind the user to check the water level regularly! A humidifier can easily use 5-gallons of water in a 24-hour day.

In operation, a water pump raises water up to the media wheel, filter belt, or evaporator pad to be evaporated. A fan then disperses the moist air into the room.

With a bad water situation, it is possible for the water impeller or the tube to become clogged. The result will require dismantling and cleaning. Figures 26-1 to 26-3 show the same machine in three stages of disassembly.

The media in another type uses a moving belt having one end immersed in a tank of water and the other end exposed to air being circulated by a fan (see Figs. 26-4 to 26-7).

Figure 26-8 shows an exploded view and parts listing of a belt medium. Figure 26-9

FIGURE 26-1. Controls indicate that this humidifier has a water-level indicator, a humidistat, and a two-speed fan.

FIGURE 26-2. The media pad has been lowered to show the humidistat, fan, hose from the pump, and the water distributor. The water distributor drips water, supplied by the pump, onto the media pad. As air is drawn through the pad, it picks up moisture, which is distributed into the room by the fan.

FIGURE 26-3. The pump is set on the tank bottom. It is coupled to the fan motor by a piece of flexible tube. The square white object, the float, indicates the water level.

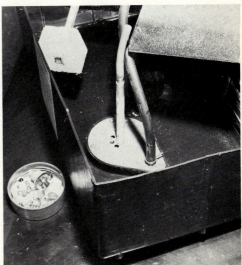

FIGURE 26-4. Front view showing controls for a portable humidifier.

FIGURE 26-5. Rear view showing media belt, fan, and controls protective cover. The motor that drives the media belt is mounted near the top left side and connects to an outlet on the side of the cabinet. This entire assembly tilts back and lifts out of the cabinet.

FIGURE 26-6. When lime is allowed to accumulate, it may join the roller and stall the motor. Remove as much as possible by scraping and remove the rest with an acid wash. Observe all the precautions printed on the acid container.

FIGURE 26-7. In hard-water areas the evaporator pad will accumulate a lime deposit that must be cleaned off. It should be removed by turning the nylon bearing about one-quarter turn, then pulling it out. This will release the roller, and the pad can be removed to be washed in either vinegar or mild acid solution that is available for removing scale from humidifiers.

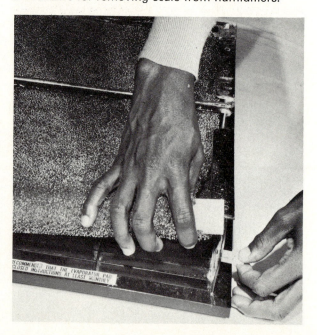

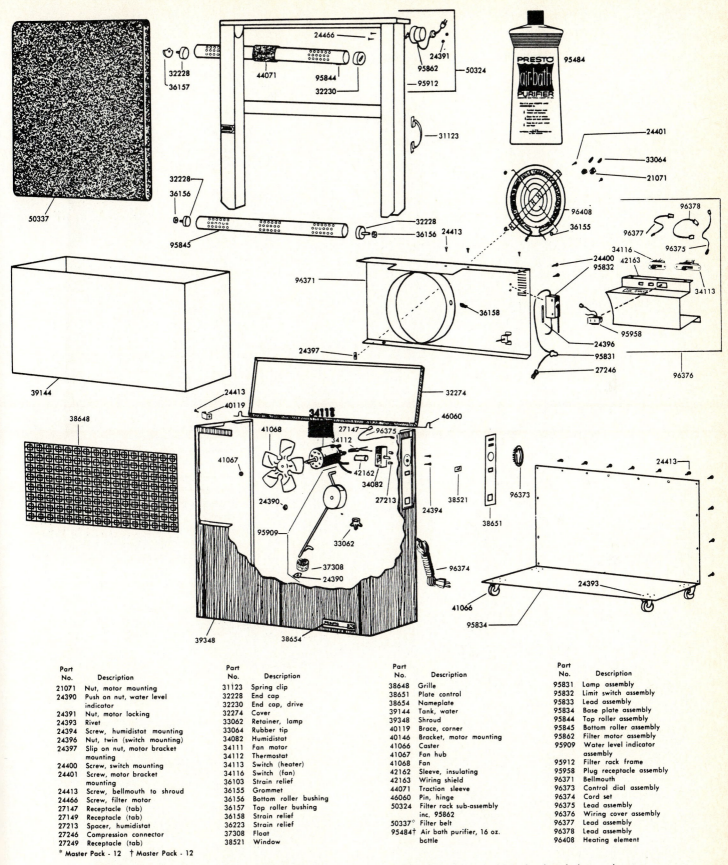

Part No.	Description	Part No.	Description	Part No.	Description	Part No.	Description
21071	Nut, motor mounting	31123	Spring clip	38648	Grille	95831	Lamp assembly
24390	Push on nut, water level indicator	32228	End cap	38651	Plate control	95832	Limit switch assembly
		32230	End cap, drive	38654	Nameplate	95833	Lead assembly
24391	Nut, motor locking	32274	Cover	39144	Tank, water	95834	Base plate assembly
24393	Rivet	33062	Retainer, lamp	39348	Shroud	95844	Top roller assembly
24394	Screw, humidistat mounting	33064	Rubber tip	40119	Brace, corner	95845	Bottom roller assembly
24396	Nut, twin (switch mounting)	34082	Humidistat	40146	Bracket, motor mounting	95862	Filter motor assembly
24397	Slip on nut, motor bracket mounting	34111	Fan motor	41066	Caster	95909	Water level indicator assembly
		34112	Thermostat	41067	Fan hub		
24400	Screw, switch mounting	34113	Switch (heater)	41068	Fan	95912	Filter rack frame
24401	Screw, motor bracket mounting	34116	Switch (fan)	42162	Sleeve, insulating	95958	Plug receptacle assembly
		36103	Strain relief	42163	Wiring shield	96371	Bellmouth
24413	Screw, bellmouth to shroud	36155	Grommet	44071	Traction sleeve	96373	Control dial assembly
24466	Screw, filter motor	36156	Bottom roller bushing	46060	Pin, hinge	96374	Cord set
27147	Receptacle (tab)	36157	Top roller bushing	50324	Filter rack sub-assembly inc. 95862	96375	Lead assembly
27149	Receptacle (tab)	36158	Strain relief			96376	Wiring cover assembly
27213	Spacer, humidistat	36223	Strain relief	50337*	Filter belt	96377	Lead assembly
27246	Compression connector	37308	Float	95484†	Air bath purifier, 16 oz. bottle	96378	Lead assembly
27249	Receptacle (tab)	38521	Window			96408	Heating element

* Master Pack - 12   † Master Pack - 12

FIGURE 26-8.   This unit differs from the one shown in Figures 26-4 to 26-7, in that it has a heater element in the fan airstream. (Courtesy of Presto)

WIRING DIAGRAM — MODEL PM06A

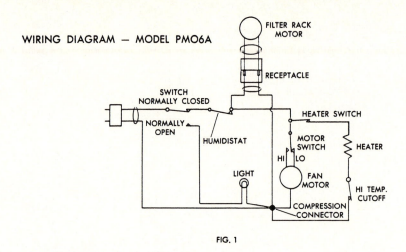

FILTER RACK MOTOR

RECEPTACLE

SWITCH NORMALLY CLOSED

NORMALLY OPEN

HUMIDISTAT

HEATER SWITCH

MOTOR SWITCH

HI LO

HEATER

LIGHT

FAN MOTOR

HI TEMP. CUTOFF

COMPRESSION CONNECTOR

FIG. 1

VIEW OF WIRING ARRANGEMENT — MODEL PM06A

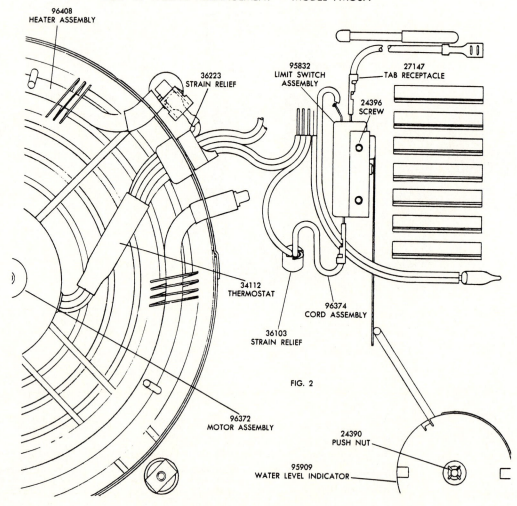

96408 HEATER ASSEMBLY

36223 STRAIN RELIEF

95832 LIMIT SWITCH ASSEMBLY

27147 TAB RECEPTACLE

24396 SCREW

34112 THERMOSTAT

96374 CORD ASSEMBLY

36103 STRAIN RELIEF

FIG. 2

96372 MOTOR ASSEMBLY

24390 PUSH NUT

95909 WATER LEVEL INDICATOR

FIGURE 26-8. (Continued)

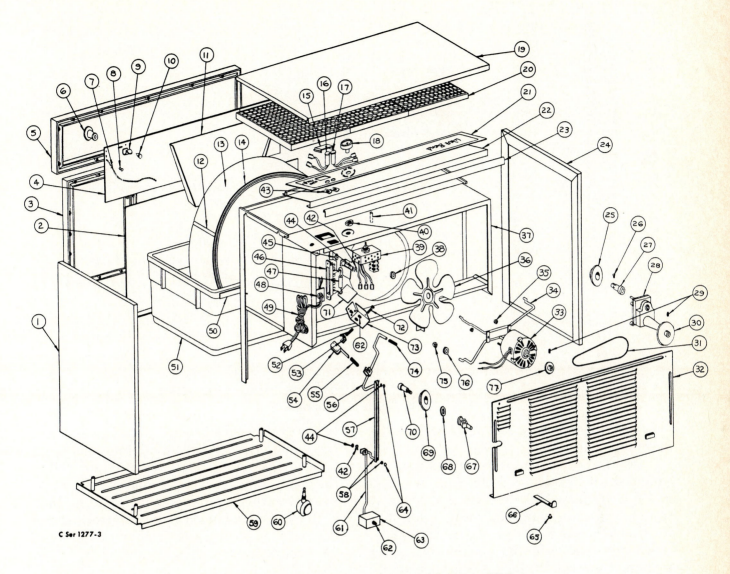

C Ser 1277-3

Parts listed below fit the West Bend Model No. 4012 Humidifier.

(Parts shown on drawing)

ITEM NO.	PART NO.	DESCRIPTION			
1	P13-36	End	15	P27-166	Lens
2	P199-10	Plate	16	P42-73	Lamp (amber - on)
3	P12-113	Frame	17	P43-73	Lamp (red - refill)
4	P93-2	Cover assembly	18	P289-19	Knob
5	P11-113	Frame	19	P88-2	Cover
6	P290-19	Knob	20	P5-208	Grille
7	P43-96	Retainer (chain)	21	P201-10	Plate
8	P151-87	Rivet	22	P16-8	Top
9	P45-96	Retainer	23	P22-48	Brace
10	P19-80	Pad	24	P12-36	End
11	P101-7	Spout	25	P68-79	Drive disc
12	P32-96	Retainer	26	P372-127	Screw (for P38-156 insert assembly)
13	P10-69	Filter belt	27	P38-156	Insert
14	P7-113D	Frame (water wheel)	28	P1-235	Gear box
			29	P431-127	Screw (for P73-79 & P74-79 discs)
			30	P73-79	Disc (for P1-235 gear box)

FIGURE 26-9. In this humidifier the filter belt is on a revolving drum that dips into the water and carries it into the airstream. All humidifiers that draw air through a wet medium also filter the air. (Courtesy of West Bend)

ITEM NO.	PART NO.	DESCRIPTION
31	P3-236	Belt (for drive motor)
32	P86-2	Cover
33	P24-201	Fan motor
34	P10-113	Frame
35	P181-81	Nut (for P24-201 motor assembly)
36	P12-38	Blade (fan)
37	P5-197G	Chassis
38	P61-23	Ring (for P12-38 fan assembly)
39	P2-212	Control - Without exchange
*	P2-212	Control - With exchange
40	P207-81	Nut (for P2-212 control assembly)
41	P16-80	Pad
42	P38-96	Retainer
43	P174-81	Nut (for lamp assembly)
44	P181-47	Washer
45	P6-209	Grommet
46	P33-67	Terminal
47	P1-237	Capacitor
48	P37-5	Clamp
49	P51-74	Cord
50	P13-200	Support
51	P3-211D	Reservoir
52	P43-134	Spring
53	P8-33	Lug
54	P7-86	Switch
55	P20-80	Pad
56	P13-54	Arm
57	P21-137	Lever
58	P5-209	Grommet
59	P170-9D	Bottom

ITEM NO.	PART NO.	DESCRIPTION
60	P3-205	Caster
61	P14-54	Arm
62	P14-80	Pad (for P2-206 float & P121-42 bracket)
63	P2-206	Float
64	P171-81	Nut (for P21-137 lever assembly)
65	P149-87	Rivet
66	P19-137	Lever
67	P187-81	Nut
68	P185-47	Washer
69	P69-79	Disc
70	P375-127	Screw
71	P171-10	Plate
72	P175-127	Screw
73	P121-42	Bracket
74	P17-80	Pad
75	P188-81	Nut
76	P4-209	Grommet
77	P74-79	Disc (for P24-201 fan motor)

(Parts not shown on drawing)

P252-127	Screw (used on P290-19 knob)
P308-127	Screw (used on P45-96 retainer)
P339-127	Screw (used for general assembly)
P82-90	Lead (used on P7-86 switch)
P185-81	Nut (used on P7-86 switch)
P366-127	Screw (used on P7-86 switch)
P373-127	Screw (used on P121-42 bracket)
P433-127	Screw (used on P33-67 terminal)

* Defective control can be returned for $ 5.00 reimbursement or credit.
Return to the attention of the <u>Customer Service Department</u>.

FIGURE 26-9. (Continued)

shows an exploded view and parts listing for a wheel-type unit.

Do not be too hasty in condemning the humidistat if the unit runs too long or cycles on too frequently. Make certain that the unit has enough water, and then assume that it is doing its best to put the correct amount of moisture into the air. About the only time to suspect the humidistat is if the humidity is persistently low with the control set to maximum and then the unit is ON for only short intervals. Even then, make certain that water is pumped to the media and is being evaporated into the room. In other words, be sure that the limit switch has not assumed cycling control and that the water supply has been amply maintained for a considerable period of time.

Troubleshooting Chart **HUMIDIFIERS**

Problem	Possible Cause	Corrective Action
Fan does not run.	See troubleshooting chart for fans (Chapter 17)	
Heater element remains cool.	See troubleshooting chart for water heaters (Chapter 36)	
Does not add moisture to air, but motor runs.	No water in tank.	Fill tank with water.
	Cold heater element.	Replace burned-out element.
	Water tube from pump is clogged.	Clean out tube.
	Impeller in pump is clogged.	Clean out pump impeller.
	Screen is clogged.	Vacuum screen and brush off any foreign matter.
	Media wheel not moving.	Clear away obstruction.

27

Incinerators

Many different makes and models of incinerators have appeared on the market and many have gone out of business because their quality of construction was poor. Their early "burnout" did little to instill customer confidence in the product.

It is interesting to see that the major survivors are built like stoves, complete with fire brick liners. Thus once a company has a successful design, it is loath to change it.

This accounts for the sameness of construction noted over a comparison of service literature 10 years old with that of current production. Essentially only the parts pricing has changed.

GAS INCINERATORS

Incinerators should consume all combustible trash and garbage completely without any smoke or odor. Any performance that does less is reason to check the unit.

Naturally you will first look inside the in-cinerator for noncombustible materials, such as tin cans, glass, bottles, and aluminum foil. These items should never have reached the chamber, but if they do they must be removed. An accumulation of this material will affect combustion and may well jam the grate.

The next point to check is the ash load. When shaking down the ashes, do it gently, with the loading door slightly open to relieve pressure within the combustion chamber (Fig. 27-1). Failure to observe these precautions may smother the flame. The ash drawer must be emptied before it gets full (Fig. 27-2) or else it will cut down the secondary air to the burner, causing a weak flame and/or poor drying action. *There must be free air motion for correct operation!*

Incidentally, these ashes make an excellent fertilizer for the lawn or garden. Spread them thin.

So much for the general housecleaning and routine maintenance of incinerators. Now for how they work. Basically, two different types are in common use. One has a low-gas input

FIGURE 27-1. To prevent pilot outage when shaking down ashes, do it gently with the loading door propped open.

FIGURE 27-2. Empty the ash tray when it is two-thirds full.

burner that is turned on whenever the incinerator is used.

The low-gas input unit provides a small (about 2000 Btu), continuous burner flame (protected from the refuse by a shield) that slowly bakes out the waste until it is dry enough to ignite and burn to ash. Much more heat comes from the materials being burned than from the burner itself. Thus, progressively (from bottom

upward), the balance of the contained refuse becomes heated, dried, ignited, and reduced to ash. Some trash may not reduce completely the first day, perhaps not even the second, but you may continue to add "packages"; and as it accumulates within the incinerator, it will dry out and eventually burn. Calcinator and Waste King have models that are representative of this type of incinerator.

The high-gas input unit has a small pilot light (about 1000 Btu) and a main burner (30,000 to 37,000 Btu) controlled by a timer. Both burners are protected from contact with the refuse by a shield. In use, this type of incinerator is loaded and the timer is set by the user for a predetermined time, depending on the load. If the load contains a large amount of wet garbage or hard-to-burn items like melon rinds, a longer burning time will be required. A short period will be sufficient for loads composed of paper only. The owner's manual will help you with specific methods of operation of a particular burner. Setting the timer automatically starts the main burner, which dries and then ignites the refuse.

Most incinerators now include an "afterburner" to consume (or eliminate) any smoke or odor that may be generated during the primary burning process.

The Locke Stove Company and the Calcinator manufacture models that include this feature (Figs. 27-3 to 27-5).

The "smokeless-odorless" aspect is required by some local building ordinances and has much merit in protecting the neighborhood from the smells of incompletely burned refuse.

The "after-burner" principle is based on a second flame so located that all smoke and odor exhaust must pass through it and thus be consumed. This process is demonstrated by a cigarette and match (Fig. 27-6).

A word to the wise: On using any incinerator, remind the owner to drain wet refuse; then wrap it in lots of dry paper to help the burning. Keep the bundle small—about the size of a loaf of bread. Accumulate refuse in the incinerator, not in the kitchen! Load the machine gently! Throwing or dropping a bag of waste into the chamber might blow out the pilot flame (Fig. 27-7).

AIR DILUTER
COOLS OUTER
JACKET

FIRE BRICK
LINED
AFTERBURNER
CHAMBER

JET TYPE
BURNER

AUTOMATIC
TIMER

FOOT OPERATED
DOOR OPENER

FIGURE 27-3. Calcinator gas incinerator. (Courtesy of Calcinator)

With the high-gas-input-type unit, make certain that enough time is allowed to burn the load completely. An accumulation of "half-burned" refuse, plus added fresh loads, will decrease the burning efficiency. More gas will be used without receiving the normal benefits of complete combustion. This is the condition that most frequently creates an odor problem.

The most common trouble with this appliance is that the pilot light will blow out. It should be the first thing checked after noticing a failure to operate. Instructions on pilot lighting and relighting are usually found on a metal plate mounted near the control.

When the timer does not advance (on models that have one), the trouble is probably within the timer itself, and it should be replaced. These burners can be shut off manually

(Fig. 27-8) by turning the timer knob counterclockwise to the OFF position.

Proper installation is important! You may assume that the installation was initially correct, but it does no harm to make a few quick checks.

1. The chimney or flue must be clear, in good condition, and free from downdrafts. A common test for chimney drafts is to light a sheet of crumpled newspaper and place it in the chimney's vent. Its smoke and small ash should go up the flue on being lighted.

2. A 6-inch-diameter pipe should connect from the incinerator to the chimney. It must be free of cracks or holes. Rusted-out sections must be replaced.

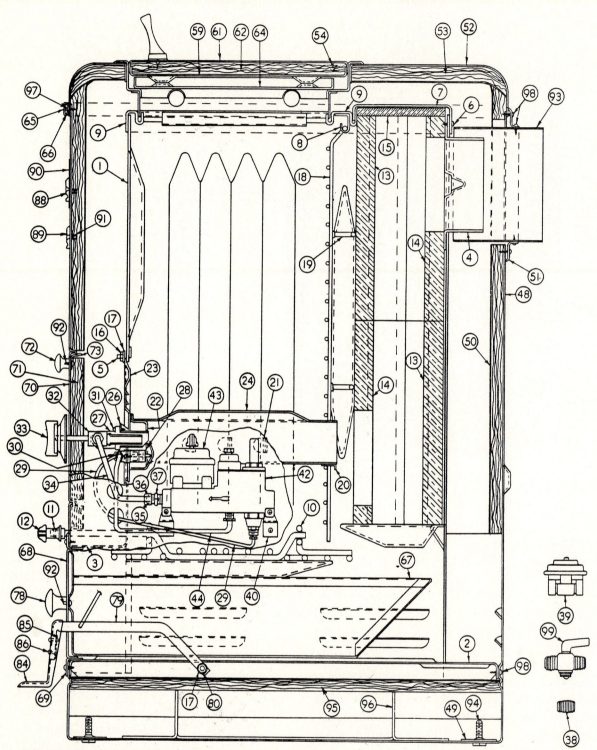

FIGURE 27-4. Construction and parts listing for the Calcinator gas incinerator. There is little change in the appliances from year to year. (Courtesy of Calcinator)

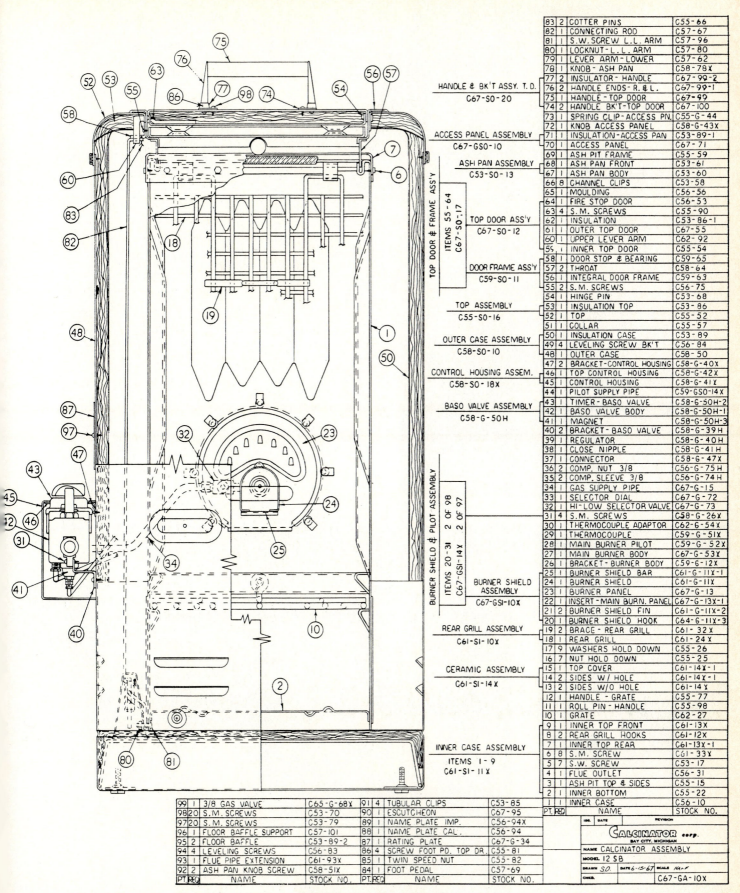

PT.	REQ.	NAME	STOCK NO.
83	2	COTTER PINS	C55-66
82	1	CONNECTING ROD	C57-67
81	1	S.W. SCREW L.L. ARM	C57-96
80	1	LOCKNUT-L.L. ARM	C57-80
79	1	LEVER ARM-LOWER	C57-62
78	1	KNOB - ASH PAN	C58-78X
77	2	INSULATOR - HANDLE	C67-99-2
76	2	HANDLE ENDS- R. & L.	C67-99-1
75	1	HANDLE - TOP DOOR	C67-99
74	1	HANDLE BK'T-TOP DOOR	C67-100
73	1	SPRING CLIP-ACCESS PN.	C55-G-44
72	1	KNOB ACCESS PANEL	C58-G-43X
71	1	INSULATION-ACCESS PAN.	C53-89-1
70	1	ACCESS PANEL	C67-71
69	1	ASH PIT FRAME	C55-59
68	1	ASH PAN FRONT	C53-61
67	1	ASH PAN BODY	C53-60
66	8	CHANNEL CLIPS	C53-58
65	1	MOULDING	C56-56
64	1	FIRE STOP DOOR	C56-53
63	4	S.M. SCREWS	C55-90
62	1	INSULATION	C53-86-1
61	1	OUTER TOP DOOR	C67-55
60	1	UPPER LEVER ARM	C62-92
59	1	INNER TOP DOOR	C55-54
58	1	DOOR STOP & BEARING	C59-65
57	2	THROAT	C58-64
56	1	INTEGRAL DOOR FRAME	C59-63
55	1	S.M. SCREWS	C56-75
54	1	HINGE PIN	C53-68
53	1	INSULATION TOP	C53-86
52	1	TOP	C55-52
51	1	COLLAR	C55-57
50	1	INSULATION CASE	C53-89
49	4	LEVELING SCREW BK'T	C56-84
48	1	OUTER CASE	C58-50
47	2	BRACKET-CONTROL HOUSING	C58-G-40X
46	1	TOP CONTROL HOUSING	C58-G-42X
45	1	CONTROL HOUSING	C58-G-41X
44	1	PILOT SUPPLY PIPE	C59-GSO-14X
43	1	TIMER - BASO VALVE	C58-G-50H-2
42	1	BASO VALVE BODY	C58-G-50H-1
41	1	MAGNET	C58-G-50H-3
40	2	BRACKET- BASO VALVE	C58-G-39H
39	1	REGULATOR	C58-G-40H
38	1	CLOSE NIPPLE	C58-G-41H
37	1	CONNECTOR	C58-G-47X
36	2	COMP. NUT 3/8	C56-G-75H
35	1	COMP. SLEEVE 3/8	C56-G-74H
34	1	GAS SUPPLY PIPE	C67-G-15
33	1	SELECTOR DIAL	C67-G-72
32	1	HI-LOW SELECTOR VALVE	C67-G-73
31	4	S.M. SCREWS	C58-G-26X
30	1	THERMOCOUPLE ADAPTOR	C62-G-54X
29	1	THERMOCOUPLE	C59-G-51X
28	1	MAIN BURNER PILOT	C59-G-52X
27	1	MAIN BURNER BODY	C67-G-53X
26	1	BRACKET-BURNER BODY	C59-G-12X
25	1	BURNER SHIELD BAR	C61-G-11X-1
24	1	BURNER SHIELD	C61-G-11X
23	1	BURNER PANEL	C67-G-13
22	1	INSERT-MAIN BURN. PANEL	C67-G-13X-1
21	2	BURNER SHIELD FIN	C61-G-11X-2
20	1	BURNER SHIELD HOOK	C64-6-11X-3
19	2	BRACE - REAR GRILL	C61-32X
18	1	REAR GRILL	C61-24X
17	9	WASHERS HOLD DOWN	C55-26
16	7	NUT HOLD DOWN	C55-25
15	1	TOP COVER	C61-14X-1
14	2	SIDES W/ HOLE	C61-14X-1
13	2	SIDES W/O HOLE	C61-14X
12	1	HANDLE - GRATE	C55-77
11	1	ROLL PIN - HANDLE	C55-98
10	1	GRATE	C62-27
9	1	INNER TOP FRONT	C61-13X
8	2	REAR GRILL HOOKS	C61-12X
7	1	INNER TOP REAR	C61-13X-1
6	8	S.M. SCREW	C61-33X
5	1	S.W. SCREW	C53-17
4	1	FLUE OUTLET	C56-31
3	1	ASH PIT TOP & SIDES	C55-15
2	1	INNER BOTTOM	C55-22
1	1	INNER CASE	C56-10
PT.	REQ.	NAME	STOCK NO.

PT.	REQ.	NAME	STOCK NO.	PT.	REQ.	NAME	STOCK NO.
99	1	3/8 GAS VALVE	C65-G-68X	91	4	TUBULAR CLIPS	C53-85
98	20	S.M. SCREWS	C53-70	90	1	ESCUTCHEON	C67-95
97	20	S.M. SCREWS	C53-79	89	1	NAME PLATE IMP.	C56-94X
96	1	FLOOR BAFFLE SUPPORT	C57-101	88	1	NAME PLATE CAL.	C56-94
95	2	FLOOR BAFFLE	C53-89-2	87	1	RATING PLATE	C67-G-34
94	4	LEVELING SCREWS	C56-83	86	4	SCREW FOOT PD. TOP DR.	C55-81
93	1	FLUE PIPE EXTENSION	C61-93X	85	1	TWIN SPEED NUT	C55-82
92	2	ASH PAN KNOB SCREW	C58-51X	84	1	FOOT PEDAL	C57-69
PT.	REQ.	NAME	STOCK NO.	PT.	REQ.	NAME	STOCK NO.

HANDLE & BK'T ASSY. T.D. C67-SO-20
ACCESS PANEL ASSEMBLY C67-GSO-10
ASH PAN ASSEMBLY C53-SO-13
TOP DOOR ASS'Y C67-SO-12
DOOR FRAME ASS'Y C59-SO-11
TOP DOOR & FRAME ASS'Y ITEMS 55-64 C67-SO-17
TOP ASSEMBLY C55-SO-16
OUTER CASE ASSEMBLY C58-SO-10
CONTROL HOUSING ASSEM. C58-SO-18X
BASO VALVE ASSEMBLY C58-G-50H
BURNER SHIELD & PILOT ASSEMBLY ITEMS 20-31 2 OF 98 C67-GSI-14X 2 OF 97
BURNER SHIELD ASSEMBLY C67-GSI-10X
REAR GRILL ASSEMBLY C61-SI-10X
CERAMIC ASSEMBLY C61-SI-14X
INNER CASE ASSEMBLY ITEMS 1-9 C61-SI-11X

CALCINATOR corp.
BAY CITY, MICHIGAN
NAME CALCINATOR ASSEMBLY
MODEL 12 SB
DRAWN S.O. DATE 6-15-67 SCALE HALF
CHK'D.
C67-GA-10X

FIGURE 27-4. (Continued)

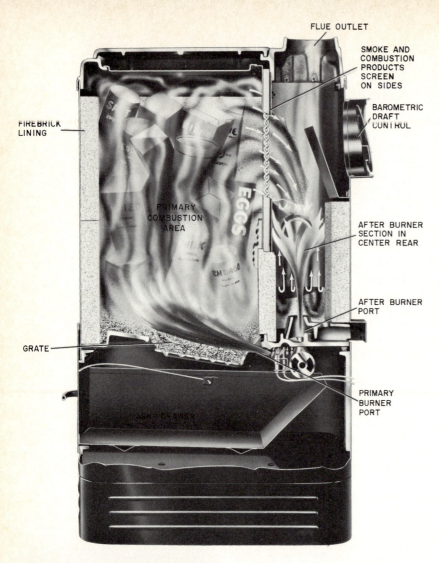

FLUE OUTLET

SMOKE AND
COMBUSTION
PRODUCTS
SCREEN
ON SIDES

BAROMETRIC
DRAFT
CONTROL

FIREBRICK
LINING

PRIMARY
COMBUSTION
AREA

AFTER BURNER
SECTION IN
CENTER REAR

AFTER BURNER
PORT

GRATE

PRIMARY
BURNER
PORT

FIGURE 27-5. Locke Stove Company's Model L-15. (Courtesy of Locke Stove Company)

FIGURE 27-6. A simple demonstration of how a flame will consume smoke and odors.

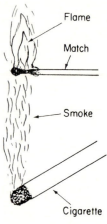

No Smoke

Flame

Match

Smoke

Cigarette

FIGURE 27-7. Warn the customer to keep packages small and not too wet, and to load the unit gently.

FIGURE 27-8. The timer must be set for enough time to consume the load. Turn the knob clockwise. Most gas incinerators use this basso valve.

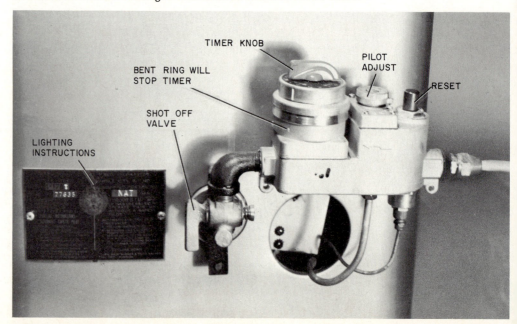

TIMER KNOB

PILOT
ADJUST

RESET

BENT RING WILL
STOP TIMER

SHOT OFF
VALVE

LIGHTING
INSTRUCTIONS

NAT

3. Other appliances, such as an oil burner or a gas water heater, that are connected to the same flue must have their dampers and doors closed while the incinerator is being used. Being careless about this point will affect the draft of the incinerator.

4. The incinerator must be level and rest solidly on the floor.

5. The gas supply line should be at least $\frac{3}{8}$-inch tubing, and if the line is 10 feet or more from the main gas line, then $\frac{1}{2}$-inch tubing should be used (or the equivalent rigid pipe). This is particularly important for the high-gas input units.

The troubleshooting chart shown has been compiled as your guide in locating and correcting improper incinerator action.

Complete installation instructions are included in Chapter 4.

Troubleshooting Chart   GAS INCINERATORS

Problem	Possible Cause	Corrective Action
Pilot outage.	Too many ashes in ash container.	Empty container when it becomes level full.
	Heavy accumulation of ashes on grate.	Shake down ashes each time incinerator is loaded.
	Dropping "package" into incinerator, which may deflect loose ash onto pilot light.	Place refuse in more gently.
	Connection between thermocouple and Baso safety valve loose and/or dirty.	Disconnect thermocouple lead and clean both contact points with lint-free cloth. Do not touch points with fingers.
	Pilot flame too small to keep thermocouple tip hot enough to hold safety valve open.	Adjust pilot for longer flame, or check for possible low gas pressure.
	Pilot air intake clogged with lint, causing a yellowish flame not hot enough to hold safety valve open.	Remove burner and clean air intake and parts thoroughly.
	Low gas pressure lowering pilot flame.	Have gas company service man check pressure; also determine if proper pilot orifice is being used.
	Buildup of waste due to incomplete burning.	Set timer for longer time and shake ashes down regularly.
Incomplete burning of wastes.	Main burner not operating long enough.	Set timer for longer period—3 hours for average mixed load, longer for hard-to-burn items, such as melon rinds and large bones.
	Accumulation of ashes and unburnables on grates keeps main flame from reaching load.	Shake grates often and drop all glass, cans, etc. into ash container before putting fresh load in unit.
	Excessive accumulation of ash interferes with action of main burner flame.	Empty ash container when level full, or sooner.
	Pilot light out.	Follow recommended procedure for relighting.

ELECTRIC INCINERATORS

The foregoing information, with the exception of those parts relating to the gas burner, is also true for electric incinerators.

Calcinator's electric unit (Fig. 27-9) has a 115-volt, 1000-watt heating element that turns on three times a day for one hour each time. An automatic timer controls these cycles. However, if a longer drying period is required to

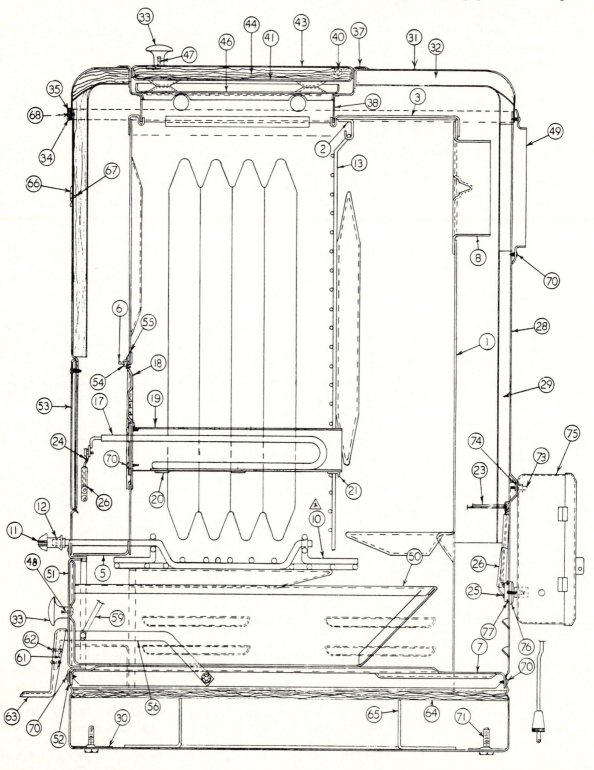

1000-WATT

FIGURE 27-9. Parts ordering list for the *electric* Calcinator. (Courtesy of Calcinator)

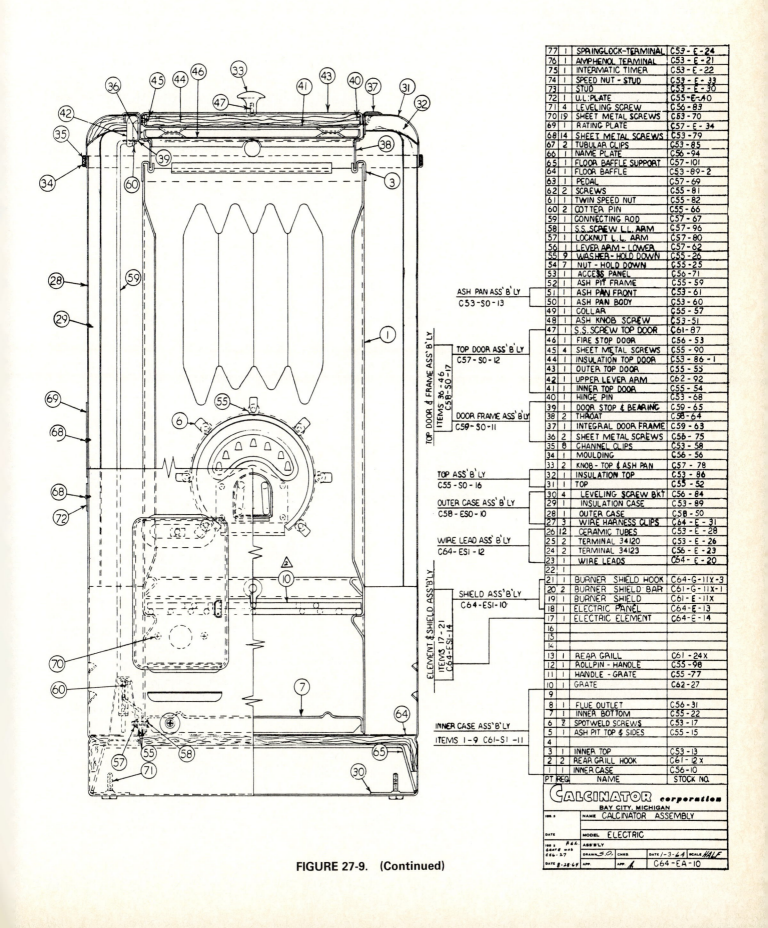

PT	REQ	NAME	STOCK NO.
77	1	SPRINGLOCK-TERMINAL	C53-E-24
76	1	AMPHENOL TERMINAL	C53-E-21
75	1	INTERMATIC TIMER	C53-E-22
74	1	SPEED NUT-STUD	C53-E-33
73	1	STUD	C53-E-30
72	1	U.L. PLATE	C56-E-10
71	4	LEVELING SCREW	C56-83
70	19	SHEET METAL SCREWS	C53-70
69	1	RATING PLATE	C57-E-34
68	14	SHEET METAL SCREWS	C53-79
67	2	TUBULAR CLIPS	C53-85
66	1	NAME PLATE	C56-94
65	1	FLOOR BAFFLE SUPPORT	C57-101
64	1	FLOOR BAFFLE	C53-89-2
63	1	PEDAL	C57-69
62	2	SCREWS	C55-81
61	1	TWIN SPEED NUT	C55-82
60	2	COTTER PIN	C55-66
59	1	CONNECTING ROD	C57-67
58	1	S.S. SCREW L.L. ARM	C57-96
57	1	LOCKNUT L.L. ARM	C57-80
56	1	LEVER ARM-LOWER	C57-62
55	9	WASHER-HOLD DOWN	C55-26
54	7	NUT-HOLD DOWN	C55-25
53	1	ACCESS PANEL	C56-71
52	1	ASH PIT FRAME	C55-59
51	1	ASH PAN FRONT	C53-61
50	1	ASH PAN BODY	C53-60
49	1	COLLAR	C55-57
48	1	ASH KNOB SCREW	C53-51
47	1	S.S. SCREW TOP DOOR	C61-87
46	1	FIRE STOP DOOR	C56-53
45	4	SHEET METAL SCREWS	C55-90
44	1	INSULATION TOP DOOR	C53-86-1
43	1	OUTER TOP DOOR	C55-55
42	1	UPPER LEVER ARM	C62-92
41	1	INNER TOP DOOR	C55-54
40	1	HINGE PIN	C53-68
39	1	DOOR STOP & BEARING	C59-65
38	2	THROAT	C58-64
37	1	INTEGRAL DOOR FRAME	C59-63
36	2	SHEET METAL SCREWS	C55-75
35	8	CHANNEL CLIPS	C53-58
34	1	MOULDING	C56-56
33	2	KNOB-TOP & ASH PAN	C57-78
32	1	INSULATION TOP	C53-86
31	1	TOP	C55-52
30	4	LEVELING SCREW BKT	C56-84
29	1	INSULATION CASE	C53-89
28	1	OUTER CASE	C58-50
27	3	WIRE HARNESS CLIPS	C64-E-31
26	12	CERAMIC TUBES	C53-E-28
25	2	TERMINAL 34120	C53-E-26
24	2	TERMINAL 34123	C56-E-23
23	1	WIRE LEADS	C64-E-20
22	1		
21	1	BURNER SHIELD HOOK	C64-G-11X-3
20	2	BURNER SHIELD BAR	C61-G-11X-1
19	1	BURNER SHIELD	C61-E-11X
18	1	ELECTRIC PANEL	C64-E-13
17	1	ELECTRIC ELEMENT	C64-E-14
16			
15			
14			
13	1	REAR GRILL	C61-24X
12	1	ROLLPIN-HANDLE	C55-98
11	1	HANDLE-GRATE	C55-77
10	1	GRATE	C62-27
9			
8	1	FLUE OUTLET	C56-31
7	1	INNER BOTTOM	C55-22
6	7	SPOTWELD SCREWS	C53-17
5	1	ASH PIT TOP & SIDES	C55-15
4			
3	1	INNER TOP	C53-13
2	2	REAR GRILL HOOK	C61-12X
1	1	INNER CASE	C56-10

CALCINATOR *corporation* BAY CITY, MICHIGAN — CALCINATOR ASSEMBLY, MODEL ELECTRIC, C64-EA-10

FIGURE 27-9. (Continued)

reduce an abnormally large or wet load, a manual switch is provided. Thus you can turn the unit on any time you wish and it will stay on until the end of the next normal drying cycle.

In order to check the heating element, push the manual switch lever on the right or the ON position. The burner (heater element) should become warm almost immediately. However, it will take about 10 minutes for it to become hot enough to ignite crumpled papers placed against it. The automatic timer will turn itself off; resume present cycling without resetting.

A good draft is necessary for efficient service. A quick check for good draft is to blow smoke from a lighted cigarette into the open combustion door. The smoke should be drawn into the combustion chamber when the door is wide open. However, there is such a thing as too much draft. This will cool the heater element and cause erratic operation.

The troubleshooting chart has been compiled as a guide in locating most problems that occur with an electric incinerator.

Complete installation instructions are included in Chapter 4.

Troubleshooting Chart ELECTRIC INCINERATORS

Problem	Possible Cause	Corrective Action
Element will not heat.	Ashes cover heater element. Faulty element connection. Defective element. No electricity to element.	Clean out and remove. Make certain terminals are clear of shield and each other. Replace. Make certain power is available to incinerator.
Incorrect heating time.	Timer cam dial trippers have slipped.	Reset trippers for ON at 9:30 A.M., 2:30 P.M., and 10:30 P.M., and for OFF at 10:30 A.M., 3:30 P.M., and 11:30 P.M.
Smokes or emits odors.	Improper draft condition.	Check and close ash pan door, chimney clean-out door, furnace fan door. Attic fan or air conditioner can overcome the chimney draft. Turn it off. Clean a dirty chimney or flue. Check downdraft condition and correct if faulty. Change venting to improve poor draft.
Refuse may remain for several days; then flare up and burn with a roar.	Insufficient heat. Improperly wrapped packages. Refuse too wet. Timer may be set for too short a cycle.	Low line voltage. Excessive draft. Burner covered with ash or non-combustibles. Packages in rear of chamber not contacting burner. Packages must be wrapped in paper. Drain off excess water; then wrap in sufficient paper. Check cam position and reset trippers.

Troubleshooting Chart ***ELECTRIC INCINERATORS*** **(Continued)**

Problem	Possible Cause	Corrective Action
Does not dispose of small packages.	Element does not heat.	See section in text on this subject.
	Packages not in contact with burner.	Continue to add refuse.
	Improperly wrapped package.	Packages must be small—size of loaf of bread, and well enough wrapped with paper to absorb moisture and hold refuse together.
	Ashes covering heater.	Clean out.

Note: We do not recommend the disassembly of a timer to make internal repairs!

28

Electric Irons

The family iron is a hard-working appliance, used by everyone and dropped on the floor occasionally. The most frequent complaints are handle breakage and cord failure. When you can change a cord or a handle, or clean up a sticky soleplate, you will have most of the iron problems licked.

If someone used too much starch in a collar, or ran an iron that was too hot over synthetic fibers, the iron will have collected a sticky coating of cooked starch and/or melted fabric on its soleplate. This coating will make the iron sticky and hard to push, and may soil white goods.

In order to clean the soleplate, try a damp cloth first; then wipe with a soft dry cloth. If this does not work, rub gently with a fine steel wool pad or a piece of fine emery.

Do not do this to a Teflon or silverstone-coated soleplate! The coated soleplates are not supposed to stick, so treat them with care so as not to scratch the surface.

Slight scratches in the soleplate will do no harm, but burrs or any other projections above the soleplate surface will damage clothes. The fine emery will do a good job on removing burrs. Before using the iron, the soleplate must be carefully cleaned and wiped off; otherwise, it will soil or damage clothing when used the next time. When repairing an iron, it is a good idea to protect the soleplate by working on a clean pad.

A steam iron can get into trouble for no other reason than the use of hard water or water with a high iron content. As the water steams, it leaves all the minerals it was carrying behind. These minerals eventually accumulate to clog ports and small steam passages; sometimes they blow out with the steam to leave a big, rusty streak on a white shirt.

Minerals can be cleaned out of the iron by filling the tank with vinegar, which will dissolve the lime. Heating the iron will hasten the process, which, in severe cases, may have to be repeated. A semiannual vinegar rinse will keep the unit in good condition. In areas where the water is extremely hard, the vinegar rinse should be more frequent.

A water softener will eliminate the lime problem, but the water will have a tendency to froth and spit more easily. Not filling the iron quite so full will usually solve this problem. The minerals are still in the water, so periodically rinse out the tank and fill with fresh water. Allowing the water in the tank to dry out will leave a mineral deposit, which can be soaked out.

Distilled water is recommended for use in the iron and may be obtained from several sources other than a store. Rainwater and melted snow do very nicely in most areas. (You should not use it in a smoky or dusty locale. Watch the air pollution.) The water obtained from defrosting the refrigerator or scraping down the freezer usually is fine for this purpose, and there is more than enough to supply the needs. (It can be used for your car battery, too.) In the summertime, another source of clean water is the condensation from a dehumidifier. Water will collect dust and similar pollutants from the air, and if it looks dirty, it should be filtered through a clean cloth.

If this water is to be saved for a while, it would be wise to put about a tablespoon of Chlorox (or some equal chlorine bleach) in a quart of water as a preservative.

What has been said about water does not apply to the nonsteam or dry iron, of course. But what is to be said about testing and repair will apply directly to the dry iron as well (Fig. 28-1). Electrically and mechanically they have the same problems. The steam iron merely has some extras of its own.

As irons acquire more features, the possibilities of more service problems increase. The Proctor-Silex shown in Fig 28-2 has an automatic shutoff module (reference number 8 on the parts list) which will automatically turn the iron off if it is left standing on its heel rest for a reasonable amount of time. The module will also turn the iron off if it is allowed to stand on its soleplate without moving for several seconds. This feature is a great safety device and its operation must be considered when the iron's electrical circuit is being tested.

There are two basic types of steam iron, which differ in the method of steam generation. The boiler type has a water tank that is heated

by the soleplate heating element to generate steam within the tank. This steam is then directed, by a valve, through holes in the soleplate, to the fabric being ironed.

The flash type has a water tank and valving that permit the water to drip onto a hot plate to generate steam, which escapes through holes in the soleplate as shown in Fig. 28-3.

Testing and troubleshooting will be rather similar for either type. Only the mechanics of assembly and disassembly of the parts will vary.

The West Bend cordless iron consists of two units: a power base and a steam iron. To operate, the heel of the iron is set on the base making certain that the contacts of each section mate to each other securely. The iron will get warm with the thermostat limiting the temperature to the preset heat. When the iron is removed from the base and used, it will cool down to be reheated when placed back on the base. To use as a spray or steam iron, water must be added to the iron, *not to the base* (Fig. 28-4). The exploded view, parts list, and wiring diagrams are included for location of parts and troubleshooting. The system sectional should help explain what happens in the steam section. In the wiring diagrams note that there is both

FIGURE 28-1. The dry iron has no water problems. Its effectiveness depends on the clothes being properly sprinkled and on the correct moisture content for good, wrinkle-free ironing. (Courtesy of Sunbeam)

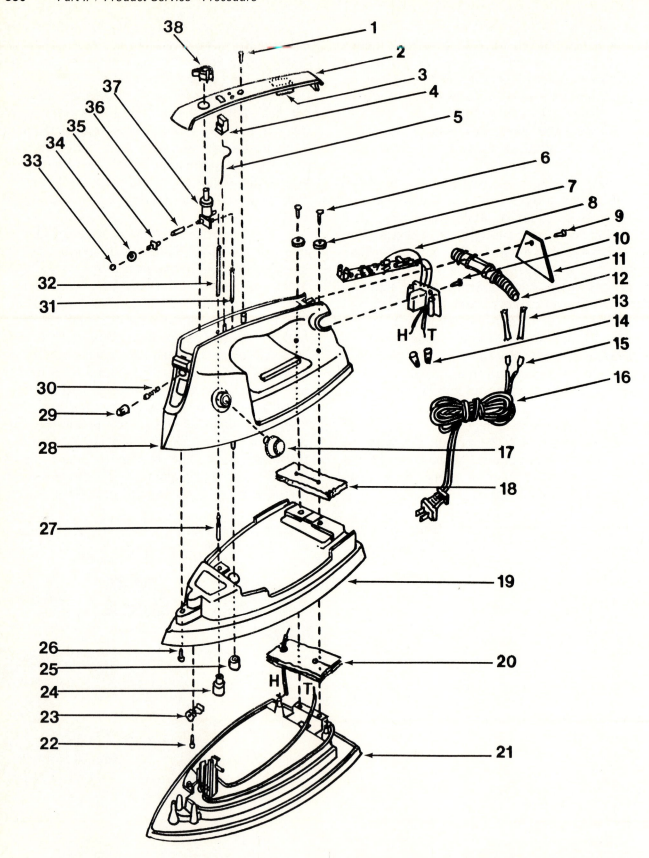

FIGURE 28-2. Proctor-Silex 12010 Series A Extra steam–spray–steam/dry. (Courtesy of Proctor-Silex)

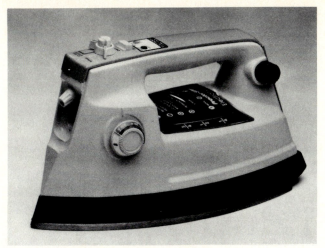

Item Number	Description
1	Screw, oval head Phillips Self tapping
2	Cover, top
3	Pad, foam (part of item 2)
4	Button, (steam/dry)
5	Extension button
6	Screw/washer (handle to soleplate 2 required)
7	Washer (item 6, 2 required)
8	Module-automatic shutoff
9	Screw (item 1)
10	Screw (module to rear handle)
11	Cover, rear
12	Guard, cord
13	Sleeve (2 required)
14	Nut, wire (2 required)
15	Terminal, female (2 required)
16	Cord, power (item 15)
17	Knob
18	Baffle
19	Cover, base
20	Baffle
21	Solepate and thermostat, (item 20)
22	Screw
23	Bracket, slide
24	Seal, blast valve
25	Seal, drip valve
26	Screw
27	Valve, blast
28	Handle (item 25)
29	Nozzle, spray
30	Plunger, spray
31	Hose, intake
32	Tube, blast
33	Gasket
34	Seal, fill
35	Body, nozzle
36	Tube, spray
37	Pump ass'y
38	Knob, spray/extra steam

FIGURE 28-2. (Continued) Proctor-Silex 12010 series A Extra steam-spray-steam/dry (Courtesy Proctor-Silex)

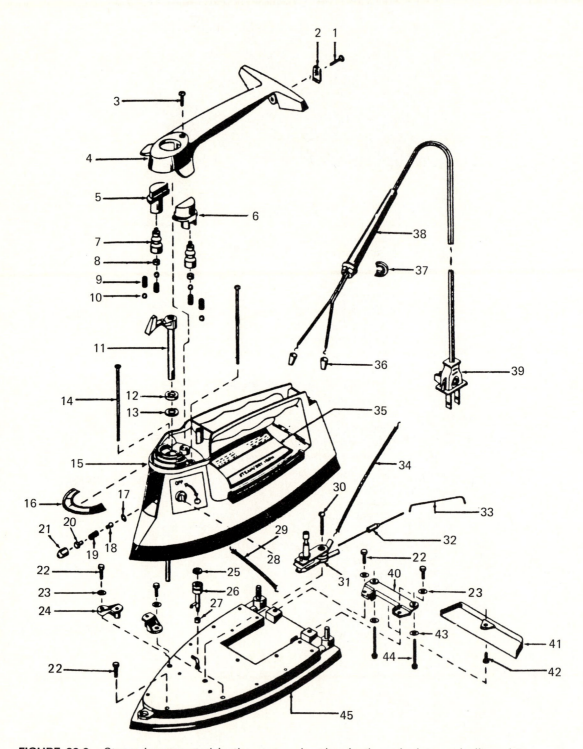

FIGURE 28-3. Steam is generated in the steam chamber in the soleplate and allowed to escape through the vents. (Courtesy of Sunbeam)

Sunbeam
Shot of Steam
Spray Mist
Comfort Iron

12319
Gray

• 51-vent soleplate for even, all-over steam coverage—SilverStone non-stick surface for effortless ironing • Large, easy to fill water reservoir—uses most tap water • Lightweight design with tough plastic shell • Comfortable, contoured stay-cool handle • Instant extra steam for stubborn wrinkles plus fast reference fabric guide • Push-button spray for quick spot dampening • Self-cleaning action—prolongs steaming life

Diagram Number	Description	Diagram Number	Description
1	Screw	23A	Tinnerman clip (2 req'd)
2	Clip	24	Support bracket (metal)
3	Screw	25	O-ring
4	Top handle	26	Injector ass'y
5	Pump button spray	27	Rubber seal
6	Pump button steam	28	Steam knob
7	Bellows	28A	O-ring used with round steam knobs
8	Pump injector seat	29	Lead wire
9	Spring	30	Thermostat screw
10	Ball	31	Thermostat
11	Lever and shaft ass'y	32	Fuse ass'y
12	Seal	33	Lead wire
13	Bearing	34	Lead wire
14	Screw (2 req'd)	35	Saddle escutcheon
15	Handle and lower housing ass'y	36	Connector (2 req'd)
16	Temperature dial	37	Strain relief
17	Pump washer	38	Cord strain relief (boot)
18	Rubber ball valve	40	Shell mounting bracket (metal)
19	Spring	41	Cover plate
20	Spray plug	42	Cover plate screw
21	Nozzle	43	Handle washer (2 req'd)
22	Bracket screw (2 req'd)	44	Handle screw (2 req'd)
23	Washers (4 req'd)	45	Soleplate and cover ass'y

FIGURE 28-3. **(Continued)**

EXPLODED ASSEMBLY DRAWING — IRON

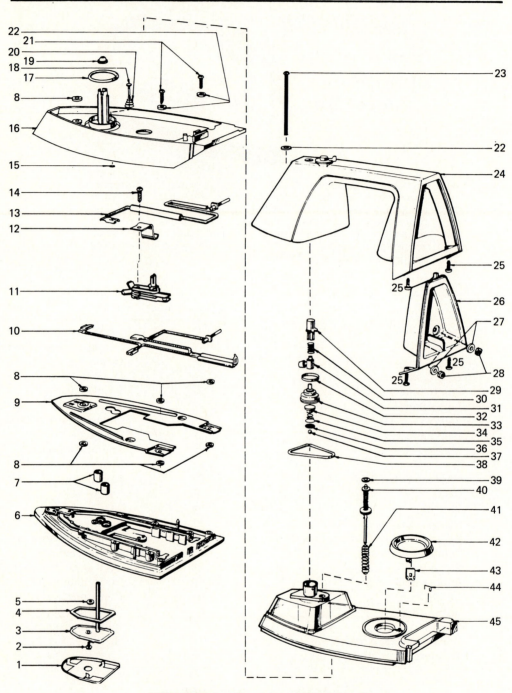

FIGURE 28-4. (Courtesy of West Bend)

Parts list—Iron

Item No.	Part Number	Description
1	P355-2D	Cover assembly
2	P663-127	Screw
3	P571-10D	Plate assembly
4	P122-72	Gasket
5	P121-72	Gasket
6	P570-10D	Plate assembly
7	P151-12	Tube
8	P271-47	Washer
9	P574-10	Plate
10	P98-73D	Lamp assembly
11	P120-97	Thermostat
12	P285-42	Bracket
13	P16-94D	Fuse assembly
14	P661-127	Screw
15	P120-72	Gasket
16	P41-199	Housing
17	P127-72	Gasket
18	P153-49	Stem
19	P133-72	Gasket
20	P77-134	Spring
21	P674-127	Screw
22	P270-47	Washer
23	P664-127	Screw
24	P41-168D	Body assembly
25	P660-127	Screw
26	P103-14	Socket
27	P274-47	Washer
28	P240-81	Nut
29	P9-174	Button
30	P79-134	Spring
31	P1-245D	Nozzle assembly
32	P60-5	Clamp
33	P104-89	Cap
34	P1-253	Piston
35	P78-134	Spring
36	P27-69	Filter
37	P3-148	Ball
38	P124-72	Gasket
39	P126-72	Gasket
40	P46-150D	Rod assembly
41	P81-134	Spring
42	P428-19D	Knob assembly
43	P48-156	Insert
44	P82-134	Spring
45	P19-211D	Reservoir assembly

FIGURE 28-4. (Continued)

EXPLODED ASSEMBLY DRAWING – POWER BASE

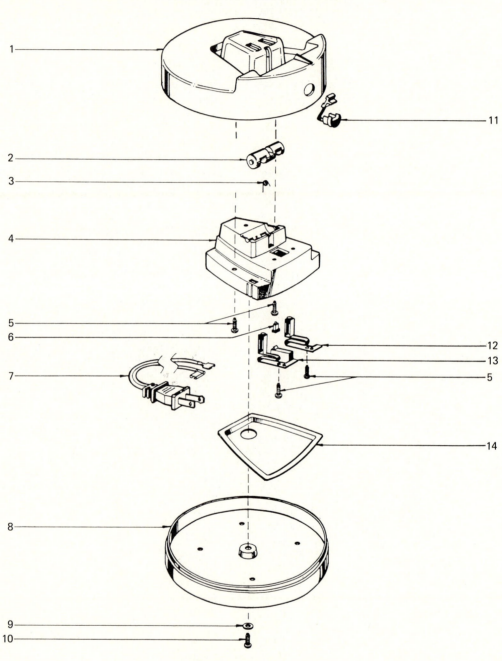

FIGURE 28-4.   (Continued)

Parts list—Power base

Item No.	Part Number	Description
1	P40-199	Housing
2	P12-52	Rotor
3	P75-134	Spring
4	P52-200D	Support assembly
5	P660-127	Screw
6	P10-111	Slide
7	P149-74	Cord
8	P522-3D	Base assembly
9	P269-47	Washer
10	P661-127	Screw
11	P37-5	Clamp
12	P74-134D	Spring assembly
13	P74-134E	Spring assembly
14	P569-10	Plate

Electrical schematic diagrams

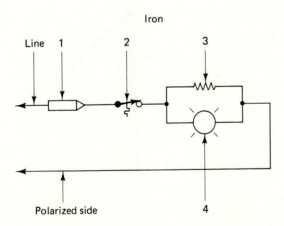

Iron

Item no.	Description	Part number
1	Fuse	P16-94
2	Thermostat	P120-97
3	Unit	P290-109
4	Lamp	P98-73

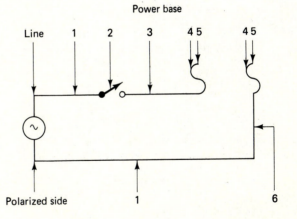

Power base

Item no.	Description	Part number
1	Cord	P149-74
2	Switch	P41-87
3	Terminal	P52-67
4	Spring	P74-134
5	Contact	P11-194
6	Terminal	P56-67

FIGURE 28-4. (Continued)

SYSTEM SECTIONAL VIEWS

Section 1. Water Reservoir, Water Path and Steam Path

The unique steam generation principle develops a slight amount of pressure during the use cycle in either the normal steam or super steam position. A pictorial of the water reservoir, water path and steam path is shown below. If the Iron leaks or fails to produce steam, all of the seals, water paths and steam paths shown below should be checked for leaks, defective parts or obstructions.

Section 2. Spray Feature

All parts of the spray section are shown below. If the Iron fails to spray or provides insufficient spray, check all of the parts below for defects or obstructions.

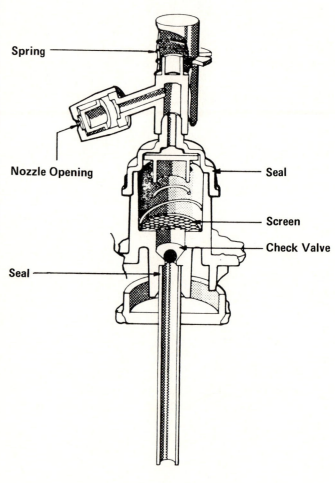

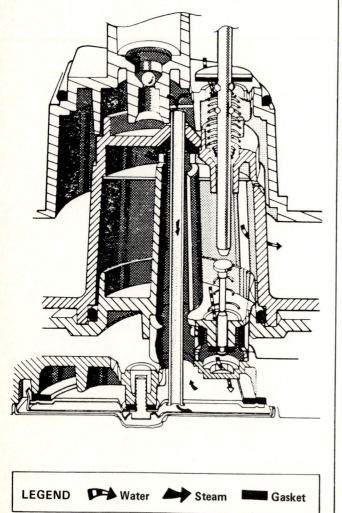

LEGEND ▷ Water   ➡ Steam   ▬ Gasket

FIGURE 28-4.   (Continued)

a fuse and a thermostat in the iron circuit. When the iron is cold (room temperature), there should be continuity between its terminals.

The power base includes a switch that is activated when the iron is placed on it properly. There will be no continuity from prong to prong of the cord unless the iron is in position. Also, be aware that the iron cord has a polarized attachment plug.

TESTING THE IRON

Aside from obvious physical damage, such as a broken handle, dented shell, a leaky tank, or scratched soleplate, you will have to make electrical tests to locate the trouble spot. All electrical tests can be made with an appliance tester or ohmmeter.

When an iron does not heat or when it operates erratically, the first part to suspect is the cord. Test it by plugging it into outlet A of the appliance tester (Fig. 28-5). Turn the thermostat all the way on, and the bulb should light. If it does not, then connect the jumper lead across the two terminals of the cord to short them. If the cord is good, the bulb will light and remain lighted while you shake, wiggle, or work over the cord with your hands. If the light flickers or goes out while you are working over the cord, it means that there is an intermittent break in the cord and it should be replaced. Having checked the condition of the cord, we move on to the iron itself.

The next part to test is the thermostat. Remove the jumper from the cord terminals and connect it across the heating element terminals. Turn the thermostat from OFF to ON and rotate its entire range. The test bulb light should come on somewhere in the first third of the control's rotation and remain ON steadily throughout the balance of turning. It should go OFF at about the same point it came ON.

Look the thermostat over carefully. Note the condition of the contact points, leaf springs, and insulators. Clean out any lint or other foreign matter that may have collected. Make certain that the terminals are tight and that contact areas are clean.

A defective thermostat must be replaced.

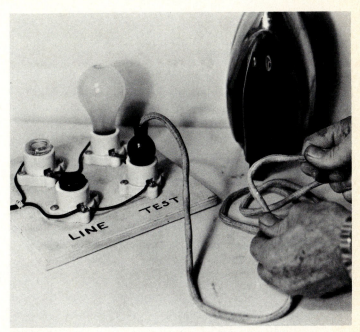

FIGURE 28-5. An iron cord is always the first thing to suspect in case of electrical troubles. It should be connected to a tester or ohmmeter and wiggled (pushed and pulled)—be particularly careful at each end. If the light flickers or if the ohmmeter needle moves, you have found the trouble.

Contact points can be cleaned to some extent, but generally repairs are not satisfactory.

The next step in electrical testing is the heater element. Remove the jumper lead from the heater element terminals and turn the thermostat all the way ON. The bulb should light. If it does not and everything else up to this point is good, the heater is defective.

To double check this diagnosis, unplug the iron cord from your tester and plug in the tester leads. Connect the test leads, one to each heater terminal. The bulb will light if the element is all right. No light, or a dim light, will indicate a defective element. There is no repair for this problem; the heater element must be replaced.

Finally, all parts and the assembled iron must be ground tested before you can say that the iron is electrically workable.

REPAIRING THE IRON

All modern irons have the heater element cast into an aluminum soleplate. This is the single most expensive part of the iron, and it is often more economical to buy a new iron rather than replace the part. Most damaged soleplates result from a thermostat that has allowed the unit to overheat.

Repairing or changing the cord is no problem. A cord can be repaired if the trouble is at either end, by simply cutting the cord at the break. And if the trouble is at the plug end, attach a new plug. This is a common cause of failure due to the cord motion while ironing and the method of "pulling the plug" when ironing is completed. The same holds true at the iron end. Just shorten the cord and reconnect. Make certain that the asbestos insulation is wrapped well and that all strands of wire are secure under the connecting screw (Fig. 28-6).

If the outer braid of the cord is grayed, burned, or shows signs of abuse, it is a lot safer to discard it and replace it with a new one.

On an iron that uses a plug-in cord, this change is no problem. But make certain that the iron terminals are tight and clean; other-

FIGURE 28-6. Access to the cord terminals is usually easy. Make certain that when reassembling, you have left no bare wires or stray strands that might touch the metal shell and give the user a shock.

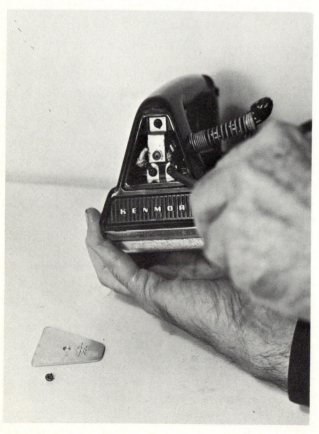

wise there will be a high-resistance connection that will overheat and damage the connector on the new cord.

Permanently attached cords follow much the same general pattern for all irons, so the procedure used on this Kenmore iron will apply to many others.

1. Remove the metal plate at back of handle.
2. Remove the two screws attaching cord to iron.
3. Untie the knot and pull out the cord. (Many kinds of strain reliefs are used in irons and they have to be removed before the cord can be removed.)

Reverse the procedure to install a new cord.

To get at the thermostat and heater of this Kenmore iron, first disconnect the cord from the iron, as explained. (It need not be removed from the handle.) Next, pry the metal cover off the thermostat dial to expose two screwdriver clearance holes in the knob. Align these holes with the two screws holding the shell to the soleplate. Remove the screws.

Lift the handle and shell assembly and take the tank assembly from the soleplate to expose the thermostat and heater terminals.

This iron has a two-piece thermostat consisting of a bimetal element and an adjustable switch. Each is mounted with its own screw; these screws must be secure to maintain the correct position of the bimetal element in relation to the switch element. Make certain that the free end of the bimetal element is free to move up and down so that it can open and close the switch with temperature changes.

Leads to the heater element are welded. If one of the leads breaks off, it will have to be silver soldered in place. Be sure to keep a $\frac{1}{8}$-inch clearance between any current-carrying lead and the exposed metal of the soleplate. A grounded wire can be dangerous.

When reassembling, make certain that all gaskets are in good condition and in place; otherwise you will have water and steam leaks.

When replacing the thermostat, make sure that the soleplate surface is clean and flat to ensure the thermostat's expansion member being in good contact with the soleplate. Also, with your test light, make certain that the thermostat has an OFF position. Adjust until it does.

In reassembly, be sure that all shims and gaskets are properly positioned and in good condition. If there is any doubt about the condition of a gasket, it should be replaced.

THERMOSTAT CALIBRATION

Many iron thermostats are set to provide a maximum temperature of 550°F (288°C) and should be adjusted for this. Some, such as the Sunbeam Comfort iron use a preset thermostat for which the maximum setting is 380 to 420°F (193 to 216°C) at the linen setting.

Steaming temperature is usually about 250°F (121°C). If an iron is set for the beginning of the steam range and cycles at 250°F (121°C), no calibration is required. Water is not used for this test.

A temperature test can be made with a thermocouple (or thermistor) in direct contact with the bottom of the soleplate, centered about one-third of the way back from the point. The soleplate must be covered to prevent excessive heat radiation. The easiest way to do this is

to place a piece of asbestos on a metal plate raised above the bench. Allow enough space for air circulation under the metal sheet or you will burn a hole in the bench top.

Fold the thermocouple through the hole in a washer, place it in position on the stand, and set the iron on top of it.

Turn the thermostat full ON and allow the iron to stabilize in temperature cycling. This may take 5 to 10 minutes. This setting should produce the maximum safe operating temperature between 540 and 560°F (282 to 293°C); if the temperature tends to creep over this range, adjust the thermostat downward.

A word of caution: An excessive temperature will melt the soleplate, so pull the plug when you see the temperature approaching 600°F (316°C).

Thermostat calibration methods will vary by type, but most of them have a small set screw in the center of the control shaft that will allow enough adjustment to bring the control into the normal operating range.

After calibration recheck control lever for a positive OFF position. Also test for leakage current from electrical conducting parts to the exposed nonconducting parts. Leakage should not exceed 0.2 milliampere and the resistance should be over 600,000 ohms.

Troubleshooting Chart   DRY AND STEAM IRONS

Problem	Possible Cause	Corrective Action
No heat.	No power at outlet.	Check outlet for power.
	Defective cord or plug.	Repair or replace.
	Broken lead in iron.	Repair or replace lead.
	Loose connection.	Clean and tighten.
	Loose thermostat control knob.	Replace knob and tighten on shaft.
	Defective thermostat.	Replace thermostat.
	Defective heater element.	Replace heater if separate. If cast in, replace soleplate assembly.
	Open thermal fuse.	Replace.
Insufficient heat.	Low line voltage.	Check voltage at outlet.
	Incorrect thermostat setting.	Adjust and recalibrate thermostat.
	Defective thermostat.	Replace thermostat.
	Loose connection.	Clean and tighten connections.
Excessive heat	Incorrect thermostat setting.	Adjust and recalibrate thermostat.
	Defective thermostat.	Replace thermostat.

Troubleshooting Chart   DRY AND STEAM IRONS   (Continued)

Problem	Possible Cause	Corrective Action
Blisters on soleplate.	Excessive heat.	After repair of control, replace or repair soleplate, depending on its condition.
Water leakage.	Careless filling. Defective seam or tank weld. Inadequate tank sealer. Damaged gasket.	Allow spilled water to dry. Be more careful. Replace tank. Replace with proper sealer. Replace gasket.
No steam.	No water in tank. Thermostat set too low. Valve in OFF position. Dirty or plugged valves or holes.	Fill tank. Set control higher. Turn valve to correct position. Clean out.
Spitting.	Incorrect thermostat setting. Excessive mineral deposit. Overfilling.	Reset thermostat, usually too low. Clean out. Be more careful.
Bad spray (spray irons).	Defective plunger.	Replace.
Iron stains clothes.	Starch stuck on soleplate. Foreign matter (iron, lime) in water. Sediment in tank.	Rub soleplate with damp cloth; then polish with dry cloth. Use distilled water. Clean tank. Use vinegar or lime.
Tears clothes.	Rough spot, nick, scratch burr on soleplate.	Remove with fine emery; then buff or polish area.
Shocks user.	Electric circuit grounded to iron.	Check 1. Cord and wiring for bare spot. 2. Thermostat for insulation breakdown. 3. Heater element for ground. Repair or replace as indicated.
Cord sparks.	Loose connection. Broken wire.	Clean and tighten. Repair or replace.
Sticks to clothes.	Dirty soleplate. Excessive starch in clothes. Iron too hot for fabric being ironed.	Clean. Iron at a lower temperature. Use less starch next time. Lower thermostat setting.

29

Knife Sharpeners; Can Openers

KNIFE SHARPENERS

Once upon a time the "Scissors Grinder" used to make the rounds to repair umbrellas, sharpen knives, grind scissors, skates, etc. His was a humble but necessary service to the housewife but he, like the iceman, has practically faded out of the picture.

The electric sharpener has been developed to help the housewife keep her knives and scissors in their best working condition. Basic sharpeners use a small shaded pole motor to drive one or two grinding wheels. (Sunbeam models use two wheels, General Electric models use one.) Used correctly either will do the job it was designed for (Fig. 29-1).

Sharpening a knife on a properly operating sharpener is no problem. Simply set the knife blade gently on the wheel(s)—as near the handle as possible and draw the knife toward you for its complete length. On long knives, light finger pressure should be applied to the tip of the blade to help maintain an even, gentle pres-

sure on the wheel(s) while drawing the blade through the sharpening stroke. This may be repeated several times.

Excessive pressure will stall the machine. Insufficient pressure may cause poor sharpening. Uneven sharpening or "scalloping" of the blade edge can be caused by uneven sharpening pressure or by a wheel that has excessive wobble.

If the customer's complaint is scalloping, test the grinder yourself before condemning it and use a good knife—an unevenly ground knife can not be used to test a grinder. Excessive wheel wobble will usually be self evident. To cure try loosening and retightening the mounting nuts, first. If this does not work or if the wheel looks damaged or unevenly worn, replace it. Also check the shaft and bearings.

Grinding wheel location and speed are important. Two-wheel grinders require the "V" between the grinding wheels to be centered in the knife notch of the housing. Their speed is approximately 675 rpm. A single-wheel grinder

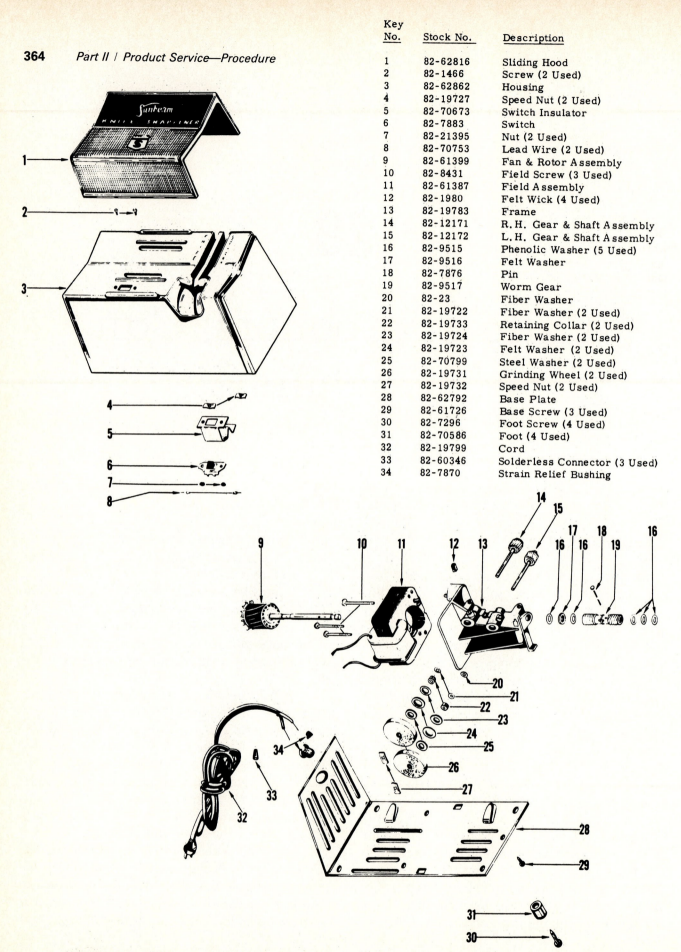

Key No.	Stock No.	Description
1	82-62816	Sliding Hood
2	82-1466	Screw (2 Used)
3	82-62862	Housing
4	82-19727	Speed Nut (2 Used)
5	82-70673	Switch Insulator
6	82-7883	Switch
7	82-21395	Nut (2 Used)
8	82-70753	Lead Wire (2 Used)
9	82-61399	Fan & Rotor Assembly
10	82-8431	Field Screw (3 Used)
11	82-61387	Field Assembly
12	82-1980	Felt Wick (4 Used)
13	82-19783	Frame
14	82-12171	R.H. Gear & Shaft Assembly
15	82-12172	L.H. Gear & Shaft Assembly
16	82-9515	Phenolic Washer (5 Used)
17	82-9516	Felt Washer
18	82-7876	Pin
19	82-9517	Worm Gear
20	82-23	Fiber Washer
21	82-19722	Fiber Washer (2 Used)
22	82-19733	Retaining Collar (2 Used)
23	82-19724	Fiber Washer (2 Used)
24	82-19723	Felt Washer (2 Used)
25	82-70799	Steel Washer (2 Used)
26	82-19731	Grinding Wheel (2 Used)
27	82-19732	Speed Nut (2 Used)
28	82-62792	Base Plate
29	82-61726	Base Screw (3 Used)
30	82-7296	Foot Screw (4 Used)
31	82-70586	Foot (4 Used)
32	82-19799	Cord
33	82-60346	Solderless Connector (3 Used)
34	82-7870	Strain Relief Bushing

FIGURE 29-1.   Parts view of Sunbeam sharpener illustrating two-wheel (No. 26) construction. (Courtesy of Sunbeam)

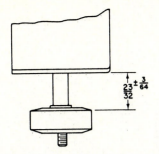

FIGURE 29-2. This is the correct location of the grinding wheel (XM25X614) on the General Electric 14EC12 can opener–knife sharpener. Spacer washers are used to obtain the proper distance. (Courtesy of General Electric)

location is shown in Fig. 29-2, which will place it in correct alignment with the knife slots in the housing shown in Fig. 29-3.

General Electric includes a pencil sharpener in its knife sharpener. This consists of an abrasive disc and guide bushing. If a pencil takes too long to sharpen (20 to 30 seconds for a new, inexpensive pencil) it may be that the abrasive disk has an accumulation of crayon, etc., which must be removed. This can be done by removing the scrap box, placing the unit in an upside down position, turning it on and

FIGURE 29-3. Parts view of the General Electric sharpener–can opener illustrating single-wheel construction. (Courtesy of General Electric)

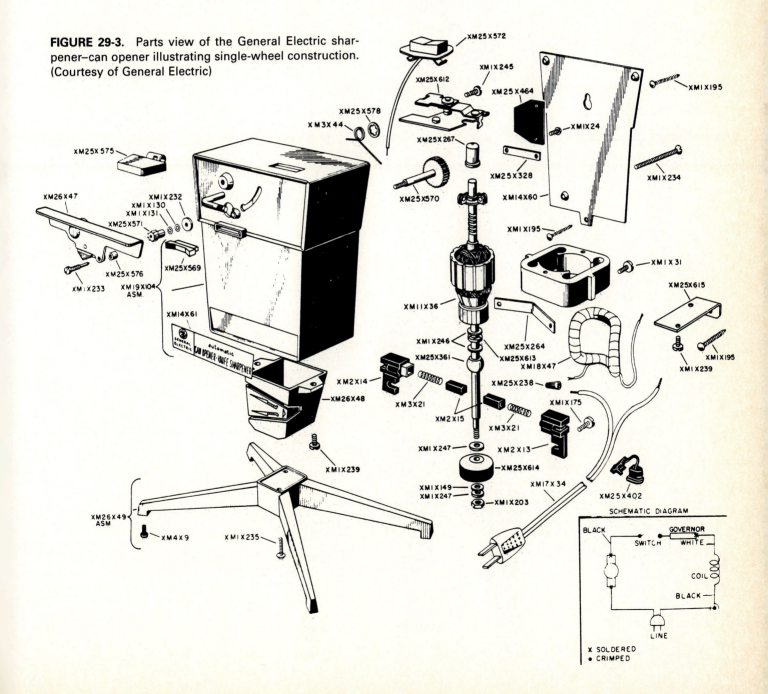

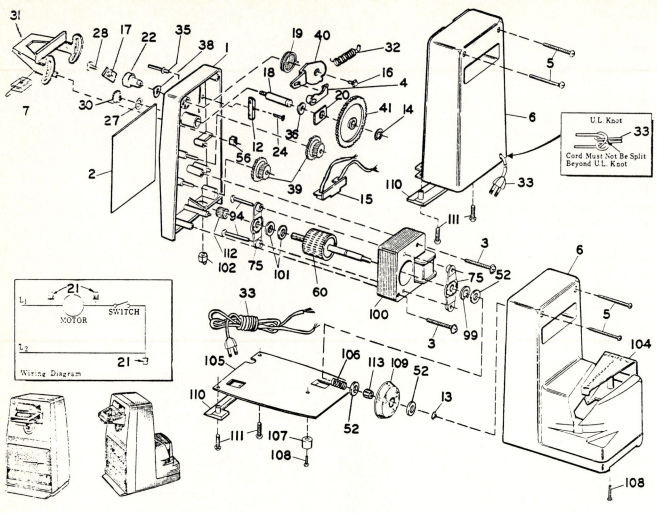

Index No.	Part No.	Description	Quan. Req'd.
1	201127	Front Housing Assembly - Chrome	1
2	200462	Nameplate - "Touch-A-Matic Imperial". .	1
3	63455	Screw - #6-32 x 1 5/16 - Motor to Front Housing	2
4	69624	Link - Plastic - For Cam Index #40. . .	1
5	76597	Screw - Upper Housing	2
6	75082	Housing - Rear - White.	1
7	67638	Magnet Assembly - For Lid Lifter. . . .	1
12	69625	Pivot Block - Lid Lifter.	2
14	74150	Retaining Ring.	1
15	69187	Switch Assembly	1
16	69163	Screw - #10-24 - For Blade Adjustment	1
17	69064	Cutter Blade.	1
18	69830	Shaft - L.H. Thread - For Index #30 . .	1
19	200209	Spring - Lever Return	1
20	69060	Clip - For Drive Gear Index #41	1
21	60346	Connector	2
22	69166	Shaft - For Blade Index #17	1
24	474600	Screw - #4-40 x 5/8 - For Index #12 . .	2
27	71673	Washer - Behind Drive Wheel	A.R.
28	69162	Screw - #10-24 - For Blade.	1
30	75658	Drive Wheel - L.H. Thread	1
31	67479	Lid Lifter	1
32	69724	Spring - For Cam Index #40.	1
33	72676	Cord & Plug Assembly - White.	1
35	200736	Pin - Can Guide	1

FIGURE 29-3. (Continued)

366

brushing the disk with a stiff-bristled toothbrush.

Incorrect adjustment of the guide bushing and abrasive disk could also affect the pencil sharpening time and quality. These should be checked if the disk seems alright.

The trouble shooting chart is for your guidance in fault location. Specific details on parts replacement are contained in the manufacturers' service manuals relating to the make and model number of the machine being serviced.

Originally, can openers and knife sharpeners were separate appliances. Now they are mostly built as a single combination unit and can be treated as such.

The major problems are cutter wheels and grinding wheels needing replacement. See Fig. 29-4 for their location and approach to changing.

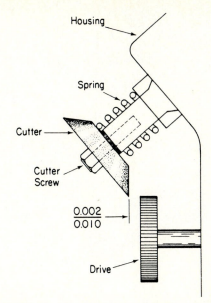

FIGURE 29-4. The spring may be a spring-type washer on some models. The cutter *must* be free to turn, but must not wobble on its shaft. When replacing the cutter, make certain that the bevel is on the inside.

Troubleshooting Chart **KNIFE SHARPENERS**

Problem	Possible Cause	Corrective Action
Motor does not run.	No power at outlet. Defective cord. Defective switch. Defective motor coil.	Replace fuse. Repair or replace cord. Replace switch. Replace field winding.
Motor hums, but wheels do not turn.	Armature rotor or shaft bound. Jammed grinding wheel. Excessive pressure while sharpening.	Realign field assembly. Lubricate bearings. Clear from bind. Caution user on light touch.
Motor is slow.	Dry bearings. Bound gears.	Relubricate. Check clearance. (If too loose, may strip gears.)
Machine is noisy.	Worn or dry bearing. Loose or worn gears. Foreign material in case.	Relubricate or replace. Replace or adjust mesh. Remove foreign object.

CAN OPENERS

When electric can openers (Fig. 29-5) first appeared on the market, many old-timers commented on the decadence of civilization and the ultimate in housewifely laziness. But for one who has cut his hands on sharp, ragged can lids and fished slivers of metal out of the contents, the neat, clean, rolled, out-of-the-way can edges resulting from the current can opener design is a real safety feature.

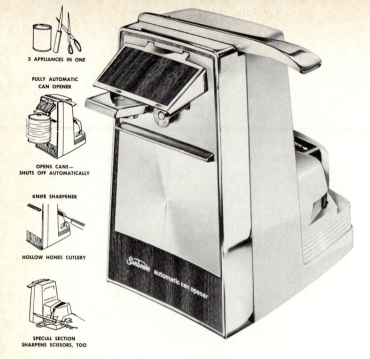

3 APPLIANCES IN ONE

FULLY AUTOMATIC
CAN OPENER

OPENS CANS—
SHUTS OFF AUTOMATICALLY

KNIFE SHARPENER

HOLLOW HONES CUTLERY

SPECIAL SECTION
SHARPENS SCISSORS, TOO

FIGURE 29-5. Sunbeam Model CS3. The modern can opener is three appliances in one: can opener, knife sharpener, and scissors sharpener. (Courtesy of Sunbeam)

FIGURE 29-6. Most can openers start automatically when the handle is pressed down to puncture the lid. The magnet keeps the lid from falling into the can. (It does not work on aluminum lids.)

FIGURE 29-7. The cutter wheel is readily removable for cleaning or replacement. Suspect this to be dull when the cutting action is poor.

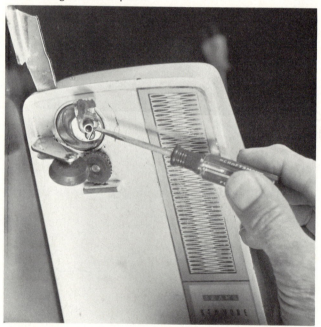

Basic design is such that a motor-driven wheel, with serrated edges, clamps under the can rim, forcing the cutter wheel to puncture the can top (Fig. 29-6). Then the can is rotated while the cutter wheel does its work. Note that it is the wheel under the rim that rotates the can, not the cutter wheel.

The cutter wheel is readily removable so that it and its shaft can be cleaned easily. It does accumulate foodstuffs and should be cleaned regularly (Fig. 29-7). If it is not cleaned, it will soon bind on its shaft, then not turn, thus increasing the motor loading and shortening its life. When reassembling the cutter wheel lubricate its shaft lightly with vegetable or mineral oil.

Adjustment of the cutter wheel should be such that the clearance between it and the driver will be between 0.002 and 0.010 inch when in operating position. Correction to this clearance is made by changing the driver-wheel spacer washers.

Motor switching is usually automatic so that operation of the clamping handle will activate a switch after the can has been punctured. Then the can will rotate to be cut. Some Sunbeam models have an automatic shutoff switch that turns the motor off when the lid is cut.

Some of the Sunbeam models have a knife sharpener included in the can opener stand which is started by pressing down the same control arm that operates the can opener. The sharpening information that is included in the knife sharpener section applies here, thus will not be repeated.

The motors used in can openers may be either the shaded pole or the series style. Some of the series motors are speed controlled by a governor (Figs. 29-8 and 29-9). All use gear reduction to reduce the motor speed to a safe drive wheel speed. Too fast a drive wheel speed will cause fast rotation of the can and spillage of its contents.

The drive wheel speed at no load will vary by make, and the model can be checked by a tachometer. Table 29-1 lists some typical speeds.

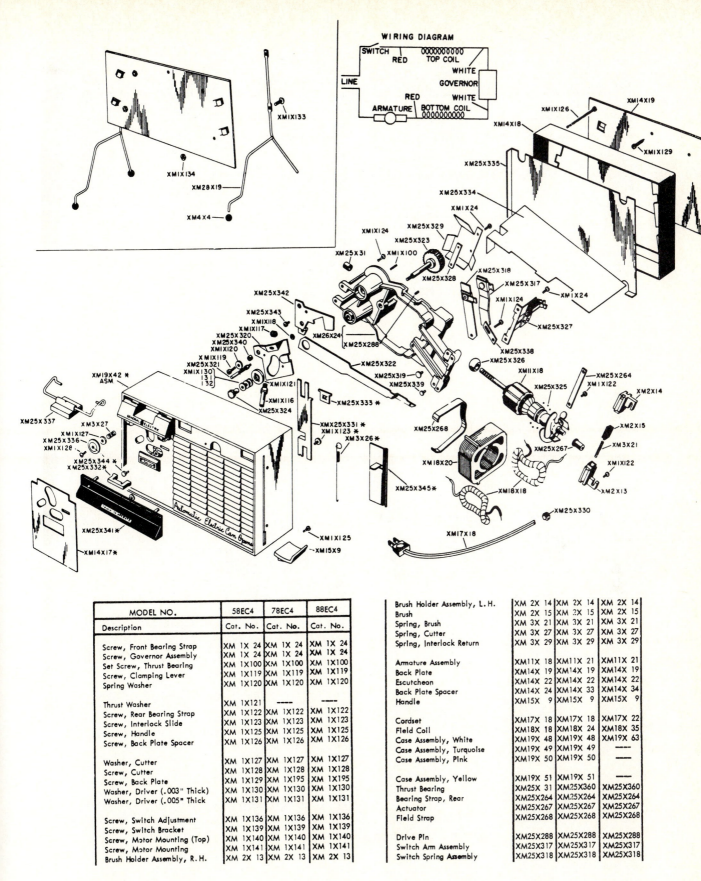

MODEL NO.	58EC4	78EC4	88EC4
Description	Cat. No.	Cat. No.	Cat. No.
Screw, Front Bearing Strap	XM 1X 24	XM 1X 24	XM 1X 24
Screw, Governor Assembly	XM 1X 24	XM 1X 24	XM 1X 24
Set Screw, Thrust Bearing	XM 1X100	XM 1X100	XM 1X100
Screw, Clamping Lever	XM 1X119	XM 1X119	XM 1X119
Spring Washer	XM 1X120	XM 1X120	XM 1X120
Thrust Washer	XM 1X121	----	----
Screw, Rear Bearing Strap	XM 1X122	XM 1X122	XM 1X122
Screw, Interlock Slide	XM 1X123	XM 1X123	XM 1X123
Screw, Handle	XM 1X125	XM 1X125	XM 1X125
Screw, Back Plate Spacer	XM 1X126	XM 1X126	XM 1X126
Washer, Cutter	XM 1X127	XM 1X127	XM 1X127
Screw, Cutter	XM 1X128	XM 1X128	XM 1X128
Screw, Back Plate	XM 1X129	XM 1X195	XM 1X195
Washer, Driver (.003" Thick)	XM 1X130	XM 1X130	XM 1X130
Washer, Driver (.005" Thick)	XM 1X131	XM 1X131	XM 1X131
Screw, Switch Adjustment	XM 1X136	XM 1X136	XM 1X136
Screw, Switch Bracket	XM 1X139	XM 1X139	XM 1X139
Screw, Motor Mounting (Top)	XM 1X140	XM 1X140	XM 1X140
Screw, Motor Mounting	XM 1X141	XM 1X141	XM 1X141
Brush Holder Assembly, R.H.	XM 2X 13	XM 2X 13	XM 2X 13
Brush Holder Assembly, L.H.	XM 2X 14	XM 2X 14	XM 2X 14
Brush	XM 2X 15	XM 2X 15	XM 2X 15
Spring, Brush	XM 3X 21	XM 3X 21	XM 3X 21
Spring, Cutter	XM 3X 27	XM 3X 27	XM 3X 27
Spring, Interlock Return	XM 3X 29	XM 3X 29	XM 3X 29
Armature Assembly	XM11X 18	XM11X 21	XM11X 21
Back Plate	XM14X 19	XM14X 19	XM14X 19
Escutcheon	XM14X 22	XM14X 22	XM14X 22
Back Plate Spacer	XM14X 24	XM14X 33	XM14X 34
Handle	XM15X 9	XM15X 9	XM15X 9
Cordset	XM17X 18	XM17X 18	XM17X 22
Field Coil	XM18X 18	XM18X 24	XM18X 35
Case Assembly, White	XM19X 48	XM19X 48	XM19X 63
Case Assembly, Turquoise	XM19X 49	XM19X 49	----
Case Assembly, Pink	XM19X 50	XM19X 50	----
Case Assembly, Yellow	XM19X 51	XM19X 51	----
Thrust Bearing	XM25X 31	XM25X360	XM25X360
Bearing Strap, Rear	XM25X264	XM25X264	XM25X264
Actuator	XM25X267	XM25X267	XM25X267
Field Strap	XM25X268	XM25X268	XM25X268
Drive Pin	XM25X288	XM25X288	XM25X288
Switch Arm Assembly	XM25X317	XM25X317	XM25X317
Switch Spring Assembly	XM25X318	XM25X318	XM25X318

FIGURE 29-8. Can openers having a series motor generally reduce speed through a worm drive from the armature shaft. This exploded view shows a good example of the construction. (Courtesy of General Electric)

MODEL NO.	58EC4	78EC4	88EC4
Description	Cat. No.	Cat. No.	Cat. No.
Rivet, Mounting Strap	XM25X319	XM25X319	----
Rivet, Switch Assembly	XM25X319	XM25X319	XM25X319
Driver Shaft Assembly	XM25X323	XM25X323	XM25X323
Bearing	XM25X326	XM25X361	XM25X361
Governor Assembly	XM25X327	XM25X327	XM25X327
Bearing Strap, Front	XM25X328	XM25X328	XM25X328
Grommet	XM25X330	XM25X330	XM25X402
Interlock Slide	XM25X331	XM25X331	XM25X331
Bumper	XM25X332	XM25X332	XM25X332
Switch Bar Hinge	XM25X333	XM25X333	XM25X333
Motor Guard	XM25X335	XM25X335	XM25X335
Cutter	XM25X336	XM25X336	XM25X336
Magnet Frame Assembly	XM25X337	XM25X337	XM25X337
Mounting Strap	XM25X338	XM25X338	----
Bushing, Clamping Lever	XM25X340	XM25X340	XM25X340
Motor Mounting Plate	XM25X342	XM25X342	XM25X342
Rivet, Motor Mounting Plate	XM25X343	XM25X343	XM25X343
Rivet, Case Support	XM25X344	XM25X344	XM25X344
Switch Bracket Assembly	XM25X351	XM25X351	XM25X351
Cutter Plate Assembly	XM25X352	XM25X352	XM25X352
Clamping Lever	XM25X353	XM25X353	XM25X353
Driver	XM25X354	XM25X354	XM25X354
Switch Bar Assembly, Red	XM25X355	----	----
Stuffing Box Liner	XM25X356	XM25X356	----
Case Support Assembly	XM25X366	XM25X366	XM25X366
Switch Bar Assembly, White	XM25X369	----	----
Motor Frame Assembly	XM26X 25	XM26X 25	XM26X 25
Washer, Driver		XM 1X191	XM 1X191
Nut, Driver		XM 1X192	XM 1X192
Lead, 4 1/2"		XM25X440	XM25X440
Connector			XM25X238
Lead, 2 5/8"			XM25X450

FIGURE 29-8. (Continued)

Key No.	Part No.	Description
1	4817	Motor, Complete
2	9392	Rear Motor Bracket
3	3248	Rear Motor Bracket Screw
4	22A1-5	Fiber Washer
5	22A1-5	Fiber Washer
6	22A1-4	Spacer Washer
7	3225	Rotor
8	2210-F	Field Coil Assembly
9	9393	Front Bracket
10	22A1C-8	Compound Gear (477)
11	3004	Drive Gear
12	8031	Front Plate w/Can Guide
13	8036	Handle
14	8942	Carrier
15	3016	Carrier Screw
16	3008	Cutter
17	3010	Cutter Spring
18	3009	Cutter Screw
19	8923	Lid Lifter w/Magnet
20	3011	Sprocket
21	3016	Carrier Screw
22	4961	Plunger
23	4964	Switch Spring
24	4974	Switch Screw, 6-32 x 1/4"
25	4820	Switch
26	22A1S-12	Solderless Connector (4991)
27	4977	Strain Relief
28	9279	Snap Bushing
29	4954	Screw for Wire Clip, 6-32 x 3/8"
30	4978	Wire Clip
31	4993	Motor Mount Screw
32	8049	Handle Spring
33	8456	Rear Housing
34	8554	Rear Housing Screw, 8-32 x 3"
35	4954	Screw for Pads, 6-32 x 3/8"
36	22A1S-8	Foot Pad
37	8556	Bottom Plate
38	9134	Cord and Plug
39	4957	Rear Housing Screw, 6-32 x 1/2"
40	2202	Decal

FIGURE 29-9. Parts list.

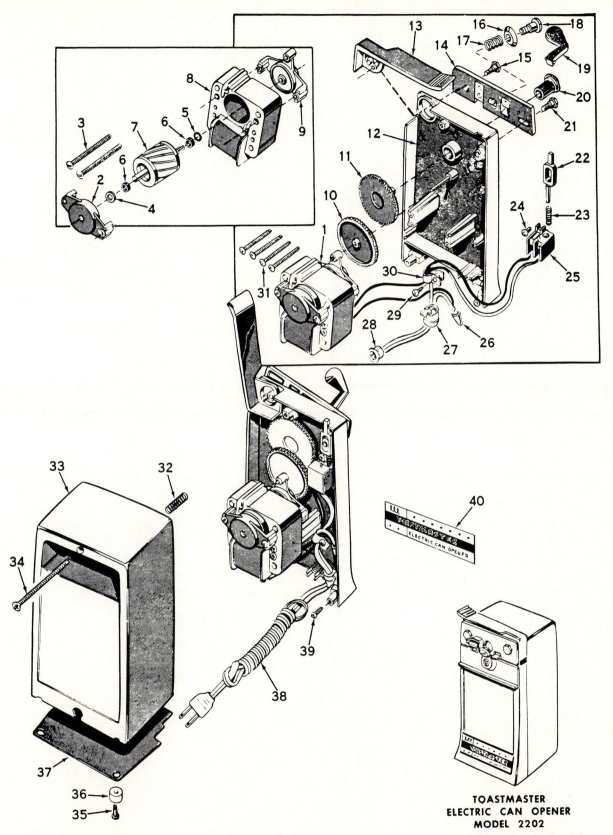

TOASTMASTER
ELECTRIC CAN OPENER
MODEL 2202

FIGURE 29-9. Can openers having a shaded-pole motor usually reduce speed through a series of gears. This exploded view of a Toastmaster is a good example of this construction. (Courtesy of Toastmaster)

TABLE 29-1

Drive-wheel speeds at no load

Make and Model	Speed (rpm)	Motor
Sunbeam DCO	34–40	Shaded pole
Sunbeam HCO	105 ± 12	Series
General Electric EC4B	175 ± 25	Series (governor)
General Electric 98 EC4B	350	Series (no governor)

Troubleshooting Chart   CAN OPENERS

Problem	Possible Cause	Corrective Action
Motor does not run.	No power at outlet.	Check fuse.
	Defective cord.	Replace.
	Switch not closing.	Repair or replace.
	Open circuit in motor.	Repair or replace.
	Armature bound.	Free bind.
	Gear train bound.	Clear jam-up.
Motor does not shut off.	Switch not opening.	Adjust switch.
	Short across switch.	Clear circuit.
Motor runs but lacks power.	Drive gear may be binding.	Oil bearing. Clean shaft.
	Poor brush contact.	Replace brushes and clean commutator.
	Defective armature.	Replace armature.
	Dull or damaged cutter wheel.	Replace cutter wheel.
Can does not turn.	Driver jammed.	Clean and oil shaft.
	Driver teeth worn.	Replace driver.
	Driver teeth clogged.	Clean teeth.
	Driver does not turn.	Tighten or replace gear.
Drive wheel runs off can.	Loose or weak spring.	Tighten or replace spring.
	Cutter not pivoting freely.	Clean and lubricate shaft.

30

Microwave Ovens

Microwaves are radio waves in the UHF classification. They are higher in frequency than the highest-frequency TV channel, number 83, whose range is 884 to 890 MHz (megahertz). The Federal Communication Commission has assigned 2450 megahertz for microwave heating applications.

THEORY

It is not necessary to know the frequency generated by the Magnetron tube but you should know that these radio waves are usually conducted by a wave guide to the oven cavity. Once inside the oven cavity, they are reflected from the metal walls to the food, which absorbs the energy in the radio waves. This energy is converted to heat by the molecules of food trying to align themselves with the fast-moving radio waves. Actually, with a frequency of 2450 megahertz the food molecules are changing direction four billion nine million times a second, and a great deal of molecular friction is developed. Friction generates the heat to cook the food both inside and outside at the same time. The penetration is about 2 inches.

BROWNING

It should be noted here that although the food is completely cooked and ready to serve, some foods need further treatment in order to make them more appealing to the diner—specifically, meats and seafood, which ordinarily have a seared, browned, or broiled surface finish when cooked in regular ovens.

Browning is accomplished by either prebroiling the meat before completing the cooking in the microwave oven or by cooking in the oven, then browning to the desired appearance.

Some models include a browning element with its own controls in the microwave oven.

WHAT GETS HOT AND WHAT STAYS COOL

Whether microwaves are reflected or absorbed depends on several factors:

1. The dielectric coefficient of the material
2. Its shape
3. Its mass
4. Its moisture content

Generally metals reflect the wave energy. Glass, paper, and most plastics allow the waves to pass through them without absorbing the energy.

For this reason, the oven and glass or paper plates remain cool and only the food gets cooked. Since the food cooks all the way through at once, it can be placed in the oven and can be ready to serve in seconds—but don't try it with TV dinners in the aluminum foil wrapper and plate. *Never use metal containers for the food in the oven.* They will reflect the radio waves away from the food. *Always use glass, plastic, or paper.*

Never operate the oven without some load—the energy has to be absorbed. A test load need be only a cup of coffee, *but* the load must contain a liquid or fat since it is the liquid molecules that readily convert the energy. Failure to have a proper load can damage the Magnetron, which is one of the most expensive parts of the unit.

SAFETY FEATURES

Microwaves are no respecters of persons. They would just as soon cook a hand or an arm as a steak. To prevent injury to the user, manufacturers have incorporated a number of safety interlock switches that make it virtually impossible for the user or serviceman to get hurt or for the oven to get seriously damaged by careless usage.

In addition to user safety, the manufacturers have done much to contain the radio waves *inside* the oven. The glass window has a built-in screen to prevent the radio energy from escaping, and the door gasket must be conductive all around the door to complete the shielding. Whenever any work is done on door or hinge adjustment, always check around the gasket with a strip of paper (a dollar bill is good). There must be a noticeable drag on the paper when you pull it from under the gasket.

SAFETY NOTES

1. Always make certain that the main OFF/ON switch is OFF and that power to the oven has been disconnected before reaching into any access openings.
2. Do not tamper with the interlocks; the high voltages in the power pack are dangerous.
3. Discharge the large capacitor by shorting its terminals before reaching into the rear of the oven.
4. Have the wiring diagram and service diagnosis procedure on hand for the particular make and model number of the oven you are working with and *follow it exactly.*
5. IF you *cannot* read or *do not* understand the diagram, *do not* attempt to service the oven. Call in someone with more experience.

INSTALLATION

The microwave oven with electronic controls, as with other electronically controlled appliances, must be connected to a correctly wired outlet. This should be a separately fused 20-ampere circuit correctly grounded and polarized. Figure 30-1 illustrates a procedure for checking polarization with a voltmeter.

Installation should provide adequate power, security, and space for the oven (Fig. 30-2). Frequently, the electronic oven is part of the range or another built-in oven. A quick check on the operation of the rest of the range will indicate whether power is available to the unit.

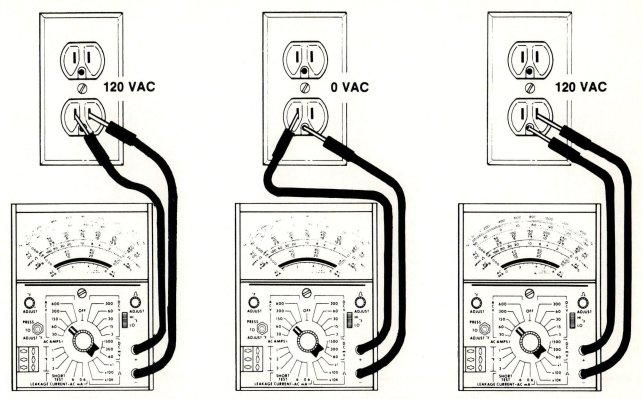

FIGURE 30-1. The electrical outlet should be checked for compliance to the National Electrical Code® standards at the time of installation. On a service call, if you are not absolutely certain of its condition, check the outlet for correct wiring. (Courtesy of Whirlpool)

SERVICE

If everything else is okay, the trouble must be in the oven circuitry. The microcomputer circuitry is very sensitive to transient voltages, especially static. The service technician is advised to "ground" himself on an exposed metal part of the oven in order to drain off any accumulated static (Fig. 30-3).

When servicing the microwave, don't be too

FIGURE 30-2. A microwave oven may be small and can be conveniently set on a kitchen counter.

FIGURE 30-3. After the service technician has discharged his own static and discharged the high-voltage capacitor, he can begin to search for the trouble in the unit.

quick to blame an electronic control unit. It is much more complex and performs many more duties than the simple clock timer, which is easy to test and replace. Usually, a wiring diagram for the unit being serviced is included somewhere in or on the back of the unit.

The control module shown in Fig. 30-4 contains both the touch panel (Fig. 30-5a) and the microprocessor unit (Fig. 30-5b). It is not serviceable and must be replaced as a complete unit. This is one model. Others may have a touch panel, electronic control panel, and their power supply as separate units, each independently replaceable. None of the electronic control units are repairable in the field (Figs. 30-6 to 30-8).

FIGURE 30-4. Location of the control module in respect to the high-voltage power supply, magnetron, and fan.

FIGURE 30-5. (a) Control module removed to show the touch pad; (b) rear view of control unit.

(a)

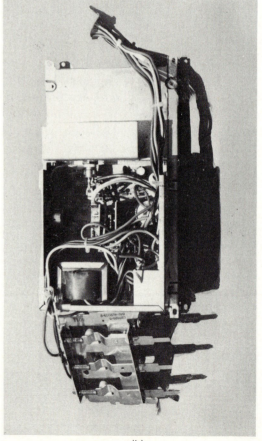

(b)

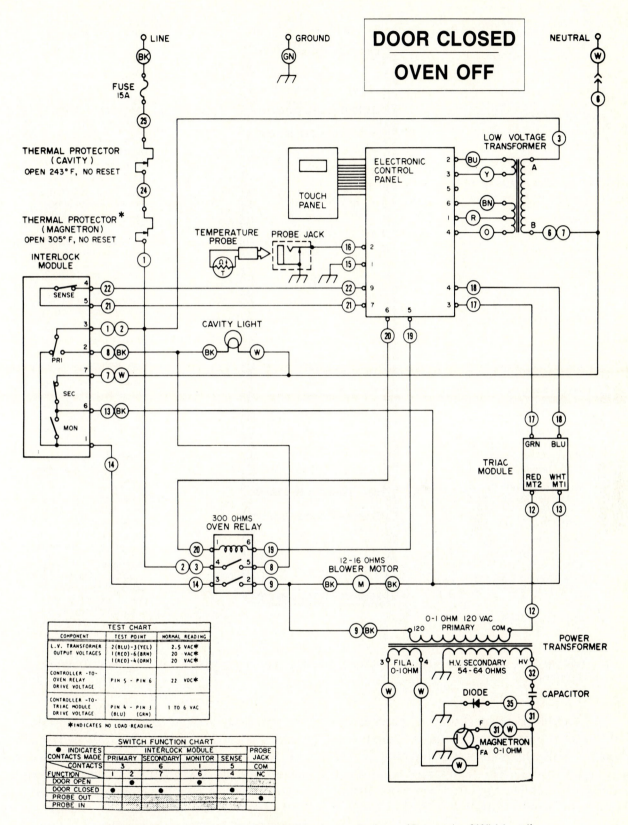

FIGURE 30-6.   Diagram for oven with electronic timer. (Courtesy of Whirlpool)

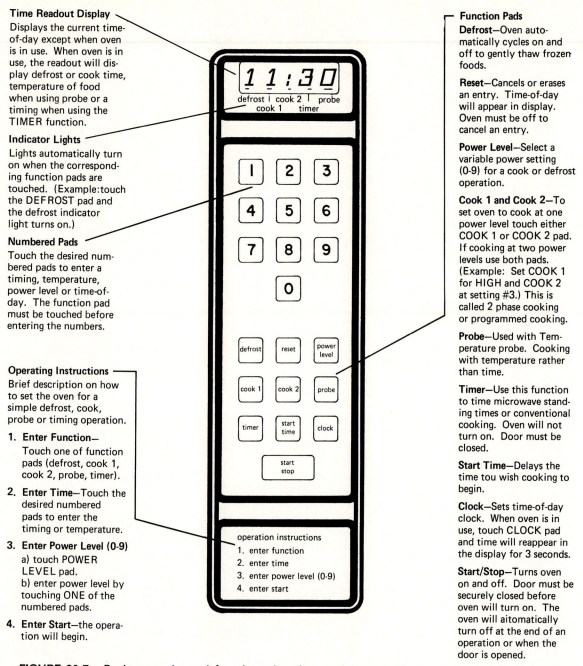

Time Readout Display
Displays the current time-of-day except when oven is in use. When oven is in use, the readout will display defrost or cook time, temperature of food when using probe or a timing when using the TIMER function.

Indicator Lights
Lights automatically turn on when the corresponding function pads are touched. (Example:touch the DEFROST pad and the defrost indicator light turns on.)

Numbered Pads
Touch the desired numbered pads to enter a timing, temperature, power level or time-of-day. The function pad must be touched before entering the numbers.

Operating Instructions
Brief description on how to set the oven for a simple defrost, cook, probe or timing operation.

1. Enter Function—
Touch one of function pads (defrost, cook 1, cook 2, probe, timer).

2. Enter Time—Touch the desired numbered pads to enter the timing or temperature.

3. Enter Power Level (0-9)
a) touch POWER LEVEL pad.
b) enter power level by touching ONE of the numbered pads.

4. Enter Start—the operation will begin.

Function Pads
Defrost—Oven automatically cycles on and off to gently thaw frozen foods.

Reset—Cancels or erases an entry. Time-of-day will appear in display. Oven must be off to cancel an entry.

Power Level—Select a variable power setting (0-9) for a cook or defrost operation.

Cook 1 and Cook 2—To set oven to cook at one power level touch either COOK 1 or COOK 2 pad. If cooking at two power levels use both pads. (Example: Set COOK 1 for HIGH and COOK 2 at setting #3.) This is called 2 phase cooking or programmed cooking.

Probe—Used with Temperature probe. Cooking with temperature rather than time.

Timer—Use this function to time microwave standing times or conventional cooking. Oven will not turn on. Door must be closed.

Start Time—Delays the time tou wish cooking to begin.

Clock—Sets time-of-day clock. When oven is in use, touch CLOCK pad and time will reappear in the display for 3 seconds.

Start/Stop—Turns oven on and off. Door must be securely closed before oven will turn on. The oven will aitomatically turn off at the end of an operation or when the door is opened.

FIGURE 30-7. Basic control panel functions, location, and usage will vary with models. (Courtesy of Whirlpool)

All troubleshooting should be preceded by a discussion with the user as to "what happened." Often, this will serve as a clue to start looking for trouble. The following troubleshooting chart is a general guide. If service information for the oven you are servicing is available, *use it.*

After service work is completed and the oven is operative, one must be certain to check the oven for microwave energy leakage. *Check*

particularly around the door. Use an approved meter such as the *Holaday 1500 Survey Meter* (Fig. 30-9). Make sure that the microwave leakage does not exceed 5 megawatts per square centimeter at 2 inches from the oven. As with all other electrical appliances, the microwave oven must be tested for electrical leakage from an exposed metal part of the frame to a *good* earth ground. Leakage must not exceed 2 milliamperes.

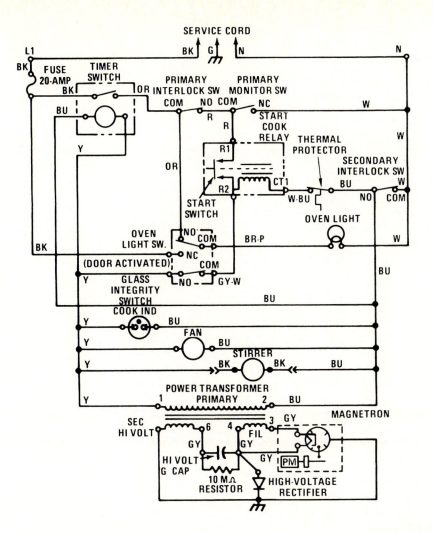

FIGURE 30-8. Diagram for oven with clock timer. (Courtesy of Whirlpool)

FIGURE 30-9. This Holaday 1500 Survey Meter is approved by the Bureau of Radiological Health. The test is made in compliance with the HEW No. 21CRF278 performance standard for microwave ovens.

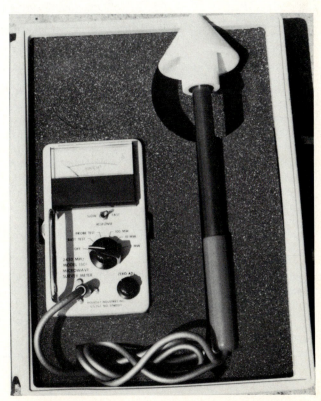

Troubleshooting Chart **MICROWAVE OVENS**

When troubleshooting a microwave oven, it is helpful to follow the sequence of operation in performing the checks. **Important:** If the oven becomes inoperative because of internal fusing, the three interlock switches must be checked.

Problem	Possible Cause	Corrective Action
Off Condition		
Line fuse blows when power cord is plugged into wall receptacle.	Shorted wire in power cord or wire harness or overloaded circuit.	Replace cord or check wiring. Put microwave oven on separate circuit.
Idle Condition		
Oven completely inoperative with controls set.	No power from service entrance. Open wire in power cord or wiring harness. Defective controls.	Check wall outlet. Replace same or repair wiring. Check touch panel, microcomputer, timer. Replace if necessary.
	Blown fuse. Defective thermal fuse.	Replace fuse. Replace thermal fuse.
Oven cavity light does not illuminate (models so equipped).	Burned-out bulb. Defective timer contacts or controls. Defective lamp socket. Defective interlock switch. Open wiring between the above components. Defective thermal fuse.	Replace. Repair or replace. Replace. Repair or replace. Check wiring. Replace.
Cooking Condition		
Oven does not go into cook cycle when door is closed or start pad is activated (timer or controls turned to Selective Time setting).	Defective contacts on timer switch or controls. Defective thermal fuse. Interlock switches defective. Defective oven relay. Open wiring between the above components. Defective power selector. Defective touch panel. Defective probe jack assembly.	Repair or replace. Repair or replace. Replace. Replace. Check wiring. Replace. Check touch panel and replace if necessary. See "Probe Jack Assembly Test."
Oven goes into cook cycle but does not complete cycle. Heat produced in oven load, but cook light does not illuminate.	Defective timer motor or controls. Open wiring in circuit to timer motor or controls. Low line voltage (should be at least 110 volts ac).	Replace. Check wiring. Check voltage with microwave oven in operation.

Troubleshooting Chart **MICROWAVE OVENS** (Continued)

Problem	Possible Cause	Corrective Action
Oven cooking light indicates cycle, but little or no heat is produced in oven load.	Shorted high-voltage circuit between voltage-doubler circuit and magnetron.	Repair or replace.
	Defective power transformer.	Repair or replace.
	Defective diode.	Repair or replace.
	Defective high-voltage capacitor.	Repair or replace.
	Defective Magnetron.	Repair or replace.
	Power selector defective (models so equipped).	Replace power selector.
Stirrer blades do not turn at top of oven cavity during cook cycle, or oven goes into cook cycle, but extreme uneven heating is produced in oven load.	Defective blower motor.	Check motor—replace if defective.
	Stirrer blades binding.	Replace stirrer blades or repair source of binding.
	Broken or loose wire connection.	Check wiring, repair if defective.
	Blocked air passage.	Check air passages for obstructions.
Oven fuse blows when door is opened.	Primary interlock switch, or secondary interlock switch defective.	Replace interlock module.
	Shorted harness wire.	Repair wiring.
Cavity lamp illuminates with door open, but light goes out when door is CLOSED with controls ON. Oven will not go into cook cycle when start switch is pushed.	Defective timer contacts (on models so equipped).	Check timer. Replace if defective.
	Broken or loose wire connection.	Repair wiring if defective.
	Primary interlock switch defective.	Replace interlock module.
	Oven relay defective.	Replace oven relay.
Oven goes into cook cycle but shuts down before end of cycle.	Thermal fuse opened.	Replace.
	Circuit overloaded.	Place microwave on separate 15-ampere circuit.
Power source fuse blows when the cook switch is depressed.	Defective power transformer.	Replace.
	Secondary circuit of power transformer is shorted.	Check wiring.
	High-voltage capacitor shorted.	Replace.
	Blower motor winding shorted.	Replace if defective.
	Cook light shorted.	Replace.
	Shorted wiring between above components.	Check wiring.
	Defective blower motor.	Check motor—replace if defective.

Troubleshooting Chart MICROWAVE OVENS   (Continued)

Problem	Possible Cause	Corrective Action
Blower motor inoperative.	Defective blower motor. Open or loose wiring in circuit to blower motor.	Replace if defective. Check wiring.
Cycle selected—unit starts without pushing start pad (electronic model only).	Start switch failed in closed position. Oven relay contacts failed in closed position. Miswired.	Replace touch panel. Replace oven relay. Correct wiring.
No power selection (high power only).	Defective triac (on models so equipped). Defective power selector. Miswired.	Replace. Remove wires and check for fused contacts in pulser timer. Check wiring; correct if miswired.
No power selection (low power only).	Defective selector. Miswired.	Check pulser timer for contacts that remain open. Check wiring. Correct if miswired.
Power selector will not stay in selected position.	Defective power selector.	Replace power selector.
Oven heats too fast.	Line voltage too high. *Note:* Check variable power control out to see if it is working properly (on models so equipped).	Line voltage should be 105 to 130 volts.
Oven will not heat.	Defective variable power control (on models so equipped).	Check the operation of the variable power control using one of the following methods: (a) Set variable power control to HIGH setting. Set timer to 5.0 minutes. Put 1 cup of water (at 60 to 70°F (16 to 21°C) initial temperature) in the center of the oven. Close door and start oven. Water should be boiling vigorously just before oven turns off. *or* (b) Put 1 cup of water in the oven to act as a load and a small fluorescent lamp (Whirlpool Part 21914). Set the power control to HIGH, set time on timer, close door, and start oven. The fluorescent lamp should

Troubleshooting Chart **MICROWAVE OVENS** (*Continued*)

Problem	Possible Cause	Corrective Action
		light and remain lit continuously while the oven is running. (Some flickering of light is normal.) If the oven seems to be receiving full power as outlined above, instruct the user regarding proper cooking times.
Oven cooks too slowly.	Low line voltage. Miswired.	The line voltage should be between 105 and 130 volts. Check wiring and correct if needed. *Note:* Be sure to discharge the high-voltage capacitor before checking the wiring.
Oven goes into cook cycle but end-of-cycle buzzer sounds and unit shuts down.	Defective temperature control or buzzer.	Replace microcomputer assembly.
Temperature control operates normally, but fails to shut off (on models so equipped).	Defective temperature control. Ground wire integrity not intact.	Replace. Check wiring.
Temperature control operates but end-of-cycle buzzer fails to sound and unit does not shut off (on models so equipped).	Defective temperature control.	Replace microcomputer assembly.
Oven goes into temperature control cycle, operates, but temperature indicator fails to move (on models so equipped).	Defective temperature control. Defective temperature probe. Defective probe jack assembly. Loose or broken wiring.	Change microcomputer assembly. Replace probe. Replace. Check wiring.
Oven light and probe indicator fail to light when probe is plugged in (on models so equipped).	Defective probe jack assembly.	Replace.
Oven does not go into cook cycle when probe is plugged in (on models so equipped).	Defective probe jack assembly, if oven operates normally in time cook cycle.	Replace.

Troubleshooting Chart   ELECTRONIC CONTROL MODELS

Problem	Possible Cause	Corrective Action
	Idle Condition	
Clock display is not blank after power-on reset.	Open or incorrect wiring.	Check and correct wiring.
	Unplugged or incorrectly aligned clock display.	Make sure that clock display is properly aligned and seated.
	Defective power supply assembly.	Replace low-voltage transformer.
	Broken or missing terminal on micro-computer.	Replace microcomputer assembly.
Control will not accept commands.	Unplugged or damaged touch panel ribbon cable.	Check touch panel. Repair ribbon cable as necessary.
	Service cord is not connected to proper supply voltage.	Check and correct service voltage.
	Chassis is not grounded properly.	Chassis must be grounded.
	Power supply voltages are out of specification limits.	Check low-voltage transformer voltages and replace low-voltage transformer if necessary.
Control does not recognize commands correctly.	Improper supply voltages.	Replace microcomputer assembly.
	Operating Condition	
Oven does not go into cooking cycle when START pad is touched or door is closed; Magnetron and blower motor do not start.	Control is not programmed properly.	Instruct customer.
	Misadjusted or inoperative interlock switches.	Check interlock switches and adjustments. Adjust or replace defective switches as needed.
	Open in wiring harness.	Check wiring and correct.
	Defective or miswired oven relay.	Check wiring and relay. Replace oven relay only if necessary.
	Inoperative driver circuitry in microcomputer assembly.	Check microcomputer assembly for proper voltages and replace if necessary.
Oven lamp lights and blower motor starts, but oven does not heat.	Open miswired, or defective power transformer.	Check wiring and transformer. Replace transformer if necessary.
	Shorted diode.	Check and replace.
	Defective triac or faulty connections to triac.	Check connections and triac. Replace triac if necessary.
	Inoperative driver circuitry in microcomputer assembly.	Check microcomputer assembly for proper voltages. Replace if necessary.
Oven cooks, but display does not countdown in cooking cycle.	Miswired or defective wiring harness.	Check harness connection.
	Defective microcomputer assembly.	Replace microcomputer assembly.

Troubleshooting Chart **ELECTRONIC CONTROL MODELS** *(Continued)*

Problem	Possible Cause	Corrective Action
Oven cooks, but turns off before end of cycle. Display randomly shows information other than remaining cook time.	Transient voltages into electronic circuitry.	Dress low-voltage harness wiring away from high-voltage circuity.
Oven stops cooking, but control continues counting with door open.	Incorrectly wired, misadjusted, or defective door switch. Defective microcomputer assembly.	Check wiring and door switch. Adjust, repair, or replace as necessary. Replace microcomputer assembly.
Speaker does not sound at conclusion of cooking sequences or after elapse of time in minute timer.	Open connection or defective speaker.	Replace microcomputer assembly.
Oven terminates prematurely or will not start in COOK TEMP. cycle.	Probe switch inoperative or miswired. Incorrect or defective probe. User not using probe correctly. Inoperative temperature control circuitry on microcomputer assembly.	Check probe switch and wiring. Replace jack if necessary. Check probe resistance. Instruct user. Check microcomputer assembly and replace if necessary.
Oven will not cook at less than full power. Display shows variable power, but oven heats too fast.	Improperly wired high-voltage circuitry. Shorted or defective triac. Defective relay driver circuitry in microcomputer assembly.	Correct wiring. Check triac relay and replace if necessary. Check and replace microcomputer assembly if necessary.
Power source fuse blows when cooking starts.	Defective power transformer. Secondary of power transformer is shorted. High-voltage capacitor shorted. Blower motor winding shorted. Timer motor winding shorted. Inadequate source or overloaded circuit.	Check and replace if necessary. Repair short or replace shorted component. Check and replace if necessary. Check and replace if necessary. Check and replace if necessary. Operate on separate 15-ampere circuit.
Oven is noisy, makes thumping sound or buzz.	Surge resistor not wired properly in circuit. Loose laminations low voltage or power transformers.	Check wiring and correct if necessary. Replace noisy component.
Blower motor will not run or turns off intermittently during variable-power cooking.	Defective blower motor. Open or loose wire. Insufficient delay time in control relay drive circuitry in microcomputer assembly.	Replace if necessary. Check voltage or blower terminals. Correct wiring if necessary. Check microcomputer assembly. Replace if necessary.

31

Ranges and Ovens

ELECTRIC RANGES

One of the most heated controversies in the appliance field, has been on the merits of gas versus electric ranges. Each has fine features. Frequently the final choice by the homeowner is influenced by a personal preference that is stronger than the actual difference between the two types of stoves.

Some characteristics of the electric range, such as slow heat-up and slow cool-off, are regarded as being an advantage or a disadvantage, depending on one's point of view. Similarly, the gas stove is alternately pointed to with pride and viewed with alarm. It has, however, been wisely pointed out that a meal cooked on one type is just as good as one cooked on the other. For the repairman, each type represents a source of income.

The electric range can be regarded as a hot plate that has become sophisticated. Basically, the operating principle is simple: heating elements are switched on and off. It is the ensemble of gadgets that makes the difference—

lights, timers, temperature sensors, and all the rest of the accessories that go into the modern range. An overall look at a range circuit diagram is usually a sure way to quick confusion, but circuit by circuit it is relatively simple (Fig. 31-1).

The range-heating element is the product of many years of design effort. It heats and cools quickly, is mechanically rugged, and relatively impervious to food and water spillage. It is an hermetically sealed unit that consists of a *Nichrome* resistance coil embedded in an insulating powder (usually magnesium oxide) that is contained in a stainless-steel tubing. Because of the insulating material, no part of the coil contacts the inner surface of the protective sheath. Such a contact would constitute an unintentional ground, in addition to producing a hot spot. Both the spiral surface unit and the curved D-shaped oven unit are constructed this way.

Depending on the type of range on which it is used, the spiral-shaped surface unit contains either a single heating coil or two such

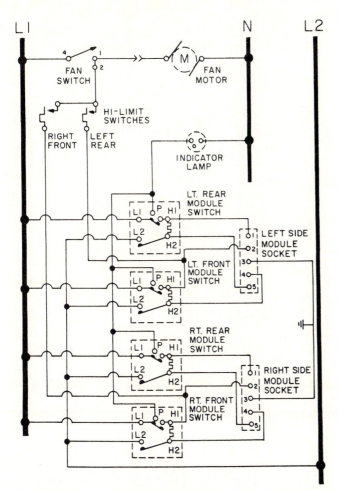

FIGURE 31-1. Schematic diagram for Whirlpool's 36-inch modular cooktop with self-ventilating downdraft exhaust. (Courtesy of Whirlpool)

FIGURE 31-2. Makeup of an electric range surface heating unit with two coils. This can be used with an infinite heat switch by connecting the two coils in parallel.

coils. Ranges with multiposition snap switches or push-button heat switches need two-coil surface units (Fig. 31-2). The switch applies 115 or 230 volts to the two elements in series and in parallel, thereby providing various cooking heats. Ranges with the "infinite" or "cycle timer" control will have single-coil surface units. The oven unit, either bake or broil, is generally a single-coil, thermostatically controlled unit.

During any particular time, the heat produced by a *nichrome* element is directly proportional to the wattage. From the familiar formula for wattage (watts $= E^2/R$), it is seen that the *wattage depends* on the *square of the voltage.* Therefore, if a heating element is switched from 115 to 230 volts, the heat-producing wattage is not just doubled, but increases by a factor of four. The wattage formula also shows

that *increasing the resistance* (as, for instance, by placing two surface unit elements in series) *lowers the heat produced.* With two voltages and two heating elements, as many as eight different heats can be obtained by using a controller designed to switch the voltages and elements in and out (see Figs. 31-3 and 31-4).

Infinite switching is a totally different arrangement (Fig. 31-5). The surface unit contains a single heating element operating at only one voltage. Heat variation is obtained by a switch (or, more accurately, a controller) that "cycles" the element ON and OFF. A set of contacts in the controller repeatedly open and close (Fig. 31-6). By turning the controller knob, the ON time can be varied from as low as 5 percent to as much as 100 percent of the ON/OFF cycle. Although the wattage remains unchanged, the heat is varied by controlling the length of time the element is energized.

In the oven, a thermostat is used for temperature control (Fig. 31-7). It is the fluid-expansion type of thermostat, not a bimetal type. A switch offers selection of upper or lower heating units, but the same thermostat exercises temperature control for either "bake" or "broil." Upper and lower heating units are rated at 2000 to 3000 watts. When continuously energized at 240 volts, they can produce the highest

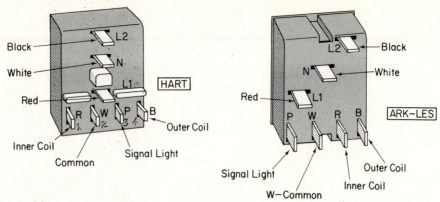

FIGURE 31-3. Terminal location for two different sources (Hart, Ark-less) of seven heat surface unit switches. Although the terminal locations are different, the heat control sequence is the same.

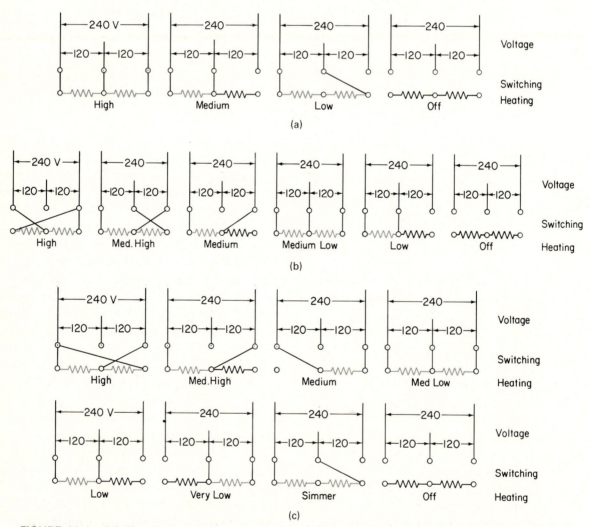

FIGURE 31-4. (a) The three-heat (four-position) switch is found on hot plates and older ranges. Each unit must have *two 120-volt* elements (coils) that may or may not be of the same wattage. (b) The five-heat (six-position) switch requires each unit to have *two 240-volt* coils, which may or may not be of the same wattage. (c) The seven-heat (eight-position) switch requires each unit to have *two 240-volt* coils of *unequal wattages.*

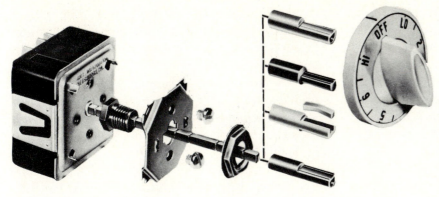

FIGURE 31-5. Universal electric range replacement control. It can replace three-heat, five-heat, seven-heat, and infinite-heat controls on most ranges. It operates as an infinite-heat control. (Courtesy of Chromalox)

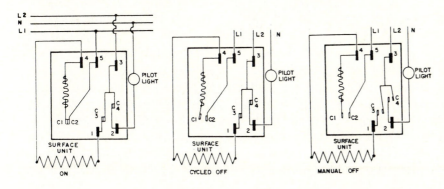

FIGURE 31-6. In a typical infinite-heat switch the internal heater causes the bimetal to open and close the switch that controls the range heater element. This bimetal heater is *connected in series* with the range (*cooking*) *heater* and must have the *correct resistance* for the element being controlled.

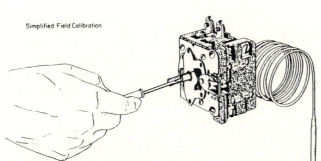

Simplified Field Calibration

FIGURE 31-7. The Wilcolator type G1-A control for an electric range oven provides calibration from the front simply by removing the dial.

temperature marked on the control knob. Lower temperatures are obtained by thermostatic cycling.

Oven timers are quite popular. Most ranges are equipped with some sort of device for automatically turning the oven on and off. Two types of timers are used: one is a clock-operated switch, the other an electric alarm bell. The switch timer actually turns the oven element on and off at a preset time. The bell merely notifies the user that a preset time has elapsed. It does nothing to control the operation of the range.

Electrical troubles in the range can generally be classified as heating element, controller, and wiring faults. In addition, such accessories as lights and timers can become defective. However, troubles that interfere with the basic action of the range are most likely to cause the owner to call for an appliance technician.

Although it is ruggedly constructed, the heating element can, with time or because of rough usage, become open-circuited. Then it will no longer heat. The heating element is easily checked with an ohmmeter. Normal resistance is somewhat less than 100 ohms. If open-circuited, it will measure ∞. During the resistance check, the heating element must, of course, be disconnected from the electric circuit. An open-circuited element will remain cold even with full voltage applied, but *testing the resistance in a live circuit means instant destruction* to the ohmmeter. If it is a two-coil surface unit, both elements should be tested.

Of all range repairs, heating element replacement is the simplest, primarily because all parts are easy to reach. As in any range repair, the first step in element replacement is to turn off the power, preferably at the fuse box.

Pull the fuses and put them in your pocket until the repair is completed. If you are carrying the fuses, someone else is less likely to energize the range while you have a fistful of range wires. If the power panel contains circuit breakers, turn off and cover them with tape to discourage anyone else from turning them on.

For ease of cleaning, the range-surface unit is designed to hinge up or lift out. Either arrangement offers easy access to the connector for surface-unit replacement. To remove the swing-up unit, the hinge hardware must be loosened.

A two-piece porcelain insulator, held together by screws or clips, surrounds the wire attachment to the surface unit. With the insulator removed, it can be seen that the wires are fastened to the heating element leads by screw-type terminal lugs. If a two-element unit is to be replaced, care should be taken to ensure exact relocation of the line wires. If necessary, prepare a simple sketch to ensure correct reattachment to the replacement unit. On a single-coil unit, with only two wires attached, no such problem exists.

After lead attachment, the insulators should be carefully reassembled around the connections and the surface unit relocated in the range *before* the fuses are restored and the surface unit energized.

Except for working in more cramped quarters, testing and replacing oven coils are no more difficult than servicing surface units. In fact, some oven coils are plug-in units that can be replaced without tools. As with the surface unit, some identification system should be used to ensure correct wire connections when a replacement is made. Using an exact replacement is preferable; however, some excellent universal-type replacement units are also available. When such a type is used, *the installation sheet supplied* with the unit *should be carefully followed to ensure a correct fit and proper attachment.*

If the heating element resistance is normal, yet the unit remains cool when switched ON, the controller is next in order as a probable source of the trouble. Some amount of range

disassembly is necessary to make the switch available to test. For example, four screws must be removed to allow pulling out the push-button switch assembly.

A voltage measurement is the most reliable test for a switch. When the switch is turned ON, a measurement across the switch terminals should register zero volts. When the switch is turned OFF, a measurement across the switch terminals should register full-line voltage, either 120 or 240 volts. A measurable voltage across a closed switch indicates a fault.

The infinite switch gives a somewhat different voltage indication. Normally, in the closed condition, the switch voltage measurement will fluctuate from zero to full-line voltage in a manner that depends on the switch setting. At a high-heat setting, the voltmeter will register zero for a longer time interval than the time it registers full-line voltage. At a low setting, the reverse is true. A faulty controller either registers zero at all times (contacts "stuck" cannot open) or registers some voltage at all times (contacts "burned" and are unable to close satisfactorily).

Switch replacement can involve the disconnection and reattachment of many wires. As mentioned before, some scheme of wire marking should be used to ensure correct relocation of each and every wire. A sketch or identifying tags are recommended.

The attaching hardware for rotary and thermostatic switches can usually be found under the knob. After the front screws are loosened, the controller is removed at the rear side of the control panel. Thermostat removal introduces a new problem, routing the fluid bulb and tubing. Removing and installing a new thermostat involves running the sensing element along a path that depends on the brand of range. The manufacturer's instruction normally gives the exact method of replacement.

Although it seldom happens, a defective range may be the result of faulty wiring, either a severed wire or a broken conductor under the insulation. So if the heating unit remains cold even though the element and switch test out satisfactorily, the wiring in the inoperative circuit should be checked. Usually arcing damage is evident as a telltale blackening of the wire.

But if the faulty location is not visible, a test is necessary.

Because of the particular circuit that is inoperative, one or possibly two conductors will, through a process of elimination, turn out to be the faulty ones. To test a wire, disconnect both its ends from the circuit; then check it with an ohmmeter. A good wire should check 0 ohms; a faulty one, ∞.

While checking the wire, jiggle it, watching for an "intermittent" break that opens and closes, causing the meter needle to bounce back and forth from 0 to ∞. If the ends of the wire are too far apart for the reach of the ohmmeter leads, temporarily fasten one end to the frame of the range and test between the other end and a nearby bare-metal part of the frame.

Replace a faulty conductor with an 105°C insulated wire. Other types of insulation are not intended to withstand heat and are not suitable. Solder is not used, for it would melt. Instead, either the spade-tongue or quick-connect terminals are used. Because of the ready availability of connector kits, no problem should be encountered in attaching suitable end connectors.

Any appliance parts supplier can usually supply the high-temperature insulated wire for range rewiring.

In addition to the electrical circuitry, ranges can have other problems: oven doors may not close properly or the oven vent may clog, each of which may affect the oven operation.

The downdraft self-ventilating cooktop (Fig. 31-8) includes a grille and an exhaust fan that is ducted to the outside of the house. If the motor fails, the filter needs cleaning, or the ductwork is faulty or too long, there is a possibility of smoke and cooking odors getting into the house. It may be necessary to review the installation instructions for the particular range you are servicing.

It should never be assumed that a range must operate satisfactorily just because a part has been replaced. Even new parts have been known to be faulty. Also, an error can be made in rewiring or reassembly. Therefore any repair job should be validated by a performance check.

Fortunately, a range is easy to check. That the unit heats up satisfactorily is quite evident without the aid of instruments. However, the cooking temperatures should be tested and checked with a thermometer.

The testing of surface and oven temperatures has been covered earlier and will not be repeated here.

However, when repairs and/or replacements that involve either surface-heating units, oven-heating units, or any part of a thermostat have been made, a check should be made. Verify that the temperature shown on the control knob coincides, within acceptable tolerances, with the actual temperature of the unit or oven. If it does not, additional work will be necessary to discover and correct the difficulty. As a rule, it is a matter of thermostat recalibration.

Upon completion of repairs and satisfactory operation of the appliance, you must make a leakage-to-ground test. Some multimeters have a leakage scale. If this is not available, use an ac milliammeter. There must not be over 2 milliamperes of leakage from a *good* earth ground to bare metal part of the appliance frame in both the OFF and ON switch position.

FIGURE 31-8. This modular cooktop will fit into a space on a countertop. Make certain that the calculated exhaust ductwork is less than 26 feet from the outside vent. Try to choose an area free of any competing draft. (Courtesy of Whirlpool)

Troubleshooting Chart ELECTRIC RANGES

Problem	Possible Cause	Corrective Action
Oven will not heat.	Selector switch is OFF. Blown fuse. Inoperative oven control. Open circuit in oven element. Loose connection. Timer inoperative.	Set selector switch. Check fuses. Check circuit continuity. Check circuit continuity. Tighten all connections. Check timer setting.
Oven too hot or too cold.	Thermostat calibration. Improper oven door fit.	See "Thermostat adjustment." See "Door seal and fit."
Oven will not turn off.	Inoperative selector switch. Inoperative timer.	Check selector switch. Check timer setting.
Oven interior light does not light.	Loose or inoperative bulb. Inoperative light switch. Loose connections.	Tighten or replace blub. Light switch replacement. Tighten.
Oven door opens under heat.	Door needs adjustment. Loose or worn pin.	See "Door seal and fit." Replace hinge.
Oven door drops down.	Worn hinge bracket.	Replace bracket.
Timer does not operate properly.	Incorrect setting. Loose connection. Inoperative motor. Inoperative mechanism.	Refer to owner's manual: see "Timer operation." Tighten. Replace motor. Replace timer.
Timer will not control oven.	Incorrect connection. Inoperative timer. Selector switch not correctly set.	Check wiring diagram. Replace timer. Set selector switch.
Oven drips water or sweats.	Oven not preheated with door open. Oven temperature excessive. Door does not seal at the top. Clogged oven vent.	Check oven operation. Check thermostat calibration. Adjust oven door. Clean vent.
Surface unit does not heat.	Blown main fuse. Loose connection. Inoperative switch. Open unit. Incorrect connection. Broken wire.	Check fuse. Tighten. Replace switch. Replace unit. Check wiring diagram. Continuity check.

SELF-CLEANING OVENS

Even the most careful housewife will find that grease, starch, and sugar will spill in an oven and often bake on to the bottom or sides. Fortunately these soils, all hydrocarbons, will decompose into water vapor and gases when heated to a temperature between 850 to 925°F (454 to 495°C).

This action is called pyrolysis—chemical decomposition by heat. Any residual ash will be loose and can be wiped out.

The oven is not an incinerator, so do not expect it to burn large quantities of soil. As much excess as possible should be wiped up before the self-cleaning process is started.

In the self-cleaning cycle the oven door must be tightly sealed to limit the entrance of oxygen and it must be securely locked so that people will not try to use it. At these temperatures, burning would be severe and an excess of oxygen into the oven could cause an explosion.

Exact operating sequence for the self-cleaning cycle will vary from model to model. Therefore, you must familiarize yourself with the manufacturer's procedure used in the various ovens you will service.

A typical basic sequence for starting the cleaning cycle is as follows:

1. Clean out excess soil, and, if necessary, remove racks (older ovens).
2. Where required, close the door window shutter.
3. Close door and move lock handle, usually to the right.
4. Set the timer, for 2 hours of operation, longer for heavier soils. If there is no timer, the operation must be manually terminated after 2 hours.
5. Set the oven temperature control to clean.

When the oven heats to about 550°F (288°C) a lock thermostat will positively lock the door so that it cannot be opened until after the cleaning cycle has been completed and the oven has cooled down to 550°F (288°C). This could be an hour after the end of the cleaning cycle. When the lock light has turned off, the oven door can be opened by moving the lock lever back to its original position.

Figure 31-9 is a schematic diagram of the wiring and parts involved in the self-clean cycle

FIGURE 31-9. In the self-clean cycle the bake and broil elements are operated at 120 volts. Once the 900°F (480°C) clean temperature has been reached, the heaters will cycle at this temperature until the clock switch opens the circuit.

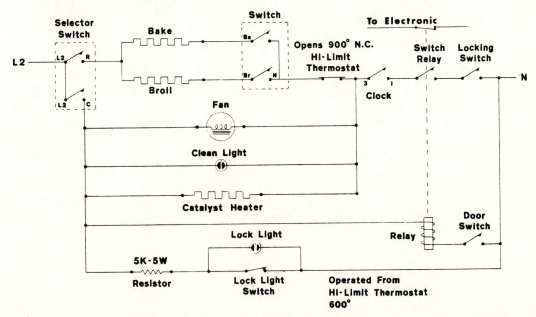

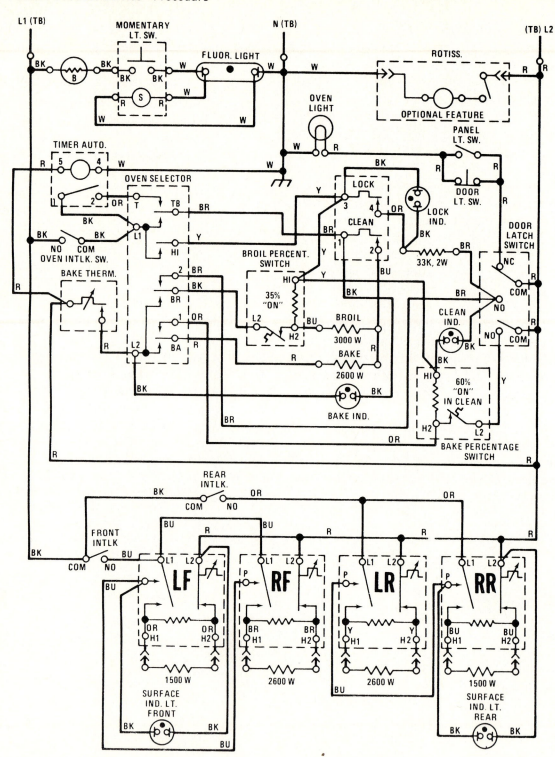

FIGURE 31-10. Basic circuitry for an electric range with a self-cleaning oven. Usually, a diagram is attached to the appliance. Use that if it is available. (Courtesy of Whirlpool)

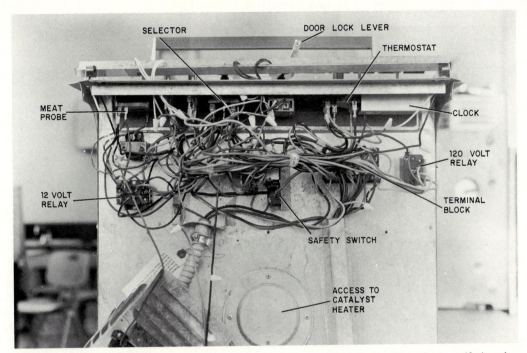

FIGURE 31-11. Locations of some of the major controls for the Chambers self-cleaning oven. Most ovens have their controls either on the top or on one side. They are accessible by pulling the oven part forward out of its mounting opening.

of a Tappan self-clean electronic range.

In addition to the bake-and-broil elements, there is a catalyst heater whose job is to heat the platinum paladium catalyst to above 800°F (427°C). At this temperature, the catalyst consumes the smoke.

SERVICE PROCEDURE

Electrical

Problems in the electrical circuitry can be traced with a volt-ohmmeter as in any other appliance. Start at the power input and trace the circuit through the control to the heater element. Use the diagram attached to the unit as a guide (Fig. 31-10). If there is no diagram, the ones included in this section are typical of many in use and should serve to guide you to the trouble area.

Mechanical

There is probably more variation in the mechanics of construction and assembly than there is electrically from one range to another (Fig. 31-11). It is always best to be guided by the manufacturer's instructions.

A quick clue to oven-cleaning effectiveness is the color of the remaining soil at the end of a cycle.

1. Brown and soft—no cleaning heat.
2. Dark brown—incomplete cleaning.
3. Loose gray ash—high enough cleaning temperature.
4. Excessive smoke during cleaning—inoperative catalyst.

The troubleshooting chart shown should serve as a guide to locating the problem in most self-cleaning ovens.

Troubleshooting Chart *SELF-CLEANING OVENS*

Problem	Possible Cause	Corrective Action
No heat for cleaning.	Incorrect control setting.	Set controls for clean cycle for a minute or so to determine if all units are heating. *Note:* If they do, instruct user. If not, proceed.
	Open fuse.	Test and replace.
	Defective oven relay.	Replace relay.
	Defective thermostat.	Replace thermostat.
	Door not locked.	Check alignment or interference with locking mechanism.
	Defective selector switch.	Replace.
Incomplete cleaning.	Incorrect control setting.	See "No heat for cleaning" above.
	Set time too short.	Time must be set for at least 2 hours.
	Low-line voltages.	Check voltage under load; if low at fuse panel, notify local electric utility. If okay there and low (100 volts) at range, have qualified electrician check wiring.
	Defective element.	Check each for operation in bake-and-broil position. Check catalyst unit for continuity.
	Excessive soil in oven.	Instruct user to remove excess before cleaning cycle.

GAS RANGES

Before getting too involved in servicing a gas range (Fig. 31-12) make certain that you know what the range is supposed to do. If there is a problem—and once you found the cause—recheck yourself to be sure that it is the correct cause; then proceed to fix it.

It is frustrating to do a lot of work and then find that you still have the original trouble. Our purpose here is to provide you with some basic knowledge on gas ranges that will assist you in keeping customers' ranges in first-class working order.

Instead of repeating information contained elsewhere, we refer you to the section on flame and burner characteristics that is included in Chapter 12. What has been said there applies to all other gas-burning appliances, just as some information in this chapter will apply to all other gas appliances.

Make certain that the automatic timer is set for *manual* while you are servicing a range.

As with the electric range, uniform baking in a gas range requires that the stove be level and solidly placed. The oven door must operate properly and the oven thermostat bypass adjustment must be correct. Too high a setting of the bypass will cause overcooking, even when the thermostat is set at LOW position. Too low a setting will cause uneven temperatures. There are several oven thermostats in current use; each has its own adjustment methods.

Before adjusting an oven control, make certain that the air shutter adjustment for the burner is clean, correctly adjusted, and that the oven burner baffle is properly set to prevent excessive drafts over the burner.

To adjust air shutter, turn the oven on and set to 350°F (177°C). Loosen the screw that holds the air mixer. Then adjust the shutter so that the inner cone is sharp, distinct, and about $\frac{1}{2}$ inch in length. Yellow tips mean that more air is needed. A flame that lifts away from

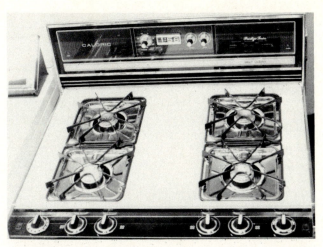

FIGURE 31-12. The cooktop of this Caloric range has controls on the front top; others have controls on a front top side.

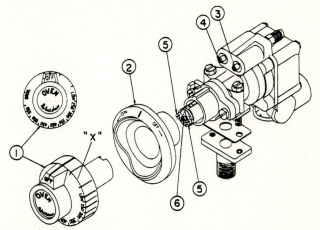

FIGURE 31-14. The Robertshaw BJ oven thermostat: 1, dial; 2, bezel (not used in some ranges); 3, pilot flame adjuster (decreases in clockwise direction); 4, bypass flame adjuster (decreases in clockwise direction); 5, locking screws; 6, calibration screws (turning toward low lowers oven temperature).

the burner means too much air is being used (Fig. 31-13).

Note: Do not adjust the air mixture for an oven burner when the burner is operated for minimum or bypass flame!

The Robertshaw control (Fig. 31-14) is a combination thermostat and oven gas valve. The oven is turned on and the temperature setting is made by a single rotation of the dial. This control is used by Norge, Tappan, Whirlpool, and others.

The bypass adjustment controls the flame that must be maintained when the oven has come up to the temperature setting on the dial. Enough gas must be bypassed by the thermostat to keep the entire burner lighted. There must be approximately $\frac{1}{8}$-inch flame around the entire burner. Should adjustment be needed, adjust as follows:

FIGURE 31-13. The air shutter adjusts the amount of primary air mixed with the gas. Open or close it until the flame is clean and free of yellow tips.

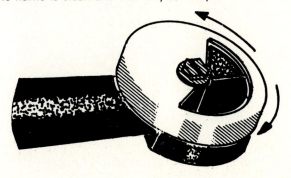

1. Turn the control dial fully on, being sure that the pilot light is on and that the gas is fully ignited.
2. After 5 minutes, turn the dial back to midway between the OFF and the LOW [or 140°F (60°C) ON] position, depending on how the particular dial is marked.
3. Remove the dial and bezel. Usually these pull off.
4. Adjust the bypass adjusting screw as needed to produce the $\frac{1}{8}$-inch flame over the entire burner.

The right-hand screw is the pilot flame adjustment. This should be set to allow about a $\frac{1}{2}$-inch flame—or until the yellow tip disappears.

Calibration for the oven control is carefully set at the factory and should not be changed unless cooking results definitely prove that oven temperatures do not match the dial setting and *no other cause can be found!*

If you feel that the oven temperature is inaccurate, place an accurate mercury thermometer or thermocouple in the oven. Set the dial for 300°F and allow the temperature to stabilize for 5 minutes after the oven flame has cut down to the bypass flame. A tolerance of ±5°F is standard.

Should calibration be needed, pull off the dial and bezel, loosen the two outside locking screws on the calibrating dial, and turn the center screw toward High or Low as needed. If the temperature reading is higher than the dial setting, turn toward Low, and vice versa. Each graduation is about 25°F. Remember to tighten the two outside screws when you are through.

Another form of this control is used on some Norge models and differs in that calibration is made by rotating the dial to the temperature reading. Proceed as for the control just described and remove the knob. At this time you will note the locking screw.

Remove the metal insert from the dial center and reinstall the dial. Hold the dial and loosen the locking screw; then turn the dial to the actual oven temperature (thermometer) reading. Tighten the locking screw. Recheck at other temperatures; then replace the dial insert when calibration is satisfactory.

Several adjustable thermostats with directions for their use are shown in Fig. 31-15.

Burner cleanliness is a factor that is often overlooked. Foreign matter, such as spilled grease, and foods can clog burner parts. Burner heads can usually be removed for cleaning. Scrubbing with soap and water will clean them. Caustic compounds will discolor the cast aluminum parts and steel wool will scratch them. Toothpicks or fine wire can be used to clean the burner ports. The lighting slots can be cleaned with a razor blade. Dust must be kept from accumulating at the primary air intake.

When ordering parts, be sure to give the range model and serial numbers. Generally, the rating plate on which these are found can be located under one of the burners, often the right front. Other models have it on the inner front frame just below the oven door when it is open.

The checklist should be reviewed before looking for mechanical failures.

IGNITION

Burners can be lighted as required manually, by a gas pilot light which is on all the time, or by an electronic ignition system. Manual ignition is the use of a match or sparking device. The standing pilot light, once lighted, will ignite any burner any time it is turned on. The action

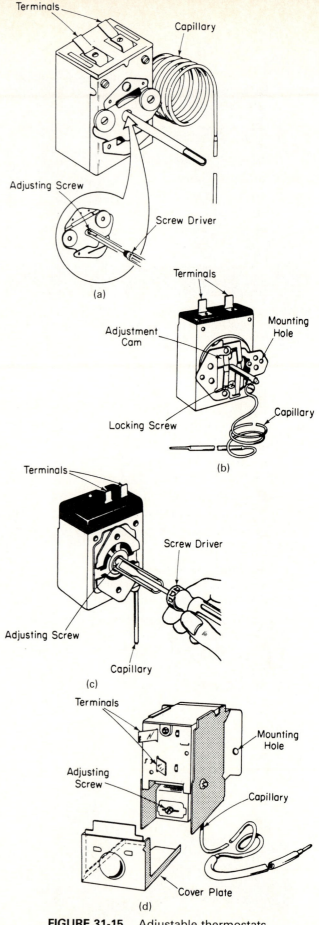

FIGURE 31-15.   Adjustable thermostats.

here is that when a burner is turned on, gas flows to the burner head through its holes. Some will flow into the flash tube toward the pilot flame, where it is ignited. The flame travels back through the flash tube to the burner head, where it ignites all the gas flowing out of the burner head holes. Neither of these systems requires electricity.

The solid-state electronic ignition system provides a pulsing spark when a gas valve is turned on. In operation, when the valve is turned on all the way, it closes a switch that activates the igniter module to produce a spark. (Fig. 31-16). When the gas has ignited, the valve is turned back slightly until you can hear that the sparking has stopped. The burner can then be adjusted to the desired flame height.

Electronic ignition for the oven is different. In one system, when the thermostat is set, pilot gas will flow, the ignition will spark, and a pilot

light turns on. The pilot light will remain on until the thermostat is turned off. In another system, when the thermostat is turned on, a switch will allow current to flow to a heater element which will remain ON while the thermostat cycles the gas flow. The heater turns off when the thermostat is turned off (Fig. 31-17).

In the event of a power failure, the surface burners can be lighted manually by holding a lighted match near the burner and turning the control knob to the start position.

The troubleshooting chart is in three sections: solid-state ignition, gas ranges, and gas-oven baking problems (Fig. 31-18).

FIGURE 31-16. Igniter control and sparking modules with high-voltage insulated wire. The side view of the sparking element is shown to illustrate the clearance of about $\frac{1}{10}$ inch between the electrode and the ground.

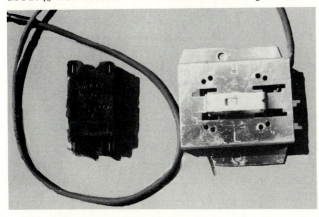

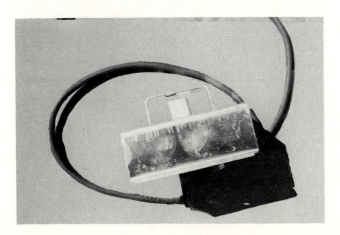

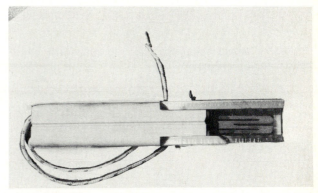

FIGURE 31-17. An oven gas igniter heater element.

FIGURE 31-18. A schematic diagram for a gas range with electronic spark ignition. (Courtesy of Whirlpool)

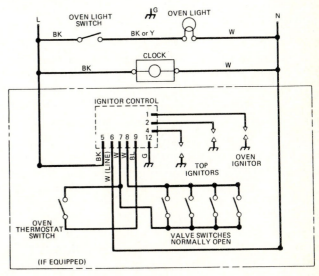

Troubleshooting Chart SOLID-STATE IGNITION

Problem: No sparks appear at any ignitor gap when *any* top burner valve is turned on.

Possible Cause	Check	Solution
No power to range.	Do accessory lights on the range work?	Check power supply to wall outlet; be sure range cord is securely plugged into outlet.
No power to module.	With voltmeter, check for line voltage across terminals N and L of module.	Correct faulty wiring or connections so that voltage is present.
High voltage not being generated because of faulty module.	If module has input but no output, it is faulty.	Replace module. Wires must be reconnected to same terminals of new module.

Problem: No sparks appear at any ignitor when certain gas valves/thermostat(s) are turned on (at least one valve may function properly).

Possible Cause	Check	Solution
Faulty switch (contacts don't close).	Check each switch which is not sparking properly by disconnecting wires from switch and connecting them to each other. If continuous sparking occurs, switch is faulty.	Replace switch.
Open circuit in wiring harness.	Check for open circuit between common terminal and all manual or re-ignition switch terminals.	If you find an open circuit, replace that section of wire.
Reignition ignitor which is controlled by switch is grounded at gap or through high-voltage wire.	With module off, disconnect high-voltage wire at module and check for continuity from the quick disconnect on the wire to ground (range frame).	If ignitor gap is grounded, readjust to $\frac{1}{10}$ inch (width of two dimes). If high-voltage wire is grounded, replace it.
Reignition ignitor which is controlled by switch has high resistance leak to ground at gap or through high-voltage wire.	With power supply disconnected, disconnect ignitor at module. Substitute a test ignitor which has short (about 12-inch) high-voltage wire lead and position tip $\frac{1}{10}$ from ground. If sparking occurs when module is reconnected to power original ignitor is faulty.	Replace original ignitor.

Troubleshooting Chart SOLID-STATE IGNITION (Continued)

Possible Cause	Check	Solution
Failure in trigger circuit because of faulty module.	If procedures above fail to determine cause, module is probably faulty.	Replace module. Wires must be reconnected to same terminals of new module.
High resistance or open connection between module and ignitor.	Disconnect ignitor from module. Connect jumper from spark output to ground. If sparking stops, ignitor high-voltage wire is probably open.	Replace ignitor.
Ground lead to module disconnected or range chassis not properly connected to ground by ground lead or through third prong of power cord plug, combined with reversed power supply polarity.	Check ground connection of range chassis and ground lead connection to module.	Connect range to earth ground. Connect ground lead to module.

Problem: Sparking occurs at ignitors when all valves/thermostat(s) are off.

Possible Cause	Check	Solution
Faulty switch; contacts are closed in OFF position.	Disconnect leads from each switch, one at a time, leaving switches disconnected as they are checked. If sparking stops when leads to any switch are disconnected in this fashion, that switch is probably faulty.	Replace faulty switch. Reconnect switches one at a time; if problem recurs when any switch is reconnected, that switch is also faulty and must be replaced. If sparking does not stop with all switches disconnected simultaneously, see below.
Short circuit from any switch lead to ground or to ground, neutral or common leads.	Disconnect switch leads from module, one at a time, leaving them disconnected as they are checked. If sparking stops when any lead is disconnected, that lead is probably faulty.	Replace faulty lead(s).
Short in trigger circuit of module.	If procedures above fail to determine cause, module is probably faulty.	Replace module. Wires must be reconnected to same terminals on new module.

Problem: Sparks are present but not at all points, nor at full strength.

Troubleshooting Chart SOLID-STATE IGNITION (Continued)

Possible Cause	Check	Solution
Arcing to range at undesired place.	One of ignitors may have very strong arc.	An ignitor may be faulty or a flaw may be present in the high-voltage wire. If possible, observe range in dark to help locate breakdown. Replace faulty wiring or ignitor. Presence of grease near ignitor tips can also sometimes cause accidental grounding to range.
Interruption in high-voltage circuit.	Additional symptom: All sparks are very weak.	Connection loose or ignitor gap distances too great. All gaps should be approximately $\frac{1}{10}$ inch or the thickness of two dimes.
Insufficient high voltage generated.	Additional symptom: All sparks are very weak.	Line voltage should be at least 94 volts ac. If at least this voltage is present, replace module.

Problem: All burners that are turned on have lighted but ignitors are still sparking.

Possible Cause	Check	Solution
For manual ignitors *only:*		
Top burner valve remains in START position.	Check top burner valve knob positions.	Turn valve knob from START position.
For flame sensing (automatic reignition) ignitors:		
Insufficient flame in spark gap.	Incorrect positioning of flame sensing ignitor.	Position ignitor so gap is approximately $\frac{1}{10}$ inch (thickness of two dimes).
	Drafts, which lift or blow out flame in spark gap.	Eliminate drafts.
	Unstable flame in spark gap.	Adjust burner air shutter to stabilize flame.
	Insufficient constant (oven) pilot flame.	Be sure oven pilot selector is turned all the way to either the natural or LP gas setting (depending on the gas on which the range is to be used).
	Unstable constant (oven) pilot.	Check for blockage in pilot orifice and/or tubing and, if necessary, correct.

Special Note Regarding Solid State Ignition

Ignitors failed to spark on some of the early models in high humidity areas or when water spilled on the wiring or ignitor. Operation could be restored by drying the system with a hair dryer, or waiting for the weather to change. Sometimes spraying the wiring with an automotive ignition spray will correct the electrical leakage. Otherwise the wiring had to be replaced.

In the meantime advise manual ignition.

***Troubleshooting Chart* GAS RANGES**

Problem	Possible Cause	Corrective Action
Pilot outage.	Draft.	Check location in respect to open doors, windows, fans, etc. Locate range or baffle air from striking pilot.
	Improper oven door fit.	Check door fit at top; tighten springs.
	Improper pilot adjustment.	Readjust to proper flame. Minimum tinge of yellow on steady flame.
	Improper burner adjustment.	Readjust main burner so that its ignition will not extinguish pilot. May be too much air or gas.
	Splatters from broiler put pilot light out.	Lower food slightly.
	Oven pilot will not light.	Check safety valve and thermocouple.
	Clogged air filter.	Clean or replace.
Top burner will not light.	Pilot light out.	See pilot outage.
	Excessive air in gas mix.	Adjust air shutter.
	Burner not in its proper position.	Reposition burner.
	Incorrect height of pilot flame.	Adjust to satisfactory flame.
	Pilot incorrectly positioned.	Reposition pilot.
Yellow tips on gas flame. Smoky flame.	Too much gas in proportion to air.	Adjust air shutter. Clean air shutter. Dirty burner. Clean burner thoroughly.
Flames lift from burner.	Excessive primary air.	Adjust air shutter.
Clogged burner carbon forms on utensils.	Incorrect air-gas ratio.	Adjust air shutter. Clean spilled food from burner.
Gas odor.	Leak.	Locate and correct.
	Pilot flame impinging on burner casting.	Relocate pilot flame so it does not touch.
	Oven burners out of adjustment.	Adjust primary air shutter.
Condensation or sweating in oven.	Oven not preheated with door open.	Open door during preheat.
	Oven door does not seal at top.	Adjust oven door.
	Clogged oven vent.	Clear vent.
	Temperature setting too high.	Operate oven at lower temperature.
	Incorrect thermostat calibration.	Recalibrate oven control.
No even heat control.	Improper bypass adjustment.	Adjust bypass to lowest possible flame.
	Foreign matter in heat control.	Clear out control.
	Defective control or capillary tube.	Replace control.

Troubleshooting Chart   GAS RANGES   (Continued)

Possible Cause	Check	Solution
Porcelain top crazing.	Utensil being used is too large for burner. Cleaning with moist cloth when hot.	Use large burners for large-bottom pans. Wait until stove is cool before wiping.

Troubleshooting Chart   GAS OVEN BAKING PROBLEMS

Problem	Possible Cause	Corrective Action
Baked goods burn bottom.	Oven flame too soft. Dark utensils. Improper circulation. Utensil too large for amount of batter used. Baffle improperly positioned, smothering flame. Flue blocked.	Increase primary air at air shutter. Use bright utensils. Use bright utensils. Use small utensil. Reposition baffle. Open flue.
Baked goods burn on one side.	Utensil in direct line of heat circulation. Range or utensil not level.	Allow $1\frac{1}{2}$ inches at sides and 1 inch front and back. Level range. Replace warped dented utensils.
Baked goods burn on top.	Oven flame too hard. Baffle upside down. Flue blocked.	Decrease primary air. Properly position baffle. Open flue.
Baked goods crack open.	Oven too hot. Batter too stiff.	Check calibration; use lower temperature. Check recipe.
Baked goods soggy.	Oven not hot enough. Improper ingredients or failing to remove promptly from pan.	Check calibration; use higher temperature. Check recipe.

32

Refrigerators

When a refrigerator breaks down, the owner's immediate concerns are: Can it be fixed before the food spoils and how much is it going to cost? Fortunately, refrigerators are much more dependable than they used to be and most of the service calls are for minor problems.

In any event, **promptness** in making the service call is important.

A COLD CABINET

If you consider the refrigerator to be a food storage cabinet connected to some machinery that keeps the inside cool, it will make your service approach easier.

Some cabinets have one door; others have two or more (Fig. 32-1). Some will have a single compartment at about 40°F (4.5°C) temperature; others will include a second compartment at about 0°F (−17°C) temperature. This second compartment is called the *freezer section*.

Obviously, if there is something wrong

with the cabinet either in how it was used, located, or damaged, the cooling machinery is not going to be able to do its job efficiently.

LOCATION AND USAGE

The location of a refrigerator is important to its proper operation. Details on this are included in Chapter 4. You should note the following:

1. Surrounding temperature.
   (a) Room temperature.
   (b) Location in respect to sunlight.
   (c) Location in respect to radiator, range, hot air duct.
   (d) Anything that will raise the cabinet temperature above a normal room temperature.
2. Surrounding space.
   (a) Is there enough air space to allow removal of the hot air coming

FIGURE 32-1. Refrigerators are offering more special-ized storage compartments. The refreshment service door is used for snacks. The dark area in the freezer door is the location of the ice and cold water dispensers.

from the condenser? Is the con-denser clean?
   (b) Is this space clear of litter, paper, etc.?
   (c) Is there enough space to allow the doors to swing freely? Constant banging of a door against another cabinet can spring the door hinges.
3. Physical condition of the cabinet.
   (a) Do the doors close and seal prop-erly?
   (b) Are the breaker strips intact?
   (c) Is the cabinet rusty or otherwise damaged?
4. Customer usage
   (a) Is the refrigerator overloaded or overcrowded with foodstuffs?
   (b) Is the door opened and closed a great deal?
   (c) Is there an abnormal ice cube de-mand?

Any of these could cause the compressor to run all the time and yet fail to keep the refrig-erator cool enough. The unit is overloaded and working beyond its capacity. Correction of the offending condition and an explanation to the customer will usually suffice.

RUNNING TIME

Typically, a refrigerator is designed so its com-pressor will run 55 to 65 percent of the time at a room temperature of 70°F (21°C). As the room temperature increases, or as an abnormal amount of warm food is placed in the refrigera-tor, or as the door is opened and closed an ab-normal number of times the compressor has to work harder and longer to transfer the extra heat from the inside of the box to the outside (see Fig. 32-2).

An increase in room temperature to 90 or 100°F (32 to 38°C) could cause the compressor to run 100 percent of the time. Under these conditions continuous operation of the compres-sor does not indicate a failure in the refrigerator system. So don't condemn the compressor or the rest of the refrigeration system until you are certain the trouble is not due to local condi-tions or usage.

Traditionally, refrigeration service calls in-crease in hot weather.

LOW-CAPACITY SYSTEM

A refrigeration system that has lost some of its refrigeration capacity will exhibit somewhat the same symptoms as those for the overworked compressor. But the causes are different and the following list should be checked with great care.

1. A loss of refrigerant is due to a leak from the sealed system to the atmos-phere. A small or slow leak will gradu-ally cause the compressor to run for a longer period of time to produce the required low temperature. Eventually the compressor will run continuously and the refrigeration compartments will not get as cold as they should. On refrigerators having an exposed evapo-rator the frost pattern will be "short." In other words, frost will not cover the entire evaporator. The pattern will continue to get "shorter."

A large leak will cause the com-

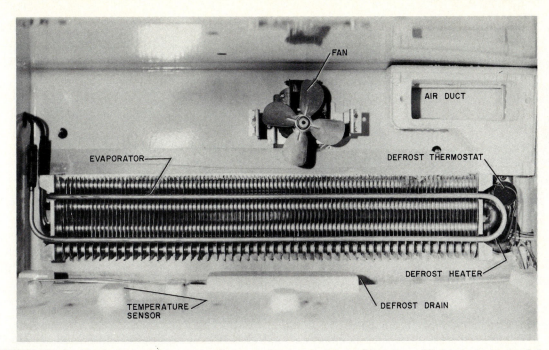

FIGURE 32-2. Typical evaporator area for a modern two-door refrigerator. All this is in back of the freezer compartment. The fan circulates 0°F (−18°C) air over the evaporator into the freezer section and some down into the refrigerator compartment. The temperature sensor controls the compressor.

pressor to run continuously, but there will be no refrigeration and the condenser will not be hot.

Obviously the leak must be located and repaired before the unit can be made to operate satisfactorily.

2. *A defective compressor* may run, but not be efficient in pumping the refrigerant through the system. A broken or stuck valve within the compressor will not allow the pressure differential between the low side and the high side of the system to build to the proper values. Typically, under these conditions, the low side will be operating at a higher than normal pressure.

3. A *restriction* will usually occur in the capillary tube. A partial restriction will allow some refrigerant flow. Generally a gage placed in the suction line will have a lower pressure reading than normal vacuum and the high-side gage will have a higher pressure reading than normal. When the compressor is stopped, equalization of pressures will be slow or will stop depending on the degree of restriction.

Ice resulting from moisture in the system is a restriction that will cause erratic operation. When the unit is warm the refrigerant will flow in a normal pattern but when the moisture freezes—usually at the point where the capillary tube enters the evaporator—all flow will stop. The quickest test for this is to place a hot, wet cloth against the evaporator where the ice has formed. The heat will melt it and allow a flow of refrigerant again until it refreezes.

4. *Air* in the system usually becomes trapped in the condenser causing a higher than normal condenser pressure and inefficient operation.

5. A refrigerant overcharge is usually indicated by a frost pattern appearing on the suction line outside of the refrigerator (see Fig. 32-3).

FIGURE 32-3. The white coating on the suction line is frosty ice that formed because an overcharge of Freon allowed the gas to continue cooling all the way back to the compressor.

FIGURE 32-4. This wiring diagram of a Whirlpool refrigerator-freezer includes both schematic and pictorial representations of its electrical system. The ice cube maker diagram is shown separately. (Courtesy of Whirlpool)

Neither air nor ice nor an overcharge should occur unless someone has previously worked on the unit and didn't know what he was doing.

Instructions on patching leaks and replacing compressors appear in other chapters.

ELECTRICAL PROBLEMS

Most refrigerators have an electrical wiring diagram attached (pasted) to the back of the cabinet. This diagram should be consulted when making electrical tests. Fig. 32-4 is a typical diagram taken from a freezer-refrigerator; it includes most features found in many different units.

This diagram presents the wiring in both the schematic and pictorial form and should be referred to in the following discussion of electrical problems.

WIRING DIAGRAM

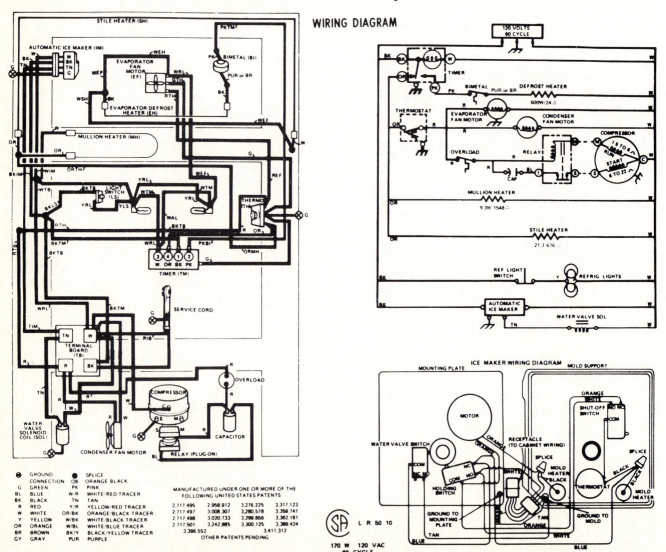

Obviously the first electrical test is to make certain that power is available in the refrigerator. Usually, the door light is a good indicator. If it lights, electricity is available; but if it does not, you will have to check elsewhere. A good place to start is the electrical outlet. Then, following the diagram, work back to the compressor. Voltage should be available at each electricity-using part—compressor, fan, heater, light. If it is not, the trouble is in either the wiring or a control—defrost timer, thermostat, starting relay, or switch. See Fig. 32-5 for general parts location and function.

At the time you are testing for voltage, make certain that the defrost timer is in the run cycle in order to insure voltage being available to the compressor circuit. The defrost timer is normally in this position since the defrost cycle is only in operation about 20 minutes every 12 hours. This will vary in some models.

When checking the defrost heaters, make certain that the timer is in the defrost position and that the defrost thermostat is in its closed position. Typically, a defrost thermostat will cut-in at 20°F (−6°C) and cut-out at 50°F (10°C). However, it is always best to verify these readings from the manufacturer's literature for that particular model.

Aside from these precautions, the electrical circuit testing is done in the usual manner, using a voltmeter to determine the existence of a proper voltage where it should be.

The ohmmeter is used to determine circuit continuity, and the word of caution here is to make certain that the power cord has been disconnected before testing.

THE AUTOMATIC ICE MAKER

The freezer compartment is used to freeze and store frozen foods. It is also used to make ice cubes, either in compartmented trays or with an automatic ice maker (Fig. 32-6). The trays require no service, but the ice maker may. In addition to electrical power, the ice maker requires a connection to the household water supply.

The connection to the house supply can be made to a cold water pipe by using a saddle valve and tubing, which may be plastic or copper. An installation kit (Fig. 32-7) is available for this type of installation. Instructions are on the back of the package. Connection to the refrigerator is made to a solenoid-controlled fill valve. This valve can be mounted either near the base of the cabinet (Fig. 32-8a) or near the top if there is a provision for it (Fig. 32-8b). Ice-maker kits, including the fill valve, are available for many different refrigerator models. Figure 32-9 illustrates an installed unit.

After the ice maker is connected and the water turned on, allow at least 12 hours before expecting ice cubes. Then throw out the first batch or two because they may contain chips or other contaminants from the new water supply lines. From then on, the rest can be used. A wiring diagram for the ice maker is given in Fig. 32-10. The water valve is special in that it is designed to operate at 105 volts ac. The other 10 volts is used by the mold heater. The water valve also includes a flow control washer. This limits the flow rate to 125 or 135 cubic centimeters per cycle, depending on how many cavities that particular model has.

OTHER FEATURES

Some refrigerators have a separate opening in the freezer section. It contains a cold-water and ice dispenser. Figure 32-11 is the schematic wiring diagram for the electrical circuitry of one of these models. The fill valve has a single input connection and a dual output with two solenoid valves. One is for the ice maker, and one is for the drinking water. Each solenoid is different. Do not interchange them or substitute one for the other. The drinking water valve is identified on the diagram and the other is located as the water valve supplying the ice maker.

The drinking water flows to a tank in the lower rear section of the food compartment. From there it is piped under the refrigerator to enter the freezer section door at the bottom hinge area. Then it goes up to the dispensing valve. When water is drawn, the valve lets a fresh supply into the tank to cool down. The ice maker produces ice in the normal manner. This ice feeds into a bin from which it is dispensed, on demand, by a screw-type auger that is activated whenever the ice dispenser switch

RCA **Whirlpool**

1962 "NO-FROST" REFRIGERATOR MODELS TOP FREEZER WITH MULLION EVAPORATOR

REFRIGERATION SYSTEM

NOTE: SYSTEM SHOWN HERE HAS A FAN COOLED CONDENSER. MODELS HAVING A STATIC CONDENSER WOULD NOT HAVE CONDENSER FAN, AND MAY NOT HAVE A CAPACITOR.

- **SIMPLE SYSTEM**
  Single capillary tube. No solenoid valves. Single evaporator (1) cools both compartments (2) (3).
- Unit thermostat (4) (air sensing) controls compressor (5) and condenser fan (6).
- Timer (7) stops refrigeration and starts defrost.

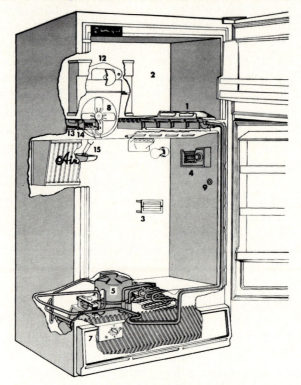

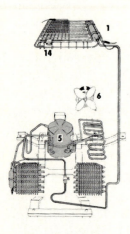

AIR SYSTEM

- **One single speed evaporator fan (8) delivers cold air to both the freezer compartment (2) and the refrigerator compartment (3).**
- Evaporator fan is controlled by unit thermostat (4) and refrigerator door switch (9). It runs when the unit runs and the refrigerator door is closed.
- Air is drawn in through openings (10 and 11) and passes over finned evaporator (1) where it is cooled.
- Cooled air is forced up air duct at rear of cabinet. Part of it goes up into freezer compartment and part of it goes through air deflector, past damper (12) and down air duct to refrigerator compartment.
- Damper controls proportion of cold air going to each compartment. Control dial set on position =1 gives maximum air flow to refrigerator compartment and minimum air flow to freezer compartment. Turning dial clockwise decreases air flow to refrigerator compartment and increases it to freezer compartment.

DEFROST SYSTEM

- Timer (7) initiates defrost (each 12 hours). Refrigerator turned off 21 minutes.
- Evaporator defrost heater (13) on until thermo disc (14) on evaporator warms to 70° or 21 minutes, maximum.
- Thermo disc resets at 30° (approx.).
- Defrost water drains from drain pan and evaporator through drain hole (15) to unit compartment pan.

REFRIGERATION CIRCUIT

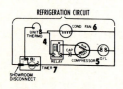

DEFROST CIRCUIT

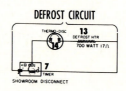

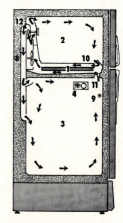

COMPONENT TESTING

Check Voltage at Outlet — Unplug to Make Electrical Checks

REFRIGERATION SYSTEM

1. Compressor does not run.
   A. Check unit thermostat for proper setting and continuity.
   B. Check compressor with starting cord.
   C. Check overload for continuity.
   D. Check relay — replace with known good relay.
   E. Check timer — timer motor must run — remove wires, check continuity, terminals 1 and 4 when timer is on run cycle.

2. Compressor runs, insufficient cooling.
   A. Check condenser fan motor by direct connection.
   B. Clean condenser.
   C. Partial restriction — frost at restriction; check for moisture by applying heat.
   D. Complete restriction (hot compressor, cool condenser, no frost).
   E. Leak or low charge — see Service Manual.

DEFROST SYSTEM

1. Timer — motor must run — turn to defrost (audible click at 2 o'clock); check continuity, terminals 1 and 2.
2. Heater — make preliminary check by turning timer to defrost position. Check wattage or feel for heat. If above test is negative — check heater direct, using ohmmeter.
3. Thermo disc (14) — must contact evaporator firmly — opens 70° and resets 30° (approx.).

AIR SYSTEM

1. Fan motor — make direct connection to check.
2. Timer — motor must run — remove wires, check continuity, terminals 1 and 4 when timer is on run cycle.
3. Thermostat — check proper setting and continuity.
4. Fan should run at constant speed when unit is on and door switch is depressed.
5. Refrigerator door switch — check continuity, closed door position.

EVAPORATOR FAN CIRCUIT

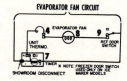

FIGURE 32-5. (Courtesy of Whirlpool)

(a)

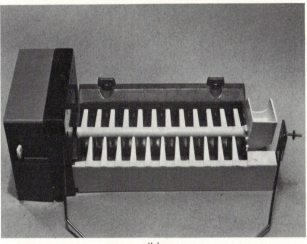

(b)

FIGURE 32-6. The automatic ice maker accepts water, freezes it, ejects the cubes into a bin, and when the bin is full, stops the process.

FIGURE 32-7. A complete water supply kit for connecting to existing plumbing may be used for purposes other than supplying water to an ice maker.

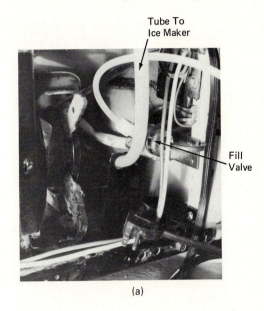

(a)

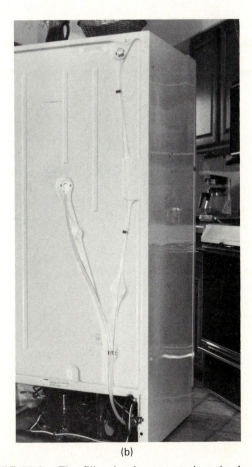

(b)

FIGURE 32-8. The fill valve is mounted at the base of the refrigerator and a plastic tube runs up the back to the existing tube.

(a)

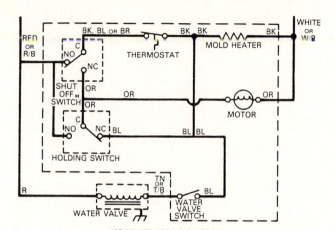

ICE MAKER WIRING DIAGRAM

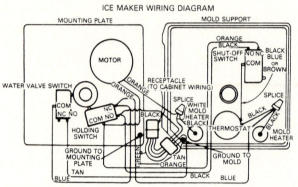

FIGURE 32-10. This diagram includes both the schematic and pictorial wiring diagrams for the ice maker. The pictorial shows the relative locations of parts and their wiring. The schematic shows the electrical location of the parts to each other and is much easier to follow when checking the circuit.

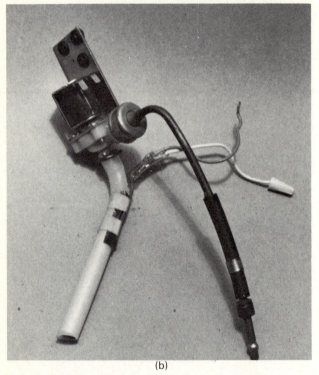

(b)

FIGURE 32-9. (a) The ice maker is an option that can be mounted in the space provided for it. This one has a control for ice cube size. (b) A fill valve of this type would be mounted on the back of the refrigerator with the fill tube extending into the freezer compartment.

is turned on. The ice dispenser switch, like the water dispenser switch, is activated by pressure on the dispensing lever.

When servicing these units, make certain that the dual water valve is properly connected and functioning. A complaint of no water could be an open solenoid coil, a defective microswitch, or frozen water in the tank due to an open tank heater. A complaint of no ice could be an ice maker problem of not producing ice or a storage bin problem of not delivering ice. This can be determined by looking in the stor-

age bin. If these two areas seem okay, the dispensing system must be checked. Electrically, the front microswitch, the control-lever interlock switch, the top freezer interlock, and the drive motor are all in series. Any break in this system will keep ice from being dispensed.

Some refrigerators have a built-in diagnostic system. These have built-in sensing units that will signal a malfunction in some areas. Figure 32-12 is a pictorial wiring diagram that includes all the electrical parts of both the refrigerator and the diagnostic system. It shows their approximate location. (Figure 32-11 is a schematic showing the working parts of this same refrigerator.) Figure 32-13 shows the location of the sensing system parts.

The purpose of these extra parts and circuitry is to inform the user of a problem in the use and operation of the refrigerator. The signal from the sensing unit passes through a microprocessor to a display panel (Fig. 32-14). This

WARNING: DISCONNECT FROM ELECTRICAL SUPPLY BEFORE SERVICING UNIT.

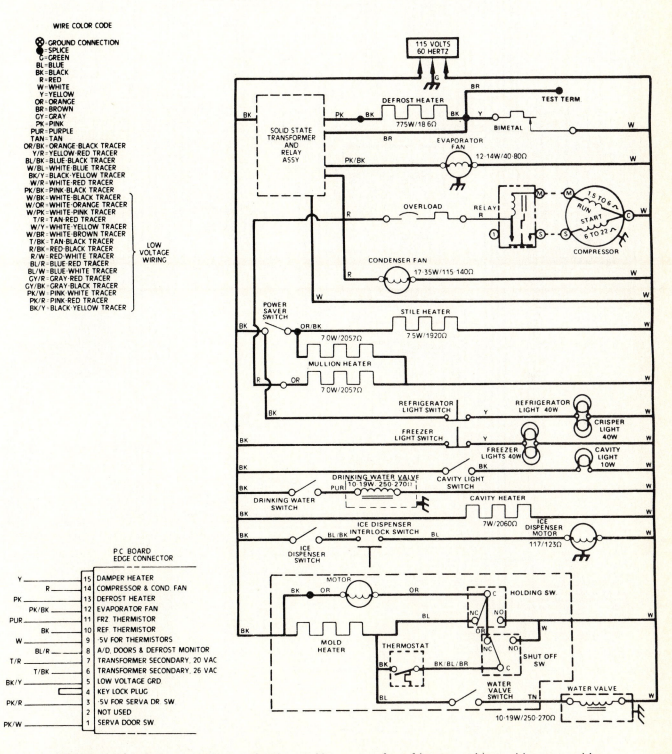

FIGURE 32-11. Schematic diagram for the working parts of a refrigerator with a cold-water and ice dispenser. The systems sentinel wiring has been omitted. (Courtesy of Whirlpool)

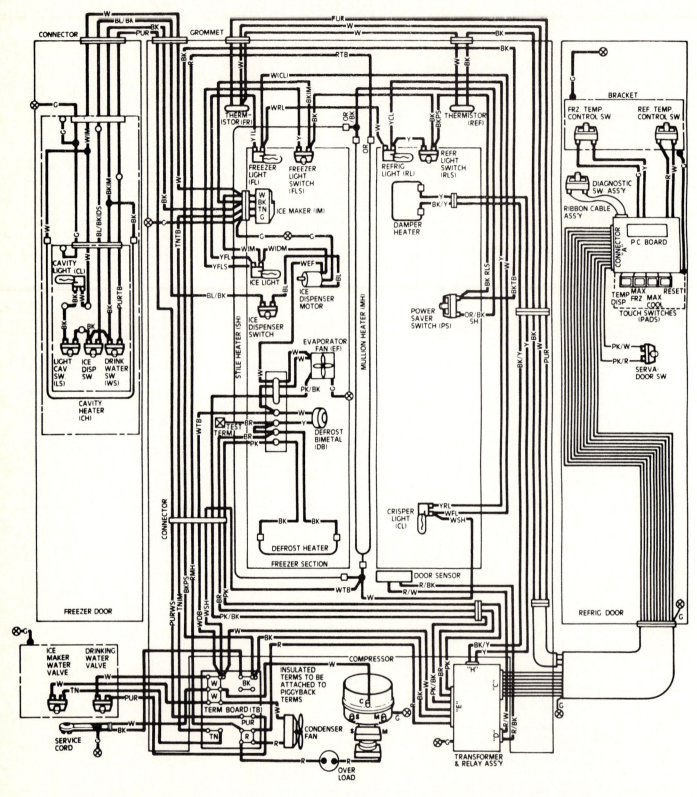

FIGURE 32-12.   This pictorial wiring diagram will help you find the approximate location of all electrical components in the Systems Sentinel II refrigerator. (Courtesy of Whirlpool)

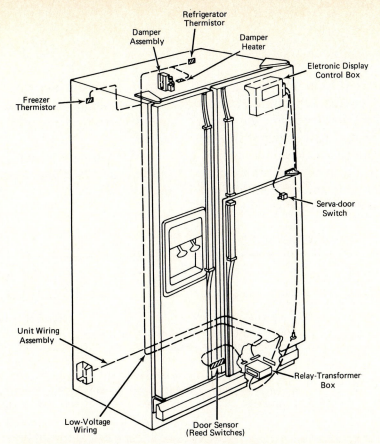

FIGURE 32-13. Location of Systems Sentinel II components. (Courtesy of Whirlpool)

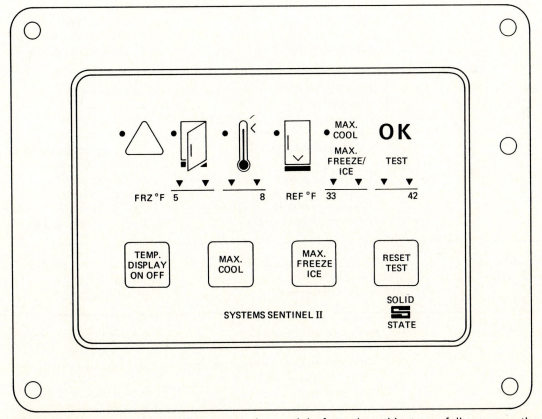

FIGURE 32-14. Display panel. To remove the module from the cabinet, carefully remove the vinyl trim and screws from the front escutcheon. It is sometimes possible to replace the PC board without removing the entire module, but a ribbon connection from the touch switch panel is difficult to reinstall with the module in place. (Courtesy of Whirlpool)

display module flashes a light indicating the faulty condition and makes a beeping sound.

There are refrigeration problems that are obvious and can be located and serviced in the regular manner. The display module will notify the owner of only specific conditions that may exist. Failure of a sensing unit or some other part in the systems sentinel could send a false signal. It is advised that unless you have had hands on experience with these models, you do not work on any area in which you are inexperienced.

The following troubleshooting chart should guide you in locating most common refrigerator problems.

Troubleshooting Chart REFRIGERATORS

Problem	Possible Cause	Corrective Action
Compressor will not run.	No power at outlet.	Replace fuse.
	Loose electrical connection.	Repair.
	Over heated compressor.	Clean condenser, check fan, clearance around cabinet.
	Thermostat stuck open or turned off.	Replace if needed.
	Defective relay, overload or compressor.	Use starting cord to check compressor. If it runs, replace defective overload on relay. If not, replace compressor.
Compressor runs but no refrigeration.	Restriction in system, moisture or permanent (damaged tube).	Repair as outlined in refrigeration chapter.
	Low charge.	Locate and repair refrigerant leak. Evacuate and recharge.
	Inefficient compressor.	Replace as outlined.
Compressor kicks out on overload.	Abnormal usage or high room temperature.	Instruct customer.
	High or low voltage, outside 10% tolerance.	Notify customer.
	Impeded air circulation.	Clean condenser, check fan if used. Provide space for air circulation.
	Defective capacitor.	Replace capacitor.
	Defective start relay.	Replace start relay.
	Defective overload protector.	Replace overload protector.
	Defective motor winding.	Replace compressor.
Refrigerator too cold.	Thermostat set too cold, or contacts stuck.	Check setting. Replace if contacts are bad.
	Abnormal location.	Instruct customer.
Refrigerator too warm.	Restricted air circulation.	Clean condenser, check fan and cabinet clearance.
	Abnormal usage, location, or high room temperature.	Instruct customer.
	Poor door seal.	Adjust door closing.
	Light switch stuck on.	Replace switch.
	Defrost heaters on all the time.	Check defrost timer and thermostat.
	Operating thermostat set too warm, or has defective sensor.	Readjust or replace if defective.

Troubleshooting Chart REFRIGERATORS (Continued)

Problem	Possible Cause	Corrective Action
	Defective compressor.	Replace compressor.
	Cooling fan not running.	Check for voltage at fan. If none, check wiring. If okay, check fan.
	Restricted air duct.	Clear air duct.
	Leaking air duct.	Reseal joints.
	Excessive frost on evaporator.	Defrost evaporator.
	Open defrost heater will allow ice to restrict duct.	Replace heater.
Noisy unit.	Loose parts.	Tighten loose parts.
	Tubing rattle.	Move tubing.
	Abnormal fan noise.	Check fan blade security, motor mounts and blade clearance from shroud.
	Abnormal compressor noise.	Check mounting pads for clearance. If noise is internal, replace compressor.
Sweating outside cabinet.	Damp or confined location.	If possible relocate refrigerator.
	Low side tubing too close to cabinet shell.	Reposition tubing.
	Mullion or stile heaters may be open.	Replace defective heaters.
	Void in insulation.	Replace insulation.
	Wet insulation.	Correct cause, replace insulation.
Sweating inside cabinet.	Abnormal usage.	Instruct user to cover foods and defrost as needed.
	Poor door seal.	Make necessary door adjustments.
	Wet insulation due to poor cabinet seal.	Remove liner, replace insulation and reseal cabinet.
Incomplete defrosting or high temperature during defrost cycle.	Defrost thermostat may be loose or defective.	Remount or replace thermostat.
	Inoperative defrost timer.	Replace timer.
	Open defrost heaters.	Replace heaters.
Water in bottom of cabinet or ice in bottom of freezer.	Clogged drain.	Clear drain.
	Misaligned drain fitting.	Reposition fitting.
	Cabinet not level.	Level cabinet.
	Defrost timer inaccurate.	Replace timer.
	Drain heater defective or not making good contact.	Retape heater in position or replace if open. Make certain it has proper voltage before replacement.
Odor.	Food.	Instruct customer to cover foods.
	Fouled drain system.	Clear drain system and flush with baking soda and water solution.
	Inefficient filter if used.	Check for correct installation. Change if over 1 year old.

Note: A warm refrigerator will always have more odor than a cold one. See notes on odor removal.

33

Toasters and Toaster Oven–Broilers

TOASTERS

Short of being knocked off the table by someone tripping over the cord, or having a heater element broken by someone fishing a piece of toast out with a fork, a toaster leads a reasonably trouble-free life.

Most automatic toasters operate the same way (Fig. 33-1). A slice of bread is dropped in the slot and the toast carriage handle is pressed down until it locks in place. This closes a switch, turning on the heat and starting the timer. After a predetermined time, adjustable to provide the desired color of toast, the carriage latch releases and a spring lifts the carriage and finished toast up from the oven.

The timing and latching devices will vary on toasters made by different manufacturers. There are two basic types, the clock timer and the thermostat timer. Both have the same job to do. That is, they control the length of time the heater elements are "on" to toast the bread (Fig. 33-2).

Most toasters have a removable plate on

the bottom for cleaning purposes. The crumbs that collect there should be cleaned out regularly. A soft brush can be used to remove those that are stuck around the edge and inside the oven.

Erratic heating of a toaster is often due to a defective cord. Check this factor by plugging the cord into a tester; then push the toast carriage handle down until it latches. The tester bulb should light. Work the cord with your hands, gradually, along its length, alternately pulling and wiggling it. When the light on the tester goes out or flickers, you have come to the bad spot. If it is near the end, it could be shortened and reused. If the cord is frayed and worn, it should be replaced.

In order to get at the cord end in the toaster, the shell will have to be removed. In fact, it has to be removed to gain access to the timer and heater elements as well.

Disassembly of the toaster is similar for many models. Thus the procedure used here can be followed for others besides the one illustrated in Fig. 33-3.

(a)

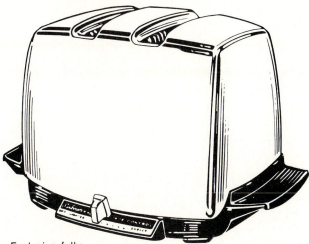

- Exclusive, fully automatic toaster lowers your bread and then raises your toast when done—all automatically
- Unique Radiant Control feature senses the amount of moisture in the bread to give you uniform toast slice after slice
- Easy-to-read front toasting control
- Extra wide wells with hinged crumb tray for easy cleaning
- Timeless design in a rich chrome finish

®SUNBEAM ©SUNBEAM APPLIANCE COMPANY. 1984   (b)

FIGURE 33-1. [(a) Courtesy of Proctor-Silex; (b) courtesy of Sunbeam]

First, the toast carriage handle must be removed. It is attached with a set screw on the underside of the handle. It is necessary to loosen the screw only far enough to permit handle removal. Next, the cord strain relief must be removed. Squeeze the strain relief insert into

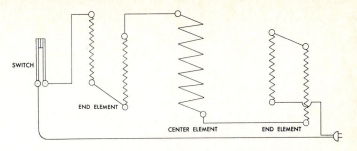

FIGURE 33-2. Note that the schematic wiring diagram for this toaster shows all the heating elements to be connected in series. If one burns out, they are all out. This must be considered in your troubleshooting procedure.

FIGURE 33-3. When working on a toaster, always take care to keep the bread guard wires out of the heating element.

the cord to allow it to unlock from the case; then pull outward until it is clear from the case.

Lay the toaster on its side on a soft, padded surface to avoid marring the chrome finish. Determine how the case is attached. Most manufacturers use screws. Some have projecting tabs that are twisted to lock them in position. These devices can be unlocked, at least once, by careful straightening. More handling is apt to break them.

The Proctor-Silex model (Fig. 33-1a) has four screws that must be removed to separate the base from the case. The case is in three parts: two side panels and a wraparound center section. When the center section is removed, the bread guard wires will fall out. Remove them immediately before an end from one damages the heater. Slide this center section up the cord as far as necessary to allow a clear working area.

All the working parts and connections are now available for test and repair.

The cord is attached under two nuts that must be securely tightened after cord replacement.

Most of the current production toasters use a disconnect female terminal as a cord connector, (Fig. 33-4). Figure 33-5 shows a representative exploded view and parts list.

If you must replace this terminal, make certain to use one that is approved for high-temperature service. The brass ones are apt to lose their spring tension and a poor contact will result.

The next step in the electrical circuit is the switch. The contact should close firmly and be held in position by a latch. Sticky, burned, and uneven points can be cleaned with fine sandpaper or a point file. They must be tight in their mountings or else they will overheat and burn out.

The holding latch in the timer assembly (Fig. 33-5) will maintain contact position until the toast is done. Then it will release and the toast carriage will lift the toast up from the oven. Make certain that this latch is free in its movement and is not bent out of position.

FIGURE 33-4. Toaster chassis showing the parts relationship.

FIGURE 33-5. Representative parts list and exploded drawing to serve as a general guide on construction. For accurate servicing you should be guided by the manufacturer's information on that particular model. (Courtesy of Toastmaster)

Key	Description	Part Number
	Case assembly (B160)	B160-186
	Case assembly w/decor plate (B161)	B161-186
1	Decor plate (B161)	B140-814
2	Case, front (B160-B161)	B140-351
3	Case, end (B160-B161)	B140-352
4	Case, center (B160-B161)	B140-350
5	Front stat. under cover (B160)	B160-300
5	Front stat. under cover (B161)	B161-300
6	Hinged crumb tray assy (B160-B161)	S101-301
7	Rear handle and buffle assy (B160)	B160-140
7	Rear handle and buffle assy (B161)	B161-140
8	Front handle (B160)	B160-801
8	Front handle w/decal (B161)	B161-143
9	Nameplate decal (B161-B160)	B161-806
10	Operating handle (B160-B161)	B140-809
11	Control knob (red) (B160)	B160-808
11	Control knob (B161)	B161-807
12	Insert, control knob (B160)	B140-804
12	Insert, control knob (B161)	B150-801
13	Timer assembly	B140-108
14	Bimetal arm assembly	D125-63-5
15	Carriage assembly w/op. lever	B140-156
16	Negator spring	B140-811
17	Vertical shaft	B140-502
18	Bread rack	B102-415
19	Frame and vert. shaft brkt. assy.	B140-167
20	Base plate	B102-304
21	Main switch lever	S101-501
22	Switch busbar and ins. assy.	B140-122
23	Element, center	B140-30
24	Element, outside	B140-35-2
25	Busbar, element connection	S101-502
26	Rivit	1B125-29
27	Guard wires	B102-511
28	Cord and terminal assy.	B140-128
29	Cord retainer cover	B140-300
30	Screw, cord ret. cover	S127-821
31	Screw, 3-48 x 3/16 front stat. to frame	B102-821
32	Screw, 4-40 x 3/16 timer to frame	B102-878
33	Screw, 5-40 x 7/16 busbar to frame	1B11S-4
34	Screw, 5-40 x 1/4 plastic handles to frame	1B14S-8
35	Screw, 3-48 x 1/8 timer to frame	1B12S-12

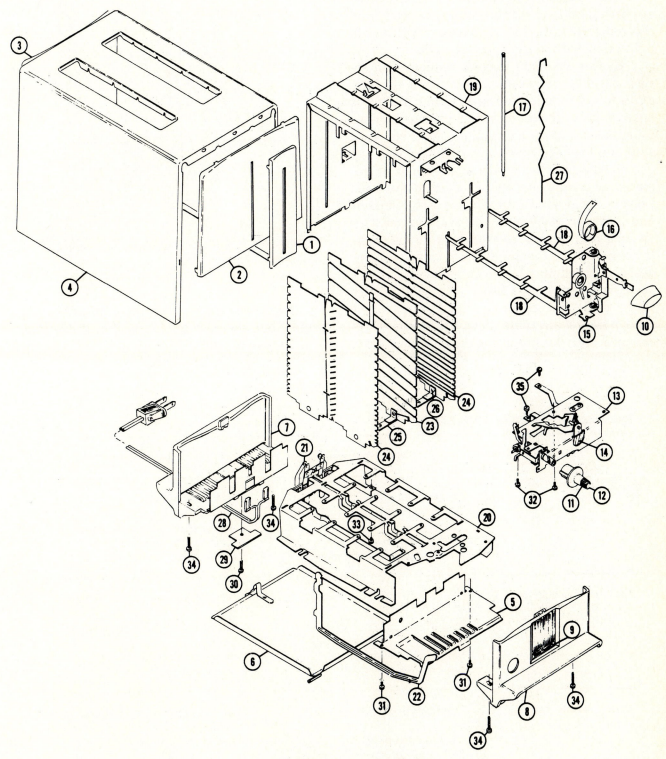

FIGURE 33-5.   (Continued)

The toast carriage guide rod must be clean and straight; otherwise the carriage will be slow in lifting. Carbon tetrachloride or lighter fluid are both good to clean the gummy deposit from this rod. Lubricate the rod sparingly with a good light oil or with silicone grease stick.

Heater elements may be connected in series, parallel, or series parallel. Use the diagram for the model you are testing. If this is not available, inspect each element carefully for a break or a loose connection. See Figs. 33-2 and 33-6 for two typical wiring diagrams.

A defective heater element should be replaced as a complete unit. A word to the wise: Don't try to unwind a turn or two of the element and reconnect it because this will lower its resistance and cause it to overheat and burn out. If the rest of the toaster element is in good condition, a break may be silver soldered.

When replacing a heater element, be sure to order the correct one because they are not always interchangeable. Both the electrical resistance and the mounting method will vary.

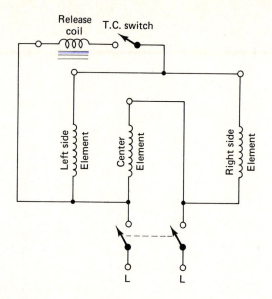

FIGURE 33-6. Typical wiring diagram in series–parallel. (Courtesy of Proctor-Silex)

Electrical testing may be done with your tester, but the rest of the mechanical testing will be visual.

Troubleshooting Chart TOASTERS

Problem	Possible Cause	Corrective Action
Fails to heat.	No power at outlet.	Check house fuse.
	Defective cord.	Repair or replace cord.
	Loose connection.	Clean and tighten connection.
	Switch not making contact.	Repair or replace switch.
	All elements burned out.	Replace heater elements.
Toast will not stay down.	Hold down latch not locking.	If bent, straighten to correct position.
		If binding, clear bind to allow free operation.
	Bind in toast carriage.	Clear cause of bind.
	Broken latch spring.	Replace spring.
Toast will not "pop up."	Toast carriage binds.	Clear bind to allow free operation of carriage.
	Release latch binds.	Check latch find and clear.
	Broken power spring.	Replace spring.

Troubleshooting Chart TOASTERS (Continued)

Problem	Possible Cause	Corrective Action
Toast lifts slowly.	Bind in slide rods. Weak spring.	Clean and lubricate rods. Adjust or replace spring.
Toast lifts too rapidly.	Excessive spring tension. Dashpot not effective (where used).	Adjust tension on lift spring. Repair piston or washer in dashpot.
Toast too light or too dark.	Incorrect adjustment of timer mechanism. Defective thermostat or timer.	Adjust timer setting. Replace defective parts.
One side untoasted.	Defective heater element.	Replace heater element.
Shocks user.	Ground wire.	Locate ground and take corrective action.

TOASTER OVEN–BROILERS

The toaster oven–broiler is a handy kitchen item for the small household. It is portable, convenient, and subject to the same cleanup, cord, and plug problems as those of any of the small portable appliances.

The models discussed here are representative, differing slightly in size and styling. Basically, they are side opening with a tray that moves outward as the door is opened. Controls are at each or either end of the door side. They consist of a thermostat, pilot light, toast control, and may have a toast, bake, broil switch.

The Proctor-Silex (Fig. 33-7) is typical of the styling. Its exploded view and parts list show the relative parts locations and their names. The wiring diagram shows the switching and heater element relationship. The four main heater elements are thermostatically controlled. The others, slow cook or keep warm, are an extra option.

The West Bend wiring diagram (Fig. 33-8) includes the main heaters, but uses a thermostatic control to operate a release coil to turn off the unit when the toast is ready.

Note that in this diagram, as in the Proctor-Silex, the main heater elements are connected in series–parallel.

SERVICING

Short of dropping the unit and warping the chassis or springing the door, there is little to service in the mechanical system.

Electrically: check for cord damage and damage to the thermal fuse or the controls before condeming the heater elements. Usually, removing the end panels will expose the heater element terminals so that they can be tested separately. Always check the electrical outlet for power before trying to service the appliance.

Failures can also occur when someone misuses any appliance. This can happen accidentally or from not knowing the limitations of the appliance. Toastmaster, in the user's manual for their oven-broiler, includes a list of safeguards that all service personnel should know and be able to pass on to customers (Fig. 33-9).

(a)

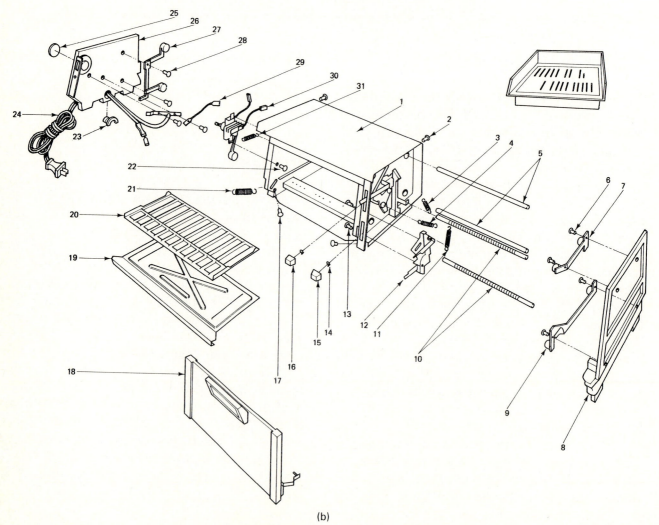

(b)

FIGURE 33-7. (Courtesy of Proctor-Silex)

Item Number	Part Number	Description
1	942940	Chassis (includes bottom cover and items 3, 4, 11, 12, 13, and 31)
2	002733	Screw, Phillips self tapping (2 required)
3	105343	Spring, color control
4	105350	Spring, Counterweight
5	610758	Element (300W), upper and lower rear (2 required)
6	002275	Screw, Phillips, self tapping, upper and lower busses to panel (4 required)
7	968308	Bus, upper (right panel)
8	969221	Panel, right (does not include bus bars)
9	968002	Bus, lower (right panel)
10	610757	Element (450W), upper and lower front (2 required)
11	105344	Spring latch
12	601594	Latch ass'y (item 1)
13	004005	Rivert, latch ass'y (item 1, 2 required)
14	105005	Spring, color knob and on/off (2 required)
15	156787	Knob, on/off
16	156787	Knob, on/off
17	002275	Screw, Phillips self-tapping panel to chassis bottom (4 required)

Item Number	Part Number	Description
18	924076	Door ass'y
19	647700	Tray, crumb
20	621830	Grill rack
21	105356	Spring, door
22	002004	Screw, Phillips, self-tapping, oven thermostat to chassis (2 required)
23	121028	Grip, cord
24	926767	Cord, power (includes connected bus bar)
25	143105	Cover, oven control knob
26	926558	Panel, left (includes oven control knob) *Note: Does not include power cord or bus bars*
27	908692	Bus ass'y (left panel)
28	002275	Screw, Phillips, self tapping, busses to panel (4 required)
29	968321	Connector
30	620853	Thermostat, oven
31	105343	Spring extension
32	935737	Pan, broiling, includes grid

(c)

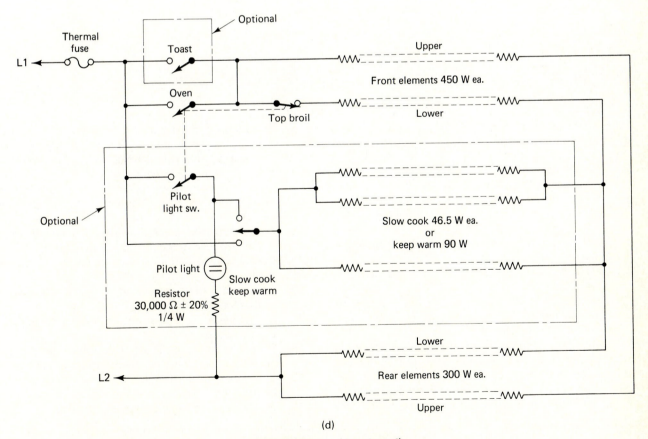

(d)

FIGURE 33-7. (Continued)

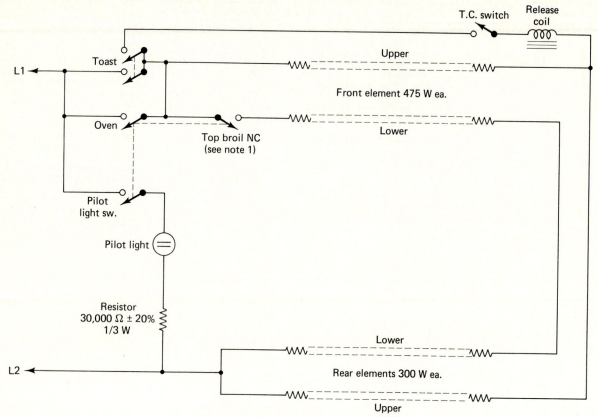

FIGURE 33-8. (Courtesy of West Bend)

Important safeguards

When using electrical appliances, basic safety precautions should always be followed, including the following:

1. Read all instructions.
2. Do not touch hot surfaces. Use handles or knobs.
3. To assure electrical safety, **do not** place this oven broiler or electric cord in water or other liquid. See cleaning instructions.
4. Close supervision is necessary when any appliance is used by or near children.
5. CAUTION: Never leave oven broiler unattended when broiling.
6. To assure electrical safety, do not clean with metal scouring pads. Pieces can break off the pad and touch electrical parts.
7. Unplug from outlet when not in use and before cleaning. Place all controls in "OFF" position before unplugging. Allow to cool before putting on or taking off parts, and before cleaning the appliance.
8. Do not operate any appliance with a damaged cord or plug or after the appliance malfunctions, or has been damaged in any manner. Return appliance to the nearest authorized service facility for examination, repair or adjustment.
9. When using this appliance, provide 4 to 6 inches air space above, behind and on both sides for air circulation.
10. For safe operation, use accessory attachments only if they are recommended by manufacturer.
11. Do not use outdoors or while standing in a damp area.
12. Do not let cord hang over edge of table or counter, or touch hot surfaces.
13. Do not place on or near hot gas or electric burner, or in a heated oven.
14. Extreme caution must be used when moving an appliance containing hot oil or other hot liquids.
15. Use extreme caution when removing tray or disposing of hot grease.
16. Do not use appliance except as intended.

FIGURE 33-9. (Courtesy of Toastmaster)

Troubleshooting Chart   TOASTER OVEN–BROILERS

Problem	Possible Cause	Corrective Action
No heat.	Blown fuse.	Replace fuse.
	Open thermal fuse.	Replace.
	Defective cord.	Repair/replace.
	Defective switch.	Replace.
Upper elements do not heat.	Defective heater.	Replace.
Lower elements do not heat.	Defective heater.	Replace.
	Defective top broil switch.	Replace.
Toaster does not turn off.	Defective thermostat.	Replace.
Toaster does not start.	Defective switch.	Replace.
Oven does not hold its temperature.	Defective thermostat.	Replace.

34

Vacuum Cleaners

A vacuum cleaner, mechanically speaking, is a simple appliance that creates a partial vacuum within itself by means of a motor-driven exhaust fan. Whenever this partial vacuum exists, atmospheric pressure causes the air surrounding the inlet to rush into the cleaner. The inrushing air is collected through a nozzle. Any small, loose matter (dust, dirt, sand) near the nozzle is picked up and carried along in the airstream. The airstream passes through a porous dust receptacle that filters the air before it is expelled from the cleaner (Fig. 34-1). Since the fan inside the vacuum can exhaust the air inside the machine faster than it can rush in, a partial vacuum remains within the cleaner as long as the motor and fan are running.

A vacuum cleaner is the most "accident-prone" appliance in the home. The number of people who use it, plus its portability, sometimes result in complaints of poor soil pickup, complaints that are not the fault of the machine. Before you pick up a screwdriver to fix a machine, check its recent usage and the external factors that might cause poor cleaning.

In order to operate properly, any vacuum cleaner must have vacuum (or suction) at the nozzle. The major factors controlling this aspect are given below.

1. *Clear intake and outlet.* There must be a clear passage for airflow, and its load of soil, from the nozzle to the dust bag. Any obstruction within the nozzle, hose, or intake chamber will reduce airflow and cause poor cleaning. Partially clogged hoses can usually be cleared by connecting the hose to the exhaust port and blowing any foreign matter out. More stubborn cases may require the use of a broom handle or similar object (with a blunt end) to push out any obstructions.

2. *Clean dust bag.* All the air coming into a vacuum cleaner is exhausted through the dust bag. An overloaded dust bag will reduce cleaning efficiency. Similar conditions will exist when using the

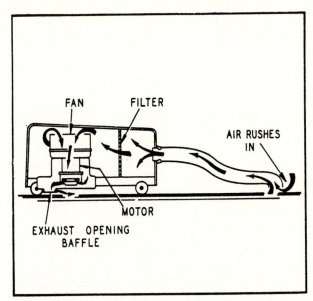

FIGURE 34-1. Airflow in a typical canister-type vacuum cleaner is through the hose, porous bag, motor, and out. The dust remains in the bag, and the filtered air cools the motor.

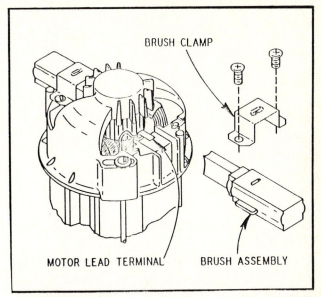

FIGURE 34-2. Motor repair that involves disassembly is *not* recommended. The motor operates at such high speed that unless the repair shop has special balancing equipment, it will develop excessive vibration, and this will cause further damage. Merely checking brushes and cleaning up the commutator is satisfactory.

older-type (cloth-type) dust bag having clogged pores.

3. *Proper adjustment at nozzle*. Poor contact of the nozzle with the surface being cleaned will allow air to flow into the cleaner nozzle without properly picking up the soil. Too tight a contact will decrease airflow into the nozzle and will also prevent proper soil intake. Correct choice of nozzle attachments for the surface being cleaned will give best results.

4. *Proper sealing at intake*. Any air leaks within the system will decrease the cleaning abilities of the vacuum cleaner. Check the hose connections at the nozzle and intake port. Look for breaks in the hose, poor fit at intake flange, and faulty seal around the motor mounting for air leakage.

5. *Proper motor and fan operation*. The amount of available suction is limited by the ability of the motor and fan to create it. The motor and fan are designed as a balanced unit to produce a predetermined suction (Fig. 34-2). *Any change in components will alter that relationship.* Therefore, in servicing, use the *correct* replacement parts.

A vacuum gage that will read to 150 inches of water column, is a handy tool for telling how well a cleaner is doing. Most modern canister cleaners should show about an 80-inch vacuum at the cleaner inlet port and about a 70-inch vacuum at the hose end. A good central vacuum should read about 100 inches at the outlets. Loss of vacuum will indicate leakage in the suction side of the unit, a slow motor, or a partially clogged fan.

Of course, water pickup by a vacuum cleaner *not* designed for this purpose will damage the motor. If this accident occurs, open the machine and dry it out thoroughly before trying to run it again. *A wet motor will burn out.*

An overfilled dust bag is another danger to the motor life of the canister cleaner. By design, all the air drawn through the cleaner passes through the motor to cool it. As the dust bag fills, the amount of air that can be drawn through the motor decreases, thus making the motor run hotter. The exhaust air temperature should not be more than 25°F (15°C) above room temperature after the machine has run for a half hour. It should be about 20°F (11°C) after 15 minutes. Also, the fuller the dust bag, the less efficient the machine. A bag should be

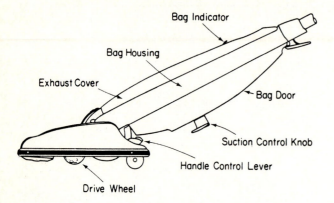

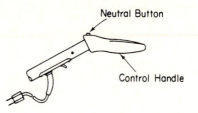

FIGURE 34-3. Outline of an upright or stick-type vacuum cleaner illustrates the differences between it and the tank type.

changed, or emptied, when it is about two-thirds full.

Worn, slipping, jammed, or broken belts are problems inherent in the upright type (stick) of vacuum cleaner (Fig. 34-3). This belt drives the brush that agitates the rug fibers to loosen the soil for the suction to pick up. It also loosens lint, thread, and hair that may have accumulated on the rug. Sometimes these threads and hairs will collect in the belt pulley grooves and in the brush assembly bearings.

Some cleaners have an electronic senser that will warn the user that the brush is turning at a reduced speed or that the dust bag is full (Fig. 34-4).

The Sears vacuum cleaner shown in Fig. 34-5 combines features of both the canister and upright cleaners. The canister has its normal collection of attachment parts. Then, when the motor-driven sweeper is added, the unit performs as an upright cleaner. Exploded views and parts listings are included to provide a basic reference for parts location.

A word of caution: Remind the user to check the nozzle, brush, and belt of her cleaner regularly to prevent an accumulation that will cause future trouble.

The method of belt changing is explained in the owner's manual. Make certain that the twist in the belt is correct; otherwise it will not stay on the motor shaft pulley.

If the revolving brush bristles have worn so they do not project through the nozzle opening, they should be replaced.

Electrically, the circuit consists of a series motor, a switch, and a cord. Some models may have a two-speed motor, which means another switch would be required for speed control (Fig.

34-6). The most frequent failure is a wire break occurring in the cord where it enters the machine.

The troubleshooting chart shown contains all the basic information you need to check out any vacuum cleaner.

FIGURE 34-4. (a) Sears power mate with cover off, showing senser on the left; (b) close-up of the senser on a power mate.

(a)

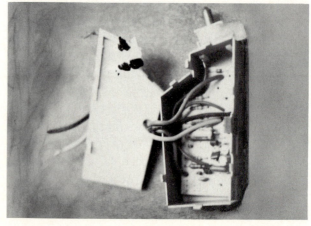

(b)

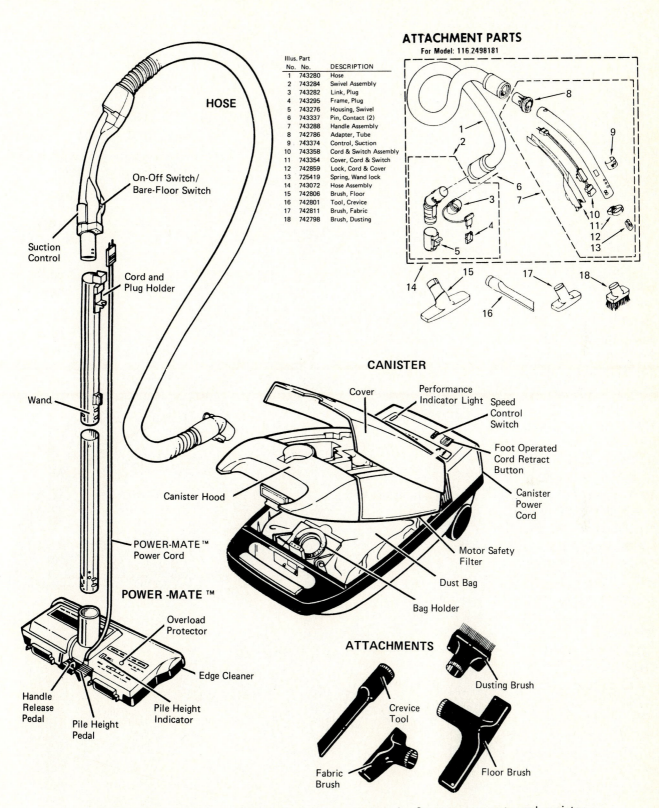

HOSE

On-Off Switch/
Bare-Floor Switch

Suction
Control

Cord and
Plug Holder

Wand

POWER-MATE™
Power Cord

POWER-MATE™

Overload
Protector

Edge Cleaner

Handle
Release
Pedal

Pile Height
Pedal

Pile Height
Indicator

Illus. No.	Part No.	DESCRIPTION
1	743280	Hose
2	743284	Swivel Assembly
3	743282	Link, Plug
4	743295	Frame, Plug
5	743276	Housing, Swivel
6	743337	Pin, Contact (2)
7	743288	Handle Assembly
8	742786	Adapter, Tube
9	743374	Control, Suction
10	743358	Cord & Switch Assembly
11	743354	Cover, Cord & Switch
12	742859	Lock, Cord & Cover
13	725419	Spring, Wand lock
14	743072	Hose Assembly
15	742806	Brush, Floor
16	742801	Tool, Crevice
17	742811	Brush, Fabric
18	742798	Brush, Dusting

ATTACHMENT PARTS
For Model: 116.2498181

CANISTER

Cover

Performance
Indicator Light

Speed
Control
Switch

Foot Operated
Cord Retract
Button

Canister
Power
Cord

Canister Hood

Motor Safety
Filter

Dust Bag

Bag Holder

ATTACHMENTS

Dusting Brush

Crevice
Tool

Fabric
Brush

Floor Brush

FIGURE 34-5. Drawing, exploded views, and parts lists for Sears power mate and canister.

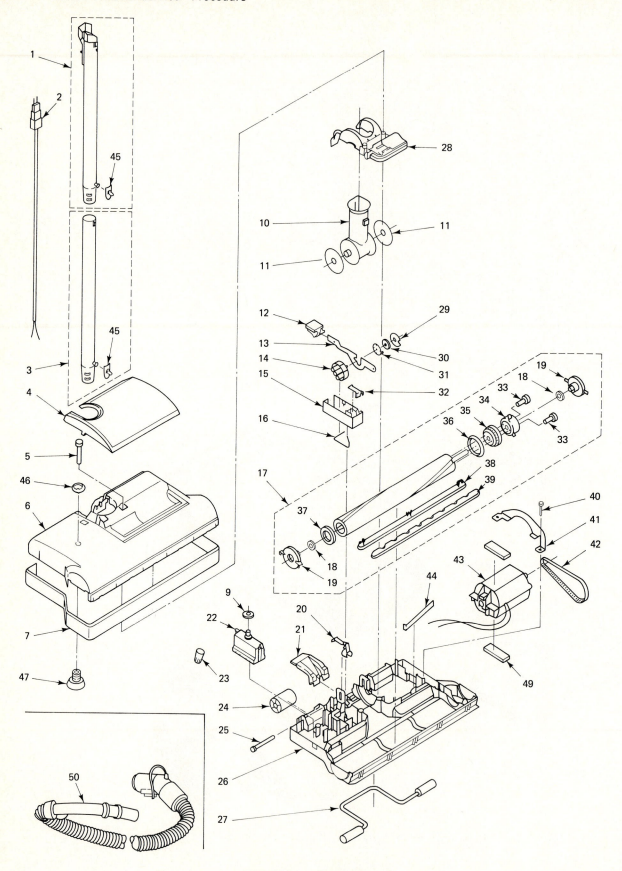

FIGURE 34-5 (Continued)

Illus. no.	Part no.	Description
1	742772	Wand and cord mount assembly
2	743131	Cord and plug
3	742488	Wand assembly (lower)
4	743637	Cover, nozzle
5	681550	Screw, 8-18 × 1-7/8 (2)
6	743626	Nozzle
7	742767	Bumper
9	742968	Seal, overload
10	742758	Swivel
11	742019	Swivel seal (2)
12	742763	Pad foot pedal
13	742017	Indicator lever
14	725725	Indicator cam
15	742015	Indicator block
16	725775	Spring assist
17	742454	Brush assembly
18	681424	Washer, metal (12 × 1/2) (2)
19	743395	Bearing and housing assembly (2) (includes illus. 18)
20	742013	Lifter
21	742755	Handle release
22	742332	Overload protector
23	596564	Wire connector
24	742765	Roller, rear (2)
25	742006	Rear axle (2)
26	743037	Base
27	744029	Front axle assembly
28	742760	Swivel cap
29	722909	Washer thrust
30	725635	Washer
31	722955	Catch, indexing
32	742016	Ratchet catch
33	681605	Screw type 25 #6 pan (2)
34	742509	End brush tufted
35	742506	Brush sprocket
36	742031	Brush sprocket flange
37	742172	String guard
38	742030	Beater brush (2)
39	742029	Beater bar
40	681297	Screw, 8-18 × 5/8 (2)
41	742422	Motor bridge
42	742024	Belt, cogged, *sears no. 20-5285
43	700853	Motor assembly
44	742002	Motor adjustment spring
45	725419	Spring wandlock
46	741261	Bezel
47	741260	Switch guide
49	742782	Gasket, motor (2)
50		Hose assembly

Refer to your canister model number for ordering the complete replacement hose or individual components.

*Available at any Sears retail or catalog store.

FIGURE 34-5. (Continued)

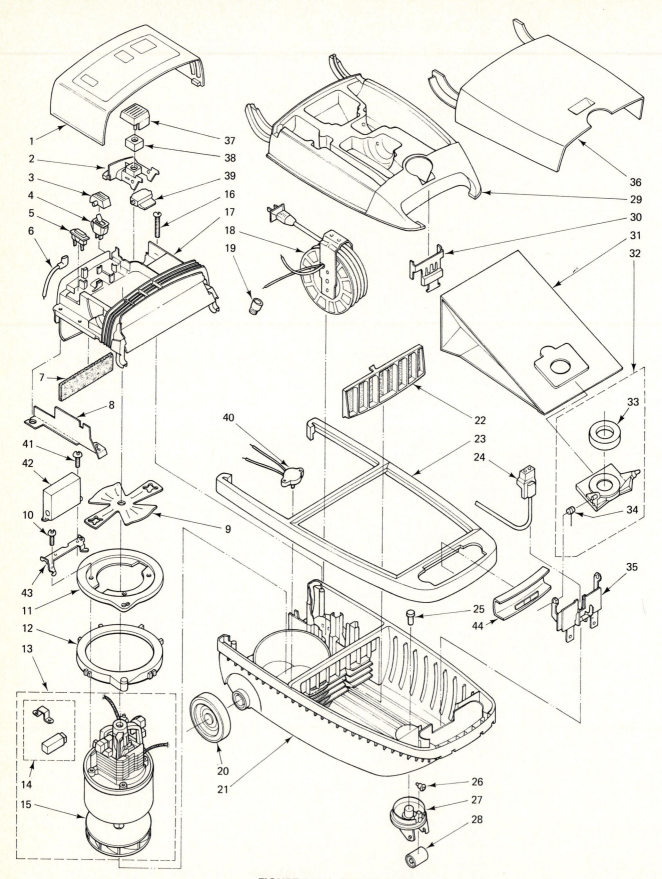

FIGURE 34-5. (Continued)

Illus. no.	Part no.	Description
1	706483	Cover, canopy
2	706520	Frame, cord lock
3	706399	Button, switch
4	706580	Switch, toggle
5	706630	Light, indicator
6	706826	Lead, indicator
7	706389	Filter, rear exhaust
8	706827	Aegis, motor
9	703515	Shield, motor
10	681546	Screw, 10-24 × 1/2 (4)
11	706011	Plate, motor mount
12	706504	Mount, motor
13	700592	Motor assembly
14	588867	Brush, motor
15		Fan motor
	588866	Upper
	588708	Lower
16	681550	Screw, 8-18 × 1-7/8 (2)
17	706026	Canopy
18	700895	Cord reel
19	596564	Connector, wire (4)
20	706007	Wheel, rear
21	706002	Base
22	706061	Filter, safety
23	706008	Bumper
24	706398	Receptacle
25	706414	Pin, caster retainer
26	706006	Bearing, caster
27	706005	Frame, caster
28	706004	Wheel, caster
29	706018	Hood
30	706019	Release, lid
31	706014	Bag, dust *sears no. 20-5055
32	676246	Mount, bag
33	706017	Gasket, part
34	706024	Spring, bag mount
35	706415	Insert, strike
36	706021	Lid, tool cover
37	706027	Button, cord reel
38	706395	Foam, button return
39	706519	Lock, cord
40	706763	Switch, vacuum
41	680968	Screw, 8-18 × 3/8
42	706397	Control, electronic
43	706577	Bracket, electronics
44	706028	Grip, handle

Following Parts not Illustrated

Lit 706842		Owner's manual
Lit 706858		Repair parts list

*Available at any Sears retail or catalog store.

FIGURE 34-5.   (Continued)

Typical Wiring Diagrams

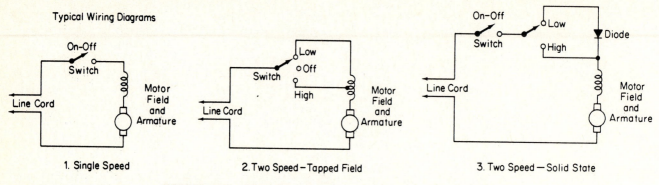

FIGURE 34-6.   Typical wiring diagrams for vacuum cleaners.

Troubleshooting Chart   VACUUM CLEANERS

Problem	Possible Cause	Corrective Action
Motor will not run.	No power.	Check outlet for power.
	Defect in cord.	Check cord plug for damaged prongs and cord for a break.
	Defective switch.	Replace.
	Worn carbon brushes in motor.	Replace with new brushes.
	Jammed fan.	Free fan. If bent or damaged, replace it.
	Frozen motor bearings.	Clean and lubricate. If too worn, replace.
	Open or shorted motor winding.	Repair or replace defective part.
	Poor contact of cord plug at wall outlet.	Check plug prongs, wire connections, and outlet.
Motor starts and stops while cleaner is being used.	Intermittent break in cord.	Shake cord to locate break.
	Loose connection in cleaner.	Check all wire connections.
	Defective switch.	Replace switch.
	A short in the wiring (will blow fuse).	Check wiring and insulate the bare spot.
Motor running too slow; no power.	Misaligned or tight motor bearings.	Realign bearings.
	Something caught on fan or armature.	Remove the foreign object.
	Burnt-out armature or field coils or poor brush contact in motor.	Repair or replace defective part.
Motor running too fast.	Fan loose on shaft and not turning with armature.	Check its balance.
	Shorted field coils.	Replace both.
	Overfilled dust bag.	Replace if disposable; otherwise thoroughly clean.
Motor sparking.	Dirty commutator (oil and/or dirt).	Clean with strip of 2/0 sandpaper.
	Worn brushes.	Replace both brushes.
	Incorrect brush seating or commutator.	Check for correct seating; allow to run free.
	Open wire in armature.	Replace or rewind armature.

Troubleshooting Chart *VACUUM CLEANERS* *(Continued)*

Problem	*Possible Cause*	*Corrective Action*
Motor noisy.	Foreign matter in motor. Fan damaged. Armature hitting. Worn bearings.	Clean out. Replace fan. Realign armature. Replace bearings.
Poor dust pickup.	Worn or damaged attachments. Leaky hose. Clogged hose. Overfilled dust bag. Clogged exhaust port. Defective motor and fan assembly.	Check attachments for cracks causing leakage and replace as needed. Check hose for air leaks and repair or replace hose. Blow or push obstruction out of hose. Replace or clean out dust bag. Clear out obstructions. See motor checkout items.
Poor pickup and cleaning (upright models, in addition to previous problems).	Broken belt. Stuck agitator brush. Incorrect nozzle adjustment for carpet nap.	Replace belt. Clean bearings free of dirt. Adjust nozzle for correct contact to carpeting.
Dust leakage into room.	Holes in dust bag. Incorrectly installed dust bag. Defective or leaky sealing gasket. Old and dirty dust bag (for machines using cloth bag).	Replace dust bag. Install per your owner's manual instructions. Replace gasket. Replace dust bag.
Cord does not retract.	Cord is dirty.	Clean and wax cord.

35

Automatic Washers

The automatic washer is an appliance designed to wash, rinse, and extract water from clothes automatically. It should remove soil from most fabrics quickly and completely. It should retain the maximum whiteness in white things and destroy bacteria that may be harmful to health—when used with the recommended water and cleaning aids, that is. All this a washer can do—IF it is in proper mechanical condition.

The location and/or installation of the machine has some minimum requirements that should be remembered:

1. An adequate source of hot [145°F (63°C)] and cold water.
2. An electrical supply of 117-volt, 60-hertz branch circuit protected with a 15-ampere slow blow-type fuse.
3. The drain must be a standpipe or tub at least 32 inches above the floor.
4. The floor must be solid and the machine set level.

Another good idea would be to keep the machine protected from weather and freezing temperatures. There is always some water left in the pump and hoses, and cold weather never helped oils and greases!

OPERATING PRINCIPLE

The basic method of washing action is a reciprocating agitator that swirls the clothes back and forth within a tub of water and such washing aids as may be used. Figure 35-1 shows the tub and agitator. Figure 35-2 shows the mechanism that drives the agitator, and Figure 35-4 shows the gearing arrangement that converts the rotary motion from the motor to the reciprocating action required for moving the load.

CONTROLS

Once set and started, an automatic washer will continue through its selected sequence of actions until completion of the cycle. The available sequences of action and the placement of controls will be different for each make and model.

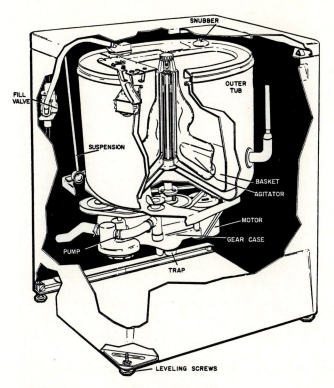

FIGURE 35-1. (Courtesy of Whirlpool)

The owner's manual for any particular machine contains specific information on its control, uses, and settings. *Be sure you study it.*

The timer is the brains of an automatic

FIGURE 35-2. Automatic washer gear case assembly. (Courtesy of Whirlpool)

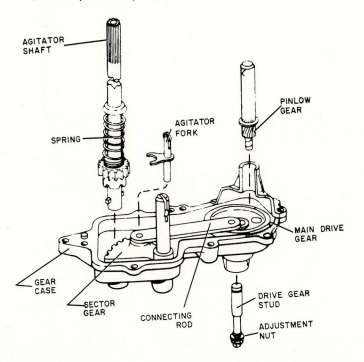

washer (Fig. 35-3). It may be a clock motor driving a sequence of cam switches, or it may be an electronic module that controls the timing of the various valves and motors that make things happen in the machine. So before trying to fix an automatic washer, make certain that you know:

1. How the machine should operate
2. What the controls do
3. Which noises are right
4. What actions to expect

You should know that in many washers, when water is running into the machine, the timer motor is stopped. The amount of water used is "watched" by the water level control. When the right amount of water is in the tub, this control will shut off the fill valves and start the timer motor.

The fill valves automatically mix hot and cold water to provide water at a preselected temperature. Water flow is solenoid controlled. Sometimes temperature is controlled by the thermal element in the mixing chamber, otherwise by proportional mixing. Repair parts for fill valves are available from most sources, but there is an area of judgment as to whether the valve should be repaired or replaced. The overall condition of the valve and relative costs should be the deciding factors.

The water level control may be either a float or a pressure switch (Fig. 35-5). The most common control is the latter. In machines that have an adjustable water level, the control is usually located in the console (control panel). In machines where the water level is fixed, you will have to look for the control.

ADJUSTMENTS

Having made certain that power and water are available at the machine before checking any other part of it, you can make the following adjustments and repairs with a minimum of tools and effort.

The fill (water inlet) hoses include a filter washer installed cup side up in the hose end that is attached to the faucet. Poor water flow

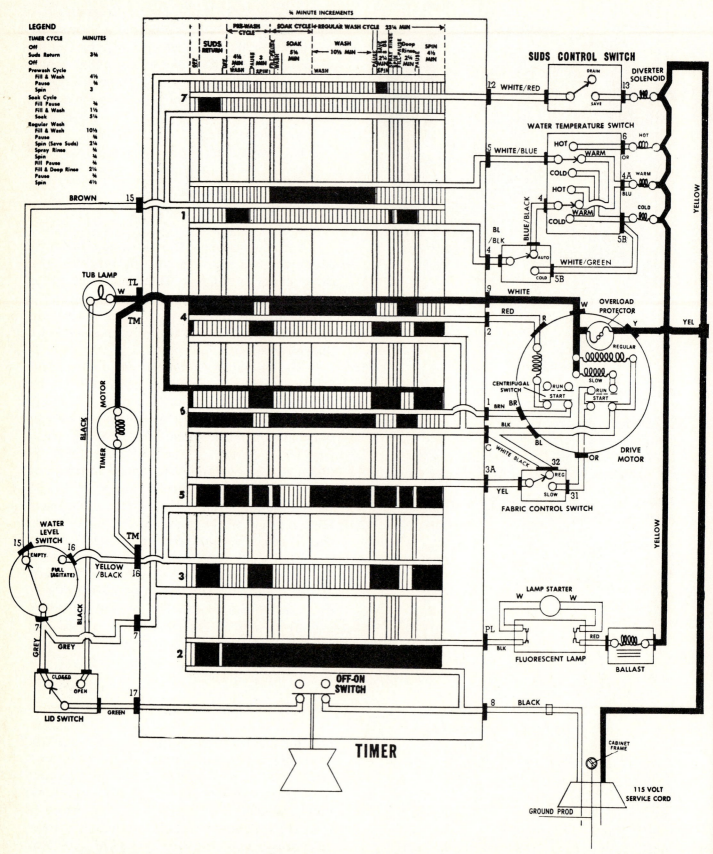

FIGURE 35-3. Typical schematic wiring diagram with cycle chart shown in timer. All wiring shown as connected to timer. (Courtesy of Maytag)

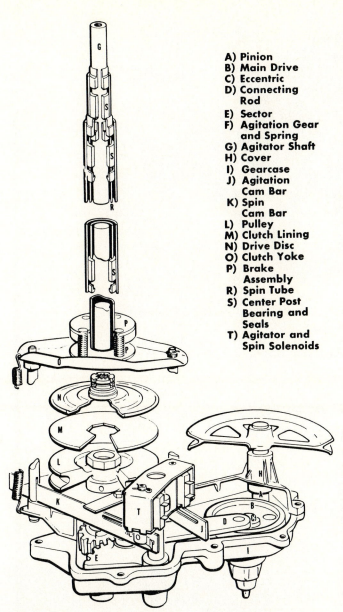

A) Pinion
B) Main Drive
C) Eccentric
D) Connecting Rod
E) Sector
F) Agitation Gear and Spring
G) Agitator Shaft
H) Cover
I) Gearcase
J) Agitation Cam Bar
K) Spin Cam Bar
L) Pulley
M) Clutch Lining
N) Drive Disc
O) Clutch Yoke
P) Brake Assembly
R) Spin Tube
S) Center Post Bearing and Seals
T) Agitator and Spin Solenoids

FIGURE 35-4. Whirlpool gear case assembly. (Courtesy of Whirlpool)

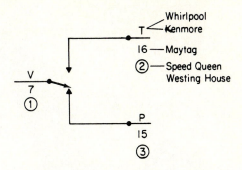

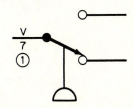

Representative Pressure Switch Terminal Identification. Shown in the No Pressure (Empty) Position

Schematic Symbol for the Pressure Switch

FIGURE 35-5. Schematic diagrams for automatic washers.

may be due to a clogged filter, which can be removed, flushed out, and replaced.

The water pump recirculates the wash water, pumps out the water, and pumps suds back into the machine (in suds saver models) (Fig. 35-6). Occasionally a button, bobby pin, toothpick, or an accumulation of lint will lodge in the pump, causing it to jam. If you are lucky, the jamming particles can be removed by removing the bottom cover plate of the pump. Do not forget to replace the gasket when reassembling this plate.

A slipping belt is the cause of approximately 60 percent of the mechanical failures in washers (Fig. 35-7). To adjust the belt tension in the Kenmore and Whirlpool machines,

1. Remove rear service panel.
2. Loosen the nut holding the motor mounting bracket (Fig. 35-8). Grasp the motor mounting bracket with pliers or a Crescent wrench and pull the motor assembly against the belt to increase the tension. While holding this position, tighten the nut.
3. Correct belt tension allows about $\frac{3}{4}$-inch deflection with firm pressure.

To replace belt: remove the rear service panel. It can be replaced through the rear service opening. It is easier to lay the machine on its front, exposing the underside. It would be a good idea to place a rug or pad on the floor, before tipping the machine, to protect the machine's finish. The following step-by-step procedure applies specifically to Whirlpool and Kenmore machines (Fig. 35-9).

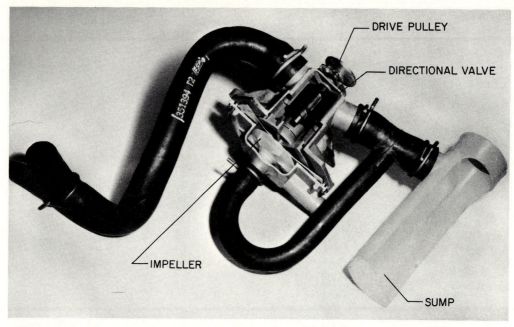

FIGURE 35-6.   Sectional view of the bidirectional pump and sump used in Whirlpool washers. Note the flapper valve used to divert the water flow from "recirculate" to "drain."

FIGURE 35-7.   Proper belt tension will allow about a $\frac{3}{4}$-inch deflection.

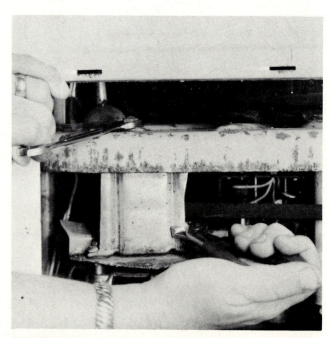

FIGURE 35-8.   To adjust the belt tension on Whirlpool and Kenmore washers, loosen the $\frac{9}{16}$-inch nut, grasp the motor mounting bracket, and pull against the belt. Then loosen the $\frac{9}{16}$-inch nut.

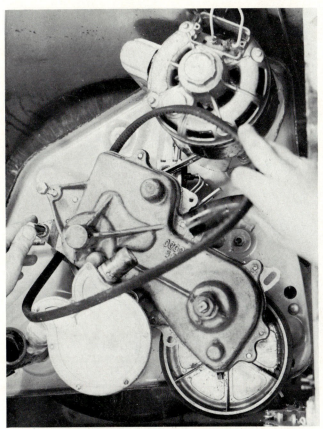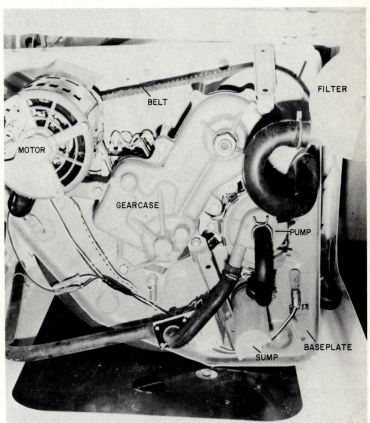

FIGURE 35-9. (a) Belt replacement on a Whirlpool washer. This is an older machine. (b) A view similar to that of part (a) on a current-model Whirlpool. The belt replacement procedure is essentially the same, although the pump design and hose routing have changed.

1. Loosen the nut holding the motor-mounting bracket to relieve belt tension.
2. Remove belt from pulleys.
3. Remove gear case mounting stud and take out spacers; allowing belt to pass through this space.
4. Unhook the brake shoe yoke spring.
5. Slide the cam bar out of the slot in the clutch yoke stud by pushing the bottom of the stud to free pressure on the cam bar. Rotate pulley to move cam bar out of slot; then push the cam bar.
6. After cam bar is free, the stud can be shifted to allow the belt to pass between the top of the stud and the clutch yoke.
7. Disconnect hoses from the pump.
8. The belt is now free to be removed.

9. Insert new belt and reverse the disassembly procedure to reinstall.

Figures 35-10 to 35-12 show the adjustment procedures for some other machines.

Agitation and spin, for Whirlpool and Kenmore, are both controlled by the solenoids and cam bars mounted on top of the gear case. The solenoid plungers are each linked to its cam bar by a rivet or a cotter pin. If one of these devices breaks, the solenoid assembly will continue its oscillation but the cam bar will not move. To replace a broken cotter pin:

1. Rotate pulley until solenoid is in a convenient position.
2. Remove broken ends of rivet or pin.
3. Install new cotter pin (or rivet), making sure that the guides are in position—one on each side of cam bar.

FIGURE 35-10.   Belt tension on this Kelvinator can be adjusted with this nut. *Don't overdo it.*

FIGURE 35-11.   To replace the Kelvinator belts: (1) Lay the washer on its front; (2) remove the four bolts holding the cross brace; (3) loosen the pump from the cross brace; (4) remove the cross brace. Now all belts are accessible. Reverse the process to reassemble.

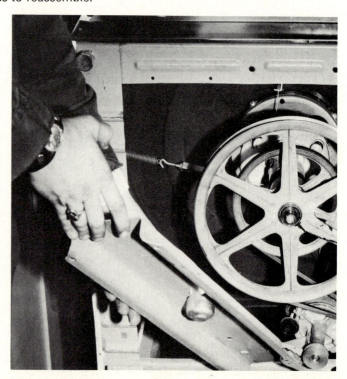

To adjust the belt tension on a General Electric washer, loosen the four motor-holding nuts, pull the motor firmly against the belt, and tighten securely (Fig. 35-13).

To change belts on this machine, detach the flexible coupling, loosen the motor-holding nuts, remove the old belt, install the new belt, and tighten the motor securely against the belt. Replace the flexible coupling that drives the two-way pump.

If this flexible coupling breaks or comes loose, there will be no water recirculation or water pumped out.

In the GE the suds-return pump and the distribution valve assembly are serviced by removing the four screws holding the mounting bracket to the suspension and by sliding the entire assembly out the back. It is easier to do this if you disconnect the hose from the distribution valve at the two-way pump end. A pair of Corbin pliers is the correct tool to use with these clamps. A defective solenoid at this point or a stuck distribution valve will affect recirculation, pumpout, and suds saving. The suds-return pump can be cleaned out by removing the four Phillips head screws holding the pump housing to the frame.

Service problems can be avoided, to a large extent, by preventative maintenance and proper, periodic lubrication.

1. Remove the agitator and clean both its underside and the bracket center post once a month.
2. Apply a few drops of oil to the two holes above the control magnet plungers and around its pivot studs at least once a year.
3. Twice a year, soak the water pump oil wick.
4. Lubricate the cam bars annually.

Figures 35-14 and 35-15 present a series of views showing parts locations of a number of different automatic washers.

Schematic wiring diagrams and a cycle of operations chart can be found on the back of each machine (Fig. 35-16). These should be followed when troubleshooting, to determine which part is not functioning.

FIGURE 35-12. The belts on this Maytag are readily accessible from the underside. It may be necessary to loosen the three water pump mounting screws to adjust the belt tension.

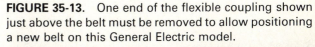

FIGURE 35-13. One end of the flexible coupling shown just above the belt must be removed to allow positioning a new belt on this General Electric model.

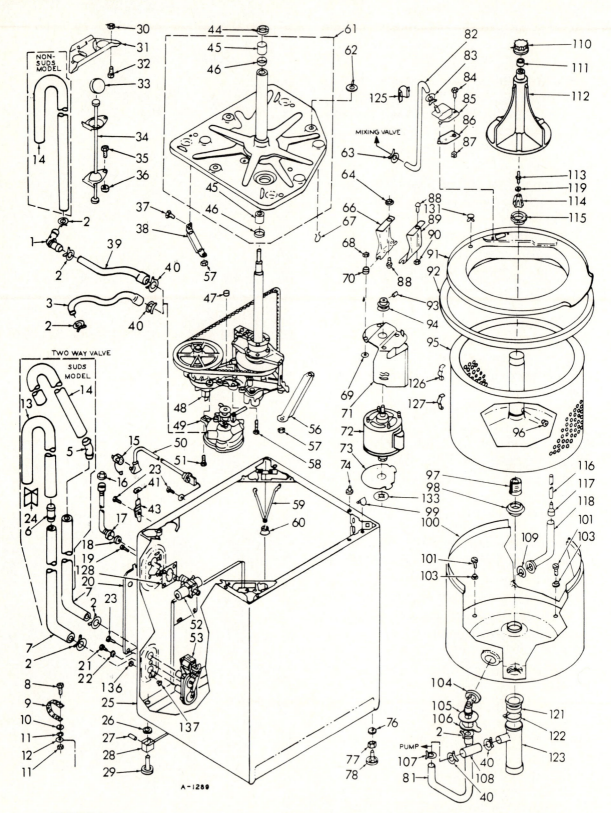

FIGURE 35-14. Complete automatic washer parts listing. (Courtesy of Sears Roebuck & Co.)

Key Number	Part Number	Description
		MACHINE SUB-ASSEMBLY
1	95440	Connector (110.6804500 and 110.6804550)
2	272856	Clamp, Hose
3	21115	Hose (110.6805500 and 110.6805550)
5	96756	Connector (110.6805500 and 110.6805550)
6	93229	Connector (110.6805500 and 110.6805550)
7	96746	Hose, Outlet (110.6805500 and 110.6805550)
8	*132848	Screw, No. 10-24 x 1
9	22965	Strap, Ground
10	*446152	Washer, No. 10
11	*120361	Nut, No. 10-24
12	16324	Wire, Ground
13	91273	Hose, Gooseneck (110.6805500 and 110.6805550)
14	96743	Hose, Gooseneck
15	92925	Adaptor, Power Cord
16	21092	Washer, Filter (2)
17	81339	Hose, Inlet (2)
18	16123	Washer, Rubber (2)
19	90864	Screw, No. 10-24 x 3/8
20	85279	Panel, Rear
21	*177359	Screw, 1/4-28 x 1/2 (110.6805500 and 110.6805550)
22	*121637	Lockwasher, 1/4 (110.6805500 and 110.6805550)
23	*98605	Screw No. 10-16 x 1/2
24	16268	Strainer, Hose (110.6805500 and 110.6805550)
25		Cabinet
	82470	(110.6804500 and 110.6805500)
	82473	(110.6804550 and 110.6805550)
26	98411	Washer, Friction
27	98439	Pin, Knurled
28	75587	Leveling Link Assembly
29	95879	Foot, Rear (2)
30	*90984	Nut, 5/16-18
31	94114	Cover, Suspension Rod (3)
32	*126358	Bolt, 5/16-18 x 1 (4)
	*126211	Bolt, 5/16-18 x 5/8 (2)
33	91073	Ball, Suspension (6)
34	82584	Rod, Suspension (3)
35	*425594	Bolt, 5/16-24 x 1 1/4
36	*138485	Lockwasher, 5/16
37	*98062	Screw, 5/16-24 x 3/4
38	82466	Brace
39	21114	Hose (110.6804500 and 110.6804550)
40	272858	Clamp, Hose
41	18849	Pad Hinge
43	92657	Hinge—LH
	18775	Hinge—RH
44	91938	Seal
45	54207	Bearing
46	54209	Seal
47	92429	Spacer
48	84470	Gear Case Assembly (Refer to Pages 6 and 7)

Key Number	Part Number	Description
49	87659	Pump Assembly (Refer to Page 10)
50	92667	Cord, Power 120 Volt
	83742	Cord, Power 240 Volt
51	96867	Screw
52	86793	Valve, Mixing—Alternate (Refer to Page 5)
	86794	Valve, Mixing—Alternate (Refer to Page 5)
	86829	Valve, Mixing—Alternate (Refer to Page 5)
53	75618	Two Way Valve Assembly (110.6805500 and 110.6805550) (Refer to Page 10)
56	93670	Brace—Long
	93671	Brace—Short
57	*96101	Nut, 5/16-24
58	*90877	Screw, 3/8-24 x 2
59	93434	Spring, Snubber
60	93061	Pad, Snubber
61	76527	Baseplate Assembly
62	17235	Gasket, Tub
63	272846	Clamp, Hose
64	*96016	Nut, 3/8-16
66	85286	Bracket, Adjusting
67	16515	Clip, Harness
68	*98023	Nut, No. 10-32
69	92217	Washer
70	19138	Grommet (4)
71	93181	Shield, Motor
72		Motor
	84602	60 Cycle—Alternate
	84811	60 Cycle—Alternate
	84813	60 Cycle—Alternate
	84829	60 Cycle—Alternate
	84986	60 Cycle—Alternate
	86508	60 Cycle—Alternate
	86805	60 Cycle—Alternate
	82052	50 Cycle
73	93182	Shield, Bottom
74	98669	Locator, Top (2)
76	98684	Washer (110.6804550 and 110.6805550)
77	*120238	Nut, 1/2-13
78	95878	Foot, Front (2)
81	92272	Hose, Recirculating
82	96625	Hose, Water Inlet
83	272847	Clamp, Hose
84	*98709	Screw, No. 8-32 x 1/2
85	93325	Inlet, Water
86	92748	Gasket, Inlet
87	98688	Nut, Snap-In
88	*98537	Bolt 3/8-16 x 3/4
89	93549	Bracket, Pivoting
90	*95721	Nut, 3/8-16
91	93263	Ring, Tub
92	91188	Rubber, Tub Ring
93	*90879	Screw, 5/16-24 x 1/2 Set
94	92216	Pulley—60 Cycle
	93219	Pulley—50 Cycle
95	82446	Basket
96	21265	Stopper

FIGURE 35-14. (Continued)

Key Number	Part Number	Description
97	96384	Block, Drive
98	21050	Gasket, Tub
99	18776	Lock, Front Top (2)
100	81393	Tub
101	19021	Screw, Tub (4)
103	21365	Washer, Rubber (4)
104	52721	Funnel, Drain (Alternate)
	92555	Funnel, Drain (Alternate)
105	50909	Gasket
106	50926	Lockwasher
107	272853	Clamp, Hose
108	92231	Hose
109	17228	Ring Retainer
110	75064	Cap, Agitator
111	17626	Sleeve
112	85174	Agitator
113	95122	Stud, Shaft
114	95107	Collar, Drive
115	21366	Nut, Lock
116	21046	Tubing, Pressure
117	97723	Connector
118	95338	Hose
119	120393	Washer
121	93553	Hose, Tub Outlet
122	9421321	Clamp, Hose
123	82621	Manifold (Alternate)
	84091	Manifold (Alternate)
125	93828	Clip, Hose
126	98103	Clip, Tub Ring (3)
127	98196	Clip, Tub Ring (1)
128	94164	Barrier (Use W/86793 and 86794 Valves)
131	98187	Plug, Tub Ring
133	98892	Clip, Motor Shield
136	95096	Grommet (110.6805500 and 110.6805550)
137	95097	Bushing (110.6805500 and 110.6805550)

Following Parts Not Illustrated
95931	Instructions, Installation
351287	Manual, Owner's
351289	List, Parts

The Following Parts Used With 240 Volt Installations
85502	Bracket, Transformer
94291	Shield, Transformer
94368	Transformer

*Standard Hardware Item—
May Be Purchased Locally

FIGURE 35-14.   (Continued)

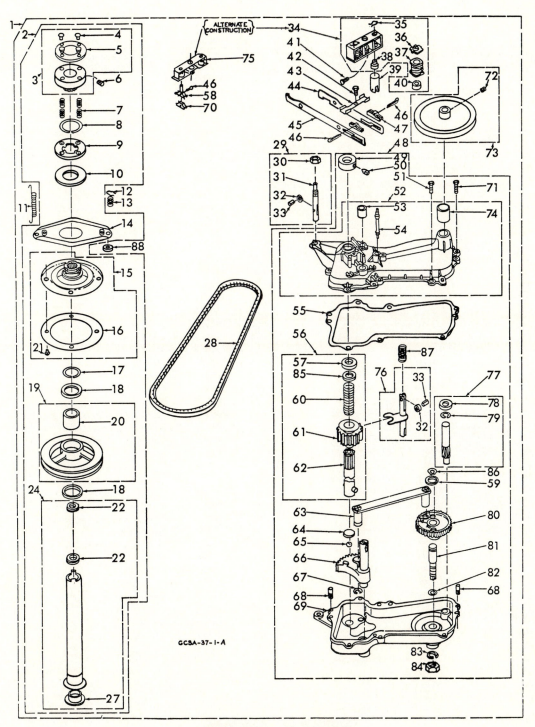

GCSA-37-1-A

FIGURE 35-14. (Continued)

Key Number	Part Number	Description
		GEAR CASE AND SUPERSTRUCTURE ASSEMBLY
1	84470	Gear Case and Superstructure Assembly
2	84533	Basket Drive and Brake Assembly
3	75078	Brake Drum Assembly
4	98281	Rivet, Tubular
5	16039	Lining, Brake
6	*90880	Screw, 5/16-18 x 3/4 Set
7	19049	Spring, Drive
8	16049	Ring, Snap
9	18929	Drum, Lower Brake
10	21094	Lining
11	17261	Spring
12	52097	Spring, Retainer
13	52446	Spring, Control Rod
14	92903	Yoke, Brake (Alternate)
	18744	Yoke, Brake (Alternate)
15	87319	Disc and Lining Assembly
16	95162	Lining
17	90368	Ring, Retainer
18	16304	Washer, Thrust
19	82949	Drive Pulley Assembly
20	16033	Bearing
21	90683	Rivet
22	91939	Seal, Oil
24	76674	Drive Tube and Bearing Assembly
27	19438	Bearing, Thrust
28	95405	Belt, "V"
29	76220	Shaft Assembly
30	*90715	Nut, 1/2-20 Self Locking
31	18930	Shaft, Clutch
32	92119	Roller, Shaft (Alternate)
	16010	Roller, Shaft (Alternate)
33	90532	Pin, Roller
34	76483	Magnet, Control
35	90515	Spring, Retainer
36	85501	Spring, Support
37	75791	Solenoid (Alternate)
	81433	Solenoid (Alternate)
38	85499	Core
39	16241	Plunger
40	85500	Sleeve
41	16135	Screw, Taper End
42	*98001	Screw, 1/4-28 x 1/2
43	94726	Spring, Brake
44	18642	Cam, Agitator and Pump
45	16051	Cam, Basket
46	90456	Pin, Cotter
47	93869	Guide, Cam Bar

Key Number	Part Number	Description
48	84473	Gear Case Assembly
49	16044	Collar, Support
50	*90207	Screw, 5/16-24 x 1/2 Set
51	*120741	Bolt, 5/16-24 x 3/4
52	76531	Gear Case Cover Assembly (Alternate)
	87889	Gear Case Cover Assembly (Alternate)
53	5676	Bearing—11/16" O.D.
	96853	Bearing—7/8" O.D.
54	93850	Support, Brake Shoe Yoke
55	95089	Gasket, Gear Case
56	84468	Agitator Shaft Assembly
57	92814	Seal
58	84866	Plunger
59	95083	Washer, Retainer
60	92815	Spring
61	17553	Gear, Agitator Shaft
62	84474	Shaft, Agitator
63	76577	Rod, Connecting
64	16018	Bearing, Thrust
65	85529	Ball, Thrust
66	75073	Gear, Sector
67	52083	Ring, Snap
68	5018	Pin, Dowel
69	95095	Case, Gear
70	95462	Guide, Cam Bar
71	*98048	Bolt, 5/16-24 x 1 1/8
72	*90461	Screw, 5/16-18 x 5/16 Set
73	75528	Pulley, Agitator Drive
74	16308	Bushing, Drive Pinion
75	84867	Wig Wag
76	80398	Fork, Agitator Gear
77	84467	Drive Pinion Assembly
78	16024	Washer, Tension
79	90094	Ring, Retaining (Alternate)
	90369	Ring, Retaining (Alternate)
80	95085	Gear, Drive
81	95084	Stud, Drive Gear
82	95080	Ring, "O"
83	*103336	Lockwasher, 7/8
84	*219754	Nut, 7/8-14
85	94276	Shield, Agitator Spring
86	95082	Ring, Retaining
87	16015	Spring
88	97329	Washer

Following Part Not Illustrated

	89122	Oil, Gear Case (Pint Can)

*Standard Hardware Item— May Be Purchased Locally

FIGURE 35-14. (Continued)

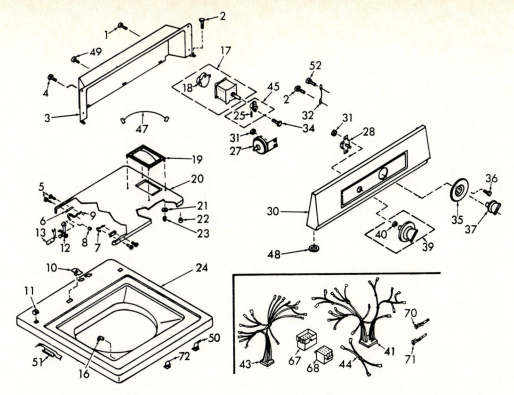

Key Number	Part Number	Description
		TOP AND CONSOLE ASSEMBLY
1	*98559	Screw, No. 10-16 x 3/8
2	*98605	Screw, No. 10-16 x 1/2
3	97931	Cover, Console
4	*98209	Screw, 8-32
5	18724	Screw and Washer
6	21097	Pad, Lid Hinge
7	91770	Hinge, Front
8	19119	Bumper
9	91367	Hinge, Rear
10	92129	Clip, Console
11	98219	Nut, Snap-In
12	96410	Bracket, Lid Switch
13	87437	Switch, Lid (Alternate)
	87438	Switch Lid (Alternate)
16	21258	Bearing (2)
17	87439	Timer Assembly—60 Cycle (Alternate)
	87321	Timer Assembly—60 Cycle (Alternate)
	87502	Timer Assembly—50 Cycle
18	81124	Motor, Timer (Use W/87439 Timer)
	87114	Motor, Timer (Use W/87321 Timer)
	81123	Motor, Timer (Use W/87502 Timer)
19	96188	Handle, Lid
20	96374	Lid (110.6804500 and 110.6805500)
	96376	Lid, Porcelain (110.6804550 and 110.6805550)
21	14351	Washer, Fibre
22	93770	Bumper, Rubber
23	98591	Nut, Acorn
24	97956	Top (110.6804500 and 110.6805500)
	360052	Top, Porcelain (110.6804550 and 110.6805550)

Key Number	Part Number	Description
25	98654	Screw, No. 10-32 x 3/16 Set
27	88930	Switch, Water Level (Alternate)
	88931	Switch, Water Level (Alternate)
28	97979	Switch, Water Temperature
30	351018	Escutcheon
31	98788	Self-Threading Nut
32	54570	Wire, Ground
34	*98614	Screw, No. 10-32 x 5/16
35	351087	Skirt, Timer Dial (110.6804500 and 110.6804550)
	351073	Skirt, Timer Dial (110.6805500 and 110.6805550)
36	*155028	Screw, No. 8-32
37	351090	Knob, Timer
39	360004	Water Temperature Knob Assembly
40	97084	Clip, Compression
41	360242	Harness, Wiring—Cabinet
43	360243	Harness, Wiring—Console
44	360461	Harness, Wiring—Two Way (1106805500 and 110.6805550)
45	360088	Timer Extension Assembly
47	86820	Wire, Jumper (110.6805500 and 110.6805550)
48	96870	Plug
49	*98605	Screw, No. 10-16 x 1/2
50	19107	Clip, Harness
51	97513	Bracket (Steel)
	351329	Holder (Nylon)
52	98209	Screw
67	94612	Cap, Terminal—Console Harness
68	94611	Plug, Terminal—Cabinet Harness
70	94613	Terminal—Male
71	94614	Terminal—Female
72	90016	Clip, Harness

*Standard Hardware Item—
May Be Purchased Locally

FIGURE 35-14.   (Continued)

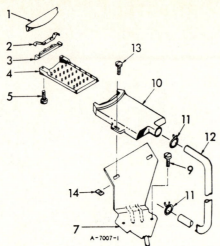

Key Number	Part Number	Description
		FILTER ASSEMBLY
1	94401	Handle
2	94403	Spring
3	94402	Bar, Release
4	94400	Cartridge, Filter
5	*176360	Screw, No. 8-32 x 5/8
7	91602	Bracket, Filter
9	*273556	Screw, No. 10 x 5/8
10	94399	Housing, Filter
11	272853	Clamp, Hose
12	94749	Hose, Recirculating
13	*98881	Screw, No. 10-16 x 5/8
14	*90416	Nut, Spring—No. 10

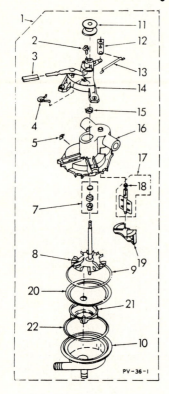

Key Number	Part Number	Description
		PUMP ASSEMBLY
1	87659	Pump Assembly
2	*97787	Screw, 10-16 x 3/4
3	96175	Sleeve
4	96266	Spring, Toggle
5	96262	Clip, Retainer
7		Seal Assembly
	87394	Alternate
	87525	Alternate
	87535	Alternate
8	87617	Impeller and Shaft Assembly
9	96988	Gasket
10	96754	Housing
11	16113	Pulley
12	92833	Lever, Valve
13	96264	Rod, Connecting
14	87290	Bearing Block Assembly
15	96265	Wick, Oil
16	96259	Body, Pump
17	87288	Valve and "O" Ring Assembly
18	21102	Ring, "O"
19	96271	Insert
20	97525	Plate, Divider
21	92344	Impeller
22	97316	Gasket

*Standard Hardware Item—
May Be Purchased Locally

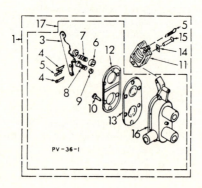

Key Number	Part Number	Description
		TWO-WAY VALVE ASSEMBLY
1	75618	Valve Assembly
3	16292	Lever, Operating
4	*90222	Pin, 3/32 x 1/2 Cotter
5	*90228	Pin, 1/8 x 3/4 Cotter
6	16299	Washer, Cup
7	16291	Spring
8	16287	Spring, Valve
9	16295	Washer, Valve Spring
10	98666	Screw, No. 8-32 x 5/8
11	86798	Solenoid
12	16293	Cover, Valve
13	75183	Diaphragm
14	90296	Clip, Stud
15	451637	Rivet
16	85205	Body, Valve
17	99830	Valve Sub-Assembly

*Standard Hardware Item—
May Be Purchased Locally

FIGURE 35-14. (Continued)

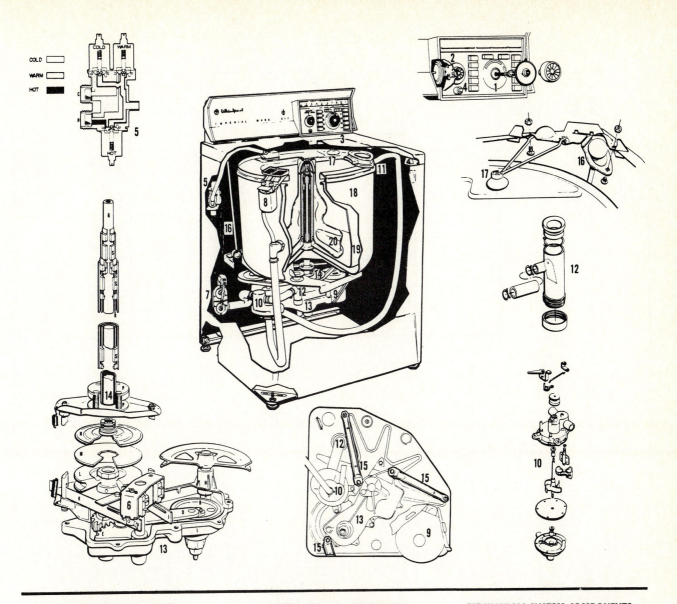

ELECTRICAL COMPONENTS

- **TIMER ASSEMBLY (1)**
  Push Button Assembly - Timer Switch
  Assembly - Timer Motor
  4 Cycle, 12 Selections
  Controls...
  > On-Off - Line switch
  > Circuits to operate solenoids
  > Water temperature - 4 Wash, 2 Rinse -
  > All automatic
  > Motor speed - 2 (1725 RPM - 1140 RPM)

- **THREE SWITCHES**
  PRESSURE (2) - Single pole, double throw
  Controls thru timer...
  > Water level - Manual infinite 7" - 11"
  > Agitator solenoid
  > Spin solenoid
  > Motor

  SPIN SWITCH (3) - Mercury type - Mounted to
  lid hinge
  Controls thru timer...
  > Spin solenoid

  LIGHT SWITCH (4) - Ballast
  Controls independent to timer...
  > Fluorescent light circuit

- **EIGHT SOLENOIDS**
  THREE - WATER INLET MIX VALVE (5)
  CONTROL MAGNET (6)
  > One - Agitation
  > One - Spin
  ONE - TWO-WAY VALVE (7)
  TWO - DISPENSERS (8)

- **MOTOR ½ HP 2-SPEED-CAPACITOR (9)**

WATER SYSTEM COMPONENTS

- **WATER INLET MIXING VALVE (5)**
  3 SOLENOID
  Non-Thermostatic
  Controls Water Temperature
  > Hot - Hot water solenoid energized
  > Cold - Cold water solenoid energized
  > Warm - Warm water solenoid energized
  > Medium - Hot and Warm solenoids
  > energized

- **TWO-WAY VALVE (7)**
  Control Wash Water Flow To:
  > Suds storage
  > Drain
  > Suds return

- **UNI-DIRECTIONAL PUMP (10)**
  Pump Operates When Motor is Running
  Flow Controlled Mechanically by Agitation
  Cam-Bar Controls...
  > Pump Out - Not Agitating
  > Suds Return - Agitating
  > Filter Recirculation - All times with
  > minimum seven inches of water

- **FILTER (11)**
  Brush Type
  Detergent Dispenser

- **DISPENSER (8)**
  Manual Fill
  Automatic Solenoid Operation
  Bleach-Dispensed 2 Increments Before End of
  Wash Cycle
  Conditioner - Dispensed during Deep Rinse Fill

- **TUB OUTLET (12)**
  Manifold and Trap Assembly

MECHANICAL SYSTEM COMPONENTS

- **GEAR CASE ASSEMBLY (13)**

A) Pinion	E) Sector	I) Gearcase
B) Main Drive	F) Agitation Gear	J) Agitation
C) Eccentric	and Spring	Cam Bar
D) Connect in	G) Agitator Shaft	K) Spin
Rod	H) Cover	Cam Bar

- **BASKET DRIVE & BRAKE ASSEMBLY (14)**

L) Pulley	P) Brake	S) Center Post
M) Clutch Lining	Assembly	Bearing and
N) Drive Disc	R) Spin Tube	Seals
O) Clutch Yoke		

- **BASEPLATE BRACES (15)**

- **SUSPENSION (16)**
  New Rods
  New Upper Mounting

- **SNUBBER (17)**
  New Mounting and Spring

- **OUTER TUB (18)**

- **BASKET (19)**

- **AGITATOR (20)**

FIGURE 35-15. Automatic washer components. (Courtesy of Whirlpool)

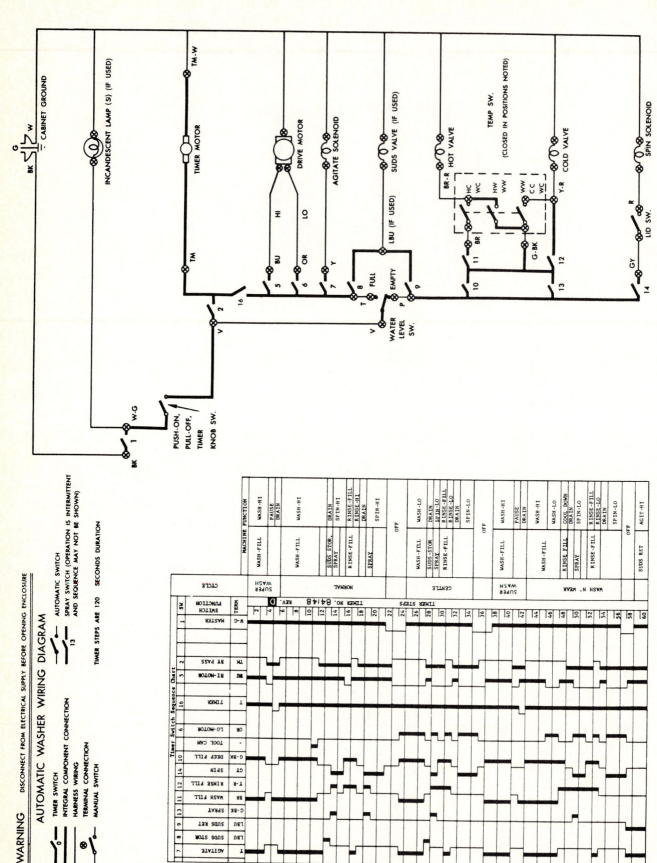

FIGURE 35-16. Typical schematic wiring diagram and cycle chart found on the back of Whirlpool and Kenmore washers. All manufacturers supply similar information. (Courtesy of Whirlpool)

454

We cannot stress enough that washability depends on: loading, the type of water, its temperature, the washing aids used, as well as on the mechanical workings of the machine. *Do not condemn the machine if these other factors are not correct!*

ELECTRONIC CONTROL SYSTEMS

The electronic controls used in an automatic washer replace the clock timer in that they control the timing and sequence of machine operation. The mechanics and duties of the other parts of the machine remain the same.

To operate the washer the water level is selected, power switch is ON, fabric selection is made, and water temperature and other options are selected. After the control system receives its input, it proceeds to provide output to the motor relays and solenoids in proper sequence and time. Figure 35-17 is a cycle sequence chart of these functions.

Control systems have gone through several stages of development as the electronic industry has developed new and more sophisticated circuits and components. This is reflected in system changes in automatic washers and other appliances. Although the actions of the washer will remain the same, the individual controls may vary. Try to use the wiring diagram and information for the specific machine you are servicing. The wiring diagram and associated information in Fig. 35-18 is included to serve as a guide to the relationship of the control system to the rest of the machine.

The control system is wired so that the part being controlled is connected to the black (hot) line, and the switching occurs at the white (neutral) line. Washers with clock timers are connected with parts being controlled to the white (neutral) line and the timer switches to the black (hot) line.

All switched circuits except the start relay are controlled by a triac. The start relay is controlled by a transistor. This supplies 8 volts dc to the start relay to hold it closed. All the other switched circuits are 120 volts ac. The switching triacs and their control circuitry are in the microcomputer assembly. The LED

(light-emitting diode) display assembly and touch control module are connected to each other and to the microcomputer by flexible circuit connectors (Fig. 35-19).

MICROCOMPUTER

The parts in the control assembly are fragile and sensitive to handling, static electricity, and excessive voltage changes.

1. Drain your body static before working on the controls.
2. Do not kink or stress the flexible circuit connector.
3. Check all connectors for secure contacts.
4. Keep your fingers off the microcomputer terminals.
5. Check the power outlet for polarity and a valid ground.
6. Check the electrical components in the washer before checking the microcomputer. Unplug the power cord, disconnect the edge connector, and test continuity for open or shorts.
7. If the washer fails to start and there is power at the outlet and the power cord is good, check the push-to-start relay for poor switch contact or an open coil. A good coil should read 150 ± 30 ohms of resistance. With the switch held in place, there must be 120 volts ac at terminals V and W at the step-down transformer and 9.5 volts ac at the transformer edge connector between W-OR and BK-OR.

If the transformer is being supplied with 120 volts ac and is not providing the correct secondary voltage, the transformer is defective. If the transformer is providing the correct secondary voltage, check for the correct dc volts between the "main relay" and the "triac reference" at the microcomputer edge connector. Make certain that the edge connector terminals are okay.

When the correct ac voltage is being supplied to the microcomputer and no or incorrect

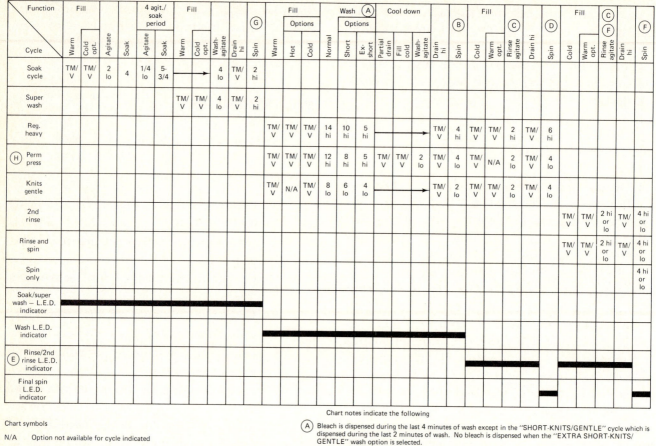

Chart symbols

N/A — Option not available for cycle indicated

TM/V — Time variable – sensed function

→ — Cycle skips function

▬ — L.E.D. indicator "on"

Time indicated is in minutes and motor speed is indicated as "hi" or "lo"

Example: $\frac{4}{hi}$ — 4 minutes hi motor speed

○ — Chart note

Chart notes indicate the following

(A) Bleach is dispensed during the last 4 minutes of wash except in the "SHORT-KNITS/GENTLE" cycle which is dispensed during the last 2 minutes of wash. No bleach is dispensed when the "EXTRA SHORT-KNITS/GENTLE" wash option is selected.

(B) Includes four spray rinses at selected rinse temperatures.

(C) Fabric conditioner dispensed during last rinse.

(D) Includes four spray rinses at selected temperature in "REG. HEAVY" cycle only.

(E) Rinse L.E.D. indicator remains on which "2ND RINSE" is selected;

(F) Lo motor "RINSE SPIN" and "SPIN ONLY" are available when "KNITS/GENTLE" is selected.

(G) Spin is omitted in "SUPER WASH" when "SUPER WASH" is followed by a fabric selection.

(H) Cool down portion of "Permit Press" cycle is skipped when cold wash is selected.

Control operation notes

If a selection is correct, the corresponding L.E.D. comes "on" if a selection is incorrect the L.E.D. does not come "on".

Cycle operation will begin eight seconds after the final selection is made if the "HOLD" options is used operation will be delayed until "RUN" is selected "RUN" and "SPIN ONLY" selections cause immediate operation.

FIGURE 35-17. Cycle sequence chart of washer functions. Third-generation control system. (Courtesy of Whirlpool)

dc voltage appears at the output, you will have to assume that the microcomputer is defective.

Use the wiring diagram for the specific model you are servicing and check circuit by circuit.

Some machines have a built-in test cycle which should be used to check the computer board. Let the test cycle *run* until it reaches the part to be tested, then *hold* it there. Connect a voltmeter between the V terminal on the start relay and the edge connector for the suspected circuit. This places the meter in parallel with the part. A reading of the line voltage would indicate that the microcomputer is supplying the correct voltage. No line voltage would indicate a faulty board.

THE TOUCH MODULE

Normally, the touch module can be checked by pushing each button because the LED above the button should come on. There are three possible failures that could occur on a touch

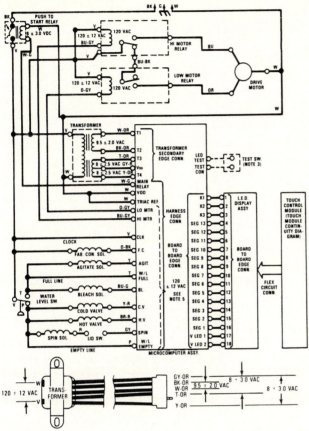

FIGURE 35-18. Wiring diagram showing the relation of the control system to the rest of a washing machine. (Courtesy of Whirlpool)

FIGURE 35-19. Typical harness connectors. (Courtesy of Whirlpool)

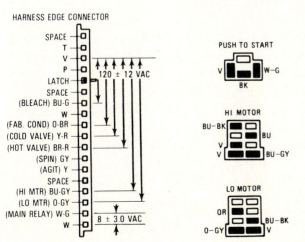

module. Whirlpool's service literature suggests this approach:

1. An open switch will not make contact when pressed.
2. A stuck switch will not break contact when released.
3. A faulty connection in the flexible connector either at its terminals or within the connector.

To check for an open switch, make a selection, then touch the suspected faulty pad to change the selection. If the LED does not light, your guess was right. If there is no change, the suspected switch is defective.

To check for a switch that is stuck or shorted, the washer will not fill with water after the 8-second waiting period. A different LED will light than the one you intended.

To check for a faulty connection, inspect the flexible connector and its pins for cleanliness and security of contact. It may be necessary to use a contact cleaner.

To completely check out the touch module, remove it from the console and disconnect the flexible circuit connector from the LED display. The conductor strips are listed from 1 through 13 in Fig. 35-20. Each pad can be checked for continuity or shorts with an ohmmeter that is set for its highest resistance scale. If any switch shows signs of leakage with the switch open, the touch module should be changed. The same holds true if any switch shows other than a short (low-resistance) reading when the switch is depressed.

The LED display assembly indicates the selections made and where the machine is during its cycle. In operation, when a pad is pushed on the touch module, it inputs a voltage through the LED display assembly to the microcomputer. The microcomputer processes the voltage and turns on the correct LED. With all this voltage traveling back and forth, it is necessary that all the connections and contacts be working properly. Make certain that all the edge connector wires are pushed all the way into the contact notches on the connector.

The Whirlpool Third-Generation Controls can activate an LED test cycle that will se-

TOUCH MODULE CONTINUITY DIAGRAM.

Example of use: When "HOLD" selector is pressed and held, continuity will be observed between 10 and 13 on flexible circuit connector. When "HOLD" selector is released, there should be an open circuit from 10 to 13 on the flexible circuit connector.

NOTE: Flexible circuit connector must be disconnected from L.E.D. display assembly for this test.

FLEXIBLE CIRCUIT CONNECTOR

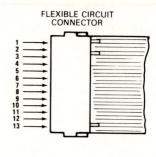

CONTACT	1	2	3	4	5	9	10
11	EXTRA SHORT	NOT USED	2ND. RINSE	SPIN ONLY	NOT USED	COLD WASH	NOT USED
12	NOT USED	SOAK CYCLE	RINSE SPIN	NOT USED	PERM'T PRESS	WARM RINSE	RUN
13	HOT WASH	SUPER WASH	SHORT WASH	NOT USED	KNITS GENTLE	REG. HEAVY	HOLD

FIGURE 35-20.   With the flexible circuit connector disconnected, an ohmmeter test from terminal to terminal, as illustrated, will determine the condition of each touch pad.

quence all the lights for 1 second at a time (Fig. 35-21). This is done by pushing a power switch and then touching "Perm't/Press" and "Knits/ Gentle" at the same time. Then touch "Hot Wash" and the test will begin. Doubtless, other manufacturers build their own test codes into their machines. Read the owner's manual for specifics.

If the LED indicators do not light or light incorrectly, it could be a faulty touch module, LED display assembly, microcomputer board, or board-to-board connector.

If one LED does not light, the LED assembly is faulty.

FIGURE 35-21.   The LED test will light each indicator above the touch pads for 1 second in a logical sequence when the "Perm't/Press" and "Knits/Gentle" pads are pressed at the same time. Then press "Hot Wash."

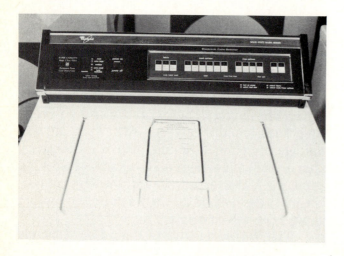

If two or more LEDs do not light, do the following:

1. Check the board-to-board connector for continuity. Unplug the machine and use an ohmmeter.
2. Check the secondary outputs of the transformer with the machine turned on.

If these tests are good, showing proper continuity and correct transformer voltages to the microcomputer, the microcomputer board is faulty.

The troubleshooting chart is offered as a guide or checklist to some of the common problems, causes, and ways to fix them. You can service many minor troubles if you can recognize and locate them. When a defective timer is suspected, you might consider the electronic control system.

NOTE: The detergent manufacturers consider 68°F (20°C) water as *cold water*. Some of the materials in cold water detergents will not dissolve in colder water and will deposit in the fabric of the clothes being washed. They may appear as light spots or streaks in the clothes. Advise the user to use some warm water to assure complete dissolving of the detergent. DON'T BLAME THE MACHINE.

Some models of Sears washers have a thermostat built into the fill valve which will allow some hot water to mix in with the cold to raise its temperature to a satisfactory level.

Troubleshooting Chart **AUTOMATIC WASHERS**

Problems	Possible Cause	Corrective Action
No water fill.	Water valves closed. Hoses kinked. Screen in fill hose clogged. No power to fill solenoid. Faulty water level control. Machine not turned on. Machine did not drain out last time used.	Turn on valves. Unkink hoses. Clean out screen. Replace selenoid. Replace control. Check controls and power at outlet. See "Water will not drain from machine."
Incorrect fill or temperature	Faulty water level control. Faulty thermal element in mixing valve. Hot-water supply inadequate. Reversed hoses—hot-water hose on cold-water connection.	Replace control. Replace valve (sometimes repair kits are available). Check temperature setting and capacity. Connect hoses correctly.
No spray rinse.	No water supply. Defective timer.	Same as no water fill. Replace timer.
Water will not shut off.	Defective timer. (Time fill machines.) Defective water level control. Foreign particles in mix and fill valve. Defective valve.	Replace timer. Replace control. Clean out valve. Replace valve.
Water leakage.	Inlet hose loosely connected to valve. Drain hoses not tight on pump. Broken hose. Leaky gasket. Cracked housing.	Tighten hose connection. Tighten hose clamps. Repair hose. Replace gasket. Replace parts.
Water will not drain from machine.	Kinked or clogged drain hose. Pump does not run. Suds lock. Faulty transfer valve. Defective timer. Loose belt.	Clear drain hose. Readjust and tighten pump. Drive mechanism. Remove suds, add cold water. Replace valve. Replace timer. Adjust belt.
Motor will not run.	No power to machine. Door switch or other safety control in motor circuit. Faulty timer. Faulty water level control. Faulty motor.	Check outlet. Check controls for operation and replace if defective. Replace timer. Replace control. Repair or replace motor.

Troubleshooting Chart AUTOMATIC WASHERS (Continued)

Problems	Possible Cause	Corrective Action
No agitation.	Motor failure. Faulty timer contacts. Faulty transmission. Defective control solenoid. Broken linkage. Faulty water level switch.	Repair or replace motor. Replace timer. Repair or replace. Replace solenoids. Replace or repair linkage. Replace switch.
Slow spin.	Belt or clutch slips.	Adjust belt.
Excessive vibration.	Washer not level. Weak flooring. Unbalanced load. Rubber cups not on feet. Damaged snubber or suspension bolts.	Level machine by adjusting leg screws. Reinforce floor. Redistribute load. Install cups on feet. Replace snubber.
Torn clothing.	Improper bleach usage. Broken agitator. Defective basket.	Add bleach to water before loading clothes in tub or dilute bleach well before adding. Replace agitator. Replace basket.
Machine will not shut off.	Defective timer. Break in wiring.	Replace timer. Repair wiring.
Timer will not advance to next cycle.	Defective timer motor. Bound timer shaft or knob. Faulty water level control.	Replace timer motor. Clear knob from panel. Replace control.
No suds return (for machines with suds saver).	Faulty distributor valve. Slipping belt. Kinked hose. Defective solenoid. Faulty water switch. Jammed pump.	Replace valve or solenoid. Tighten or replace belt. Straighten hose. Replace solenoid. Replace switch. Remove pump and clean out.
No recirculation of water during agitation.	Jammed pump. Defective pump drive. Clogged hose. Defective distribution valve.	Clean out pump. Replace coupling or tighten. Clean out hose. Clean out or replace valve or solenoid.

Note: Timer does not advance during water fill period until the water level switch has been satisfied.

36

Water Heaters

As civilized people we have come to depend a great deal on a water heater, and we are quick to complain when there is no hot water (Table 36-1). A number of different kinds of water heaters are in use, but the most popular home units are the automatic storage types that use either gas or electricity for fuel. Our concern is only with these.

TABLE 36-1

Typical hot-water usage chart*

Activity	Gallons
Handwashing	0.9
Shower	5–7
Tub bath	7-12
Shaving	2
One meal per person	2-5
Automatic washer	11-34
Conventional washing	30
Dishwasher	8
Average bathroom use per day	
Adult	12
Child	24

* One kilowatt-hour will provide approximately 4 gallons of hot water under average conditions (100°F temperature rise 56°C).

An automatic heater consists of a storage tank (Fig. 36-1), thermostatic controls (Fig. 36-2), some insulation, and a heat source (Fig. 36-3). Electric units have a heat source either immersed in the tank or strapped around the outside of it. Modern gas heaters have the burner enclosed beneath the tank. Both electric and gas units use thermostats to control the water temperature.

ELECTRIC WATER HEATERS

New immersion heating elements in a water heater are easier to replace than you may think. You need a few common tools and some sandpaper.

First, turn off the switch; then remove shielding covers and test the exposed ends of the elements with an ohmmeter. If the unit is open, it is defective.

The local utilities service department or electrical supply house should be able to supply you with whatever replacement you might

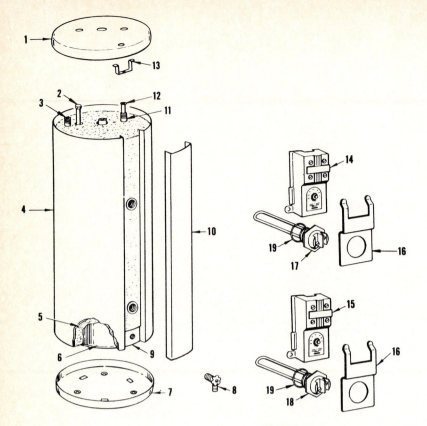

TYPICAL ELECTRIC WATER HEATER PARTS IDENTIFICATION

1. TOP PAN
2. ANODE
3. HEAT TRAP ASSEMBLY
4. JACKET
5. INSULATION
6. TANK
7. BOTTON PAN
8. DRAINCOCK
9. DRAINCOCK PANEL
10. FRONT PANEL
11. NIPPLE
12. DROP TUBE
13. JUNCTION BRACKET
14. THERMOSTAT—UPPER
15. THERMOSTAT—LOWER
16. THERMOSTAT BRACKET
17. ELEMENT—UPPER
18. ELEMENT—LOWER
19. ELEMENT GASKET

FIGURE 36-1. Typical electric water heater.

FIGURE 36-2. (a) High limit control for water heater. Set at 190°F (88°C) to open the circuit. This control cannot be held closed by blocking the reset lever, making it a safety control. (b) Adjustable thermostats that can be used on any storage-type electric water heater. This is the control that is set for the desired water temperature. (Courtesy of Chromalox)

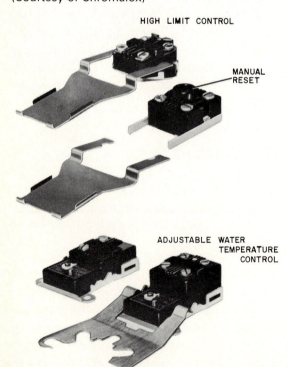

HIGH LIMIT CONTROL

MANUAL RESET

ADJUSTABLE WATER TEMPERATURE CONTROL

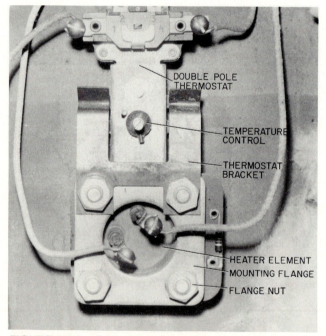

FIGURE 36-3. Removing the upper cover exposes the wiring, heater element, and the control. The lower heater element has its own thermostat.

need. To be sure of receiving the proper type, furnish the supplier with the make, model number, and capacity of the tank. The shape of the new unit might be different but the wattage should be the same (Fig. 36-4).

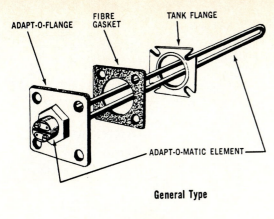

ADAPT-O-FLANGE   FIBRE GASKET   TANK FLANGE

ADAPT-O-MATIC ELEMENT

General Type

Adapt-O-Flanges

ADF-1
Square Type

2½" Bolt Circle.
Fits more than 85% of
all water heaters.

ADF-2
Square Type

2¾" Bolt Circle

ADF-3
Square Type

3¼₆" Bolt Circle.
Fits Frigidaire and
Thermogray water heaters.

ADF-4
Round Type

2½" Bolt Center.
Fits 6-bolt White
water heaters.

ADF-5
Elliptical Type

3¼₆" Bolt Center.
Fits 2-bolt flanges.

FIGURE 36-4. The Chromalox Adapt-O-Matic water heater element with its assortment of mounting flanges serves as a replacement for most of the elements in current use.

IMMERSION HEATERS

With the switch off and while draining water from the tank, save time and future guessing by making a rough sketch showing color and location of each wire (Fig. 36-5). Tape your drawing to the tank housing for ready reference. Disconnect wires at the terminals of both the element and the thermostat with a sturdy screwdriver. Bend the freed wires back from the work area.

Apply gradual force to free bolts that secure the element to the tank. After removal, wire brush corrosion off the bolt threads. Then pull out the heating element.

Scrape off every bit of the old gasket left around the tank hole with an old chisel, stiff putty knife, or a screwdriver. Be sure to sand this surface smooth because a good seal with the new gasket is vital. If a new gasket does not come with the replacement, be sure to get one. Never try to reuse an old gasket.

Prepare the new heating element by sanding smooth any burrs or rough spots on its gasket side. Set the gasket in place on the unit. Then place the element in position. Replace bolts and tighten them alternately. Scrape any dirt from the wire ends with a knife and connect to proper terminals as indicated by your sketch.

Remember to refill the tank with water before throwing the switch back on! You will burn out the new unit if you don't!

Figure 36-6 describes a typical wraparound heater element replacement.

WRAPAROUND HEATER ELEMENTS

The major advantage of the wraparound heaters is that they will not become coated with lime in a hard-water area. The major disadvantage is that unless the contact between the heater element and the tank is good for its entire length, hot spots will develop. These hot spots could cause early burnout failure of the element.

EFFECT OF HARD WATER

Immersion-type heating elements that become coated with lime and scale will be less efficient

A-4 DOUBLE ELEMENT
Limited Demand Operation

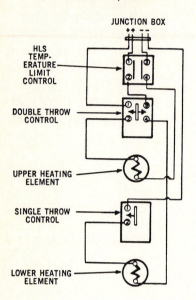

Circuit No. A-4 is a 4-wire circuit used where specifications require 4 leads at junction box; where simultaneous operation of both upper and lower element is prohibited; and where time switch operation of lower element is required. This 4 wire circuit also permits the lower element to operate off an un-metered or flat-rate circuit. (Also Note: To change to simultaneous operation of both elements, move lead from terminal 4 to terminal 3 on upper Control Thermostat.)

A-5 DOUBLE ELEMENT

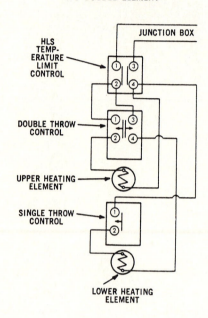

Circuit No. A-5 gives complete power disconnect through the temperature limit control. All electric water heaters produced after October 15, 1964 are required by Underwriters Laboratories to be equipped with a manual re-set temperature-limiting device, such as Chromalox HLS, wired as in Circuit A-5, so as to disconnect all ungrounded conductors. This provides complete protection against any combination of wiring, element or thermostat fault.

N.E.M.A. STD.

B-4 DOUBLE ELEMENT
Simultaneous Element Operation

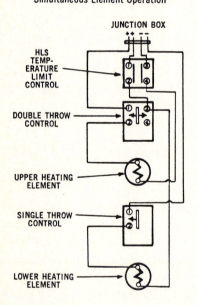

Circuit No. B-4 is a 4-wire circuit used where specifications require 4 leads at junction box; where simultaneous operation of both elements is permitted; and where it is desired to operate the lower element off an unmetered or flat-rate circuit, or time switch. (Double Throw Control Thermostat is used at top position to meet the wider differential requirement at the top of the tank.)

N.E.M.A. STD.

C-2 SINGLE ELEMENT

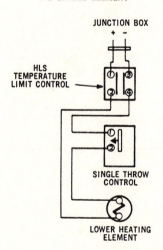

Circuit No. C-2 is a 2-wire circuit for single element heaters.

N.E.M.A. STD.

FIGURE 36-5. Standard water heater circuits.

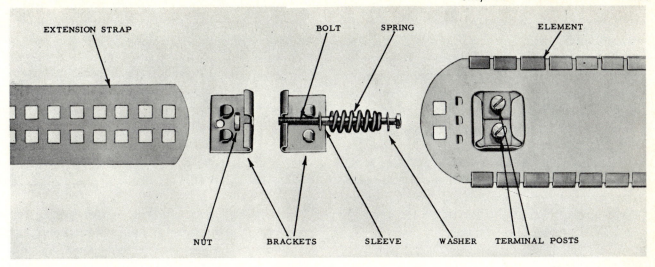

EXTENSION STRAP BOLT SPRING ELEMENT

NUT BRACKETS SLEEVE WASHER TERMINAL POSTS

This Photo Shows Parts
Not Assembled

The nut holds the bolt to
the left bracket. The sleeve,
spring, washer and bolt head
go over the right bracket.

This Photo Shows Parts
Assembled to Element

TO REPLACE A WRAP AROUND HEATER ELEMENT

1. SHUT THE POWER TO THE UNIT OFF!
2. DISCONNECT WIRES FROM OLD ELEMENT.
3. UNFASTEN OLD ELEMENT AND TEST FOR FREEDOM OF MOVEMENT AROUND TANK.
4. CHECK THE REPLACEMENT ELEMENT FOR SIZE, WATTAGE AND FREEDOM FROM KINKS.
5. ATTACH END TO OLD ELEMENT SO THAT IT WILL PULL AROUND THE TANK AS THE OLD ELEMENT IS REMOVED.
6. MAKE CERTAIN THE NEW ELEMENT MAKES GOOD CONTACT ALL AROUND THE TANK THEN SECURE IT AND CONNECT TO THE POWER LINE.

FIGURE 36-6.

in their heat transfer to the water. Eventually enough lime buildup will allow the element to overheat and burn out. In extremely hard water areas, the elements should be removed and inspected semiannually. The tank should be flushed by draining at least 5 gallons of water once a month to remove the sediment deposits from the tank bottom.

A water softener in a hard-water area will do much to prolong the life of a water heater. What has been said about hard water, lime deposits, and flushing the heater holds true for gas heaters as well as electric.

INSTALLATION NEEDS

A gas water heater requires more installation care than an electric unit. An electric water heater requires only a secure footing, accessibility to plumbing, and three No. 10 wires to a 220-volt separately fused 30-ampere power source. A gas water heater requires no electricity but does need a gas line, available air, and a vent to a chimney or some other outdoor exit for its products of combustion.

THE GAS BURNER

The gas burner in a water heater is subject to the same principles of operation, needs, and controls as the gas burners discussed in Chapter 12. Figure 36-7 illustrates the part relationship in a typical gas water heater.

The two troubleshooting charts should provide enough information to guide you through a service call on a typical water heater.

Troubleshooting Chart   **ELECTRIC WATER HEATERS**

Problem	Possible Cause	Corrective Action
No hot water.	No power.	Check fuse and replace it.
	Defective thermostat.	Replace thermostat.
	Thermostat out of calibration.	Adjust thermostat.
	Defective heater element.	Replace heater element.
	Incorrect heater element.	Check elements for wattage and voltage against parts list.
	Incorrect wiring.	Check wiring against diagram.
Water not hot enough.	Thermostat set too low.	Reset thermostat to a higher temperature.
	Incorrect heater elements.	Replace heater element.
	Excessive lime in tank.	Clean out deposit.
Water too hot.	Thermostat set too high.	Lower thermostat setting.
	Defective thermostat.	Replace thermostat.
Not enough hot water.	Under sized heater for demand.	Check water use against recovery rate.
	Thermostat set too low.	Raise thermostat setting.
	Lime coated element.	Remove and clean.
	Defective element.	Replace element.
	Incorrect wattage or voltage element.	Replace.
	Incorrect wiring to element.	Rewire.

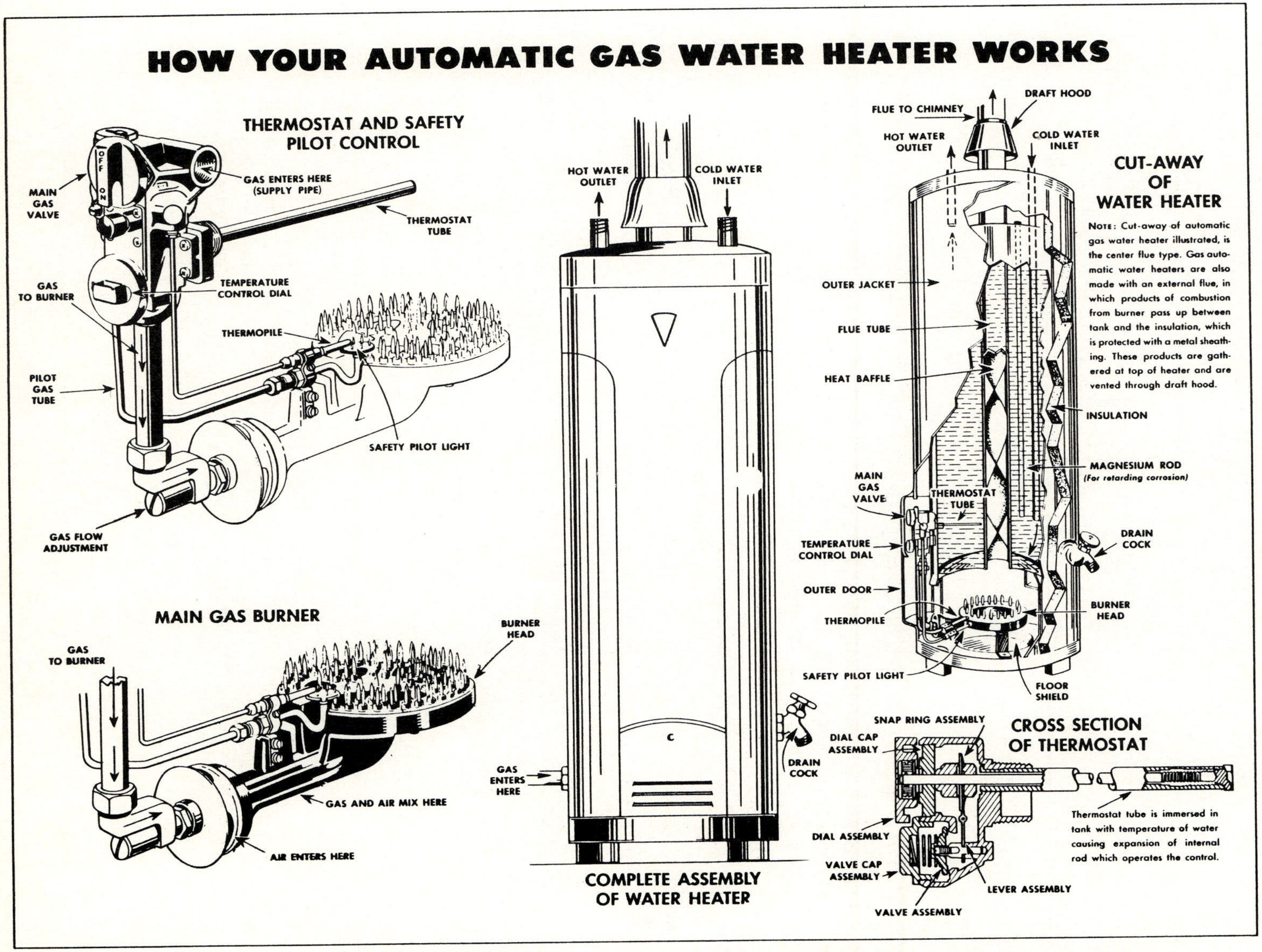

WHY YOUR GAS WATER HEATER IS CALLED "AUTOMATIC"

What is the meaning of the word "automatic"? Simply this: *it works by itself.* That is, an automatic gas water heater turns itself on and off. When the water in the tank cools below a set temperature, or when the hot water is drawn from the tank and is replaced with cool water, the thermostat cools and turns on the gas. The same thermostat also shuts the gas off when the temperature of the water in the tank becomes hot enough.

Other automatic features are built into an automatic gas water heater. The safety pilot ignites the gas at the main burner when the thermostat turns on the gas. It is called a safety pilot because if it goes out, the gas supply to the water heater is shut off.

The safety pilot also generates a small amount of electricity because its flame heats an ingenious device called a thermocouple. A thermocouple is a tiny electric generator made of two different metals joined firmly. When these metals are heated, a small electric current is generated. Although the electric energy is small, it is enough to hold the safety shut-off gas valve open. If the pilot flame becomes too small, or if it should go out, the thermocouple then does not produce enough electricity, and a spring closes the gas valve. In this way the gas automatic water heater not only works by itself, but it is entirely safe to use.

HOW YOUR AUTOMATIC GAS WATER HEATER WORKS

CUT-AWAY OF WATER HEATER

Note: Cut-away of automatic gas water heater illustrated, is the center flue type. Gas automatic water heaters are also made with an external flue, in which products of combustion from burner pass up between tank and the insulation, which is protected with a metal sheathing. These products are gathered at top of heater and are vented through draft hood.

DRAFT HOOD — COLD WATER INLET — FLUE TO CHIMNEY — HOT WATER OUTLET — OUTER JACKET — FLUE TUBE — HEAT BAFFLE — MAIN GAS VALVE — TEMPERATURE CONTROL DIAL — OUTER DOOR — THERMOPILE — SAFETY PILOT LIGHT — THERMOSTAT TUBE — INSULATION — MAGNESIUM ROD (for retarding corrosion) — DRAIN COCK — BURNER HEAD — FLOOR SHIELD

CROSS SECTION OF THERMOSTAT

Thermostat tube is immersed in tank with temperature of water causing expansion of internal rod which operates the control.

SNAP RING ASSEMBLY — DIAL CAP ASSEMBLY — DIAL ASSEMBLY — DRAIN COCK — VALVE CAP ASSEMBLY — LEVER ASSEMBLY — VALVE ASSEMBLY

COMPLETE ASSEMBLY OF WATER HEATER

COLD WATER INLET — HOT WATER OUTLET

THERMOSTAT AND SAFETY PILOT CONTROL

THERMOSTAT TUBE — GAS ENTERS HERE (SUPPLY PIPE) — TEMPERATURE CONTROL DIAL — THERMOPILE — SAFETY PILOT LIGHT — MAIN GAS VALVE — GAS TO BURNER — PILOT GAS TUBE — GAS FLOW ADJUSTMENT

MAIN GAS BURNER

BURNER HEAD — GAS ENTERS HERE — GAS AND AIR MIX HERE — AIR ENTERS HERE — GAS TO BURNER

FIGURE 36-7. Gas water heater. (Courtesy of the American Gas Association)

467

Troubleshooting Chart   GAS WATER HEATERS

Problem	Possible Cause	Corrective Action
Unable to light pilot.	Gas shut off.	Open gas cock.
	Pilot orifice or filter clogged.	Remove part and clean.
	Flame blowing too hard.	Excessive gas pressure; reset regulator.
	Pilot tube pinched or clogged.	Clean or replace.
	Air in gas line.	Purge air from line.
	Loose thermocouple.	Clean contact area and secure connection.
	Weak flame will not heat thermocouple.	Adjust pilot flame.
		Check gas pressure.[a]
	Pilot flame not hitting thermocouple.	Readjust to correct position.
		Replace pilot tip if badly burned or misshapen.
Overheats or water is too hot.	Thermostat set too high.	Normal setting is 140°F (60°C).
	Defective valve.	Replace valve.
	Defective thermostat.	Replace thermostat.
	Incorrect gas pressure.	Check gas pressure.
	Improper venting.	Repair.
	Dirt on valve seat.	Clean valve seat.
Water not hot enough.	Thermostat set too low.	Reset to 140°F (60°C).
	Thermostat defective.	Replace thermostat.
	Impatient user.	Allow time for hot water to get to the faucet.
	Cold-water inlet is not at correct distance from tank bottom.	Replace with pipe of correct length.
Not enough hot water.	Heater too small for usage.	Replace heater.
	Low gas input.	Check gas pressure.
	Incorrect burner.	Compare rating with burner.
	Thermostat out of adjustment.	Readjust thermostat.
Excessive soot accumulation. Yellow flame.	Improper primary air adjustment.	Adjust air shutter.
	Improper venting.	Check vent-to-burner requirements.
	Incorrect burner orifice.	Replace.
	Orifice clogged.	Clean orifice.
	Incorrect gas pressure.	Adjust to correct pressure.
Main burner will not fire.	Low gas pressure.	Check pressure.
	Valve sticking closed in the thermostat.	Replace valve.
	Tubing clogged.	Clean or replace.
	Defective diaphragm valve.	Replace valve.
	Defective pilot.	Adjust.
	Thermostat not operating.	Adjust or replace.

469 / Water Heaters

Chap. 36 / Water Heaters **469**

Chap. 36 / Water Heaters **469**

Troubleshooting Chart   GAS WATER HEATERS   (Continued)

Problem	Possible Cause	Corrective Action
Main burner will not shut off.	Valve sticking open in thermostat.	Replace valve.
	Dirt on valve seat.	Clean.
	Valve out of adjustment.	Readjust.
	Thermostat not working.	Adjust or replace.
	Thermostat out of calibration.	Recalibrate.

[a] Gas pressure should be 7 inches water natural, 11 inches water LP.

37

Water Softeners

Water in its travels through the soil absorbs many impurities, primarily calcium and magnesium carbonates and sulfates. These can be removed by an ion exchange in which sodium would replace the calcium in the salts. The sodium salts are soluble and make the water more useful.

The appliance that can do this is a water softener. There are two basic models (Fig. 37-1). One has two separate units: the brine tank and the resin tank. The other (compact) unit has one unit which houses both the resin tank and the brine tank.

INSTALLATION

The water softener is installed as close to the location of the incoming water supply as possible (Fig. 37-2). A water softener improperly installed will not operate at its best. Check the installation for basic conformance. The drain line must be at least $\frac{3}{4}$ inch in diameter and not over 6 feet long to avoid excessive back pressure that may affect regeneration. This drain line can be an ordinary garden hose. A longer drain line should be larger in diameter.

Following are some points to check for the unit shown in Fig. 37-2.

Installation Items to Check Out

1. *Water volume.* Before installing this water softener, make sure that your water system will pump enough water to backwash the softener.

2. *Drain.* The unit should be located close to a drain. The drain must be capable of disposing of water at the unit backwash rate for a period of 20 minutes.

3. *Water pressure.* Check pressure with a reliable pressure gage; 20 psi minimum is required. Adjust pump control if necessary. 120 psi is maximum allowable pressure. Install a pressure-reducing valve if necessary.

FIGURE 37-1. Both models will do the same job. The two-unit model has a greater salt storage capacity. (Courtesy of Sta-Rite Industries)

4. *Electrical requirement*. 120-volt, 60-hertz ac fused 15-ampere grounded circuit constant electrical power required. Outlet must be within reach of the power cord provided with the softener (6 feet).

5. *Water temperature*. Cold-water service only. Provide at least 10 feet of pipe between softener and hot water heater to prevent hot water backup.

6. *Location*. Make sure that the softener stands on a smooth, level floor. An uneven, rough floor or sharp objects may damage the brine tank. The water softener must be installed in a location where it will not be subject to freezing temperatures or subject to impact damage.

Another item to watch for in the installation is the electrical plug. Be sure that it is connected. A disconnected plug will throw the timer off and can easily cause the skip of a few regenerations.

Instructions for setting regeneration time and frequency are fully explained on the inside cover of most timers (Fig. 37-3).

SERVICING

Some automatic water softeners have a solenoid valve to control the drain during regeneration. These cycles are controlled by the solenoid valve action, which, in turn, is initiated by the timer (Fig. 37-4).

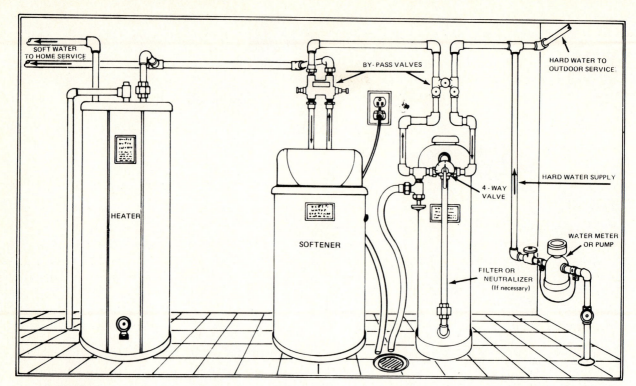

FIGURE 37-2. Typical installation site and plumbing for a single-tank model. (Courtesy of Sta-Rite Industries)

Failure of this valve to function, because it has become clogged with impurities, is one of the major causes of service in water softeners. When the valve sticks open, the softener will regenerate constantly. If it sticks closed or if the coil is bad, it cannot regenerate at all.

To service a drain solenoid valve (Fig. 37-5), you must first unplug the power; then turn off the water to the softener. Next, unplug the leads from the solenoid coil and disassemble the unit. This step will expose the valve plunger, spring, and diaphragm. Wash all foreign matter from the tube and the valve parts; then reassemble. Of course, if the solenoid coil is defective (open or burned out), it must be replaced. Remember to see that the household has enough water on hand before you start servicing, or that the manual by-pass valve has been set.

In the Water King softener the valve is operated by a motor that selectively moves the plunger through its various regeneration stages. The motor operation is controlled by the timer.

Periodically, the softening capacity of the resin bed will have to be restored (regenerated). A fresh solution of brine with a high concentration of sodium ions must displace the calcium ions that the resin has collected from the hard water.

This regeneration procedure involves five cycles, summarized below. The plunger positions itself for each of these operations (cycles) in turn:

1. *SERVICE.* Untreated water filters down through dense, high-quality resin bed which traps hardness, minerals, and sediment.
2. *BACKWASH.* Fast upflow removes the silt, sediment, and iron which have been filtered out of the water.
3. *BRINE.* Slow, brining downflow maximizes salt efficiency.
4. *SLOW RINSE.* Ensures maximum brine contact time.
5. *FAST RINSE.* Removes all traces of

The changing or resetting of the program clock is a simple matter and can be accomplished by following instructions below without professional assistance. You cannot injure the program clock by manually turning the TIME DIAL or the SKIPPER WHEEL in either direction.

Proper frequency and length of regeneration time are specified by the manufacturer or its representative. If in doubt as to the proper setting call your water softener dealer.

MANUAL REGENERATION

Push Lever "N" in direction of Arrow "P" as far as it will go.

Step No. 2:

REGENERATION FREQUENCY

1. Turn SKIPPER WHEEL until #1 Tab is opposite Arrow "F".
2. Lift SKIPPER TABS for days regeneration is to occur.

Step No. 3:

SET LENGTH OF REGENERATION CYCLE

1. Loosen Screw "H".
2. Lift Dial "M" and set Pointer "J" to total regeneration time as indicated by white numbers on Dial "K".
3. Tighten Screw "H" being sure gears engage.

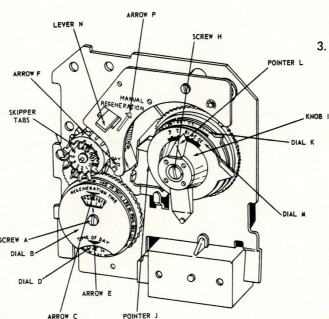

Step No. 1:

SET REGENERATION TIME

1. Loosen Screw "A" and Rotate Black Dial "B" until Yellow Arrow "C" points to time regeneration is to begin.
2. Tighten Screw "A".
3. Lift Yellow Dial "D" and turn in either direction until Arrow "E" points to actual time of day. Release dial.

Step No. 4:

SET FAST RINSE TIME

1. Lift Knob "I" and set Pointer "L" to fast time rinse as indicated by yellow numbers on Dial "M".
2. Release Knob being sure gears engage.

NOTE: Fast rinse occurs within total regeneration time setting.

EXAMPLE: If at the time the resetting of the clock is completed the TIMER DIAL "A" and the SKIPPER TABS are in the position illustrated in Figure 2a above, it would be (1) 10 P.M., (2) the softener would regenerate at 2:30 A.M. the next morning, (3) the length of the regeneration cycle will be 70 minutes and (4) it would regenerate every 4 days.

FIGURE 37-3. Typical program clock that controls both the regeneration frequency and timing needs for an automatic water softener. Electronic solid-state control units are programmed differently.

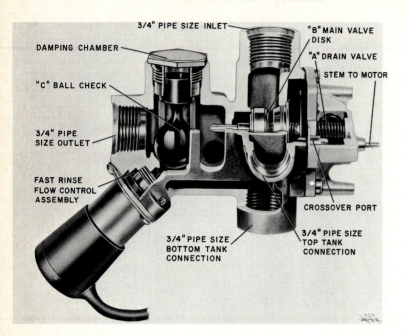

DAMPING CHAMBER

3/4" PIPE SIZE INLET

"B" MAIN VALVE DISK

"A" DRAIN VALVE

STEM TO MOTOR

"C" BALL CHECK

3/4" PIPE SIZE OUTLET

FAST RINSE FLOW CONTROL ASSEMBLY

CROSSOVER PORT

3/4" PIPE SIZE BOTTOM TANK CONNECTION

3/4" PIPE SIZE TOP TANK CONNECTION

FIGURE 37-4. Popular valve used on many automatic water softeners. Its operation is controlled by the timer. (Courtesy of Erie Mfg. Co.)

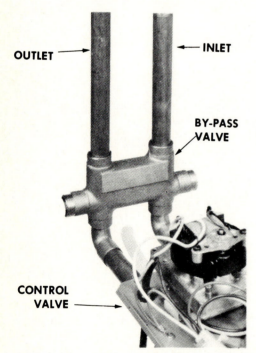

OUTLET

INLET

BY-PASS VALVE

CONTROL VALVE

FIGURE 37-5. Motor-operated drain valve shown in the *service* position (plunger all the way in). The other valve positions are: back wash: plunger all the way out; brine and slow rinse: plunger halfway in; fast rinse: plunger two-thirds of the way in. (Courtesy of Sta-Rite Industries)

Typical inlet-outlet connections for compacts and two tank models.
NOTE: Shown with by-pass valve.

Attach drain line. Fasten ½" I.D. drain tubing to "Drain hose barb". This drain tubing should not be more than 9 feet above the floor where softener stands. Fasten drain line to prevent kinks or whipping during regeneration. Use "Solid Drain Kit" if code requires solid drain line. Do not remove hose barb. Loss of parts in the hose barb will result in loss of softener mineral.

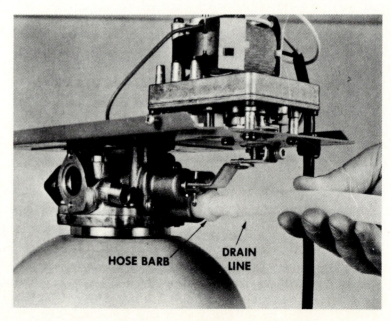

HOSE BARB

DRAIN LINE

SOFTENER VIEWED FROM BACK.
TYPICAL FOR COMPACT AND TWO TANK MODELS

brine and repacks resin bed to prepare unit for start of new softening operation.

This five-cycle operation, recommended by resin manufacturers, gives you the most thorough restoration of the resin bed, thus assuring the highest-quality softening performance throughout the life of the softener.

Be sure to check any change in the water usage habits *and* check the water for a change in hardness before condemning a softener for failing to keep the water soft.

Some people, having a manual or semiautomatic water softener, are apt to forget to add salt, and regenerate their unit on a fixed schedule. This allows hard water to fill the hot-water tank (and the rest of the plumbing). Then one is never certain to attain the full benefits of softened water. So advise softener users to set up a firm schedule for regeneration and stick to it!

THE BRINE TANK

The concentrated brine solution used to regenerate the resin in a water softener is made in the brine tank. The salt rests on a screen several inches from the bottom of the tank with only the bottom inch or so immersed in water, the level of which is controlled by a float valve. Brine is drawn from the bottom of the tank during the brining cycle of regeneration and fresh water flows back into the brine tank during the rinse cycle.

Areas of potential trouble in the brine tank could be

1. An excessive amount of salt impurities collecting on the screen
2. "Bridging" of salt over the water
3. Float-valve failure within the brine tank

The valve can stick open, thereby allowing water to overflow, or it can remain closed, thus preventing brine flow. The entire float assembly

can be lifted out of the brine tank for inspection and cleaning (Fig. 37-6).

"Bridging" occurs when wet salt pellets dry out and reform as a solid mass above the water level. Poking through the salt with a broom handle will correct the situation.

An accumulation of impurities builds up to a layer of mud that insulates the salt from the water. Even $\frac{1}{2}$ of 1 percent salt impurities could build up to a $\frac{1}{2}$-inch layer of mud within one year! The salt level should be allowed to drop to empty periodically so that the base plate can be inspected and any mud removed. Any of these problems will prevent brine flow to the softener. Thus the resin will not be regenerated and the household will have no soft water.

The exploded drawings and parts lists shown in Figs. 37-7 to 37-10 illustrate the construction and relative positioning of the parts.

FIGURE 37-6. Before removing the brine tank float valve assembly, make certain to turn the water off or water will spray when the brine tube is disconnected.

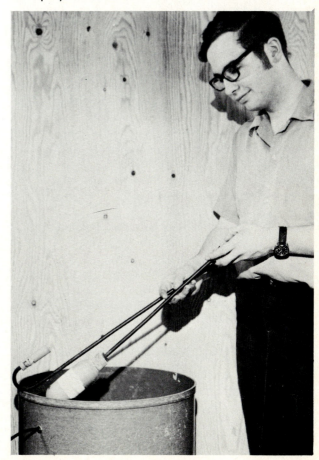

Ref. No.	Part No.	Part Description
1	607155	Shroud-Solid State Control Assy.
2	300811	Transformer - 115-24 Volt
3	400176	#6 - 32 x 5/16 Machine Screw (2)
	400720	Lockwasher - #6 Int. Tooth (2)
	400320	Nut - #6 (2)
4	401186	Caplug
5	300808	Snap Bushing
6	400187	#10 - 32 x 3/8 Machine Screw (6)
7	607148	Drive Motor Assembly
8	400173	#6 - 32 x 3/8 Self Tapping screw (4)
9	300018	Nozzle Cap
10	400602	Nozzle Cap Gasket
11	300017	Nozzle Screen
12	300016	Nozzle - #60
13	400601	Nozzle Gasket
14	300015	Throat - #50
15	300805	Valve Body
16	402012	O-Ring (2)
17	300137	End Cap
18	400127	#10 - 32 x 5/8 Machine Screw (2)
19	402013	Seal
20	—	Installation Horn
	300145	Inlet
	300146	Outlet
21	400188	#10 - 32 x 1/2 Machine Screw (4)
22	705406	Elbow Assembly - consists of:
	420289	Elbow - 45°
	200102	Brass Insert - 3/8"
	200125	Delrin Sleeve - 3/8"
	200256	Brass Compression Nut
23	300140	Inlet-Outlet Adapter (2)
25	300142	Retainer, Inlet-Outlet Adapter (2)
26	609009	Plunger Assembly - 5 3/4"
27	400501	Cotter Pin
28	400168	#10 - 32 x 5/8 Capscrew
29	300669	Drain Cap
30	400055	O-Ring
31	—	Hose Barb Flow Control
	507070	1.25 GPM (302/90)
	507080	1.50 GPM (302/165, 301/150)
	507081	2.50 GPM (302/200, 302/330, 301/200)
32	507097	Drive Link Assembly (includes 28)
34	400187	#10 - 32 x 3/8 Machine Screw (4)
	707189	Valve Control Assembly (All Above except Items 1, 19, 20, 21, 22, 23, 25, 31)

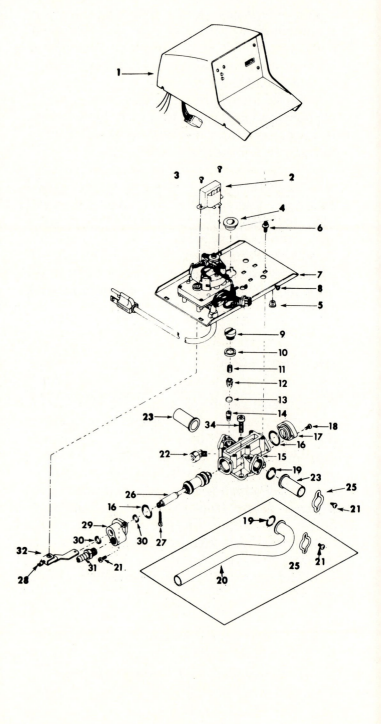

FIGURE 37-7.   Control assembly. (Courtesy of Sta-Rite Industries)

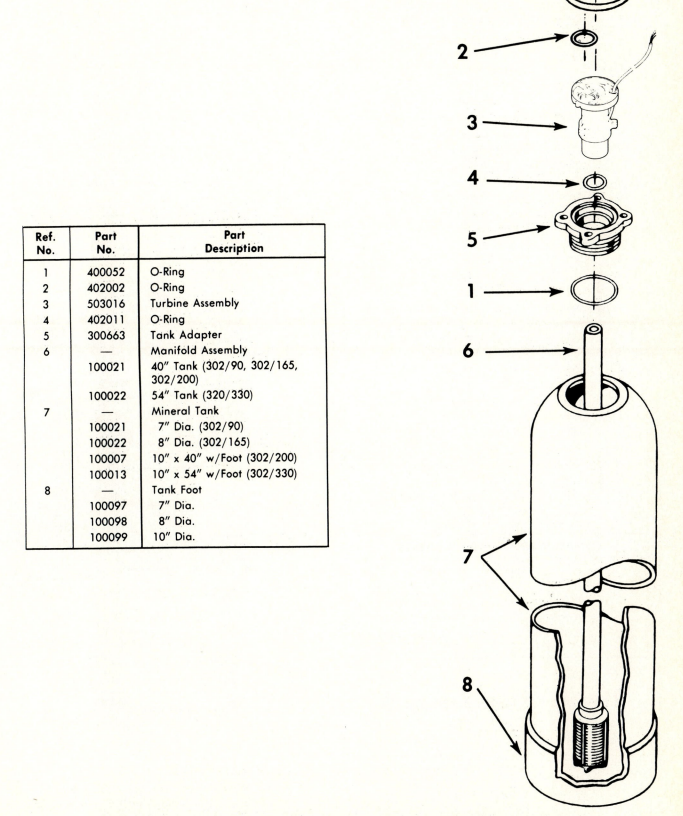

Ref. No.	Part No.	Part Description
1	400052	O-Ring
2	402002	O-Ring
3	503016	Turbine Assembly
4	402011	O-Ring
5	300663	Tank Adapter
6	—	Manifold Assembly
	100021	40″ Tank (302/90, 302/165, 302/200)
	100022	54″ Tank (320/330)
7	—	Mineral Tank
	100021	7″ Dia. (302/90)
	100022	8″ Dia. (302/165)
	100007	10″ x 40″ w/Foot (302/200)
	100013	10″ x 54″ w/Foot (302/330)
8	—	Tank Foot
	100097	7″ Dia.
	100098	8″ Dia.
	100099	10″ Dia.

FIGURE 37-8. Mineral tank assembly (two-tank models). (Courtesy of Sta-Rite Industries)

Ref. No.	Part No.	Part Description
1	420530	Elbow - ⅜ Tube to ⅜ Tube (Includes Items 2, 3, and 4)
2	420538	O-Ring OR (2)
3	420539	Nut and Spacer (2)
4	420537	Grab Ring (GR)
5	—	With Item 3
6	—	Brine Valve Pipe
	200097	(302 models)
	200541	(301 Models)
7	200553	Brine Well Cover
8	—	Brine Select Rod
	200598	(302 Models)
	200581	(301 Models)
9	200069	Float Rod
10	200119	Float Rod Guide
11	400100	#6 - 32 x ⅜ Self Tap Screw
12	200115	Float Locator (2)
13	200091	Float
14	400521	Circular Push-On ⅛
15	—	Brine Select Tube
	200574	(302/90)
	200544	(302/165, 302/200, 302/330, 301/150, 301/200)
16	506024	Large Cap & Seal Assembly
17	400082	O-Ring No. 1-021
18	200189	Brine Valve Seal
19	200188	Brine Valve Stem
20	200191	Brine Valve Insert
21	200192	Tension Washer
22	200155	Ball - ⅝"
23	400059	Tetraseal
24	200066	Brine Valve Cap, Small
25	200125	Delrin Sleeve - ⅜"
26	506015	Brine Valve Body Assembly (Includes Item 23)
	200436	Brine Line - 4 Foot

COMPLETE BRINE VALVE ASSEMBLY		
	706050	Brine Valve Assembly 301/150
	706061	Brine Valve Assembly 301/200
	706064	Brine Valve Assembly 302/90
	706063	Brine Valve Assembly 302/165
	706062	Brine Valve Assembly 302/200-302/330.

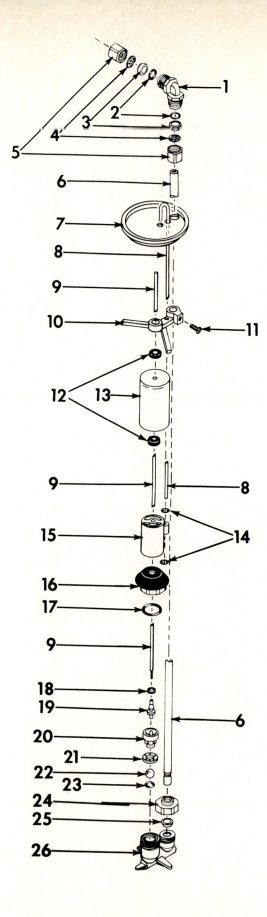

TYPICAL FITTING ASSEMBLY

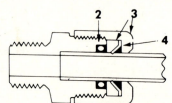

NOTE: Item 21 has one side marked "UP" in raised lettering. The "UP" side must face top of the assembly.

FIGURE 37-9. Brine valve. (Courtesy of Sta-Rite Industries)

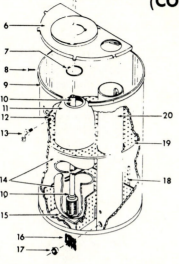

BRINE TANK ASSEMBLY
(COMPACTS)

BRINE TANK ASSEMBLY
(2 TANK MODELS)

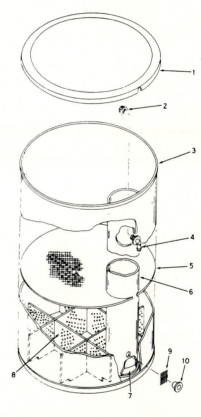

Ref. No.	Part No.	Part Description
1	402001	"O" Ring
2	402002	"O" Ring
3	503012	Tank Insert Assembly
4	402011	"O" Ring
5	100417	Tank Adaptor
6	200594	Salt Shield
	200047	Salt Shield
7	400052	"O" Ring
8	400202	#8 x ⅝ Self Taping Screw
9	200490	Brine Tank
10	703034	Manifold Assembly
11		Mineral Tank
	100021	7 x 40 For 201/90
	100022	8 x 40 For 201/150
	100226	9 x 40 For 201/200
12	200145	Overflow Fitting
13		Part of 12
15	200146	Base Plate
16	200087	Screen, Brine Well
17	200088	Bushing, Brine Well
18		Grid Plate Support (3)
	200559	6" For 201/90
	200555	8" For 201/150, 201/200
19		Grid Plate
	200573	With 7" Hole For 201/90
	200535	With 8" Hole For 201/150
	200597	With 9" Hole For 201/200
20	505017	Brine Well Assembly, 201/90 with Item 16 and 17.
	505018	Brine Well Assembly For 201/150, 201/200
Not Shown	390738	Shroud

Ref. No.	Part No.	Part Description
1	200141	Brine Tank Cover (18" Dia.)
2		Nut (Part of Ref. No. 4)
3	705005	Brine Tank 18" x 40"
4	200145	Overflow Fitting (Complete)
5	200190	Grid Plate (18")
6	505007	Brine Well Assy. (For 18" x 40")
7	200213	Brine Well Bottom Cover
8	200252	Grid Plate Support - 10¾" (4)
9	200087	Screen (3)
10	200088	Brine Well Insert (3)

FIGURE 37-10. Brine tank assembly. (Courtesy of Sta-Rite Industries)

Troubleshooting Chart **WATER SOFTENERS**

Problems	Possible Cause	Corrective Action
Failure to regenerate.	Electrical power not getting to unit.	Make sure valve cord is live at all times. Check drop cords if used. Make sure power source is not turned off by a remote switch or pull chain.
	Loose wiring.	Check all connections *with power off.*
	Control not counting. (No flow-light activity.)	
	Seized turbine.	Free turbine.
	Faulty turbine assembly.	Replace turbine assembly.
	Faulty solid state control.	Replace solid-state control.
Failure to draw brine.	Plugged nozzle.	Remove nozzle cap and clean with fine wire.
	Plugged nozzle screen.	Remove nozzle cap and clean screen.
	Worn plunger seals.	Replace plunger.
	Low pressure.	Set pump to maintain at least 20 psi at softener during brine cycle.
	Restriction at drain line or brine line.	Remove restriction (refer to installation instructions).
	Faulty brine valve.	Check float to see that it is not binding on brine well or guide. Disassemble brine valve and clean (make sure all fittings are tight during reassembly).
	Brine well fills too fast, shuts off brine valve prematurely.	Check brine well screens; clean or replace.
	Loose brine line fittings.	Tighten fittings.
Intermittent brining.	Low pressure.	Set pump to maintain 20 psi at softener.
	Partly plugged nozzle or nozzle screen.	Remove nozzle cap; clean nozzle and screen.
Water not always soft.	Unit regenerated ''last night.'' Hot water is hard and cold water is soft.	Reprogram control to next higher hardness setting.
	Brine valve float loose.	Move locators snug against float. Brine tank should refill to approximately ½ inch above grid plate.
Salt storage tank overflows.	Faulty brine valve operation.	Check freedom of float. Clean brine tank. Disassemble and clean brine valve.

***Troubleshooting Chart* WATER SOFTENERS (*Continued*)**

Problems	Possible Cause	Corrective Action
Salt in house lines.	Slow brining due to: 　Plugged nozzle and nozzle screen. 　Low pressure. 　Drain line or brine line restriction. 　Worn seals on plunger. Excessive amount of brine in brine tank.	Remove nozzle cap; clean screen and nozzle (use fine wire). Set pump to maintain 20 psi at softener during brine cycle. Remove restriction. Replace plunger. Adjust float so that the water is ½ inch above grid plate when float shuts off. Clean brine tank. Check brine valve for dirt particles in seat (clean).
Softener continues to pass hard water after regeneration cycle is complete.	No salt in brine tank. When a softener is undersized, it becomes exhausted quickly. (After regeneration the softened water will be only partly soft.) An insufficient salt dosage will produce an incomplete regeneration. Partially softened water is the result. Water in brine tank does not refill up to and above grid plate to proper level. Bypass in plumbing.	Add salt to brine tank. Check water hardness and unit spec. Replace unit with larger size if necessary. Use high brine valve setting. Adjust float so that the water is ½ inch above grid plate when float shuts off. Screens in brine well plugged. Replace or remove and clean. Trace plumbing. Make sure hard- and soft-water lines are not interconnected. Make sure bypass valve is closed and that valve seat is in good condition.

Source: Sta-Rite Industries.

Glossary

Abrasive Any rough substance (or compound) used for grinding and polishing, as sandpaper or emery.

Abreast Even with; in line with; side by side.

Abused Used badly; mistreated.

Accelerates Speeds up; goes faster.

Accumulate To collect; gather; pile up.

Accumulator A part that collects, as a storage battery *accumulates* electricity.

Activate To make active; to start going; to put into motion; to make work.

Additives Something which is added to something else to make it work better, such as putting antifreeze in a car radiator to prevent freezing.

Adequate Just enough; barely satisfactory.

Advent A coming or arrival.

Adverse Undesirable; unfavorable; unwanted; incorrect.

Aesthetic Sensitive to art and beauty; artistic.

Affix To attach; to connect one device to a dissimilar material.

Agitate To shake jerkily; to put or keep in irregular motion.

Agitation The act of being shaken or stirred roughly.

Agitator That part of a washing machine that moves the clothes in the tub.

Air gap The space between the end of a fill-hose and the tub into which water flows.

Air, primary The air necessary for combustion that is mixed with the gas in a burner before ignition takes place.

Air, secondary The air surrounding a flame that is necessary to complete combustion.

Alignment Being brought into a straight line.

Allergen Any substance or vapor which causes a person to react with skin eruptions or eyes, nose irritations.

Alternating current (AC) A flow of electricity which changes its polarity. Normally, this change occurs at a fixed rhythm.
Example: Power-line frequencies are 60 hertz.

Amber color Brownish yellow.

Ambient conditions Surrounding conditions; conditions on all sides.

Ambient temperature Temperature of fluid (usually air) which surrounds objects on all sides.

Ammeter An instrument for measuring the strength of an electric current in amperes.

Ampere Amount of current moved by one volt through a resistance of an ohm.

Amplified Made stronger; increased, such as electrical impulses; strengthened by means of electronic tubes.

Amplifier A device that increases the strength of an electronic impulse, such as the amplifier in a radio.

Analyze To separate into parts so as to find out their nature, purpose, and use; to examine the parts of a whole carefully and in detail.

Anti-syphon loop A loop in a drain hose extending above the normal water level of a tub to prevent unwanted water drainage.

Armature The part that turns (revolves) in an electric motor or dynamo or generator; made of an iron core wound with wire. The vibrating part in an electric relay or bell.

Asbestos A fire-resistant, heat-resistant mineral; a silicate of calcium and magnesium used in fireproofing and insulation.

Assurance Anything that inspires confidence, as a promise, positive statement; a guarantee; self-confidence.

Atomization Making a liquid into a mist-like spray.

Atomize To make a liquid into a fine spray.

Automatic defrost A device which works automatically by temperature to melt and do away with ice, frost, etc.; in refrigerators and freezers.

Baffle A wall or screen to stop or hinder the flow of gases, liquids, etc.

Ballast An inductance coil or transformer used with a fluorescent tube.

Behavioral sciences The sciences which study the behavior or conduct, actions, and reactions of humans and animals—such as psychology.

Bevel A tool made of a rule with a movable arm, used in measuring or marking angles.

Beveled Cut at an angle; not a right angle; sloped.

Bezel A sloping or slanting surface, as the cutting edge of a chisel.

Bimetal Two dissimilar metals joined in such a way that the assembly will change shape with a change in temperature.

Bleed To come through or show through a covering coat of paint. (To ooze sap, juice, etc., as a bruised plant.)

Boil To be heated to a point where bubbles of vapor rise and break at the surface.

Braze To solder with a metal having a high melting point, especially with an alloy of zinc and copper.

Brine solution A saturated or full solution of water and salt, usually table salt (NaCl).

Btu British thermal unit. Quantity of heat required to raise temperature of 1 pound of water 1 degree F.

Bushing A metal lining used for making the effect of friction on moving parts less; it can be taken out.

Calcium A soft, silver-white metallic element found combined with limestone, chalk or marble, etc. Also an element that is often dissolved in water causing "hardness."

Calibrate To fix, check, or correct the graduations of an appliance or measuring instrument.

Cam A wheel or projection on a wheel which gives an alternating or irregular motion to a wheel or shaft—or one which gets its irregular motion from such a wheel. Also, an irregular shaped rotating or sliding part used to change circular motion to up and down motion.

Capable Able, skilled, competent, having the "know-how," having the necessary qualities.

Capacitor An electric condenser; something which receives and stores an electric charge. A capacitor blocks the flow of direct current and allows alternating current to pass.

Capacity Content or volume; all a container can hold; ability to hold, contain, or absorb.

Capillary tube A small diameter copper tube. A type of refrigerant control. Also, used to connect temperature control bulbs to control mechanisms.

Categories Classes or divisions in a system of classification.

Celsius Consisting of or dividing into 100 degrees, of or by the Celsius thermometer, where water freezes at 0° and boils at 100°. Abbreviation is C.

Centrifugal Moving or tending to move away from the center (of a moving circle).

Centrifugal force The force which causes objects to move away from the center of rotation due to inertia (the tendency to stay at rest or keep moving in the same direction unless acted on by an outside force).

Chafed Worn away by rubbing; rubbed so as to be made worn.

Change of state When a substance (element or compound) that is solid becomes liquid or gas. For example, when water becomes ice, it goes through a change of state.

Charge To place a charge, as a powder, fuel, gas, etc., within or upon. To add an electrical charge; to replenish. The amount of fuel for which a machine is made.

Charred Reduced to (made into) charcoal by burning; scorched.

Circuit   The path over which electricity can flow.

Clarify   To make or become clear, free of impurities—said of liquids. To make or become easier to understand.

Coincide   To take up the same place in space; to happen at the same time. To agree, match exactly, be identical.

Combustible   Able to take fire; can be easily burned up; inflammable.

Commutator   Something which changes the direction of an electric current. In a dynamo or motor, the revolving part that collects the current from or gives (distributes) it to the brushes.

Compatible   Able to get along well with one another; in agreement. For example, sugar and water will mix—are compatible—but oil and water are not.

Complex   Complicated, not simple; two or more related parts.

Complexity   Complication, anything intricate or complicated; condition of being complicated.

Components   One (or more) of the parts of a whole.

Compound gage   A device used, in refrigeration, which measures pressures below the atmospheric pressure *as well as* above the atmospheric pressure.

Compressor   A machine that compresses air, gas, etc.

Comprise   To include, contain, consist of, be made up of.

Compromise   Something midway between different things. A settlement in which each side gives in some.

Condenser   A device which changes gases or vapors to a liquid; a device which receives and stores an electric charge.

Conduction   Sending of electricity or heat, etc., by passing energy from particle to particle.

Conduit   Pipe for protecting electric wires or cable. Usually insulated on the inside; different from gas or water pipe.

Conformance   Being in agreement with one another, being in harmony; similarity; acting in agreement with rules, customs, ritual, etc.

Conjunction   Adjoining together or being joined together.

Connection   Joining of two or more parts; a union.

Connectives   Devices which connect or join two or more things together.

Consumed   Used up; wasted; destroyed, as by fire.

Contact points   Two movable points or areas that complete a circuit when pressed together. These points are usually made of tungsten, platinum, or silver.

Contaminants   Soil, dirt, gases which are where they are not supposed to be. These are what cause pollution.

Contaminate   To make corrupt so that humans cannot eat or drink. To make impure, unclean; to pollute.

Continuity   Unbroken, continuous path or line through which electricity may flow.

Contour   Outline of a figure, land, etc.

Conversely   Opposite; reversed in order or position; on the other hand.

Corbin pliers   Type of hose-clamp plier, made by Corbin.

Corrosion   Being worn away gradually, as by the action of chemicals; rust. A material made by the wearing away of a metal by chemicals.

Cracking (of valve)   Opening of valve, a tiny crack.

Crimp   To squeeze or contract.

Crocus cloth   Used as a fine sandpaper to smooth surface.

Curd   A deposit formed by action of soap with hard water.

Deceleration   Slowing down in speed or motion.

Defective   Incomplete, faulty, having a fault or imperfection.

Deflect   To swerve; to bend or turn to one side.

Defrost   To melt and do away with frost and ice.

Dehumidify   To remove moisture from the air, etc.

Dehydrate   To remove water from; to become dry; to lose water.

Derive   To get or receive from a source—as a formula; or one compound from another.

Detergent   A cleaning agent, like soap, but made synthetically, not from fats and lye.

Deterioration   Becoming or being made worse; depreciation.

Determine   To set limits, bounds; to decide upon; to be a deciding factor; to give a definite aim to, to direct.

Device   An invention; a mechanical contraption or tools or method.

Diagnosis   Deciding the nature of a trouble or difficulty by careful observation and examination.

Diaphragm   A flexible separating membrane or device.

Dictates Controls; regulates.

Dilemma A situation requiring a choice between undesirable results.

Dismantle To take apart, to strip of covering.

Disperse To break up and scatter in all directions.

Dissipate To scatter; to drive completely away; to dispell; to make disappear.

Door Switch An on-off device used in appliances that operates by the opening and closing of a door or panel.

Dormant Inactive.

Drain To make or become gradually dry or empty.

Drier A device containing a drying substance placed in the refrigerant circuit. Its main purpose is to collect and hold excess water in the system.

Dryer sensor A device in a dryer that eliminates overdrying of clothes by detecting the remaining moisture content in the clothes.

Ecological Having to do with the study of man's environment (mostly concerning air and water).

Ejector A device which throws out or expels or discharges what is not wanted.

Electrostatic Dealing with static electricity, that is, electricity at rest.

Empathy Intellectual or emotional identification with another; putting oneself in another's place and understanding what he would feel.

Encounter To meet accidentally or in opposition; meeting in conflict; a fight; unexpected meeting.

Enhance To make greater; heighten; intensify.

Enumerate To count or name one by one; to specify.

Environment Surroundings; conditions and influences surrounding and affecting development and behavior.

Epoxy (patch) Epoxy is a synthetic plastic adhesive.

Erratic Having no fixed course, irregular, wandering.

Eutectic point A eutectic alloy is that composition of two or more metals that have one sharp melting point and no plastic range.

Evacuate To withdraw from, to make empty, remove the contents from.

Evaporator That which changes a liquid or solid into vapor, or which removes moisture from a product by heating.

Excess An amount or quantity greater than is needed, surplus.

Excessive Too much; more than is wanted or needed.

Expendable Can be used up or spent or consumed.

Fahrenheit Designating or of a thermometer on which the boiling point of pure water is 212° and the freezing point is 32°. Abbreviation is F.

Feasible Possible, practical, probable; capable of being done.

Ferromagnetic A highly magnetic material such as iron, nickel, steel, etc.

Fill valve A device that allows water to enter the machine, generally operated by a pressure switch.

Flange A rim or collar on a wheel which sticks out to hold the wheel in place, give it strength, or attach it to something else.

Foreign object An object which does not belong where it is found.

Freeze To be formed into ice; be hardened or solidified by cold.

Freon A special gas used in refrigeration.

Fungi A group of plants (mildews, molds, mushrooms, rusts) that have no leaves or flowers or green coloring and reproduce by means of spores.

Fusible Able to be united as if melted together.

Fusion *See* Latent Heat.

Gaseous Lacking substance or solidity; like or in the form of gas.

Gases *Natural*—A gas that comes from the ground. *Butane*—The by-products of the production of oil and gasoline.
LP—The same as butane.
Mfg.—A gas that is manufactured from another substance.

Gas is an airlike fluid neither independent shape nor volume, tending to expand indefinitely.

Gasket A piece or ring of rubber, metal, etc., put around a piston or joint to make it leakproof.

Glycerine An odorless, colorless, syrup-like liquid; used as a solvent, skin lotion, and in explosives.

Graphite A soft, black shiny form of carbon; used for lead in pencils, lubricants, electrodes.

Hazard Chance; risk; danger.

Heat sensor A thermal resistance unit wherein the temperature changes the resistance (A thermal resistance unit).

Hermetically sealed Airtight.

Hertz (Hz) The modern unit of measurement for frequency. One hertz equals one cycle per second.

Hi-pot   Test for shock hazard. Abbreviation for high-voltage (potential) insulation test. Normally, the test voltage is 1000 volts + twice the operating voltage of the part being tested.

Humid   Damp, moist; containing much water vapor.

Humidistat   A device that senses and indicates the amount of moisture in surrounding air. It can also sense and indicate dryness.

Humidity   The amount of moisture in the air; dampness.

Hydrologic cycle   The cycle of water from evaporation, rising, condensing, and falling in the form of rain or snow.

Ice ejector   A machine which ejects ice from ice tray.

Idler   One not busy; inactive; not in use; that which causes a machine to idle.

Ignite   To set fire to, to heat to a great degree.

Ignitor, automatic spark   Electrical or mechanical means of controlling the starting of a controlled flame.

Immerse   To dip or plunge into, as in a liquid.

Immersion   The act of putting an object into a liquid so that it is completely covered by the liquid.

Impart   To give a share or part of; to make known or reveal.

Impedance   Seeming, apparent resistance in an alternating electrical current, corresponding to the true resistance in a direct current.

Impeller   A wheel-like device upon which fins are fastened. It spins to compress air a little or to pump water.

Impurities   Dirt or other foreign matter mixed in a substance so as to make it unclean.

Incinerate   To burn to ashes, burn up.

Incinerator   A furnace usually used for burning trash.

Increment   Amount of increase or addition.

Inductance   Inherent property of an electric circuit that opposes a change in the current. The property of a circuit whereby energy may be stored in magnetic field.

Inflammable   Easily set on fire; any material that would burn very quickly.

Ingredients   Any of the things that a mixture is made of.

Inhibitor   A substance added to oil, water, gas, etc., to prevent unwanted action such as rusting, foaming, etc.

Insoluble   Cannot be dissolved.

Insurance   Being sure, safe; a guarantee, as guaranteeing a replacement for a loss.

Intangibles   Cannot be touched, cannot be easily defined or grasped.

Integrated circuit   A circuit consisting of an inseparable assembly of numerous electronic components in a single package.

Integrity   Being complete; soundness, honesty, sincerity, uprightness.

Interference   Obstruction: something blocking the way.
Radio: Static, unwanted signals making a confusion of sounds.
Physics: Mutual action of two waves of vibration (sound, light, etc.) in reinforcing or neutralizing each other.

Intermittent   Starting and stopping again at different times.

Intricate   Complicated; hard to follow or understand because it is entangled, complicated.

Inventory   Itemized list of goods, property, etc.; stock listing, store of goods, etc.

Ion   An electrically charged atom that has lost or acquired an extra electron. An added electron makes the atom a negative (−) ion. A lost electron makes the atom a positive (+) ion.

Ion exchange   Ion exchange occurs when one ion or group of ions replaces another ion or group of ions in a solution.

Isolate   Set apart from others, place alone; separate.

Jam   Stuck; wedged fast; unworkable.

Jeweler's loupe   Magnifying glass with eye piece which fits over eye glasses or can be held in place by squeezing the eye.

Laced (as in well laced)   Substance containing additives to improve performance. Means of tying wires together to form a bundle arranged so that the proper ends are in position to be connected to the parts of a machine.

Ladder diagram   A schematic where all components are in a ladder form (one component after another in an up and down picture).

Laminate   To form or press into a thin sheet or layer; to separate into thin sheets or layers.
Laminations: Thin sheets or layers of steel used in cores of transformers, motors, or generators.

Latent Heat   The energy (Btu) needed to cause a change of state.

It cannot be measured by thermometer.

A. Fusion—Change of state of element or compound from solid to liquid, or reverse.

B. Vaporization—Change of state from a liquid to a gas.

C. Sublimation—Change from a solid to a gas.

Leaf springs A suspension spring made of several pieces of flat steel.

Leveler (1) A solvent used to smooth a paint; touch-up job.
(2) Adjustable feet on an appliance.

Liberally Amply, abundantly, sufficiently, unrestrictedly.

Liming up Expression used to indicate a buildup of lime deposit resulting from use of hard water.

Line Something used in electrical work that carries the main load of electrical current. Any wire, pipe, etc., or system of these, carrying fluid, electricity, etc.

Lint Fine bits of thread, ravelings, or fluff from cloth or yarn.

Liquidate The process of changing a solid into a liquid state. To settle or clear up the affairs of a business.

Lubrication The act of making slippery or smooth in order to reduce friction. This is usually done with an oil or grease.

Magnesium (Mg.) A metallic element that is a light silver-white, easy to work with, burns with a hot, white light—used in photographic flash bulbs.

Malfunction Failure to operate as it should.

Mandatory Required, necessary, commanded by authority.

Manifold A device used for refrigeration service which includes valves and gages to allow connection to a refrigeration system for analyzing its operation.

Manometer An instrument for measuring the pressure of gases of vapors.

Mediocre Ordinary, average, commonplace, neither very good nor very bad.

Mercury switch An electric switch made by placing a large globule of mercury in a glass tube having electrodes arranged in such a way that tilting the tube will cause the mercury to make or break the circuit.

Microorganism Any very tiny (microscopic) animal or vegetable bit of life, such as bacteria or spores.

Migrate To move from one area to another.

Milliampere One thousandth of an ampere.

Misalignment Not properly aligned, not straight.

Modulator A device that regulates.

Module A separable, self-contained device consisting of a group of interconnected parts, designed to do a specific job; usually part of a more complex unit.

Molykote Trade name for a molybdenum-sulfide lubricant.

Mullion element Heating element in the small bars of a refrigerator to prevent frost buildup.

Nichrome Nickel-chromium alloy used to make resistance heating elements.

Nondetergent A very weak cleaning agent.

Nuts Tinnerman—Special sheet metal nuts.

Obsolete Out of date; no longer in use or practice.

Ohm (symbol Ω) Unit of measurement of resistance. Amount of resistance that will allow a current of 1 ampere to flow under a pressure of 1 volt.

Ohmmeter Meter used to measure resistance in ohms.

Orifice A small opening in tube, etc., to control the flow of gases.

Oscillation Fluctuation, instability; a single swing of a swinging object. The variation between maximum and minimum values, as electric current.

Oven element A certain resistance used to warm and cook foods in an enclosed unit.

Overload protector A thermally operated device (thermostat) which turns off a machine when too much current goes through.

Oxidation Being united with oxygen.

Palatable Pleasing to the taste.

Parallel circuit Circuit which contains two or more paths for electrons supplied by common voltage.

Peak Maximum value of sine wave: The highest voltage current or power reached during a particular cycle or operating time.

Penetrate To make a way into and through something; pierce.

Percolate To drain or ooze through a porous substance; to pass a liquid gradually through a porous substance; filter, as water through soil.

Perpetual Indefinitely long in use, service, etc.; lasting forever; continuing indefinitely without interruption; constant.

Perpetual motion   The motion of a hypothetical machine which, once started, would operate indefinitely by creating its own energy.

Ph content   Degree of acidity of water. Neutral water has a pH of 7, acids are lower, bases higher.

Pinion gear   A small cogwheel which meshes with a larger gearwheel or rack.

Pivot   A point, shaft, etc., on which something turns; to turn as if mounted on such a point.

Pliable   Easily bent, flexible.

Potential   Voltage—the electrical energy which moves electrons in a circuit.

Precipitate   To condense (vapor, etc.) and cause to fall as rain or snow; to separate a dissolved substance out from a solution.

Predetermined   Determined or decided upon beforehand.

Pressure   Force exerted against an opposing body; the thrust distributed over a surface.

Pressure gage   A tool that measures pressure in PSI (pounds per square inch).

Pressure switch   An on-off device operated by weight or pressure.

Printed circuit board   A stiff insulating and copper foil sheet on which the foil has been etched to produce the circuit conductors. Usually it serves as the component support and circuit interconnecting means for a module.

Priority   A right to proceed earlier, to be ahead; coming before in order or importance.

Productive   Causing or having good results.

PSI   Pounds per square inch; a measurement of pressure.

Psychrometer (Sling)   A device which measures relative humidity in air.

Pulser timer   A device used in major appliances that starts and stops different operations with a throbbing effect.

Purging hose   A hose through which gas, air, water, or other liquid is forced to cleanse away unwanted matter.

Pyrolysis   Chemical change brought about by the action of heat.

Pyrometer   A thermometer for measuring unusually high temperatures.

Pyroxylin putty   Putty used to fill in nicks, scratches, etc., in the surface of a finished appliance before painting.

Race (on which bearings ride)   The inner or outer ring that provides a contact surface for the balls or rollers in a bearing.

Radiation   The process in which energy is in the form of rays of light, heat, etc.

Recurrence   The act of appearing again, or periodically; the act of turning back in the opposite direction.

Refrigerant   Any substance that produces a cooling effect by its absorption of heat while expanding or vaporizing.

Regenerate-Regeneration   To bring into existence again; reestablish; to cause to be completely reformed or improved; to renew; restore; to grow (a part) anew; to replace a part lost.
Radio: To amplify by feeding energy back from the output into the input circuit.

Relay   A magnetic switch which uses a little current in a control circuit to operate a device needing heavier current in the *operating* circuit.

Relegated   Assigned to a class or sphere; belonging to a specific group.

Reliability   Dependability; can be counted on.

Reoccurrence   Happening again.

Replenish   To make full or complete again, as with new stock or supply; supply again.

Residual   A remainder; a remaining force as in magnetism; left over.

Residue   What is left after part is taken away.

Resin beads   A sand-like plastic material used in water softeners.

Resistance   The quality of an electric circuit which opposes the flow of current through it. This produces heat in the carrier. It is measured in ohms and symbolized by R.

Resonance   The state of adjustment of an electric circuit, as of a radio receiving apparatus tuned to a particular station that permits the greatest flow of current of a particular frequency.

Restrict   To limit, confine, keep within bounds.

Rheostat   A device for regulating strength of an electric current by varying resistances without opening the circuit.

Rigid   Stiff; not bending or flexible.

SAE   Society of Automotive Engineers, or rated horsepower. A simple formula of long standing is used to determine what is commonly referred to as SAE.

$$\frac{\text{bore diameter}^2 \times \text{number of cylinders}}{2.5}$$

Safety thermostat   A thermostat that limits the temperature of an appliance.

Schematic (drawing or diagram)   A sketchy

line drawing or diagram which gives the outline and basics of an object, situation, or proposition, but does not fill in details.

Score To make a scratch mark or line to show a starting point or cutting line.

Screw extractor Tool used to remove broken bolts, screws, etc., from holes.

Seals A rubber rim used in washing machines to keep the water or fluids from getting into the motor, or transmission of the machine.

Sediment Any matter that settles to the bottom of a liquid, such as soil in still water.

Seize When two parts rub together so that they get hot and force the lubricant out of the area they "gall" and finally "freeze" or stick together.

Self-contained Having all working parts in a closed unit.

Sensible heat The heat energy required to change the temperature of any element or compound.

Series circuit An electrical path where components are connected in line. One break disrupts the whole path.

Serrate-serrated Having sawlike teeth or notches along the edge.

Shaft A bar supporting, or sending (relaying) motion to a mechanical part of an engine.

Shim A thin piece of metal used for filling space, leveling, etc. Material used for shims.

Short A side circuit of very low relative resistance connecting two points in an electric circuit of higher resistance so as to divert or turn aside most of the current. A disrupted or broken electric circuit caused by this situation is said to be "short" or "shorted."

Siphon break An air gap between hose and drain.

Sleeve A tube or tube-like part fitting around another part.

Slinger A loose fitting ring around the shaft which turns and throws aside any leaking liquid.

Slip The difference between the speed of the rotating magnetic field of an induction motor and the speed of the rotor.

Snifter valve A valve on the inlet side of a pump which will permit air to enter the system.

Snubber To check or stop with a sharp retort. A device used to limit the travel of some part.

Sodium A silver-white, alkaline, metallic element, found in nature only in combined form. Symbol: Na

Solder A metal alloy used when melted to join two metal surfaces together, or to patch metal parts or surfaces.

Solenoid A coil of wire carrying an electric current and having the properties of a magnet.

Soleplate The bottom plate of an iron that touches the clothes.

Solidify To make or become firm, hard, solid. To crystallize.

Soluble Can be dissolved, as sugar is *soluble* in water.

Solvent Something used for dissolving something else.

Spark ignitor An electrical or mechanical means of starting a controlled flame, as in an oil burner.

Specific Precise, definite, exact; specially made for; definitely stated.

Speed nuts Nuts with minimum amount of threads to aid in speedy removal or replacement.

Spike *See* Transient.

Spin To whirl rapidly in one direction as in an automatic washer.

Stabilize To make firm, steady, to keep from changing.

Stable Firm, steady, not easily changed or decomposed (chemistry).

Standpipe Used with washing machines to accept the drain hose when there is no tub.

Starter The automatic switch used in a fluorescent light.

Stipple To paint, draw or engrave in small dots, instead of lines or solid areas.

Strain relief A device which firmly attaches the power card to the machine.

Sublimation *See* Latent Heat.

Submersible Able to be placed under water.

Subsequent Following in time, place, or order; coming after.

Suds lock Condition of a washing machine water system becoming obstructed by use of excessive amount of soap or detergent.

Sufficient Enough, as much as needed.

Sulfates Salts of sulfuric acid.

Sump A pit or reservoir serving as a drain for fluids.

Surge *See* Transient.

Suspended matter Visible pollution in water or air.

Suspension bolt Bolts used to hold an object from the top to allow movement of the part being held.

Swage The process of enlarging a tube.

Switch A device for directing or controlling current flow in circuit.

Switch (Continued)
Types of switches:
Normally open (N.O.)—A contact pair which is open when the device is in the de-energized condition.
Normally closed (N.C.)—A contact pair which is closed when the device is in the de-energized condition.
Two-way/three-way—Two or three on-off devices that operate one or more resistances, such as a light bulb.

Synchronous speed   The speed of the rotating magnetic field set up by the stator of an induction motor. In a synchronous motor the rotor locks into step with the rotating magnetic field.

Tabulation   A systematic list or table. To list systematically; to put in a table as facts, statistics, etc.

Tachometer   A device which measures or records speed or velocity as of a motor in rpm.

Tacky   Sticky, as varnish, glue, paint, etc., before completely dry.

Tap (in a coil winding)   To cut threads in a hole; can be used to indicate the fluted tool used to cut the threads.

Temperature   The degree of hotness or coldness of anything, usually as measured on a thermometer.

Terminals   A connecting point in an electric circuit. (When referring to the battery it would indicate the two battery posts).

Thermocouple   Device which generates electricity, using the principle that if two dissimilar metals are welded together and the junction is heated, a voltage will develop across open ends.

Thermostat   A device uniformly sensitive to temperature changes. Used to regulate and control heat and cold producing apparatus.

Thermostat, operating   A thermostat that controls the operating temperature of a device.

Thermostat, safety   A thermostat that automatically limits the temperature of an appliance.

Timer, pulser   A device used in major appliances that starts and stops different operations with a pausing effect. Mechanism used to control on and off times of an electrical circuit.

Tinnerman nuts   *See* Nuts.

Tolerances   Amount of variation or change allowed from a standard accuracy—especially the difference between the allowable maximum and minimum sizes of some mechanical part.

Torque   Turning force exerted by the armature of a motor; usually measured in pound-inches or ounce-inches.

Transformer   Device which transfers electrical energy from a primary circuit into variations of voltage in secondary circuit, by electro-magnetic induction.

Transient   A sudden temporary increase of voltage in the power supply lines.

Transmission   A device that transmits motive force from engine to the wheels usually by means of gears or hydraulic cylinders.

Triac   An electronically controllable solid-state switch that will allow ac to flow in each direction.

Turbidity   Having the sediment stirred up, muddy, cloudy.

Turbine   An engine driven by the pressure of steam, water, or air against the curved vanes of a wheel.

Unalterable   Cannot be changed or altered.

Vacuum   An enclosed space from which practically all air has been removed; reduction in pressure below atmospheric pressure; a space with nothing at all in it; a space with most of the air or gas pumped out of it. (Note: A perfect vacuum is not attainable).

Validate   To prove to be sound, having legal force. To be well grounded on principles or evidence.

Valve, expansion   Type of refrigerant control which maintains pressure difference between high side and low side pressure in refrigerant mechanism. Valve is caused to operate by pressure in low or suction side. Often referred to as an automatic expansion valve or AEV.

Valves   Device for controlling liquid flow: a gate regulating the flow of water.
The *mix valve* lets both the hot and cold in according to the setting.
The *fill valve* allows water to enter the machine generally operated by a pressure switch.

Valves, transfer/distributor   Mechanical device for starting, stopping, or regulating the flow of liquid, gas, or loose material.

Vaporization   *See* Latent Heat.

Variable   Changeable, not constant, not fixed.

Velocity (air)   Rate of motion in a particular direction in relation to time speed.

Vent   An opening for passage or escape of fluids, smoke, or air.

Verify   To prove to be true by evidence; to confirm; to test the accuracy of, as by investigation.

Versatile   Able to do many things; to turn easily from one subject or action to another.

Viscosity   The internal fluid resistance of a substance, caused by molecular attraction. Like a sticky, syruplike solution made by treating cellulose with potassium hydroxide and carbon disulfide. Term used to describe resistance of flow of fluids.

Voltage, low line Condition of voltage being at a lower level than that for which the appliance or motor was designed.

Voltmeter Instrument used to measure voltage action in electrical circuits, an electromotive force, or a difference in electrical potential by volts.

Volts Unit of measurement of electromotive force in an electrical circuit; that force which will cause a current of 1 ampere to flow through a conductor whose resistance is 1 ohm.

VOM A multiple purpose electrical meter used for testing the voltage, current, and resistance in an electricity-using machine.

Vulnerable Can be hurt, physically injured; open to harm or injury.

Water level Surface or height of still water.

Wattmeter Meter used to measure power in an electric circuit in watts.

Watts Unit of electric power, equal to a current of one ampere under one volt of pressure.

Wet and dry bulb thermometer A device that measures the humidity as well as the temperature.

Zip tube Designed to aid in igniting a flame heating unit. A perforated pipe to convey a light to a pilot burner.

Index[*]

---

[*] Upright page numbers = Text; *Italic* page numbers = Illustrations; **Bold face** page numbers = Tables. They are shown in this order.